T0295308

IET ENERGY ENGINEERING SERIES 207

Power Electronics for Next-Generation Drives and Energy Systems

Other volumes in this series:

Volume 1 **Power Circuit Breaker Theory and Design** C.H. Flurscheim (Editor)
Volume 4 **Industrial Microwave Heating** A.C. Metaxas and R.J. Meredith
Volume 7 **Insulators for High Voltages** J.S.T. Looms
Volume 8 **Variable Frequency AC Motor Drive Systems** D. Finney
Volume 10 **SF_6 Switchgear** H.M. Ryan and G.R. Jones
Volume 11 **Conduction and Induction Heating** E.J. Davies
Volume 13 **Statistical Techniques for High Voltage Engineering** W. Hauschild and W. Mosch
Volume 14 **Uninterruptible Power Supplies** J. Platts and J.D. St Aubyn (Editors)
Volume 15 **Digital Protection for Power Systems** A.T. Johns and S.K. Salman
Volume 16 **Electricity Economics and Planning** T.W. Berrie
Volume 18 **Vacuum Switchgear** A. Greenwood
Volume 19 **Electrical Safety: a guide to causes and prevention of hazards** J. Maxwell Adams
Volume 21 **Electricity Distribution Network Design, 2nd Edition** E. Lakervi and E.J. Holmes
Volume 22 **Artificial Intelligence Techniques in Power Systems** K. Warwick, A.O. Ekwue and R. Aggarwal (Editors)
Volume 24 **Power System Commissioning and Maintenance Practice** K. Harker
Volume 25 **Engineers' Handbook of Industrial Microwave Heating** R.J. Meredith
Volume 26 **Small Electric Motors** H. Moczala *et al.*
Volume 27 **AC-DC Power System Analysis** J. Arrillaga and B.C. Smith
Volume 29 **High Voltage Direct Current Transmission, 2nd Edition** J. Arrillaga
Volume 30 **Flexible AC Transmission Systems (FACTS)** Y-H. Song (Editor)
Volume 31 **Embedded generation** N. Jenkins *et al.*
Volume 32 **High Voltage Engineering and Testing, 2nd Edition** H.M. Ryan (Editor)
Volume 33 **Overvoltage Protection of Low-Voltage Systems, Revised Edition** P. Hasse
Volume 36 **Voltage Quality in Electrical Power Systems** J. Schlabbach *et al.*
Volume 37 **Electrical Steels for Rotating Machines** P. Beckley
Volume 38 **The Electric Car: Development and future of battery, hybrid and fuel-cell cars** M. Westbrook
Volume 39 **Power Systems Electromagnetic Transients Simulation** J. Arrillaga and N. Watson
Volume 40 **Advances in High Voltage Engineering** M. Haddad and D. Warne
Volume 41 **Electrical Operation of Electrostatic Precipitators** K. Parker
Volume 43 **Thermal Power Plant Simulation and Control** D. Flynn
Volume 44 **Economic Evaluation of Projects in the Electricity Supply Industry** H. Khatib
Volume 45 **Propulsion Systems for Hybrid Vehicles** J. Miller
Volume 46 **Distribution Switchgear** S. Stewart
Volume 47 **Protection of Electricity Distribution Networks, 2nd Edition** J. Gers and E. Holmes
Volume 48 **Wood Pole Overhead Lines** B. Wareing
Volume 49 **Electric Fuses, 3rd Edition** A. Wright and G. Newbery
Volume 50 **Wind Power Integration: Connection and system operational aspects** B. Fox *et al.*
Volume 51 **Short Circuit Currents** J. Schlabbach
Volume 52 **Nuclear Power** J. Wood
Volume 53 **Condition Assessment of High Voltage Insulation in Power System Equipment** R.E. James and Q. Su
Volume 55 **Local Energy: Distributed generation of heat and power** J. Wood
Volume 56 **Condition Monitoring of Rotating Electrical Machines** P. Tavner, L. Ran, J. Penman and H. Sedding
Volume 57 **The Control Techniques Drives and Controls Handbook, 2nd Edition** B. Drury
Volume 58 **Lightning Protection** V. Cooray (Editor)
Volume 59 **Ultracapacitor Applications** J.M. Miller
Volume 62 **Lightning Electromagnetics** V. Cooray
Volume 63 **Energy Storage for Power Systems, 2nd Edition** A. Ter-Gazarian
Volume 65 **Protection of Electricity Distribution Networks, 3rd Edition** J. Gers
Volume 66 **High Voltage Engineering Testing, 3rd Edition** H. Ryan (Editor)
Volume 67 **Multicore Simulation of Power System Transients** F.M. Uriate
Volume 68 **Distribution System Analysis and Automation** J. Gers
Volume 69 **The Lightening Flash, 2nd Edition** V. Cooray (Editor)
Volume 70 **Economic Evaluation of Projects in the Electricity Supply Industry, 3rd Edition** H. Khatib
Volume 72 **Control Circuits in Power Electronics: Practical issues in design and implementation** M. Castilla (Editor)
Volume 73 **Wide Area Monitoring, Protection and Control Systems: The enabler for Smarter Grids** A. Vaccaro and A. Zobaa (Editors)
Volume 74 **Power Electronic Converters and Systems: Frontiers and applications** A.M. Trzynadlowski (Editor)
Volume 75 **Power Distribution Automation** B. Das (Editor)
Volume 76 **Power System Stability: Modelling, analysis and control** A.A. Sallam and B. Om P. Malik
Volume 78 **Numerical Analysis of Power System Transients and Dynamics** A. Ametani (Editor)
Volume 79 **Vehicle-to-Grid: Linking electric vehicles to the smart grid** J. Lu and J. Hossain (Editors)
Volume 81 **Cyber-Physical-Social Systems and Constructs in Electric Power Engineering** S. Suryanarayanan, R. Roche and T.M. Hansen (Editors)
Volume 82 **Periodic Control of Power Electronic Converters** F. Blaabjerg, K. Zhou, D. Wang and Y. Yang
Volume 86 **Advances in Power System Modelling, Control and Stability Analysis** F. Milano (Editor)
Volume 87 **Cogeneration: Technologies, Optimisation and Implentation** C.A. Frangopoulos (Editor)
Volume 88 **Smarter Energy: from Smart Metering to the Smart Grid** H. Sun, N. Hatziargyriou, H.V. Poor, L. Carpanini and M.A. Sánchez Fornié (Editors)
Volume 89 **Hydrogen Production, Separation and Purification for Energy** A. Basile, F. Dalena, J. Tong and T.N. Veziroğlu (Editors)
Volume 90 **Clean Energy Microgrids** S. Obara and J. Morel (Editors)
Volume 91 **Fuzzy Logic Control in Energy Systems with Design Applications in Matlab/Simulink®** İ.H. Altaş
Volume 92 **Power Quality in Future Electrical Power Systems** A.F. Zobaa and S.H.E.A. Aleem (Editors)

Volume 93 **Cogeneration and District Energy Systems: Modelling, Analysis and Optimization** M.A. Rosen and S. Koohi-Fayegh
Volume 94 **Introduction to the Smart Grid: Concepts, technologies and evolution** S.K. Salman
Volume 95 **Communication, Control and Security Challenges for the Smart Grid** S.M. Muyeen and S. Rahman (Editors)
Volume 96 **Industrial Power Systems with Distributed and Embedded Generation** R. Belu
Volume 97 **Synchronized Phasor Measurements for Smart Grids** M.J.B. Reddy and D.K. Mohanta (Editors)
Volume 98 **Large Scale Grid Integration of Renewable Energy Sources** A. Moreno-Munoz (Editor)
Volume 100 **Modeling and Dynamic Behaviour of Hydropower Plants** N. Kishor and J. Fraile-Ardanuy (Editors)
Volume 101 **Methane and Hydrogen for Energy Storage** R. Carriveau and D.S-K. Ting
Volume 104 **Power Transformer Condition Monitoring and Diagnosis** A. Abu-Siada (Editor)
Volume 106 **Surface Passivation of Industrial Crystalline Silicon Solar Cells** J. John (Editor)
Volume 107 **Bifacial Photovoltaics: Technology, applications and economics** J. Libal and R. Kopecek (Editors)
Volume 108 **Fault Diagnosis of Induction Motors** J. Faiz, V. Ghorbanian and G. Joksimović
Volume 109 **Cooling of Rotating Electrical Machines: Fundamentals, modelling, testing and design** D. Staton, E. Chong, S. Pickering and A. Boglietti
Volume 110 **High Voltage Power Network Construction** K. Harker
Volume 111 **Energy Storage at Different Voltage Levels: Technology, integration, and market aspects** A.F. Zobaa, P.F. Ribeiro, S.H.A. Aleem and S.N. Afifi (Editors)
Volume 112 **Wireless Power Transfer: Theory, Technology and Application** N. Shinohara
Volume 114 **Lightning-Induced Effects in Electrical and Telecommunication Systems** Y. Baba and V.A. Rakov
Volume 115 **DC Distribution Systems and Microgrids** T. Dragičević, F. Blaabjerg and P. Wheeler
Volume 116 **Modelling and Simulation of HVDC Transmission** M. Han (Editor)
Volume 117 **Structural Control and Fault Detection of Wind Turbine Systems** H.R. Karimi
Volume 118 **Modelling and Simulation of Complex Power Systems** A. Monti and A. Benigni
Volume 119 **Thermal Power Plant Control and Instrumentation: The control of boilers and HRSGs, 2nd Edition** D. Lindsley, J. Grist and D. Parker
Volume 120 **Fault Diagnosis for Robust Inverter Power Drives** A. Ginart (Editor)
Volume 121 **Monitoring and Control using Synchrophasors in Power Systems with Renewables** I. Kamwa and C. Lu (Editors)
Volume 123 **Power Systems Electromagnetic Transients Simulation, 2nd Edition** N. Watson and J. Arrillaga
Volume 124 **Power Market Transformation** B. Murray
Volume 125 **Wind Energy Modeling and Simulation Volume 1: Atmosphere and plant** P. Veers (Editor)
Volume 126 **Diagnosis and Fault Tolerance of Electrical Machines, Power Electronics and Drives** A.J.M. Cardoso
Volume 128 **Characterization of Wide Bandgap Power Semiconductor Devices** F. Wang, Z. Zhang and E.A. Jones
Volume 129 **Renewable Energy from the Oceans: From wave, tidal and gradient systems to offshore wind and solar** D. Coiro and T. Sant (Editors)
Volume 130 **Wind and Solar Based Energy Systems for Communities** R. Carriveau and D.S-K. Ting (Editors)
Volume 131 **Metaheuristic Optimization in Power Engineering** J. Radosavljević
Volume 132 **Power Line Communication Systems for Smart Grids** I.R.S. Casella and A. Anpalagan
Volume 134 **Hydrogen Passivation and Laser Doping for Silicon Solar Cells** B. Hallam and C. Chan (Editors)
Volume 139 **Variability, Scalability and Stability of Microgrids** S.M. Muyeen, S.M. Islam and F. Blaabjerg (Editors)
Volume 142 **Wind Turbine System Design: Volume 1: Nacelles, drive trains and verification** J. Wenske (Editor)
Volume 143 **Medium Voltage DC System Architectures** B. Grainger and R.D. Doncker (Editors)
Volume 145 **Condition Monitoring of Rotating Electrical Machines** P. Tavner, L. Ran, C. Crabtree
Volume 146 **Energy Storage for Power Systems, 3rd Edition** A.G. Ter-Gazarian
Volume 147 **Distribution Systems Analysis and Automation 2nd Edition** J. Gers
Volume 151 **SiC Power Module Design: Performance, robustness and reliability** A. Castellazzi and A. Irace (Editors)
Volume 152 **Power Electronic Devices: Applications, failure mechanisms and reliability** F. Iannuzzo (Editor)
Volume 153 **Signal Processing for Fault Detection and Diagnosis in Electric Machines and Systems** M. Benbouzid (Editor)
Volume 155 **Energy Generation and Efficiency Technologies for Green Residential Buildings** D. Ting and R. Carriveau (Editors)
Volume 156 **Lithium-ion Batteries Enabled by Silicon Anodes** C. Ban and K. Xu (Editors)
Volume 157 **Electrical Steels, 2 Volumes** A. Moses, K. Jenkins, Philip Anderson and H. Stanbury
Volume 158 **Advanced Dielectric Materials for Electrostatic Capacitors** Q. Li (Editor)
Volume 159 **Transforming the Grid Towards Fully Renewable Energy** O. Probst, S. Castellanos and R. Palacios (Editors)
Volume 160 **Microgrids for Rural Areas: Research and case studies** R.K. Chauhan, K. Chauhan and S.N. Singh (Editors)
Volume 161 **Artificial Intelligence for Smarter Power Systems: Fuzzy Logic and Neural Networks** M.G. Simoes
Volume 165 **Digital Protection for Power Systems 2nd Edition** Salman K Salman
Volume 166 **Advanced Characterization of Thin Film Solar Cells** N. Haegel and M.Al-Jassim (Editors)
Volume 167 **Power Grids with Renewable Energy** Storage, integration and digitalization A.A. Sallam and B. OM P. Malik
Volume 169 **Small Wind and Hydrokinetic Turbines** P. Clausen, J. Whale and D. Wood (Editors)
Volume 170 **Reliability of Power Electronics Converters for Solar Photovoltaic Applications** F. Blaabjerg, A.I Haque, H. Wang, Z. Abdin Jaffery and Y. Yang (Editors)
Volume 171 **Utility-scale Wind Turbines and Wind Farms** A. Vasel-Be-Hagh and D.S.-K. Ting
Volume 172 **Lighting interaction with Power Systems, 2 volumes** A. Piantini (Editor)
Volume 174 **Silicon Solar Cell Metallization and Module Technology** T. Dullweber (Editor)

Volume 175 **n-Type Crystalline Silicon Photovoltaics: Technology, applications and economics** D. Munoz and R. Kopecek (Editors)
Volume 180 **Protection of Electricity Distribution Networks, 4th Edition** J. Gers and E. Holmes
Volume 182 **Surge Protection for Low Voltage Systems** A. Rousseau (Editor)
Volume 184 **Compressed Air Energy Storage: Types, systems and applications** D. Ting and J. Stagner
Volume 186 **Synchronous Reluctance Machines: Analysis, optimization and applications** N. Bianchi, C. Babetto and G. Bacco
Volume 191 **Electric Fuses: Fundamentals and new applications 4th Edition** N. Nurse, A. Wright and P.G. Newbery
Volume 193 **Overhead Electric Power Lines: Theory and practice** S. Chattopadhyay and A. Das
Volume 194 **Offshore Wind Power Reliability, availability and maintenance, 2nd edition** P. Tavner
Volume 196 **Cyber Security for Microgrids** S. Sahoo, F. Blaajberg and T. Dragicevic
Volume 198 **Battery Management Systems and Inductive Balancing** A. Van den Bossche and A. Farzan Moghaddam
Volume 199 **Model Predictive Control for Microgrids: From power electronic converters to energy management** J. Hu, J.M. Guerrero and S. Islam
Volume 204 **Electromagnetic Transients in Large HV Cable Networks: Modeling and calculations** Ametani, Xue, Ohno and Khalilnezhad
Volume 208 **Nanogrids and Picogrids and their Integration with Electric Vehicles** S. Chattopadhyay
Volume 210 **Superconducting Magnetic Energy Storage in Power Grids** M.H. Ali
Volume 211 **Blockchain Technology for Smart Grids: Implementation, management and security** H.L. Gururaj, K.V. Ravi, F. Flammini, H. Lin, B. Goutham, K.B.R. Sunil and C. Sivapragash
Volume 212 **Battery State Estimation: Methods and Models** S. Wang
Volume 215 **Industrial Demand Response: Methods, best practices, case studies, and applications** H.H. Alhelou, A. Moreno-Muñoz and P. Siano (Editors)
Volume 213 **Wide Area Monitoring of Interconnected Power Systems 2nd Edition** A.R. Messina
Volume 217 **Advances in Power System Modelling, Control and Stability Analysis 2nd Edition** F. Milano (Editor)
Volume 225 **Fusion-Fission Hybrid Nuclear Reactors: For enhanced nuclear fuel utilization and radioactive waste reduction** W.M. Stacey
Volume 238 **AI for Status Monitoring of Utility Scale Batteries** Shunli Wang, Kailong Liu, Yujie Wang, Daniel-Ioan Stroe, Carlos Fernandez and Josep M. Guerrero
Volume 905 **Power system protection, 4 volumes**

Power Electronics for Next-Generation Drives and Energy Systems

Volume 1: Converters and control for drives

Edited by
Nayan Kumar, Josep M. Guerrero, Debaprasad Kastha and Tapas Kumar Saha

The Institution of Engineering and Technology

Published by The Institution of Engineering and Technology, London, United Kingdom

The Institution of Engineering and Technology is registered as a Charity in England & Wales (no. 211014) and Scotland (no. SC038698).

First published 2022

The Institution of Engineering and Technology
Futures Place
Kings Way, Stevenage
Hertfordshire SG1 2UA, United Kingdom

www.theiet.org

British Library Cataloguing in Publication Data
A catalogue record for this product is available from the British Library

ISBN 978-1-83953-468-3 (Volume 1 hardback)
ISBN 978-1-83953-470-6 (Volume 1 pdf)
ISBN 978-1-83953-469-0 (Volume 2 hardback)
ISBN 978-1-83953-471-3 (Volume 2 pdf)
ISBN 978-1-83953-496-6 (2 Volume set hardback)

Typeset in India by MPS Limited

Cover image: Wladimir Bulgar/Science Photo Library via Getty Images

Contents

About the editors

Nayan Kumar is a post-doctoral fellow at the School of Interdisciplinary Research, Indian Institute of Technology Delhi, India. His research interests include power electronics and its applications such as in photovoltaic systems, wind turbines, electric vehicles, reliability, harmonics, and adjustable speed drives. He received his PhD in electrical engineering from the National Institute of Technology Durgapur, Durgapur, West Bengal, India, in 2018.

Josep M. Guerrero is a professor with the Department of Energy Technology, Aalborg University, Denmark. His research interests include power electronics, distributed energy-storage systems, energy management systems, smart metering and the internet of things for AC/DC microgrids. He serves as an associate editor for the *IEEE Transactions on Power Electronics, IEEE Transactions on Industrial Electronics*, and *IEEE Industrial Electronics Magazine,* and as an editor for the *IEEE Transactions on Smart Grid.*

Debaprasad Kastha is a professor at the Indian Institute of Technology Kharagpur, India. Prior assignments include the Research and Development Division of Crompton Greaves, Ltd., Mumbai. He has been performing research on power electronics and drives for more than two decades and has authored or co-authored about 50 technical papers, books, and electronic teaching aids. His research interests include the areas of wind power generation, machine drives, dc power supply, and distribution systems.

Tapas Kumar Saha is a professor at the National Institute of Technology Durgapur, India. He is well known in the field of Power Electronics and Machine Drives and their applications in Renewable Energy Generation. He also serves as a reviewer of numerous IEEE transactions and conferences. His current research interests include control and implementation of renewable energy generation systems through power electronics and machine drives.

Chapter 1

Characteristics and modeling of wide band gap (WBG) power semiconductor

S. Toumi[1,2,3]

Wide band gap (WBG) materials are an interesting class of semiconductors that offer a key criterion to the modern technological applications in terms of their high efficiency, high-frequency power applications, and high operating temperature and voltage. The famous families of the WBG semiconductors are the silicon carbide (SiC), the gallium nitride (GaN), the gallium oxide (Ga_2O_3), the diamond, and the aluminum nitride (AlN). Attention has turned to work on basic electronic devices like Schottky diodes, solar cells, IGBT transistors, and HEMTs using the WBG semiconductors to perform the characteristics of the device and to reach the suitable properties for the desired technological applications.

In this chapter, we present the basic physical properties and the principle techniques for the characterization of the WBG materials. We present also the physics background of some powerful semiconductor devices based on WBG materials which allow an important achievement in the recent research and development requested by the modern technologies and the highly advanced science of such class of semiconductor materials. We conclude this chapter by a case study where we investigate the interface state of the Schottky contact based on 4*H*-SiC as a WBG material. The studied contacts are formed by a metal (molybdenum/tungsten) deposited on a WBG semiconductor (4*H*-SiC). The analysis of these contacts is made via the current–voltage (*I–V*) measurement of different temperatures. As a result of this study, the existence of the inhomogeneity of the interface metal/4*H*-SiC is evidenced by the temperature behavior of the physical parameters characterizing these structures.

[1]Couches minces et Hétérostructures, Unité de Recherche: Matériaux, Procédés et Environnement, University M'hamed Bouguara, Algeria
[2]Department of Physics, Faculty of Science, University M'hamed Bouguara, Algeria
[3]Optoelectronic Devices Laboratory, University Ferhat Abbas, Algeria

Nomenclature

A	diode area
A^*	Richardson's constant
a	lattice constant
$BFOM$	Baliga figure of merit
$BHFOM$	Baliga's high-frequency figure of merit
C	speed of light
C-SiC	cubic silicon carbide system
c	lattice constant
$CFOM$	combined figure of merit
D_h	diffusion constant of holes
E	electric field
E_B	breakdown electric field strength
E_g	band gap energy
E_{g0}	band gap at 0 K
FOM	figure of merit
H-SiC	hexagonal silicon carbide system
I	current of the charge carriers
I_B	base current
I_E	emitter current
I_C	collector current
I_s	saturation current
$JFOM$	Johnson's figure of merit
$KFOM$	Keyes figure of merit
k	Boltzmann constant
m	effective mass of the charge carrier
n	Refractive index, ideality factor
n_i	intrinsic carrier density
N_A	acceptor density
N_D	donor density
N_C	effective density of states in the conduction band
N_V	effective density of states in the valence band
P	polarization
P_{PE}	piezoelectric polarization
P_{SP}	spontaneous polarization
q	elementary electric charge
R_s	series resistance

R-SiC	rhombohedral SiC system
S	vertical quadratic error
SiC	silicon carbide
T	absolute temperature
V_B	breakdown voltage
W	width of the space charge region
WBG	wide band gap
α_{opt}	absorption parameter
α-SiC	hexagonal SiC system
β-SiC	cubic SiC system
β	inverse thermal voltage
ε_s	dielectric constant of the semiconductor
ϕ_{B0}	zero-bias barrier height
$\overline{\phi}_{B0}$	mean value of the Schottky barrier high
λ	wavelength
μ	electric mobility
μ_h	drift mobility of holes
v_d	drift velocity
v_S	sound velocity
v_{sat}	saturated velocity
ρ_2	coefficient quantifying the barrier high deformation
ρ_3	coefficient quantifying the barrier high deformation
σ_s	standard deviation

1.1 Introduction

Today, the technology of the semiconductor devices reached a very important step especially in the type of the semiconductor materials used in their fabrication. The ancient semiconductor material used in the electronic power devices is silicon. However, the physics and the technological application of the silicon reach the limits. There is another class of semiconductor materials offering important characteristics that are requested by the modern technology. Those materials are called the wide band gap (WBG) semiconductors citing the silicon carbide (SiC), the gallium nitride (GaN), the gallium oxide (Ga_2O_3), the diamond, and the aluminum nitride (AlN). These semiconductor materials give a considerable potential for the power electronic application. The fabricated devices based on the WBG materials have the following performances: high efficiency, high-frequency power applications, and high operating temperature and voltage. The power device fabricated using the SiC with a band gap of 2.86 eV could minimize the power loss significantly. It could also combine two devices at the same time like MOSFET in combination with rectifier devices like Schottky diodes. The high electron mobility semiconductor such as GaN with a band gap of 3.4 eV is used in the devices offering advantages for the power switching. The other WBG materials like the gallium arsenide (GaAs) with a band gap of 1.4 eV, diamond and oxides are an excellent choice for the technology of the electronic devices. The semiconductor materials with a WBG presented previously are significantly employed in the power devices as well as for the high-temperature applications. Those materials have interesting properties citing high saturation electron drift velocity, high thermal conductivity, high breakdown electric field, and remarkable physical and chemical stability. All these advantages could guarantee a variety of electronic devices that operate at elevated temperatures more than 600°C for a very high-power levels. The elaborated devices with the WBG semiconductors can also operate at high saturation electron drift velocity so that they can be used for devices operating at elevated frequency like RF and microwave. High breakdown electric field permits the realization of high-power devices which make the possibility to use them in integrated circuits.

The basic power device structures for electronic and photonic devices are the *p–n* junction diode. It can be realized by doping the semiconductor with acceptor or donor impurities to get the two regions of the diode (the *p* and *n* regions). We can have two classes of *p–n* junction diode: the homojunction and the heterojunction diode. The last ones are fabricated using different semiconductor materials in contrary to the first one using just a single semiconductor for both regions. Another power basic device serves as common devices in a very large type of semiconductor devices. It is the Schottky diode formed by a deposition of a metal on a semiconductor material. It serves to contact metal wires to silicon substrates. In addition to these basic electronic devices, the family of the transistor with all their types (CMOS, BJT, MOSFET, IGBT, etc.) trends to the miniaturization, the increase in operation speed, and the minimization in power consumption.

The physics of the transport mechanism of the charge carriers in the semiconductor devices is of great importance. For a specified electronic device, different mechanisms of current could exist and the understating of such currents could be principle for particular technological applications. The basic equations modeling the physics phenomenon in any semiconductor device are governed on the one hand by the theoretical *p–n* junction diode equation. On the other hand, the geometry of the semiconductor device is also crucial in the equation governing the transport mechanism of the charge carriers, most of the MOS device parameters depend on the length L and the width W characterizing the conducting channel between the source and the drain. The physics of the transport mechanisms of charge carrier in the semiconductor could be modeled by the famous relation between the energy $\boldsymbol{E}$ and the momentum $\boldsymbol{k}$ for the charge carriers in the studied material, and it is important to identify the interactions with phonons and photons. This relationship allows us to have details about the band structure in the considered semiconductor material. The energy bands for the semiconductor materials have been studied using different numerical formalism citing for example the pseudopotential, the orthogonalized plane-wave and the $\boldsymbol{k.p}$ methods.

SiC is one of the key WBG materials used in the new technology of the electronic devices. It offers applications with high power, frequency, and temperature compared to silicon as a material for the electronic devices. The SiC is a semiconductor with an elevated saturation velocity and high thermal conductivity. The crystalline structure of the SiC is very diverse and contains a lot of sequences of the atomic plans which give a different kind of polytypes. The arrangement of the SiC units defines two structures for these WBG materials, *C* for the cubic system of the SiC and *H* for the hexagonal one. From the point of view of the technological applications, the structures such as 3*C*, 4*H*, and 6*H*-SiC are the most important. Another interesting application results just from varying the stacking of the atomic plan in the SiC crystalline structure, it changes completely the band gap of the material. For example, the band gap in 3*C*-SiC is equal to 2.39 eV but it is equal to 3.023 eV for 6*H*-SiC and 3.265 eV for the 4*H*-SiC structure.

In this chapter, we introduce the WBG materials by citing the basic physical properties and the principle techniques to characterize those materials, followed by the physics governing some power semiconductors devices based on WBG materials. Recent research and development are also aimed in this chapter to open a big window to challenges and opportunities requested by the modern technologies and the highly advances science of such class of semiconductor materials. We conclude this chapter by a case study where we investigate the interface state of the Schottky contact based on WBG materials such as 4*H*-SiC. The studied contacts are formed by a metal (molybdenum/tungsten) deposited on 4*H*-SiC. The fabricated Schottky diode using the WBG semiconductor like 4*H*-SiC shows a very interesting characteristics for high temperature compared to those fabricated with silicon Si showing an acceptable characteristic for a limited temperature going to just 125°C. The characterization of these diodes is made via the current–voltage (*I*–*V*) measurement for different temperatures. As a result of this study, the existence of the inhomogeneity of the interface metal/4*H*-SiC is evidenced by the temperature behavior of the physical parameters characterizing these structures. Finally, almost all the physics that govern the WBG semiconductor materials will be treated in this chapter carefully citing the structural properties and the current mechanisms present in such materials and the electronic devices using them.

1.2 Overview

The recent technology with all their aspect of elaboration and processing of new families of semiconductor offers us a new generation of semiconductor called the WBG materials. Among this huge group of semiconductors, we divide them as the conventional WBG materials and the ultra-WBG materials. The most mature materials in terms of process and device technology are the SiC and the GaN classified as a conventional WBG materials. For the ultra-WBG semiconductors we could cite materials like diamond, gallium oxide GaO_2, cubic boron nitride BN, and aluminum GaN on AlN AlGaN/AlN. The SiC is considered as the second most important conventional WBG material after GaN. It has an extreme hardness caused by their very strong chemical bonds. Consequently, it has a remarkable melting point. Such properties and other, that we will state later in this section, make the SiC as an excellent semiconductor for protecting other materials and for mechanical cutting applications. The commercialization of the SiC technology is very cheap comparing with other WBG materials such as diamond, making it as the most used semiconductor for the recent applications [1]. In what follows we will cite the important physical properties of the different classes of the wide and ultra-WBG semiconductors starting with SiC, GaN and finiliazing by the diamond and the oxides such as β-Ga_2O_3, α-Al_2O_3, In_2O_3, SnO_2, and $CuAlO_2$.

1.2.1 Physical properties of SiC

SiC is a semiconductor material that could be crystallized in a large variety of structures more than 200 structures. Each one demonstrates exceptional electrical, optical, thermal, and mechanical properties. These physical properties are excellent subjects for academic analysis and could be used for accurate simulation of devices with SiC materials. The crystal structure of SiC is formed by four atoms of carbon nabouring one Si atom with a 4.6 eV as a bond energy which gives this semiconductor all the important properties cited below. The SiC is known as the best candidate for the polytypism. It can be adopted for several crystal structures varying in stacking sequence without changes in chemical composition. The zinc blende or wurtzite structure is the stable one. To classify the huge family of the polytypes of SiC taking in count the stacking structures, the most known SiC polytypes are illustrated by the notation of Ramsdell, Zhdanov, and Jagodzinski [2]. In Ramsdell's classification, polytypes are described by the crystal structure and by the number of Si-C bilayers in the unit cell as follows: *C* for cubic, *H* for hexagonal, and *R* for rhombohedral. Habitually, 3C-SiC is called β-SiC and the other polytypes like 4*H*-SiC and 6*H*-SiC are referred to as α-SiC. Table 1.1 illustrates the lattice constants of the famous SiC polytypes at room temperature [3].

The variation of the lattice constants is very remarkable from Table 1.1, because the crystal structure differs for each polytype. The effect of the temperature and the doping density is also observed on the lattice constants which is the case for all the semiconductor materials. Another important physical property is the band gap energy E_g characterizing the band structure of the different polytypes. In general, the semiempirical expression of the band gap energy E_g is given by

Table 1.1 Values of the lattice constants for the main SiC polytypes at room temperature

Lattice constants	3C-SiC	4*H*-SiC	6*H*-SiC	2*H*-SiC [4]	15R-SiC [4]
a	4.3596 A°	3.0798 A°	3.0805 A°	0.3076 nm	0.30817 nm
c(A°)	–	10.0820	15.1151		

Eq. (1.1) [5]:

$$E_g = E_{g0} - \frac{\alpha T^2}{T + \beta} \tag{1.1}$$

E_{g0} is the band gap at 0 K, T is the absolute temperature, and α, β are the fitting parameters.

The values of E_{g0} and the fitting parameters α, β are illustrated in Table 1.2 for the 3*C*-SiC and 4*H*-SiC materials .

Accordingly, for each SiC polytype, we define a band gap energy varying from one type to another due the band structure and stacking sequence of the polytype. For example, the value of E_g at the room temperature is 2.36 eV for 3*C*-SiC, 3.26 eV for 4*H*-SiC, and 3.02 eV for 6*H*-SiC.

The direct physical parameters resulting from the band structure are the absorption coefficient α_{opt} and the refractive index n especially when an optical technique is used in the materials characterization.

The formula of the absorption parameter α_{opt} can be approximated by Eq. (1.2) [7]:

$$\alpha_{opt} = \frac{A_{ab}}{h\nu}\left(\frac{\left(h\nu - E_g + \hbar\omega\right)^2}{e^{\left(\frac{h\nu}{kT}\right)} - 1} + \frac{\left(h\nu - E_g - \hbar\omega\right)^2}{1 - e^{\left(\frac{-h\nu}{kT}\right)}}\right) \tag{1.2}$$

$\hbar\omega$ is the energy of the involved phonon, ν is the photon energy, k is the Boltzmann constant, and A_{ab} is a parameter to be defined.

The variation of refractive index $n(\lambda)$ as a function of the wavelength is given by Eq. (1.3):

$$n(\lambda) = A + \frac{B\lambda^2}{\lambda^2 - C^2} \tag{1.3}$$

A, B, and C are parameters to be defined.

The band gap level E_g is strongly related to the effective densities of states in the conduction band N_C, in the valence band N_V and to the intrinsic carrier density n_i as shown by Eq. (1.4) [8]:

$$n_i = \sqrt{N_C N_V} e^{\left(-\frac{E_g}{2kT}\right)} \tag{1.4}$$

Table 1.2 Some values of the parameters charactering the band gap energy E_g *for the 3C-SiC and the 4H-SiC [6]*

Parameters characterizing the band gap, E_g	**3C-SiC**	**4*H*-SiC**
$E_g(0)$ (eV)	2.39	3.265
α (eV/K)	0.66×10^{-3}	3.3×10^{-2}
β (K)	1,335	1.0×10^5

One of the important physical properties giving the current of the charge carriers (electron/hole) in the semiconductor is the electric mobility μ. The electron/hole mobility in the 4*H*-SiC and in the 6*H*-SiC expressed in $cm^2\ V^{-1}\ s^{-1}$) is given by the following equations [9]:

$$\begin{cases} \mu_{electron}(4H - SiC) = \dfrac{1020}{1 + \left(\dfrac{N_D + N_A}{1.8 \times 10^{17}}\right)^{0.6}} \\ \mu_{electron}(6H - SiC) = \dfrac{450}{1 + \left(\dfrac{N_D + N_A}{2.5 \times 10^{17}}\right)^{0.6}} \end{cases} \tag{1.5}$$

$$\begin{cases} \mu_{hole}(4H - SiC) = \dfrac{118}{1 + \left(\dfrac{N_D + N_A}{2.2 \times 10^{18}}\right)^{0.7}} \\ \mu_{hole}(6H - SiC) = \dfrac{98}{1 + \left(\dfrac{N_D + N_A}{2.4 \times 10^{18}}\right)^{0.7}} \end{cases} \tag{1.6}$$

N_D and N_A are the donor and the acceptor densities, respectively, expressed in cm^{-3}.

It has to be noted that the electric mobility of the charge carrier (electron/hole) is strongly dependent on the doping density. Usually, the electric mobility versus the temperature T varies as $\mu \sim T^{-n}$. The values of n are taking in the range of [2.4–2.8] for lightly doped material with densities in the range of $[10^{14}–10^{15}]\ cm^{-3}$ for 240 K$<T<$600 K and in the range of [1.8–2.4] for moderately doped material $[10^{16}–10^{17}]\ cm^{-3}$ for 280 K$<T<$600 K.

For each power devices using semiconductor junction, we define the breakdown properties characterizing by the critical electric field strength or the breakdown electric field strength E_B and the corresponding breakdown voltage V_B. It can be extracted experimentally from the breakdown characteristics of the semiconductor device when the electric field is crowding. Since the diode junction is the principle content of any semiconductor devices, the V_B for a diode junction is defined by Eq. (1.7) [10]:

$$V_B = \frac{\varepsilon_S E_B^2}{2qN_D} \tag{1.7}$$

ε_s is the dielectric constant of the used semiconductor.

It is essential to note that the critical electric field is increasing with the increase of the doping density. Different analyses reported [11–13] that the value of E_B of 4*H*- and 6*H*-SiC cutting at <0001> direction is approximately higher by a factor of eight than for the Si for a specific doping density. However, for 3*C*-SiC cutting at <111> direction, it is higher just by a factor of three or four comparing to the one of the Si, because the band gap for the polytype 3*C*-SiC is small relative to the other polytypes. The physical explanation of this behavior is that when the doping density rises, the width of the space charge region W decreases, so the charge carriers will be accelerated in small distance. Accordingly, the mobility is reduced (caused by the scattering phenomenon) in such highly doped semiconductor. The formula of the width $W(V_B)$ is given by Eq. (1.8) [14]:

$$W(V_B) \approx \frac{2V_B}{E_B} \tag{1.8}$$

An important parameter characterizing the high field regime is the drift velocity v_d. In this regime, the electric field dependence of the drift velocity does not have a linear behavior because the charge carriers transfer more energy to the lattice so more phonons are emitted. The variation of the drift velocity as a function of electric field E is given by Eq. (1.9) [15]:

$$v_d = \frac{\mu E}{\left(1 + \left(\frac{\mu E}{v_S}\right)^{\gamma}\right)^{1/\gamma}} \tag{1.9}$$

v_S is the sound velocity and γ is a parameter.

For a particular electric field, the drift velocity is saturated. Then, the saturated velocity v_{sat} is expressed by the following equation [16]:

$$v_{sat} = \frac{8\hbar\omega}{3\pi m^*} \tag{1.10}$$

$\hbar\omega$ is the energy of the emitted phonon and m^* is the effective mass of the charge carrier.

Lophitis *et al.* [6] gave the different values for the saturation velocity for the electrons and the holes as a charges carrier in the used SiC materials. We resume those values in Table 1.3.

At the end of this under-section, we resume a maximum of the different physical properties of the most used SiC polytypes in Table 1.4.

1.2.2 Physical properties of nitrides semiconductor

Another famous class of the WBG semiconductors is known as the nitrides materials. The nitrides semiconductor based on the GaN as the principle constituent is

Table 1.3 Values of v_{sat} *for several polytypes of SiC material compared to silicon Si*

Parameter	**Si**	**4*H*-SiC**	**3C-SiC**	**6*H*-SiC**
Saturated electron drift velocity, v_{sat} ($\times 10^7$ cm/s)	1 [17]	2.2 [17]	4.38 (as a mobility model parameter) [18]	1.9 [17]

Table 1.4 Major physical properties of the most used SiC polytypes [19]

Physical properties	3C-SiC	4*H*-SiC	6*H*-SiC
Bandgap (eV)	2.36	3.26	3.02
Electron effective mass			
$m_{//}(m_0)$	0.67	0.33	2.0
$m_{\perp}(m_0)$	0.25	0.42	0.48
Hole effective mass			
$m_{//}(m_0)$	~1.5	1.75	1.85
$m_{\perp}(m_0)$	~0.6	0.66	0.66
Number of conduction band minima	3	3	6
Effective density of states in the conduction band (cm^{-3})	1.5×10^{19}	1.8×10^{19}	8.8×10^{19}
Effective density of states in the valence band (cm^{-3})	1.9×10^{19}	2.1×10^{19}	2.2×10^{19}
Intrinsic carrier density (cm^{-3})	0.1	5×10^{-9}	1×10^{-6}
Breakdown electric field (MV cm^{-1}) (at $N_D = 3 \times 10^{16}$ cm^{-3})			
E_B perpendicular to *c*-axis	1.4	2.2	1.7
E_B parallel to *c*-axis	1.4	2.8	3.0
Relative dielectric constant			
ε_S perpendicular to *c*-axis	9.72	9.76	9.66
ε_S parallel to *c*-axis	9.72	10.32	10.03
Young modulus (GPa)	310–550	390–690	390–690

mainly formed by the family of three compounds as InN, GaN, and AlN. The huge variety of alloys that could be realized by this family of nitrides gives the advantages to get the widest tunable range in wavelength and consequently maintaining their physical and chemical properties in good quality. For these raisons, the elaboration techniques of such materials are the key issue. The GaN-based semiconductors exist in different crystalline structures citing the wurtzite, the zinc blende, and the rock salt structure. For the materials AlN, GaN, and InN as a bulk, the wurtzite structure is the most thermodynamically stable one in the ambient conditions. But the zinc blende structure for GaN and InN is the stable one if they are used as thin films on substrates like Si, SiC, MgO, and GaAs grown on the direction (011). However, the rock salt structure is used for AlN, GaN, and InN in very high-pressure conditions [20]. The parameters of the wurtzite structure of the nitrides AlN, GaN, and InN are given in Table 1.5, the internal displacement parameter u is defined as the anion–cation bond length along *c*-axis, in the unit of *c*.

In the fabrication process of the nitride materials, the choice of the fabrication mechanism plays a key role in the evaluation of the band gap energy of such materials. For example, the quality of the InN films deposited by beam epitaxy gives a value of 0.7 eV as band gap energy instead of 1.9 eV [21]. This is a very important finding; it gives a new research area of narrow band gap for the nitrides semiconductor. Accordingly, the spectral region for those materials is enlarged ranging from near infrared (at $\lambda = 1.9\ \mu m$ with $E_g = 0.7$ eV) for InN to ultraviolet (at

Table 1.5 Wurtzite crystalline structure parameters for the principle nitrides materials evaluated at T = 300 K [21]

Crystalline structure parameters	GaN	AlN	InN
a (A°)	3.189	3.112	3.545
c (A°)	5.185	4.982	5.703
u^2	0.376c	0.382c	0.377c
TEC ($\times 10^{-6}$ K^{-1})	a5.59/c3.17	a4.15/c5.27	–

$\lambda = 0.36\ \mu$m with $E_g = 3.4$ eV) and deep ultraviolet (at $\lambda = 0.2\ \mu$m with $E_g = 6.2$ eV) for GaN and AlN, respectively. Taking into account Eq. (3.1), the band gap energy in the nitrides materials is calculated. For the semiconductors GaN, AlN, and InN, the band gap energies are 3.4 eV, 6.2 eV, and 0.7 eV, respectively, with the fitting parameters α (meV/K) taking values of 0.909, 1.799, and 0.245, β (K) having values of 830, 1462, and 624 correspondingly [6, 22].

In nitride semiconductors the electric polarization effects are also present. In crystal structure based on GaN (i.e. wurtzite structure), the polarization is oriented parallel to the direction along the *c*-axis. The total polarization P is calculated by Eq. (1.11) [23]:

$$P = P_{SP} + P_{PE} \tag{1.11}$$

P_{SP} and P_{PE} are the spontaneous and the piezoelectric polarization, respectively.

The first principle calculations (*ab initio*) based on the density functional theory (DFT) permit the evaluation of the two polarizations using different approximations like Berry-phase approach, local-density approximation (LDA), generalized gradient approximation (GGA), and Heyd–Scuseria–Ernzerhof (HSE) screened-exchange hybrid functional. In Figure 1.1, we illustrate the different values of the spontaneous polarization charge calculated with different approximations for several nitrides materials.

The polarization effects favorite the spin splitting properties in the nitrides materials which make them a good candidate for spintronic applications [30–32].

To describe the mobility μ in GaN and AlN WBG materials, the Masetti model [33] given by Eq. (1.12) is the more adequate to calculate it [6]. This model is a temperature independent and gives all the parameters at T= 300 K:

$$\mu = \mu_{\min 1} e^{-\frac{P_c}{N_D+N_A}} + \frac{\mu_{cont} - \mu_{\min 2}}{1 + (N_D + N_A/C_r)^{\alpha 1}} - \frac{\mu_1}{1 + (C_s/N_D + N_A)^{\beta 1}} \tag{1.12}$$

The fitting parameters used to calculate the mobility are summarized in Table 1.6.

As the SiC, the nitride materials are also known for their high critical electric field and their saturation velocity as it is shown in Table 1.7.

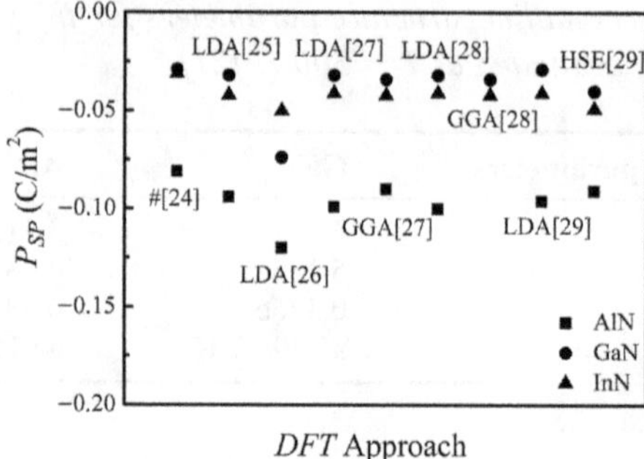

Figure 1.1 Spontaneous polarization charge of nitrides semiconductor reproduced from Ref. [23]

Table 1.6 Fitting coefficients used for evaluating the mobility μ for the electrons and holes as charges carriers for the GaN and AlN WBG semiconductors [6]

Fitting parameters	GaN		AlN	
	Electrons as charge carriers	Holes as charge carriers	Electrons as charge carriers	Holes as charge carriers
μ_{cont} (cm^2/Vs)	1,500–1,800	20	300	14
$\mu_{\min 1}$ (cm^2/Vs)	85	33	20	11
$\mu_{\min 2}$ (cm^2/Vs)	75	0	65	0
μ_1 (cm^2/Vs)	50	20	20	10
P_c (cm^{-3})	6.5×10^{15}	5×10^{15}	8×10^{17}	5×10^{18}
C_r (cm^{-3})	9.5×10^{16}	8×10^{16}	7×10^{16}	8×10^{17}
C_s (cm^{-3})	7.2×10^{19}	8×10^{20}	5.2×10^{17}	8×10^{18}
α_1	0.55	0.55	0.88	1.05
β_1	0.75	0.7	0.75	0.75

Table 1.7 Some physical properties of GaN as an epitaxial and as a bulk [34]

Physical properties	GaN	GaN (bulk)
Electric break down field E_B (V/cm)	2×10^6 (epitaxial)	3.3×10^6
Saturation velocity v_{sat} (cm/s)	3×10^7 (epitaxial)	3×10^7
Thermal conductivity (W/K cm)	1.3 [35]	2.3

Despite the great properties offered by the nitride semiconductors, they have some disadvantages compared to the SiCs. The nitrides do not have a native oxides with make the unfeasibility to fabricate the MOS structures with GaN. The technological process to grow the GaN boules is very difficult. Because of that, the pure wafers of GaN are not possible for elaboration but they are deposited on SiC or sapphire. So, the GaN wafers are very expensive compared to those fabricated with the SiC [14].

Nowadays, a new family of the WBG materials called ultra-WBG semiconductors (*next generation of the WBG semiconductors*) offered important properties like AlGaN and BN [36,37]. This category of WBG materials will not be discussed in detail in this chapter since we limit our self in this chapter just for the conventional WBG semiconductor. However, some materials of this class are listed in Section 1.7.

Recently, the development in the technology of the electronic device uses another WBG material called diamond. In general, all the applications using diamond utilize one or more of their excellent properties such as high refractive index and optical dispersion, extreme hardness, in addition to its high thermal conductivity and unique electronic properties like the large band gap and the high carrier mobilities for both electrons and holes. This *excellent* WBG material has a high saturation velocity, high breakdown field, low dielectric constant, and negative electron affinity. These properties make the diamond very interest for different device applications citing for example power transistors operating at high power levels, high frequencies, high temperatures, radiation detectors, and high-voltage diodes [38,39]. Diamond is formed by carbon atoms; its crystalline structure generally is diamond (cubic with $a = 3.5669$ Å as a lattice constant [40,41]) and graphite (hexagonal with $a = 2.461$ Å and $c = 6.708$ Å as a lattice constant). The values of the density of state of the valence and the conduction band at $T = 300$ K are 1.8×10^{19} and 5×10^{18} cm^{-3}, respectively [6]. In Table 1.8, some of the properties of diamond are listed in comparison with other WBG materials.

To fabricate power electronic devices using diamond, we have to elaborate *p*/*n*-type diamond technologies. The *p*-type diamond, additionally exists in nature, can be processed without problem by using boron as a dopant element. It forms an acceptor level at 0.37 eV over the balance band maximum [42]. However, the *n*-type does not exist in nature and represents technological difficulties to process it. Several techniques have been investigated to fabricate *p*/*n*-type diamond like

Table 1.8 Some physical properties of diamond compared to other WBG materials [14]

Properties	6*H*-SiC	4*H*-SiC	GaN	Diamond
E_g (eV)	3.03	3.26	3.45	5.45
Dielectric constant ε_r	9.66	10.1	9	5.5
Electric break down field (KV/cm)	2,500	2,200	2,000	10,000
Electron mobility (cm^2/V s)	500	1,000	1,250	2,200
	80			
Hole mobility (cm^2/V s)	101	115	850	850
Thermal conductivity (W/cm K)	4.9	4.9	1.3	22
Saturation velocity (cm/s)	2×10^7	2×10^7	2.2×10^7	2.7×10^7

Table 1.9 Properties of some WBG oxides [4,36]

Characterized parameters	α-Al_2O_3	β-Ga_2O_3	In_2O_3	SnO_2	$CuAlO_2$
Crystalline structure	Corundum	Monoclinic	Bixbyite	Rutile	Delafossite
Angle (°)	α=55.28	β=103.83	–	–	–
a_0 (nm)	0.51284	1.2214	1.0117	0.47397	0.2857
b_0 (nm)	–	0.30371	–	–	0.2857
c_0 (nm)	–	0.57981	–	0.31877	1.6939
E_g (eV)	8.3	4.8	3.6	3.6	2.54 [46]
Thermal conductivity	~15 W/mK [47]	10–30 W/mK [48]	0.58 W/mK (as a meso-porous) [49]	// to the c axis 0.98 W/cm K⊥ to the c axis 0.55 W/cm K [50]	3.5 W/mK (at room temperature, material as a bulk) [51]
Saturation velocity (m/s)	–	2×10^7 [52]	–	–	–
Breakdown electric field	320 kV/mm as a nanocomposites [53]	8 MV cm^{-1} [44]	~7.2 MV/cm (In_2O_3 on an ultrathin Solution (ZrO_x) as dielectric thin film [54])	3,990 V/cm [55]	–

microwave plasma-enhanced CVD (MPACVD) and hot filament-assisted CVD. The growth uses as substrate Si, SiC, sapphire, and tungsten carbide. For the electronic application, the MPACVD is the appropriate technique to elaborate the diamond films. The first diamond electronic device was a point contact diode working at elevated temperature reaching 573 K [43]. The research and development in the diamond electronic devices and its processing mechanisms are still new and it needs more years to investigate them very well.

In addition to the wide bad gap materials cited above, there is another class of them, the WBG oxides such as the gallium oxide β-Ga_2O_3, α-Al_2O_3, In_2O_3, SnO_2, and $CuAlO_2$ [36,44,45] classified also as ultra-WBG materials. Table 1.9 illustrates some the important properties of those oxides.

1.3 Power semiconductor devices

Power semiconductor devices have marked a double of efficiency every 3 years for the last 18 years. The principle aim in fabricating those devices is to achieve the faster, the smaller, the lighter, and the more reliable power semiconductor with keeping reductions in overall system cost [56]. In order to realize those criterions, many investigations are made such as the calculation of the figure of merit (*FOM*). The higher value of *FOM* indicates that the performances for the material are to be used as a power device. Basically, the FOM relates the physical properties with the properties of the application domain. Several figures of merit are proposed to scale different semiconductor materials. The *FOM* which gives information of the used WBG material for high-power application at high frequency is Johnson's FOM (*JFOM*) defined by Eq. (1.13) [57]. Which means that the WBG material with the two-dimensional electron gas (2DEG) will be much suitable for high-power application at high frequency [58]:

$$JFOM = \left(\frac{E_c v_{sat}}{2\pi}\right)^2 \tag{1.13}$$

E_c is the high breakdown electric field.

To choose a WBG material that operates at lower frequency with dominant conduction losses, we have to calculate another *FOM* called the Baliga *FOM* expressed by Eq. (1.14). The *BFOM* permits us to define the intrinsic limit of the material:

$$BFOM = \varepsilon_r \mu E_g^3 \tag{1.14}$$

To take into account the domination of the switching loses at high frequency, we calculate the Baliga's high-frequency *FOM* (*BHFOM*) given by Eq. (1.15) [59]. From this equation, we can conclude that the device with very high breakdown voltage cannot be applied in high-frequency operation:

$$BHFOM = \frac{\mu E_C V_G^{0.5}}{2V_B^{1.5}} \tag{1.15}$$

V_G is the gate drive voltage.

If we consider the thermal limitations due to transistor switching, we have to calculate the Keyes *FOM* (*KFOM*) defined by Eq. (1.16) [60]:

$$KFOM = \chi\left(\frac{Cv_{sat}}{4\pi\varepsilon_r}\right)^{0.5} \tag{1.16}$$

C is the speed of the light.

The last FOM is the combined *FOM* (*CFOM*) which takes in consideration the high frequency, the high power, and the high temperature applications. It is given in Eq. (1.17):

$$CFOM = \chi \varepsilon_r \mu v_{sat} E_C^2 \tag{1.17}$$

From Table 1.10, we can clearly remark the very good different *FOM* calculated for basic WBG materials compared to silicon [61].

Table 1.10 Values of FOMs for the basic WBG material used in power electronics in comparison with Si

Materials	Si	GaN	SiC
Johnson's *FOM (JFOM)*	1	270–480	324–400
Keyes *FOM (KFOM)*	1	1.4	4.5–4.8
Baliga's *FOM (BFOM)*	1	17–34	6–12
Baliga's high-frequency *FOM (BHFOM)*	1	86–172	57–76
Combined *FOM (CFOM)*	1	108–290	275–310
T_{max} (°C)	300	700	600

More details about the different power semiconductor (WBG) devices and their applications are cited in Section (1.5) of this chapter.

1.4 Characterization and modeling of WBG

The characterization of any materials is based on the determination of their physical properties and the defects present in the structure. It is crucial to understand the nature of these defects because they could affect the reliability's and the performance's device [62]. In general, all the characterization techniques used for the other materials are applicable for the case of the WBG semiconductor. But special care is done for the WBG materials like: the important value of gap for such materials and the sample structure (in the case of the SiC where the processing of the polytypes depends on the device application) taking into account the application domain of the fabricated WBG device. In what follows, we cite some of the principle techniques used in the characterization of the WBG materials.

1.4.1 Photoluminescence (PL)

The PL is an excellent technique for the evaluation of both defects and the purity of the semiconductor materials. The PL characterization is made at low temperature (2, 4.2, and 77 K). The temperature dependences of the PL spectra are used to understand the recombination paths. Principally, more the structure of the WBG material is complicated, more the PL will be, especially for SiC spectra [63–67].

1.4.2 Raman scattering

The Raman spectroscopy is a powerful tool used to characterize the physical properties of the investigated material. It is based on the interaction of the light with the solid materials leading to a different scattered wavelength that gives important information about the material under study. The most famous interaction of the light is the one with the phonons, it is called *Raman scattering.* For the reason that the Raman scattered light contains detailed information about the phonons, Raman scattering spectroscopy is an important technique for the identification of the material polytype and the characterization of the stress or the defects specially in SiC. The Raman scattering technique also permits the determination of the carrier density when the material is highly doped [68,69].

1.4.3 Hall-effect and capacitance–voltage (C–V) measurements

To determine the free carrier density and the mobility of the material under characterization, we apply the Hall-effect technique. However, the net doping density (i.e. in Schottky structure) can be measured by the variation of the capacitance as a function of voltage in this material. The use of these two techniques requires some conditions in the preparation of the sample to be characterized [70]. Since we are limited in this chapter book, we do not detailed these characterization techniques.

1.4.4 Carrier life time measurements

One of the important physical properties in the material is the carrier life time. It could be measured by several techniques citing for example: time-resolved PL (TRPL), the photo-conductance decay (PCD), and the reverse recovery (RR). In the TRPL characterization, the PL peak is proportional to the number of excess free carriers. It is mentioned that there are no conditions to be done to material under analysis in this type of carrier life time measurements. For the PCD, the carrier life time is determined by the measure of the transient electrical conductance after being pulsed by a photo excitation. In this class of carrier lifetime characterization, some simple processing preparation made to the material under investigation is required. The last technique used in the characterization of the carrier lifetime is RR. It is principally used to determine the stored carriers in the reverse regime in the diodes.

1.4.5 Detection of extended defects

The extended defects (surface irregularities) are known to have a great impact on the electronic structure of the semiconductor material and in return the performances of the electronic devices. It can be classed into two parts: the dislocation and grain boundaries. Many characterization techniques are used to measure these defect citing: the chemical etching [71], the X-ray topography [72], the photoluminescence mapping and imaging [73], and the high-resolution mapping of surface morphology [74].

1.4.6 Detection of point defects

The point defects are classified as vacancies when an atom (or an atom pair) is omitted and as interstitials (the extra atom) which are situated in a normally unoccupied structural site [75]. The point defects can be characterized by the deep level transient spectroscopy (DLTS) and by the electron paramagnetic resonance (EPR) when the point defect has a spin [76].

1.4.7 Secondary ions mass spectroscopy (SIMS)

The SIMS technique permits us to determine the unintentional/intentional impurities presence, their concentration, their depth distribution, and the changes in the depth distribution with processing. More details about this technique and its application in different WBG materials can be found in Ref. [77].

1.4.8 Modeling the WBG materials

To model some of the physics behavior of WBG semiconductor, we have to investigate principally its electronic structure (its band structure indicating the density of states in the valence and the conduction band) and the charge carrier current inside this material. To explain the transport of the charge carrier, we have to use the Boltzmann theory to build up a microscopic model for fundamental macroscopic quantities like conductivity, diffusion coefficient, and mobility. The current expression for several WBG devices will be discussed in Section 1.5.

The relationship between the energy $\boldsymbol{E}$ and the momentum $\boldsymbol{k}$ for the charge carriers in the studied material allows us to have details about the band structure in the considered semiconductor material. The electronic band structure for a semiconductor material has been studied using different numerical approximation formalism citing for example the pseudopotential, the orthogonalized plane-wave, and the $\boldsymbol{k.p}$ approach. The last one is the most used approximation to determine the band structure of three and lower dimensional materials like quantum dots wire, wells. This approach is very accurate near the bandedges. For the $\boldsymbol{k.p}$ formalism, we start the calculations with a known form of the band structure for the problem at the bandedges and after that we use the perturbation theory attempts to describe the bands away from the high symmetry points. We expand around the high symmetry points, starting from the central cell function, in terms of known functions and the problem is significantly simplified to get the solution of the studied system [78].

However, the pseudopotential method is more general than the $\boldsymbol{k.p}$ method because it can be used not only for the band structure calculations but even for the oscillator strengths of the transition calculations. This method could be preferable then the $\boldsymbol{k.p}$ approach if the optical matrix elements are used as an inputs in the band structure calculations [79].

The last method is the orthogonalized plane-wave (OPW) method, it is used to obtain approximations to the solutions of the Schrödinger equation for a periodic potential. The approach permits to have information about the state of electrons in an ideal crystal. The first application of this method is done to determine the wave function and energy eigenvalues of states near points of high symmetry on the Fermi surface of beryllium [80].

All thc citing formalism of the $\boldsymbol{k.p}$ methods are the basic of the so-called the first principle calculations based on the density functional theory (DFT) calculations. Many research studies have investigated the WBG semiconductors on the base of the DFT calculations to have the electronic and the optical properties of those materials such as the lattice constants, the band gap energy, the density of states DOS, the effectives masses of the charge carriers, the dielectric functions, the

Table 1.11 Electronic and optical parameters of some WBG semiconductors

WBG materials	Optical dielectric constant	Static dielectric constant $\varepsilon(0)$	Static refractive index $n(0)$	a (A°) with LDA approximation	c/a	Internal structural parameter u	E_g (eV)
GaN	5.0 [81]	6.15 [82]	2.95 [82]	3.152 [81]	1.631 [81]	0.376 [81]	2.6 [83], 2.14 [81]
AlN	3.9 [81]	–	–	3.090 [81]	1.602 [81]	0.382 [81]	4.41 [81]
InN	7.6 [81]	–	–	3.507 [81]	1.618 [81]	0.379 [81]	−0.18 [81]
3C-SiC	–	–	–	–	–	–	1.27 [84]
6*H*-SiC	–	–	–	–	–	–	1.96 [84]
4*H*-SiC	–	–	–	–	–	–	2.18 [84]
2*H*-SiC	–	–	–	–	–	–	2.10 [84]

absorption coefficient, the refractive index, and the energy loss. Some electronic and optical parameters are listed in Table 1.11. All these parameters are calculated by the DFT approach.

1.5 WBG power semiconductor devices: a comparison

The use of the silicon (Si) as a principle semiconductor in a huge type of semiconductor devices has served well so far. But recently, the technological applications demand to operate under higher power density to reduce the chip area, higher frequency to reduce the size of the passive components and ultimately reduce the overall copper loss and higher temperature to significantly reduce cooling needs. All those interesting properties call for WBG semiconductors to be used in the power devices [85]. In this section, we will compare the most used WBG power semiconductor devices in terms of transport mechanisms of the charge carrier (the density of the current) in each device modeling the physics phenomenon inside it. The reliability and the mobility for the most used WBG electronic devices are analyzed too in this section.

1.5.1 Schottky contacts

The use of the WBG materials in the Schottky contacts reveals of great importance since they offer the ability to control the properties of the metal contacts. The Schottky or the metal/semiconductor contacts are characterized with a barrier height (SBH) defined by the energy barrier for the charge carrier traversing the interface. This barrier determines the physics behind the electrical behavior of the Schottky contacts if it is an ohmic or a rectifying one. While the SBH is a crucial parameter to investigate the physical and the electrical properties of the contact, it is important to study the basic principles that control the Schottky barrier high [86]. A deep analysis of the SBH in these contacts using WBG materials is detailed in Section 1.8.

1.5.2 Metal/oxide/semiconductor structures (MOS)

Several semiconductor devices show surface sensitivity. To study those surface properties, the MOS device is the perfect structure to investigate the semiconductor surfaces and the oxide/semiconductor interfaces [87]. The electrical properties of the MOS structures are important to characterize them. From the *I–V* curves, one could analyze the behavior of the different parameters modeling the current across the oxides. In the MOS structures, there are four principle carrier transport mechanisms present in the oxide region when the electric field or temperature is sufficiently high. Each one of them is dominant under a specific conditions. The basic conduction processes in oxides in the MOS devices are: (a) the direct tunneling, (b) the Fowler–Nordheim tunneling, (c) the thermionic emission, and (d) the Frenkel–Poole emission, as illustrated in Figure 1.2 [88] associated with the variation of each conduction mechanism as a function of the temperature and the applied voltage.

1.5.3 Bipolar junction transistor (BJT)

The BJT was first fabricated by silicon, but this material presents limitations for high-voltage applications [89]. However, the use of the WBG materials such as SiC and diamond resolved the problem and gives a high-performances to power BJTs [90, 91].

The BJT is a switching device that can handle both electron and hole carriers. Its schematic representation is given in Figure 1.3, where the symbols *E*, *B*, and *C* are the emitter, the base, and the collector, respectively. The direction of the current of charge carriers is indicated by the arrows. Because of the switching properties of

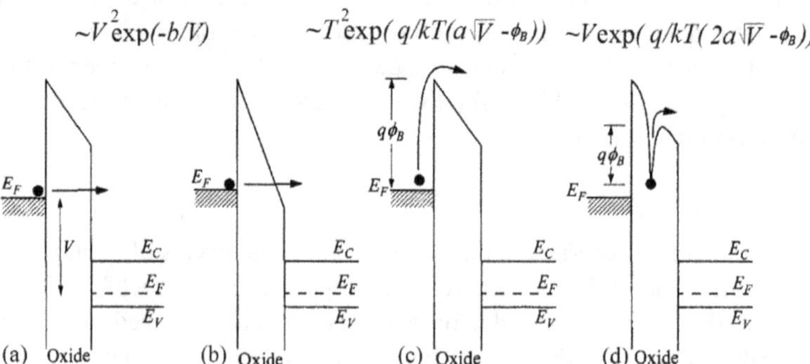

Figure 1.2 Basic conduction processes in oxides in the MOS devices: (a) the direct tunneling, (b) the Fowler–Nordheim tunneling, (c) the thermionic emission, and (d) the Frenkel–Poole emission: where ϕ_B, a, b *are the barrier high and a constant, respectively*

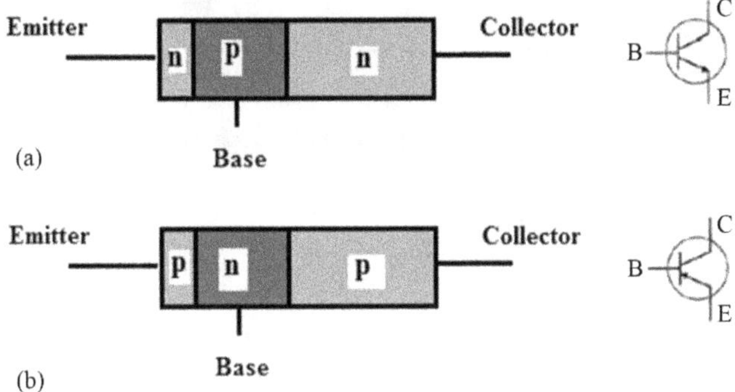

Figure 1.3 Schematic representation of the bipolar junction transistor (BJT): (a) npn type and (b) pnp type

the device, it is controlled by the currents injection (I_E, I_B, I_C) and they are inter-related as follows [92]:

$$\begin{cases} I_C = \alpha I_E \\ I_C = \beta I_B \\ I_C + I_B = I_E \\ \beta = \dfrac{\alpha}{1-\alpha} \end{cases} \tag{1.18}$$

The parameters α, β characterize the device properties.

For a *pnp* BJT, the current issued from the emitter region is given in Eq. (1.19). In this representation, it is the sum of the diffusing and the drift current through and flow the base region:

$$\begin{aligned} & I_E = I_{diffusion} + I_{drift} = qD_h \frac{dp_h}{dx} + qp_h\mu_h E \\ & I_{diffusion} > I_{drift} \end{aligned} \tag{1.19}$$

D_h and μ_h are the diffusion constant and the drift mobility of holes, respectively.

1.5.4 High electron mobility transistor (HEMT)

A HEMT is the advanced structure for the *III–V*-based transistors. They are fabricated by heterojunctions between semiconductors of different alloy composition and band gaps like AlGaAs, GaAs, GaN, and SiC. In the last three decades, a new generation of HEMT was constructed called pseudomorphic HEMT (pHEMT). The use of WBG materials gives a high quality of those devices especially in switching, RF, microwave, and millimeter-wave applications. Figure 1.4 is a schematic illustration of HEMT's device structure based on GaN material showing

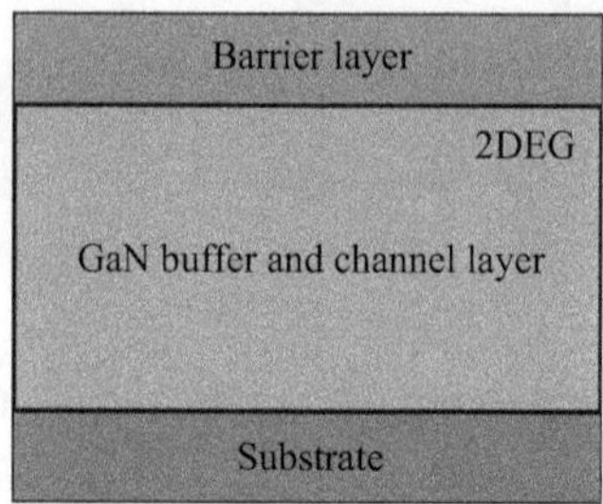

Figure 1.4 Schematic representation of HEMT's device structure

two-dimensional electron gas (2DEG) issued from electrons that are gathered in the GaN side of the heterojunction. The 2DEG phenomenon is caused by the combined effect of spontaneous and piezoelectric polarization [93].

1.5.5 Reliability analysis of WBG power semiconductors devices

In the last years, the WBG power semiconductor devices are developed to achieve the reliability keys required in many applications such as optoelectronics, hybrid electrical vehicles, RF generators, aviation applications, satellite communications, advanced military application, intelligent fire control systems, renewable energy systems (RESs) [94], radar systems [95], and other technological applications. These keys are based principally on the high switching frequency, blocking capability, operating temperature, high efficiency, and lower cost. To measure the reliability of the WBG devices, many tests (refer to Table 1.12) have to be done [96]. The use of each test depends on the type of the application domain in which the WBG device will be used.

Despite the WBG devices advantages mentioned in the sections above compared to the Si devices, there are some technological and physics problems to be solved to exploit all the advantages of such WBG devices and to perform all the reliability issues.

Basically, two processing difficulties of the SiC devices are to be solved: low inversion layer channel mobility and gate oxide reliability under high temperature and high electric field. However, for the GaN devices, many physics problems limit the reliability of such WBG devices citing for example the presence of the trap, material's defect, degradation of the metal in the Schottky device under high voltage and high current, insulated gate dielectric and surface charge. Another technical problem which is added to the precedent ones is the packaging problems of the SiC and the GaN devices in high power and temperature conditions. Those problems are major specifically in the GaN devices [95].

One of the important issues of the reliability is the mobility performances. The mobility of the charge carrier in any electronic device in general and in a WBG devices in particular is studied for many years to achieve the maximum one. Many analyses demonstrated that the mobility in the SiC or in the GaN devices is infected

Table 1.12 Some reliability tests used for the WBG electronic devices offered by the Nire company [96]

Reliability tests	Principle characteristics of the test
HTGB: high-temperature gate bias	Gate bias of up to ±70 V, run up to 200°C
SURGE: Surge Current test	Pulse the DUT with 50-Hz half-sine pulses (10 ms pulse width) up to a maximum pulse current of 1 kA
Short circuit	Maximum short circuit voltage capability is 15 kV
DI/DT	di/dt turn-on testing on DUTs with a max pulse current of 600 A
DV/DT	Accommodate DUTs rated between 600 V and 1.6 kV, at a dV/dt rate adjustable up to 300 V/ns or greater. 300 V/ns
Avalanche	Avalanche ruggedness of semiconductor switching devices rated up to 15 kV, and diodes rated up to 1.5 kV
Switching	Switching test can accommodate up to 3.3 kV devices. This test can run devices up to 1.5 kW
TDDB	Time-dependent dielectric breakdown test can use Gate bias of up to +/-70 V and can be run for as long as desired.
HTOL	High temperature reverse bias test can accommodate up to 6 kV devices and can be run up to 200°C for as long as desired

essentially by the state of the SiO_2/SiC (caused by the trapping of the electrons) or the SiO_2/GaN interface [97,98], the increase of the temperature [99], and the doping of the homoepitaxial *p*-GaN layer (with magnesium Mg) in the MOSFET devices [100]. Different performances of the mobility are done on the processing of the SiC and the GaN devices. Such processing performances are done on the new HEMT devices where a double channel is inserted to increase the mobility in these devices. It is established too that the insertion of the interlayer of GaN among the *p*-GaN and the oxide in the OG-FET (oxide GaN interlayer FET) increases the mobility meaningfully [101]. Another elaboration mechanism shows its efficiency to perform the mobility in the SiC devices like the MOSFETs, it is the introduction of N element at the SiO_2/SiC interface also called the nitridation of the interface. Additional processing technics used for different WBG electronic devices to increase the mobility of the charge carrier in these semiconductor devices citing the doping of the gate oxide by several ions like the boron, the phosphorus, alkali, and alkaline earth elements. The crystal orientation in the SiC device (MOS) influences significantly the mobility of the charge carrier. Higher values of the mobility can be found when using different off-axis angle cut. The analysis showed that the minimum of the off-axis angle guarantees the better properties of the MOS channel [102].

1.6 Recent research and developments

Recently the WBG materials have been deeply used in different technological applications, citing high power, high temperature, and high-speed devices (radar, satellite). The engineering of the SiC as several electronic and optoelectronic devices demonstrates its high benefits compared to other semiconductor materials. Most of the investigation and research works with the GaN are done largely for radio frequency electronics and optoelectronics applications. Comparing the two famous family of the WBG materials SiC and GaN, the SiC as a power device is at a more advanced stage than the GaN power devices. Because the SiC is the best suitable transition material to silicon for future power electronics devices, however, the GaN is more difficult to growth it with specific high properties requested from the modern technology. For diamond, representing as a crucial material for power devices in several applications [103], numerous researches are made recently on the structures and devices based on the synthesized diamond [104,105] to deepen the investigation and to broaden its technological applications.

1.7 Challenges and opportunities

Despite the important properties offered by the WBG materials, they are still limited in other applications in point of view of the figures-of-merit for device performance scale. This gives a birth for a new generation of WBG materials called the ultra-WBG (UWBG) semiconductors. For example, AlGaN/AlN, diamond, Ga_2O_3, cubic BN, and perhaps others not yet discovered [36]. Some physical properties of various WBG and UWBG materials are summarized in Table 1.13. The table gives the values of three important metrics required for the device application such as the quality of their substrates, the substrate diameter and their ability to be *p*-doped or to be *n*-doped. Since the present chapter book is reserved for the WBG materials, we do not go in deep for the properties and the physics of this class of WBG semiconductors. For more details about the UWBG semiconductors, the reader can consult Ref. [36].

Table 1.13 Some properties of the selected WBG and UWBG materials [36]

Semiconductor	**WBG**		**UWBG**		
	GaN	**4*H*-SiC**	**AlGaN/AlN**	**β-Ga_2O_3**	**Diamond**
Bandgap E_g (eV)	3.4	3.3	Up to 6.0	4.9	5.5
Thermal conductivity (W m^{-1} K^{-1})	253	370	253–319	11–27	2,290–3,450
State-of-the-art substrate quality (dislocations per cm^2)	$\approx 10^4$	$\approx 10^2$	$\approx 10^4$	$\approx 10^4$	$\approx 10^5$
State-of-the-art substrate diameter (inches)	8 (on Si)	8	2	4	1
Ability to be p-doped	Good	Good	Poor	No	Good
Ability to be n-doped	Good	Good	Moderate	Good	Moderate

1.8 Case study

The metal/semiconductor contacts called also the Schottky diodes with a SiC as a WBG material are mandatory for a large type of high-power switching devices. Different metals are used in SiC Schottky diodes [106,107] to investigate the physics of the interface metal/semiconductor.

The case study of this chapter in this book will focus and discuss the physics of the Schottky structure elaborated with SiC as a WBG material, using the 4*H* type. We investigate two kinds of Schottky diodes: the first one is realized using the molybdenum (Mo) as a metal contact and the second one is realized using the tungsten (W) as a metal contact. This investigation is under the high-temperature semiconducting devices application.

As cited in Section 1.5.1, the Schottky diodes are characterized by the barrier high at the interface of the metal and the semiconductor. In the last years, the inhomogeneity of the interface contact has been considered as an interpretation for the measured electrical properties expected for a spatially inhomogeneous Schottky Barrier (SB) [108–110]. In this section, the barrier high at the interface between the metal and the 4*H*-SiC is investigated.

The study of the *I*–*V* curves of the Schottky diode (SD) at the room temperature does only not give a complete understanding about the nature of barrier formation at the metal/4*H*-SiC interface. The inhomogeneity of such contact means that the measured barrier height is temperature dependent. Investigating the *I*–*V*–*T* measurements of Schottky diode usually shows an abnormal decrease in the barrier height ϕ_β and an increase in the ideality factor *n* with a decrease in temperature.

The studied diodes are elaborated by using an *n*-type 4*H*-SiC in the (0001) direction and a Si-face 8° off-oriented toward $\left(11\bar{2}1\right)$. The wafers had an *n*-type epitaxial-layer with a donor concentration in the range of 8.0×10^{15}–1.3×10^{16} cm^{-3}. For the metal contact, we have chosen the molybdenum (Mo) and tungsten (W), for their compatibility to a standard silicon process technology and for their thermal stability. The elaborated Schottky diodes had a circular geometry with a diameter of 150 μm for the Mo/4*H*-SiC and 200 μm for the W/4*H*-SiC. The *I*–*V*–*T* characteristics of the Mo/4*H*-SiC and W/4*H*-SiC structures are represented in Figure 1.5(a) and (b) in the temperature range of 298–498 K and 303–448 K, respectively.

To model those *I*–*V* curves, we used the thermionic emission (TE) theory where the current of the charge carriers as a function of the applied voltage is given by Eq. (1.20) [111]:

$$I = I_s\left(\exp\left(\frac{\beta}{n}(V - R_sI)\right) - 1\right) \tag{1.20}$$

The saturation current I_s is given by

$$I_s = AA^*T^2\exp(-\beta\phi_{B0}) \tag{1.21}$$

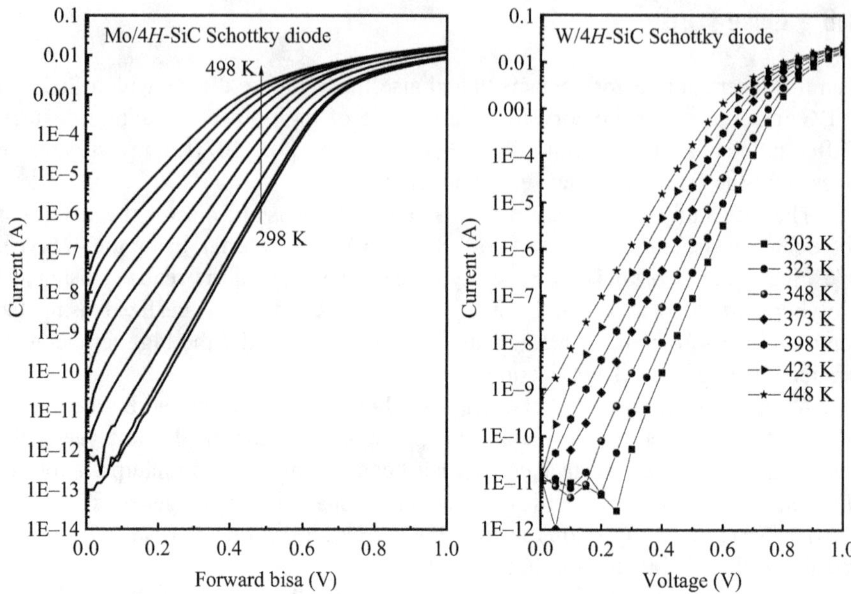

Figure 1.5 I–V–T characteristics of: (a) Mo/4H-SiC: first published in the Indian Journal of Physics 93(9), 2019 [113]; (b) W/4H-SiC structures: first published in the Applied Physics A Journal 127(9), 2021 [115]

A^*, A, $\beta=q/kT$, T, k, q, ϕ_{B0}, n, and R_s are the effective Richardson constant of 146 A/cm^2 K^2 for 4*H*-SiC [112], the diode area, the inverse thermal voltage, the absolute temperature, the Boltzmann constant, the elementary electric charge, the zero-bias barrier height, the ideality factor, and the series resistance, respectively.

The temperature dependences of the current I measured in the Schottky contacts allow us to model the inhomogeneous interface characterized by the barrier high ϕ_β using the Gaussian distribution proposed by Werner [109]. Consequently, the expression of the barrier high and the ideality factor n at an inhomogeneous Schottky contacts are illustrated by Eqs. (1.22) and (1.23), respectively:

$$\phi_{ap} = \overline{\phi}_{B0} - \frac{\beta}{2}\sigma_s^2 \tag{1.22}$$

$$n_{ap} = \frac{1}{1 - \rho_2 + \frac{\beta}{2}\rho_3} \tag{1.23}$$

$\overline{\phi}_{B0}$, σ_s, ρ_2, and ρ_3 are the mean value of the Schottky barrier high, the standard deviation, and the coefficients quantifying the barrier high deformation, respectively.

Substituting Eqs. (1.21), (1.22), and (1.23) in Eq. (1.20), we get [113]:

$$I = AA^*T^2 \exp\left(-\beta\left(\overline{\phi}_{B0} - \frac{\beta}{2}\sigma_s^2\right)\right) \times \left\{\exp\left[\beta\left(1 - \rho_2 + \frac{\beta}{2}\rho_3\right)(V - R_s I)\right] - 1\right\} \tag{1.24}$$

The simultaneous extraction of all the parameters ($\overline{\phi}_{B0}$, σ_s, ρ_2, ρ_3, R_s) that characterize the inhomogeneity of the studied contacts without the need to know any parameter initially or extract it from any graphical methods or using the *C–V* curves, can be made by optimizing the vector of parameters.

The optimization of the parameters, in our case study, is based on the fitting of the *I–V–T* data by minimizing the vertical quadratic error *S* on the vertical axis (the measured current) given by Eq. (1.25) [114]:

$$S = \sum_{i=1}^{m}\left(\frac{I_{i\exp} - I_{ifit}}{I_{ifit}}\right)^2 \tag{1.25}$$

I_{iexp} and I_{ifit} are the *i*th measured current and the *i*th fitting value of the current, respectively, and *m* is the number of measuring points.

Figure 1.6(a–e) illustrates the variation of the parameters ($\overline{\phi}_{B0}$, σ_s, ρ_2, ρ_3, R_s) as a function of temperature for the Mo/4*H*-SiC, respectively, first published in the *Indian Journal of Physics* **93**(9), 2019 [113]. For the W/4*H*-SiC Schottky contacts, the variation of the characterizing parameters ($\overline{\phi}_{B0}$, σ_s, ρ_2, ρ_3, R_s) versus temperature is shown in Figure 1.7(a–e), respectively, which was first published in the *Applied Physics A Journal* **127**(9), 2021 [115]. The different physical parameters and remarks that could be deduced from Figures 1.6 and 1.7 are resumed in Table 1.14.

If $-\rho_2 + \rho_3/\frac{2kT}{q} << 1$ and $\rho_2 << \rho_3/\frac{2kT}{q} << 1$ (which are not shown here), Eq. (1.23) becomes [109]:

$$n = 1 + \rho_2 - \rho_3/\frac{2kT}{q} \equiv 1 + \frac{T_0}{T} \tag{1.26}$$

T_0 is the temperature parameter showing the deviation of the ideality factor *n* from unity and considered as a constant by Werner *et al.* [109] and equal to $-\frac{\rho_3 q}{2k}$.

Taking into account the variation of the value of n_{ap} demonstrated by Saxena [116]:

$$n = 1 + \frac{T_0}{T} \tag{1.27}$$

As a result from Eqs. (1.26) and (1.27), we obtain

$$T_0 = \rho_2 T - \frac{q\rho_3}{2k} \tag{1.28}$$

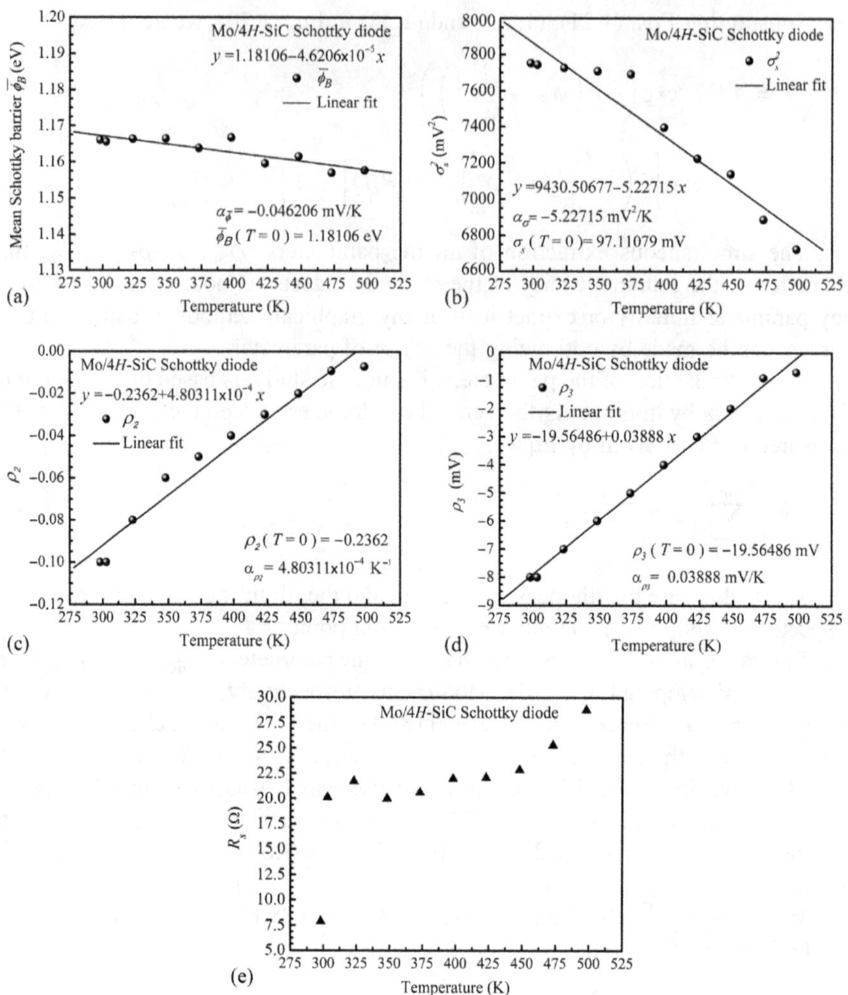

Figure 1.6 Temperature dependences of the parameters: (a) the mean barrier high ($\overline{\phi}_{B0}$), (b) the standard deviation (σ_s), (c) the coefficient (ρ_2), (d) the coefficient (ρ_3) and (e) the series resistance (R_s) characterizing the barrier high inhomogeneity for the Mo/4H-SiC contact reproduced with permission from Ref. [113]

Figure 1.8 illustrates the variation of the parameter T_0 as a function of temperature using Eqs. (1.26) and (1.28) for the Mo/4*H*-SiC contact which was first published in the *Indian Journal of Physics* **93**(9), 2019 [113]. In Figure 1.9, we represent the temperature dependences of parameter T_0 using also Eqs. (1.26) and (1.28) for the W/4*H*-SiC Schottky contact which was first published in the *Applied Physics A Journal* **127**(9), 2021 [115]. The values of the parameter T_0 represented

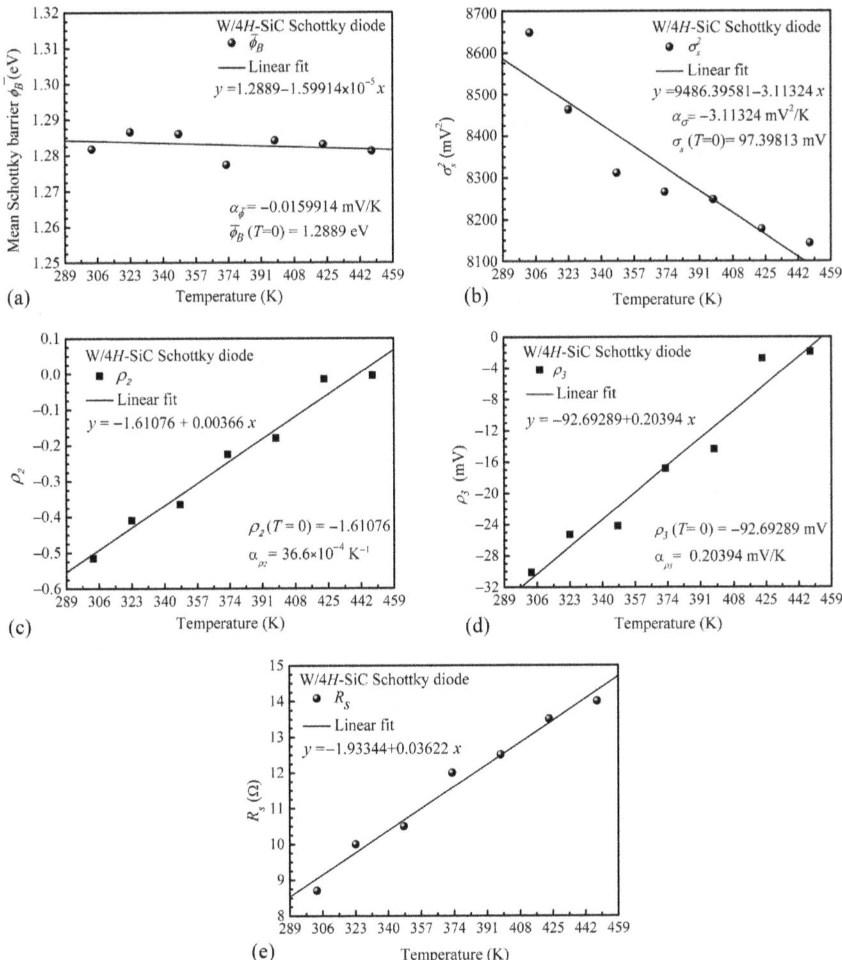

Figure 1.7 Variation of the characterizing parameters: (a) the mean barrier high ($\overline{\phi}_{B0}$), (b) the standard deviation (σ_s), (c) the coefficient (ρ_2), (d) the coefficient (ρ_3), and (e) the series resistance (R_s) versus temperature for the W/4H-SiC contact reproduced with permission from the reference [115]

in Figures 1.8 and 1.9 show the presence of an inhomogeneous interface in both contacts.

From this deep investigation of the two Schottky contacts (Mo/4*H*-SiC and W/4*H*-SiC), the inhomogeneity of the interface between the metal and the WBG semiconductor is evidenced in terms of Werner's model assuming the Gaussian distribution of the barrier high ϕ_B characterizing these interfaces. In the same research axis, many investigations [117–123] of the Schottky diode using WBG

Table 1.14 Extraction of the physical parameters deduced from Figures 1.6 and 1.7 for the Mo/4H-SiC and the W/4H-SiC Schottky contact, respectively

Parameters	**Equations modeling the parameters characterizing the inhomogeneity of the barrier high ϕ_B**	**Mo/4*H*-SiC**	**W/4*H*-SiC**
σ_s	$\sigma_s^2(T) = \sigma_s^2(T=0) + \alpha_\sigma T$ [109]	$\alpha_\sigma = -5.22$ mV^2/K, σ_s (T=0) = 97.11 mV [113]	$\alpha_\sigma = -3.11$ mV^2/K, σ_s (T=0) = 97.39 mV [115]
$\overline{\phi}_B$	$\overline{\phi}_{B0}(T) = \overline{\phi}_{B0}(T=0) + \alpha_{\overline{\phi}_{B0}} T$ [109]	$\alpha_{\phi B0}$= −0.046 mV/K, ϕ_{B0}(T=0) = 1.18 eV [113]	$\alpha_{\phi B0}$ = −0.015 mV/K, ϕ_{B0}(T=0) = 1.28 eV [115]
ρ_2	A linear fit	At T=0 K, ρ_2= 0.23 with a slope $\alpha_{\rho 2} = 4.80\times10^{-4}$ K^{-1} [113]	At T=0, $\rho_2 = -1.61$ with a slope $\alpha_{\rho 2} = 36.6\times10^{-4}$ K^{-1} [115]
ρ_3	A linear fit	At T=0 K, ρ_3 = −19.56 mV, with a slope $\alpha_{\rho 3}$ = 0.038 mV/K [113]	At T=0, ρ_3 = −92.69 mV with a slope $\alpha_{\rho 3}$ = 0.2 mV/K [115]
R_s		Increase of the series resistance with the increase of the temperature [113]	Increase of the series resistance with the increase of temperature [115]

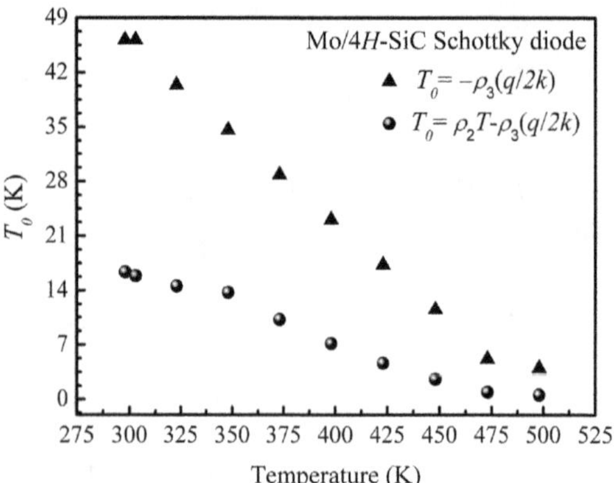

Figure 1.8 Variation of the parameter T_0 as a function of temperature for the Mo/4H-SiC contact. Reproduced with permission from Ref. [113]

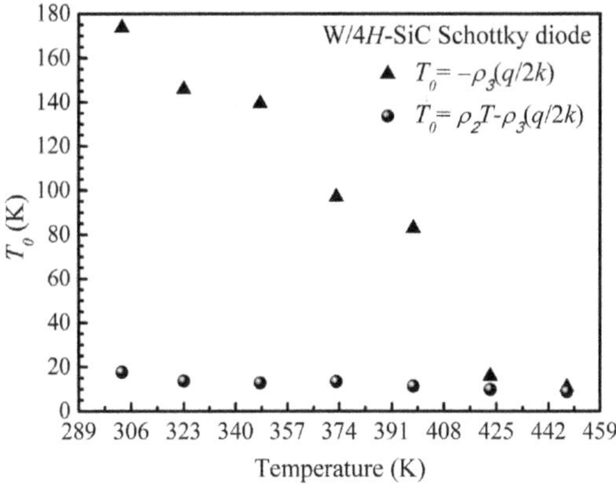

Figure 1.9 T_0 parameter versus temperature for the W/4H-SiC contact reproduced with permission from Ref. [115]

semiconductors are done and proved the presence of the inhomogeneity of the interface metal/WBG semiconductor.

1.9 Conclusion

This chapter is aimed to investigate the WBG materials in terms of their unique physical and structural properties providing a response to the technological requests. To cover approximately all the physicals and the technological aspects of these class of materials, we provided an overview for the famous categories of the WBG materials and the crucial criterions to use them in the power semiconductor devices. The characterization and modeling of WBG materials have also been treated followed by the citation of the recent researches and developments opening widows to a new generation of WBG materials which are the ultra-WBG materials. The chapter is supported by a case study involving the investigations about the interface state of the Schottky diodes based on WBG materials. The studied Schottky contacts are elaborated by deposing the molybdenum/tungsten as a metal on the 4*H*-SiC as a WBG semiconductor. To characterize them, we have measured the *I–V–T* curves with the variation of temperature. The analysis of these contacts reveals that the barrier high at the interface of the metal/semiconductor is distributed inhomogeneously. The inhomogeneity of the barrier high ϕ_B is modeled by a Gaussian distribution defined by a mean value ϕ_{B0} and a standard deviation σ_s. All the parameters that model the studied Schottky contacts and their interface barrier are extracted and investigated as a function of temperature. Finally, we may say that even there are a considerable researches in such class of materials (WBG), they still demand more investigations and deep analysis in terms of other properties such as the nanoscale applications. So, a continued development of these materials could be predictable to realize an important devices/nano-devices improvement.

Acknowledgments

The author would like to thank Pr T. Guerfi from the Department of Physics, University of Boumerdes, Algeria, for the correction of the manuscript, for his fruitful discussion of this work and helpful comments. The authors also gratefully acknowledge Pr Z. Ouennoughi from the Optoelectronic Laboratory, University of Sétif, Algeria, for the help with the *I–V–T* measurements treated in this work. Additional thanks are due to Pr L. Frey, Pr H. Ryssel, and Pr K. Murakami from Chair of Electron Devices, University Erlangen, Cauerstr. 6, 91058 Erlangen, Germany and Pr R. Weiss from University Erlangen, Cauerstr. 8, 91058 Erlangen, Germany for providing us with all the *I–V–T* measurements of the studied Schottky diodes.

The author offers a special thanks to her parents and family for their kind encouragement and support in writing this chapter in this book. Without their support and understanding, this chapter in this book would not have been published.

References

[1] R. Szweda, *Gallium Nitride & Related Wide Bandgap Materials & Devices. A Market and Technology: Overview 1998–2003*, Chapter 4, 2nd ed., Elsevier Advanced Technology, New York, NY, p. 107, 2000.

[2] A.R. Verma and P. Krishna, *Polymorphism and Polytypism in Crystals*, John Wiley & Sons, Inc., New York, NY, 1966.

[3] T. Kimoto and J.A. Cooper, *Fundamentals of Silicon Carbide Technology: Growth, Characterization, Devices, and Applications*, John Wiley & Sons Singapore Pte. Ltd., New York, NY, p. 14, 2014.

[4] D. Klimm, "Electronic materials with a wide band gap: recent developments," *International Union of Crystallography (IUCr)*, Vol. 1, Part 5, pp. 281–290, 2014.

[5] S. Adachi, *Properties of Group-IV, III-V, and II-VI Semiconductors*, John Wiley & Sons, Ltd, Chichester, 2005, p. 120.

[6] N. Lophitis, A. Arvanitopoulos, S. Perkins, and M. Antoniou, *TCAD Device Modelling and Simulation of Wide Bandgap Power Semiconductors*, Chapter 2, Intech Open, London, pp. 17–44, 2018.

[7] P.Y. Yu and M. Cardona, *Fundamentals of Semiconductors*, 3rd ed., Springer, New York, NY, p. 348, 2005.

[8] S.M. Sze and K.K. Ng, *Physics of Semiconductor Devices*, 3rd ed., John Wiley & Sons, Inc., New York, NY, p. 20, 2007.

[9] T. Kimoto and J.A. Cooper, *Fundamentals of Silicon Carbide Technology, Growth, Characterization, Devices, and Applications*, John Wiley & Sons Singapore Pte. Ltd., Singapore, pp. 24–25, 2014.

[10] B.J. Baliga, *Fundamentals of Power Semiconductor Devices*, 2nd ed., Springer Ed., Springer, New York, NY, p. 15, 2019.

[11] J.A. Cooper, M.R. Melloch, R. Singh, A. Agarwal and J.W. Palmour, "Status and prospects for SiC power MOSFETs," *IEEE Transactions on Electron Devices*, Vol. 49, no. 4, pp. 658–664, 2002, doi: 10.1109/16.992876.

[12] M. Bhatnagar and B.J. Baliga, "Comparison of 6H-SiC, 3C-SiC, and Si for power devices," *IEEE Transactions on Electron Devices*, Vol. 40, no. 3, pp. 645–655, 1993, doi: 10.1109/16.199372.

[13] M. Ruff, H. Mitlehner, and R. Helbig, "SiC devices: physics and numerical simulation," *IEEE Transactions on Electron Devices*, Vol. 41, no. 6, pp. 1040–1054, 1994, doi: 10.1109/16.293319.

[14] B. Ozpineci, L.M. Tolbert, S.K. Islam and M. Chinthavali, *Proceedings of the 10th European Conference on Power Electronics and Applications*, p. 257, 2003.

[15] C. Hamaguchi, *Basic Semiconductor Physics*, 2nd ed., Springer, New York, NY, 2010.

[16] T. Kimoto and J.A. Cooper, *Fundamentals of Silicon Carbide Technology, Growth, Characterization, Devices, and Applications*, John Wiley & Sons Singapore Pte. Ltd, Singapore, p. 27, 2014.

[17] I.A. Khan and J.A. Cooper, "Measurement of high-field electron transport in silicon carbide," *IEEE Transactions on Electron Devices*, Vol. 47, no. 2, pp. 269–273, 2000, doi: 10.1109/16.822266.

[18] M. Roschke and F. Schwierz, "Electron mobility models for 4H, 6H, and 3C SiC [MESFETs]," *IEEE Transactions on Electron Devices*, Vol. 48, no. 7, pp. 1442–1447, 2001, doi: 10.1109/16.930664.

[19] T. Kimoto and J.A. Cooper, *Fundamentals of Silicon Carbide Technology, Growth, Characterization, Devices, and Applications*, John Wiley & Sons Singapore Pte. Ltd., p. 521, Singapore, 2014.

[20] W. (Wayne) Bi, H.-C. (Henry) Kuo, P.-C. Ku, and B. Shen, *Handbook of GaN Semiconductor Materials and Devices*, Taylor & Francis Group, New York, NY, p. 4, 2018.

[21] W. (Wayne) Bi, H.-C. (Henry) Kuo, P.-C. Ku, and B. Shen, *Handbook of GaN Semiconductor Materials and Devices*, Taylor & Francis Group, New York, NY, p. 6, 2018.

[22] W. (Wayne) Bi, H-C. (Henry) Kuo, P.-C. Ku, and B. Shen, *Handbook of GaN Semiconductor Materials and Devices*, Taylor & Francis Group, New York, NY, p. 7, 2018.

[23] W. (Wayne) Bi, H.-C. (Henry) Kuo, P.-C. Ku, and B. Shen, *Handbook of GaN Semiconductor Materials and Devices*, Taylor & Francis Group, New York, NY, p. 8, 2018.

[24] F. Bernardini, V. Fiorentini, and D. Vanderbilt, "Spontaneous polarization and piezoelectric constants of III-V nitrides," *Physical Review B*, Vol. 56, no. 16, pp. R10024–R10027, 1997.

[25] S.-H. Wei and A. Zunger, "Valence band splittings and band offsets of AlN, GaN, and InN," *Applied Physics Letters*, Vol. 69, pp. 2719–2721, 1996.

[26] F. Bechstedt, U. Grossner, and J. Furthmu "Dynamics and polarization of group-III nitride lattices: a first-principles study," *Physical Review B*, Vol. 62, pp. 8003–8011, 2000.

[27] F. Bernardini, V. Fiorentini, and D. Vanderbilt, "Accurate calculation of polarization-related quantities in semiconductors," *Physical Review B*, Vol. 63, pp. 193201-1–193201-4, 2001.

[28] A. Zoroddu, F. Bernardini, P. Ruggerone, and V. Fiorentini, "First-principles prediction of structure, energetics, formation enthalpy, elastic constants, polarization, and piezoelectric constants of AlN, GaN, and InN: comparison of local and gradient-corrected density-functional theory," *Physical Review B*, Vol. 64, pp. 045208-1–045208-6, 2001.

[29] M.A. Caro, S. Schulz, and E.P. O'Reilly, "Theory of local electric polarization and its relation to internal strain: impact on polarization potential and electronic properties of group-III nitrides," *Physical Review B*, Vol. 88, pp. 214103-1–214103-22, 2013.

[30] R. Cangas and M. A. Hidalgo, "Magnetoconduction in a two-dimensional system confined in wurtzite AlxGa1xN/GaN heterostructure," *Applied Physics Letters*, Vol. 102, p. 162413, 2013.

[31] L.I. Ming, S. Gang, and F. Li-Bo, "A new method to calculate the Rashba spin splitting in III-nitride heterostructures," *Chinese Physics Letters*, Vol. 29, no. 12, pp. 127104-1–127104-6, 2012.

[32] I. Lo, M.H. Gau, J.K. Tsai, et al., "Anomalous k-dependent spin splitting in wurtzite AlxGa1xNGaN heterostructures," *Physical Review B*, Vol. 75, pp. 245307-1–245307-6, 2007.

[33] G. Masetti, M. Severi, and S. Solmi, "Modeling of carrier mobility against carrier concentration in arsenic-, phosphorus-, and boron-doped silicon," *IEEE Transactions on Electron Devices*, Vol. 30, no. 7, pp. 764–769, 1983, doi: 10.1109/T-ED.1983.21207.

[34] F. Medjdoub and K. Iniewski, *Gallium Nitride (GaN), Physics, Devices, and Technology*, Taylor & Francis Group, LLC, New York, NY, p. 3, 2016.

[35] W. (Wayne) Bi, H.-C. (Henry) Kuo, P.-C. Ku, and Bo Shen, *Handbook of GaN Semiconductor Materials and Devices*, Taylor & Francis Group, New York, NY, p. 306, 2018.

[36] J.Y. Tsao, S. Chowdhury, M.A. Hollis, et al., "Ultrawide-bandgap semiconductors: research opportunities and challenges," *Advanced Electronic Materials*, Vol. 4, pp. 1600501 (1 of 49), 2018.

[37] R.J. Kaplar, A.A. Allerman, A.M. Armstrong, et al. "Review—ultra-wide-bandgap AlGaN power electronic devices," *ECS Journal of Solid State Science and Technology*, Vol. 6, no. 2, pp. Q3061–Q3066, 2017.

[38] R. Szweda, *Gallium Nitride & Related Wide Bandgap Materials & Devices. A Market and Technology: Overview 1998–2003*, 2nd ed., Elsevier Advanced Technology, New York, NY, p. 123, 2000.

[39] L.M. Porter, K. Das, Y. Dong, J.H. Melby, and A.R. Virshup, "4.03 - Contacts to wide-band-gap semiconductors," in: P. Bhattacharya, R. Fornari, and H. Kamimura, editors, *Comprehensive Semiconductor Science and Technology*, Elsevier, New York, NY, pp. 44–85, 2011, ISBN 9780444531537.

[40] S. Shikata, T. Tanno, T. Teraji, H. Kanda, T. Yamada, and J. Kushibiki, "Precise measurements of diamond lattice constant using Bond method,"

Japanese Journal of Applied Physics, Vol. 57, no. 11, pp. 111301(1–5), 2018.

[41] D. Riley, “Lattice constant of diamond and the C–C single bond,” *Nature*, Vol. 153, pp. 587–588, 1944. https://doi.org/10.1038/153587b0

[42] S. Koizumi, “Doping and semiconductor characterizations, n-Type diamond growth and the semiconducting properties,” in: *Power Electronics Device: Applications of Diamond Semiconductors*, Woodhead Publishing, Elsevier, New York, NY, pp. 117, 2018 (Chapter 2), ISBN 978-0-08-102183-5.

[43] R. Szweda, *Gallium Nitride & Related Wide Bandgap Materials & Devices. A Market and Technology: Overview 1998–2003*, 2nd ed., Elsevier Advanced Technology, New York, NY, p. 124, 2000.

[44] H. Xue, Q. He, G. Jian, et al., “An overview of the ultrawide bandgap Ga_2O_3 semiconductor-based Schottky barrier diode for power electronics application,” *Nanoscale Research Letters*, Vol. 13, p. 290, 2018. https://doi.org/10.1186/s11671-018-2712-1.

[45] Z. Galazka, “β-Ga_2O_3 for wide-bandgap electronics and optoelectronics,” *Semiconductor Science and Technology*, Vol. 33, no. 11, p. 113001 (61 pp), 2018. https://doi.org/10.1088/1361-6641/aadf78.

[46] N.N. Som, V. Sharma, V. Mankad, M.L.C. Attygalle, and P.K. Jha, “Role of $CuAlO_2$ as an absorber layer for solar energy converter,” *Solar Energy*, Vol. 193, pp. 799–805, 2019. https://doi.org/10.1016/j.solener.2019.09.098. ISSN 0038-092X.

[47] D.G. Cahill, A.S.-M. Lee, and T.I. Selinder, “Thermal conductivity of κ-Al_2O_3 and α-Al_2O_3 wear-resistant coatings,” *Journal of Applied Physics*, Vol. 83, no. 11, pp. 5783–5786, 1998. https://doi.org/10.1063/1.367500.

[48] Y. Zheng and J.H. Seo, “A simplified method of measuring thermal conductivity of β-Ga_2O_3 nanomembrane,” *Nano Express*, Vol. 1, no. 3, p. 030010 (1–10), 2020. https://doi.org/10.1088/2632-959X/abc1c4.

[49] K. Du, S.P. Deng, N. Qi, et al., “Ultralow thermal conductivity in In2O3 mediated by porous structures,” *Microporous and Mesoporous Materials*, Vol. 288, p. 109525, 2019. https://doi.org/10.1016/j.micromeso.2019.05.050. ISSN 1387-1811.

[50] P. Turkes, Ch. Pluntket, and R. Helbig, “Thermal conductivity of SnO_2 single crystals,” *Journal of Physics C: Solid State Physics*, Vol. 13, no. 26, pp. 4941–4951, 1980.

[51] C. Ruttanapun, W. Kosalwat, C. Rudradawong, et al., “Reinvestigation thermoelectric properties of $CuAlO_2$,” *Energy Procedia*, Vol. 56, pp. 65–71, 2014. https://doi.org/10.1016/j.egypro.2014.07.132. ISSN 1876-6102.

[52] D. Madadi and A.A. Orouji, “β-Ga_2O_3 double gate junctionless FET with an efficient volume depletion region,” *Physics Letters A*, Vol. 412, p. 127575 (1–8), 2021. https://doi.org/10.1016/j.physleta.2021.127575. ISSN 0375-9601.

[53] X. Guo, Z. Xing, S. Zhao, et al., “Investigation of the space charge and DC breakdown behavior of XLPE/α-Al_2O_3 nanocomposites,” *Materials*, Vol. 13, p. 1333, 2020. https://doi.org/10.3390/ma13061333.

[54] H.F. Liu, N.L. Yakovlev, D.Z. Chi, and W. Liu, "Post-growth thermal oxidation of wurtzite InN thin films into body-center cubic In_2O_3 for chemical/gas sensing applications," *Journal of Solid State Chemistry*, Vol. 214, pp. 91–95, 2014. https://doi.org/10.1016/j.jssc.2013.10.017. ISSN 0022-4596.

[55] I.O. Mazali, W.C. Las, and M. Cilense, "Synthesis and characterization of antimony tartrate for ceramic precursors," *Journal of Materials Synthesis and Processing*, Vol. 7, pp. 387–391, 1999. https://doi.org/10.1023/A:1021822115116.

[56] C. Chris, J.H. Wort, and R.S. Balmer, "Diamond as an electronic material," *Materials Today*, Vol. 11, nos. 1–2, pp. 22–28, 2008. https://doi.org/10.1016/S1369-7021(07)70349-8. ISSN 1369-7021.

[57] F.A. Marino, N. Faralli, D.K. Ferry, S.M. Goodnick, and M. Saraniti, "Figures of merit in high-frequency and high-power GaN HEMTs," *Journal of Physics: Conference Series*, Vol. 193, no. 1, p. 012040, 2009. https://doi.org/10.1088/1742-6596/193/1/012040.

[58] W. (Wayne) Bi, H.-C. (Henry) Kuo, P.-C. Ku, and B. Shen, *Handbook of GaN Semiconductor Materials and Devices*, Taylor & Francis Group, New York, NY, p. 307, 2018.

[59] B.J. Baliga, "Power semiconductor device figure of merit for high-frequency applications," *IEEE Electron Device Letters*, Vol. 10, no. 10, pp. 455–457, 1989. https://doi.org/10.1109/55.43098.

[60] S.L. Selvaraj, A. Watanabe, A. Wakejima, and T. Egawa, "1.4-kV breakdown voltage for AlGaN/GaN high-electron-mobility transistors on silicon substrate," *IEEE Electron Device Letters*, Vol. 33, no. 10, pp. 1375–1377, 2012. https://doi.org/10.1109/LED.2012.2207367.

[61] W. (Wayne) Bi, H.-C. (Henry) Kuo, P.-C. Ku, and B. Shen, *Handbook of GaN Semiconductor Materials and Devices*, Taylor & Francis Group, New York, NY, p. 308, 2018.

[62] T. Kimoto and J.A. Cooper, *Fundamentals of Silicon Carbide Technology: Growth, Characterization, Devices, and Applications*, John Wiley & Sons Singapore Pte. Ltd., Singapore, p. 125, 2014.

[63] W.J. Choyke, *Materials Research Bulletin*, Vol. 4, 1969.

[64] R.P. Devaty and W.J. Choyke, "Optical characterization of silicon carbide polytypes," *Physica Status Solidi (a)*, Vol. 162, pp. 5–38, 1997. https://doi.org/10.1002/1521-396X(199707)162:1<5::AID-PSSA5>3.0.CO;2-J.

[65] W.J. Choyke and R.P. Devaty, "Optical properties of SiC: 1997–2002," in: *Silicon Carbide – Recent Major Advances*, Springer, New York, NY, p. 413, 2004.

[66] T. Egilsson, I.G. Ivanov, N.T. Son, et al., *Exciton and Defect Photoluminescence from SiC, in Silicon Carbide, Materials, Processing, and Devices*, Taylor & Francis Group, New York, NY, p. 81, 2004.

[67] E. Janzen, A. Gali, A. Henry, et al., *Defects in SiC, in Defects in Microelectronic Materials and Devices*, Taylor & Francis Group, New York, NY, p. 615, 2008.

[68] T. Kimoto and J.A. Cooper, *Fundamentals of Silicon Carbide Technology: Growth, Characterization, Devices, and Applications*, John Wiley & Sons Singapore Pte. Ltd., Singapore, p. 135, 2014.

[69] L. Bergman and R.J. Nemanich, "Raman spectroscopy for characterization of hard, wide-bandgap semiconductors: diamond, GaN, GaAlN, AlN, BN," *Annual Review of Materials Science*, Vol. 26, pp. 551–579, 1996.

[70] S.M. Sze and K.K. Ng, *Physics of Semiconductor Devices*, 3rd ed., John Wiley & Sons, Inc., Singapore, p. 33, 2007.

[71] T. Kimoto and J.A. Cooper, *Fundamentals of Silicon Carbide Technology: Growth, Characterization, Devices, and Applications*, John Wiley & Sons Singapore Pte. Ltd., Singapore, p. 142, 2014.

[72] D.K. Bowen and B.K. Tanner, *High-Resolution X-Ray Diffractometry and Topography*, Taylor & Francis Group, New York, NY, p. 181, 1998.

[73] D.K. Schroder, *Semiconductor Material and Device Characterization*, 3rd ed., Wiley-IEEE, New York, NY, p. 668, 2006.

[74] M. Kitabatake, "Electrical characteristics/reliability affected by defects analysed by the integrated evaluation platform for SiC epitaxial films," in: *The International Conference on Silicon Carbide and Related Materials 2013*, Miyazaki, Japan, We-1A-1, 2013.

[75] P. Hewlett, *Lea's Chemistry of Cement and Concrete*, 4th ed., Elsevier, New York, NY, pp. 195–198, 2003

[76] T. Kimoto and J.A. Cooper, *Fundamentals of Silicon Carbide Technology: Growth, Characterization, Devices, and Applications*, John Wiley & Sons Singapore Pte. Ltd., Singapore, p. 150, 2014.

[77] S.J. Pearton, *Wide Bandgap Semiconductors: Growth, Processing and Applications,* Noyes and William Andrew, LLC, New York, NY, 2000.

[78] J. Singh, *Electronic and Optoelectronic Properties of Semiconductor Structures*, Cambridge University Press, Cambridge, MA, p. 74, 2003.

[79] P.Y. Yu and M. Cardona, *Fundamentals of Semiconductors: Physics and Materials Properties*, Springer, New York, NY, p. 63, 1996.

[80] T.O. Woodruff, "The orthogonalized plane-wave method," in: F. Seitz and D. Turnbull, editors, *Solid State Physics*, Academic Press, New York, NY, Vol. 4, pp. 367–411, 1957. https://doi.org/10.1016/S0081-1947(08)60156-3. ISSN 0081-1947, ISBN 9780126077049.

[81] A. Janotti, D. Segev, and C.G. Van de Walle, "Effects of cation d states on the structural and electronic properties of III-nitride and II-oxide wide-band-gap semiconductors," *Physical Review B*, Vol. 74, no. 4, p. 045202 (1–9), 2006. https://doi.org/10.1016/S0081-1947(08)60156-3. ISBN 9780126077049.

[82] A. Said, M. Debbichi, and M. Said, "Theoretical study of electronic and optical properties of BN, GaN and BxGa1xN in zinc blende and wurtzite structures," *Optik*, Vol. 127, no. 20, pp. 9212–9221, 2016. https://doi.org/10.1016/j.ijleo.2016.06.103. ISSN 0030-4026.

[83] D. Andiwijayakusuma, M. Saito, and A. Purqon, "Density functional theory study: electronic structures of RE: GaN in wurtzite $G_{15}RE_{1}N_{16}$," *Journal of Physics: Conference Series*, Vol. 739, pp. 012027(1–9), 2016. https://doi.org/10.1088/1742-6596/739/1/012027.

[84] P. Kackell, B. Wenzien, and F. Bechstedt, "Electronic properties of cubic and hexagonal SiC polytypes from ab initio calculations," *Physical Review B*, Vol. 50, no. 15, pp. 10761–10768, 1994.

[85] W. (Wayne) Bi, K.-C. (Henry) Kuo, P.-C. Ku, and B. Shen, *Handbook of GaN Semiconductor Materials and Devices*, Taylor & Francis Group, New York, NY, p. 329, 2018.

[86] L.M. Porter, K. Das, Y. Dong, J.H. Melby, and A.R. Virshup, *Contacts to Wide-Band-Gap Semiconductors*, Elsevier, New York, NY, p. 60, 2011.

[87] E.H. Nicollian and J.R. Brews, *MOS (Metal Oxide Semiconductor) Physics and Technology*, John Wiley & Sons, Singapore, p. 9, 1982.

[88] S.M. Sze and K.K. Ng, *Physics of Semiconductor Devices*, 3rd ed., John Wiley & Sons, Inc., Singapore, p. 228, 2007.

[89] B.J. Baliga, *Fundamentals of Power Semiconductor Devices*, 2nd ed., Springer Ed., Springer, New York, NY, p. 522, 2019.

[90] T. Kimoto and J.A. Cooper, *Fundamentals of Silicon Carbide Technology: Growth, Characterization, Devices, and Applications*, John Wiley & Sons Singapore Pte. Ltd., Singapore, p. 353, 2014.

[91] H. Kato, "Technical aspects of diamond p-n junction and bipolar junction transistor formation," in: *Power Electronics Device: Applications of Diamond Semiconductors*, Elsevier, New York, NY, p. 364, 2018, Chapter 5.

[92] H. Kato, "Technical aspects of diamond p-n junction and bipolar junction transistor formation, in: *Power Electronics Device: Applications of Diamond Semiconductors*, Elsevier, New York, NY, p. 360, 2018, Chapter 5.

[93] W. (Wayne) Bi, H.-C. (Henry) Kuo, P.-C. Ku, and B. Shen, *Handbook of GaN Semiconductor Materials and Devices*, Taylor & Francis Group, New York, NY, p. 309, 2018.

[94] J. He, T. Zhao, X. Jing, and N.A.O. Demerdash, "Application of wide bandgap devices in renewable energy systems – benefits and challenges," in: *3rd International Conference on Renewable Energy Research and Applications*, Milwakuee, USA, Oct 19–22, pp. 749–754, 2014

[95] H. Jin, L. Qin, L. Zhang, X. Zeng, and R. Yang, "Review of wide band-gap semiconductors technology," *MATEC Web of Conferences*, Vol. 40, p. 01006, 2016. https://doi.org/10.1051/matecconf/20164001006.

[96] Renewable Energy Solutions NIRE Group (www.groupnire.com).

[97] P. Fiorenza, G. Greco, F. Iucolano, A. Patti and F. Roccaforte, "Channel mobility in GaN hybrid MOS-HEMT using SiO_2 as gate insulator," *IEEE Transactions on Electron Devices*, Vol. 64, no. 7, pp. 2893–2899, 2017. https://doi.org/10.1109/TED.2017.2699786.

[98] N.S. Saks and A.K. Agarwal, "Hall mobility and free electron density at the SiC/SiO_2 interface in 4H–SiC," *Applied Physics Letters*, Vol. 77, pp. 3281–3283, 2000. https://doi.org/10.1063/1.1326046.

[99] F. Roccaforte, F. La Via, V. Raineri, F. Mangano, and L. Calcagno, "Temperature dependence of the c-axis mobility in 6H-SiC Schottky diodes," *Applied Physics Letters*, Vol. 83, pp. 4181–4183, 2003. https://doi.org/10.1063/1.1628390

[100] S. Takashima, K. Ueno, H. Matsuyama, et al., "Control of the inversion-channel MOS properties by Mg doping in homoepitaxial p-GaN layers," *Applied Physics Express*, Vol. 10, pp. 1210041–12100414, 2017. https://doi.org/10.7567/APEX.10.121004.

[101] D. Ji, A. Agarwal, H. Li, W. Li, S. Keller, and S. Chowdhury, "880 V/ 2.7 $m\Omega\cdot cm^2$ MIS gate trench CAVET on bulk GaN substrates," *IEEE Electron Device Letters*, Vol. 39, no. 6, pp. 863–865, 2018. https://doi.org/10.1109/LED.2018.2828844.

[102] M. Cabello, V. Soler, G. Rius, J. Montserrat, J. Rebollo, and P. Godignon, "Advanced processing for mobility improvement in 4H-SiC MOSFETs: a review," *Materials Science in Semiconductor Processing*, Vol. 78, pp. 22—31, 2018. https://doi.org/10.1016/j.mssp.2017.10.030

[103] D. Tournier, P. Brosselard, C. Raynaud, M. Lazar, H. Morel, and D. Planson, "Wide band gap semiconductors benefits for high power, high voltage and high temperature applications," *Advanced Materials Research*, Vol. 324, pp. 46–51, 2011.

[104] F. Bénédic, M.B. Assouar, P. Kirsch, et al., "Very high frequency SAW devices based on nanocrystalline diamond and aluminum nitride layered structure achieved using e-beam lithography," *Diamond and Related Materials*, Vol. 17, nos. 4–5, pp. 804–808, 2008. https://doi.org/10.1016/j.diamond.2007.10.015. ISSN 0925-9635.

[105] F. Ke, L. Zhang, Y. Chen, et al., "Synthesis of atomically thin hexagonal diamond with compression," *Nano Letters*, Vol. 20, no. 8, pp. 5916–5921, 2020. https://doi.org/10.1021/acs.nanolett.0c01872.

[106] E. Farzana, Z. Zhang, P.K. Paul, A.R. Arehart, and S.A. Ringe, "Influence of metal choice on (010) β-Ga_2O_3 Schottky diode properties," *SN Applied Science*, Vol. 110, p. 202102, 2017. https://doi.org/10.1063/1.4983610.

[107] O. Pakma, Ş. Çavdar, H. Koralay, N. Tuğluoğlu, and Ö.F. Yüksel, "Improvement of diode parameters in Al/n-Si Schottky diodes with Coronene interlayer using variation of the illumination intensity," *Physica B: Condensed Matter*, Vol. 527, pp. 1–6, 2017. https://doi.org/10.1016/j.physb.2017.09.101. ISSN 0921-4526.

[108] Y.P. Song, R.L. Van Meirhaeghe, W.H. Laflère, and F. Cardon, "On the difference in apparent barrier height as obtained from capacitance-voltage and current-voltage-temperature measurements on Al/p-InP Schottky barriers," *Solid-State Electronics*, Vol. 29, no. 6, pp. 633–638, 1986. https://doi.org/10.1016/0038-1101(86)90145-0. ISSN 0038-1101.

[109] J.H. Werner and H.H. Guttler, "Barrier inhomogeneities at Schottky contacts," *Journal of Applied Physics*, Vol. 69, pp. 1522–1533, 1991. https://doi.org/10.1063/1.347243.

[110] R.T. Tung, "Electron transport of inhomogeneous Schottky barriers," *Applied Physics Letters*, Vol. 58, pp. 2821–2823, 1991. https://doi.org/10.1063/1.104747.

[111] S.M. Sze, *Physics of Semiconductor Devices*, Wiley-Interscience, New York, NY, p. 154, 1981.

[112] M.J. Bozack, "Surface studies on SiC as related to contacts," *Physica Status Solidi (b)*, Vol. 202, pp. 549–580, 1997. https://doi.org/10.1002/1521-3951(199707)202:1<549::AIDPSSB549>3.0.CO;2-6>.

[113] S. Toumi and Z. Ouennoughi, "A vertical optimization method for a simultaneous extraction of the five parameters characterizing the barrier height in the Mo/4H–SiC Schottky contact," *Indian Journal of Physics*, Vol. 93, pp. 1155–1162, 2019. https://doi.org/10.1007/s12648-019-01393-y.

[114] J. Osvald and E. Dobrocka, "Generalized approach to the parameter extraction from I–V characteristics of Schottky diodes," *Semiconductor Science and Technology*, Vol. 11, pp. 1198–1202, 1996.

[115] S. Toumi, Z. Ouennoughi, and R. Weiss, "Temperature analysis of the Gaussian distribution modeling the barrier height inhomogeneity in the Tungsten/4H-SiC Schottky diode," *Applied Physics A*, Vol. 127, p. 661, 2021. https://doi.org/10.1007/s00339-021-04787-0.

[116] A.N. Saxena, "Forward current-voltage characteristics of Schottky barriers on n-type silicon," *Surface Science*, Vol. 13, no. 1, pp. 151–171, 1969. https://doi.org/10.1016/0039-6028(69)90245-3. ISSN 0039-6028.

[117] H. Wu, X.W. Kang, Y.K. Zheng, et al., "Optimization of recess-free AlGaN/GaN Schottky barrier diode by TiN anode and current transport mechanism analysis," *Journal of Semiconductors*, Vol. 43, p. 0628036, 2022.

[118] H. Helal, Z. Benamara, E. Comini, et al., "A new approach to studying the electrical behavior and the inhomogeneities of the Schottky barrier height," *European Physical Journal – Plus*, Vol. 137, no. 450, pp. 1–8, 2022.

[119] I. Jabbari, M. Baira, H. Maaref, and R. Mghaieth, "Schottky barrier inhomogeneity in (Pd / Au) $Al_{0.22}$ $Ga_{0.78}N$/GaN/SiC HEMT: triple Gaussian distributions," *Chinese Journal of Physics*, Vol. 73, pp. 719–731, 2021. https://doi.org/10.1016/j.cjph.2021.08.011. ISSN 0577-9073.

[120] J.H. Lee, A.B.M. Hamidul Islam, T.K. Kim, Y.-J. Cha, and J.S. Kwak, "Impact of tin-oxide nanoparticles on improving the carrier transport in the Ag/p-GaN interface of InGaN/GaN micro-light-emitting diodes by originating inhomogeneous Schottky barrier height," *Photonics Research*, Vol. 8, no. 6, pp. 1049–1058, 2020. https://doi.org/10.1364/PRJ.385249.

[121] D.-k. Shi, Y. Wang, X. Wu, et al., "Improving the barrier inhomogeneity of 4H-SiC Schottky diodes by inserting Al_2O_3 interface layer," *Solid-State Electronics*, Vol. 180, p. 107992, pp. 1–5, 2021. https://doi.org/10.1016/j.sse.2021.107992. ISSN 0038-1101.

[122] S. Duman, A. Turut, and S. Doğan, "Thermal sensitivity and barrier height inhomogeneity in thermally annealed and un-annealed Ni/n-6H-SiC Schottky diodes," *Sensors and Actuators A: Physical*, Vol. 338, p. 113457, 2022. https://doi.org/10.1016/j.sna.2022.113457. ISSN 0924-4247.

[123] M.H. Ziko, A. Koel, T. Rang, and J. Toompuu, "Analysis of barrier inhomogeneities of p-type Al/4H-SiC Schottky barrier diodes," in: *Materials Science Forum*, Vol. 1004, Trans Tech Publications Ltd, pp. 960–972, 2020.

Chapter 2

Reliability of smart modern power electronic converter systems

Rupa Mishra[1], Nayan Kumar[2], Dibyendu Sen[3] and Tapas Kumar Saha[3]

Renewable energy sources have received considerable attention during the past two decades, owing to the advantages in terms of the absence of greenhouse gas emissions, cleanliness, and sustainability. With the growing energy crisis and environmental consciousness, the global perspectives are in the direction of promoting sustainable technologies. The wind and solar energy conversion systems under such a perspective are found very promising. Therefore, these two renewable technologies are the main focus of this chapter.

The power electronic converters interfacing play a vital role in the solar and wind energy conversion system with a maximum conversion efficiency. These converters can introduce reliability challenges to the system if not properly designed. A reliability index of a power electronics converter is calculated in terms of design, operation, maintenance, and performance assessment. Generally, the converter failure rate contributes to its overall cost of energy. In this regard, one of the significant challenges facing is reliability.

The overall purpose of a reliability analysis is to analyze failure and its impact on systems and to develop a rigid and reliable system.

This chapter focuses on the reliability of the used power electronic systems applied for wind energy conversion system (WECS) and solar energy conversion system (SECS).

2.1 Introduction

Depletion of conventional fossil fuel-based energy and its impact on climate change has been the main driving force to widen and advance renewable energy

[1]VIT Chennai, India
[2]IIT Delhi, India
[3]NIT Durgapur, India

technology. Renewable sources (RES) such as solar and wind energy, biomass, and hydropower energy are used for power generation to minimize greenhouse gas emission. Among the available RESs, wind and solar energy have emerged as significant sources for smart grid operation. Although energy production has been changing for a renewable reality, there are still many barriers to RESs development, such as the conversion cost, site selection, and distribution network.

Power electronics is the enabling technology for optimizing energy harvesting from RESs. Due to the progress in the power semiconductor configuration, the conversion efficiency of power electronics ranges above 98%. But the reliability of power electronics is becoming of serious concern [1]. Due to the huge demand for power electronics in control of semiconductor switches, particularly in solar and wind energy conversion systems, the reliability analysis (stress on power electronic components) is critical [2–4]. Therefore, it is essential to propose reliable power electronic systems to lower the risks of many failures during operation; otherwise, it will enhance maintenance costs.

In light of the above concern, this chapter briefly discusses the failure mode identification and reliability of solar and wind energy conversion systems.

2.1.1 Basic of reliability analysis

Reliability assessment is essential to the distributed generation systems when renewable energy sources and converters are studied individually or compositely [3–5]. The development of reliability is observed when the integration of wind turbine generators (WTG) and Photovoltaic (PV) are incorporated into the main electrical system. Thus, an initiative has been developed among the researchers regarding the uncertainty analysis with Renewable Energy (RE).

Reliability is defined as system's capability to adequately achieve its purpose under certain circumstances for an intended time [4]. The probability that a device would fail at t will give the failure distribution is presented as:

$$F(t) = \int_0^t f(t)dt \tag{2.1}$$

where t is the time or life, or age (in years or hours).

The reliability function $R(t)$ is formulated by subtracting the failure distribution from 1. It is presented as:

$$R(t) = 1 - \int_0^t f(t)dt = 1 - F(t) \tag{2.2}$$

Also, $f(t) - -\frac{dR(t)}{dt}$.

The mean time to failure (MTTF) is shown as:

$$MTTF = \int_0^\infty R(t)dt \tag{2.3}$$

The probability that a device is serviced until time t and fails in the next moment is conditional probability. It is also defined as a hazard rate:

$$\lambda(t) = \frac{f(t)}{R(t)} \tag{2.4}$$

The probability of failure is given as

$$\therefore \lambda(t) = -\frac{dR(t)}{dt}\frac{1}{R(t)} \Rightarrow R(t) = e^{-\lambda(t)} \tag{2.5}$$

The probability distribution function (PDF) is

$$f(t) = -\frac{dR(t)}{dt} = -\frac{d}{dt}\left(e^{-\lambda(t)}\right) \Rightarrow f(t) = \lambda e^{-\lambda(t)} \tag{2.6}$$

The mean time between failures (MTBF) defines the time span between two unplanned power cuts:

$$\mathrm{MTBF} = \frac{\sum_{i=1}^{n}\left(t_{pki} - t_{asci}\right)}{\text{Number of failure}} \tag{2.7}$$

where i is the sampling time, n is the sampling number, and t_{PK} is the starting time of unplanned power cut.

2.1.1.1 Weibull distribution function

Weibull distribution [3,4] is commonly used for analyzing the failure rate. In this case, α (scale parameter) and β (shape parameter) play a crucial role in reliability modeling. The probability distribution function is now given as:

$$f(t) = \frac{\beta t^{\beta-1}}{\alpha} e^{-\left(\frac{t^{\beta}}{\alpha^{\beta}}\right)} \tag{2.8}$$

Failure rate varies with time and is expressed as:

$\beta < 1$: Descending failure rate.

$\beta = 1$: Constant failure rate.

$\beta > 1$: Rise of failure rate.

Reliability will be the same when $t = \alpha$.

Here, to make the computation more accessible, the parameters are calculated either graphically or by linear regression.

The correlation between failure rate variation and time is commonly presented with the help of the bathtub curve. The literature reported that the failures of turbines and solar panels are commonly assumed to follow a curve [3]. Generally, the life span of the new wind turbine and solar panel is nearly 20 years. Normally, the life span of any newly installed component life span decomposed into three distinct spans. The newly installed solar panel and wind turbine equipment have a relatively

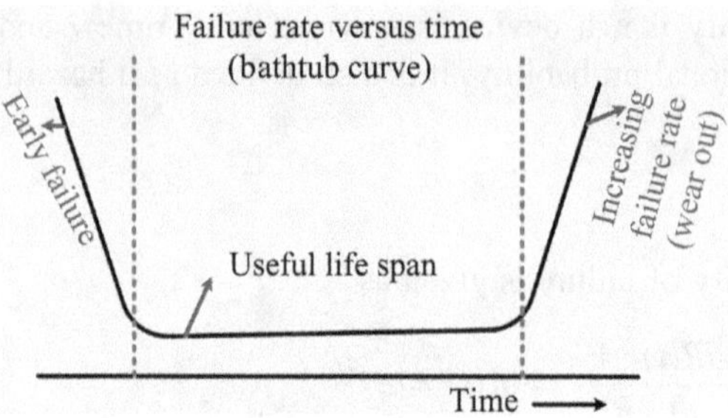

Figure 2.1 Normalized bathtub curve

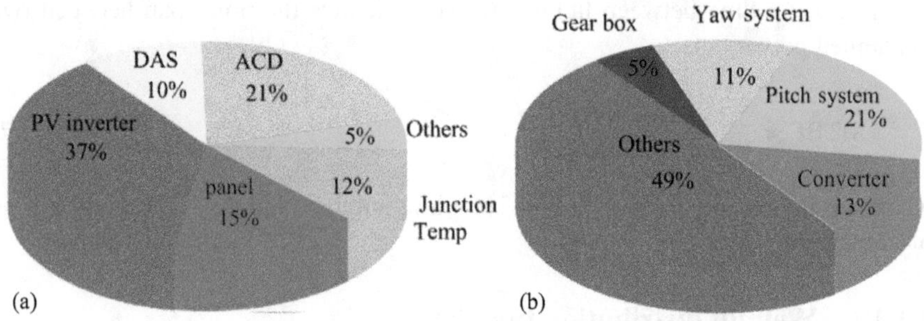

Figure 2.2 Subassemblies of (a) PV and (b) WECS to the failure rate [12]

higher failure rate for various reasons, such as manufacturing shortcomings during shipping and improper installation. In this period, the rates of failure are high, then decrease, which is the infant mortality or early failures period. The failure rates seem to vary by lower rates in the second span, known as the useful lifetime period [3–5]. The equipment will be in nearly constant service during this useful life period. The component performance will start to fall at the end of its useful life. This duration is known as the wear-out period or end of life. In this range, failure rates are increased with time. From the reliability view, the durability of various power electronic components is executed in the useful life. As shown in Figure 2.1, the bathtub curve initiated with a higher early failure rate, followed by a constant useful life and an increasing wear-out period/mortality rate [6–8].

In Figure 2.2, the failure rate for WT and PV converters accounts for 13% and 37%, respectively. Also the involvement of power electronics converter in the renewable energy system is presented in Figure 2.3. Therefore, improved power electronics reliability may effectively contribute to more energy harvesting and cost reduction.

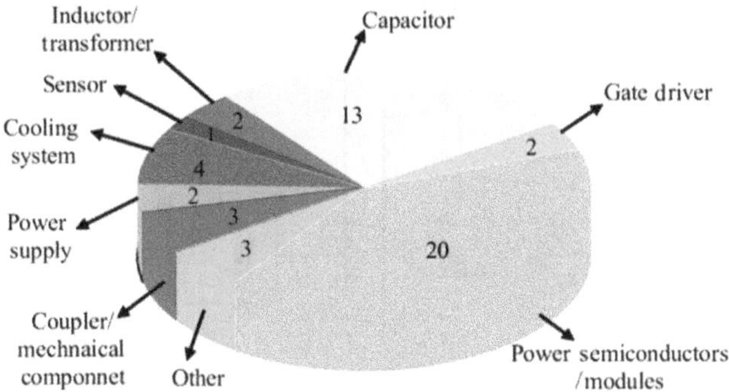

Figure 2.3 Failure rate of different power electronic components

2.2 Topology configuration of solar and wind-based energy conversion system

In the PV-based energy conversion system, ample source and battery size (system-level reliability) will not estimate the exact reliability because many other components are involved in the system. Failure of any of these components (component-level reliability) will result in the malfunction of the entire topology.

The key elements of a solar system include PV panels, batteries, inverter, charge controller, and wiring. In [7], the failure rate of the charge controller and battery can be estimation was reported. It is reported that the conventional inverter has a very high failure rate. The idea of a novel charge controller and further decreased failure rate through screening, derating, and redundancy was discussed. Normally, a multi-string inverter is suitable for the PV generator size. The failure rate of the inverter is estimated by the sum of the individual electronic components used inside the inverter. The Weibull scale and shape parameters of the inverter individual component and battery used in this work check the PV-based energy conversion system's reliability.

The solar energy conversion system is consisting of a dc–dc converter and an inverter as shown in Figure 2.4. Here, five semiconductor switches (IGBT) and the cooling system are designed to ensure the maximum junction temperature at the rated operating condition [6,7].

In this configuration, MPPT operation is implemented for generating gating signals for the DC–DC converter. Then, the PV inverter delivers the extracted power to the utility/stand-alone load by regulating the dc-link voltage [8].

The structure of the wind energy conversion system (WECS), considered for the distributed generation, is shown in Figure 2.4. The control structure includes a back-to-back converter composed of the generator and load side. The output of the inverter is fed to load with a filter designed for the system. The combined control is developed with generator and load side control, operating simultaneously in decoupled environments [9].

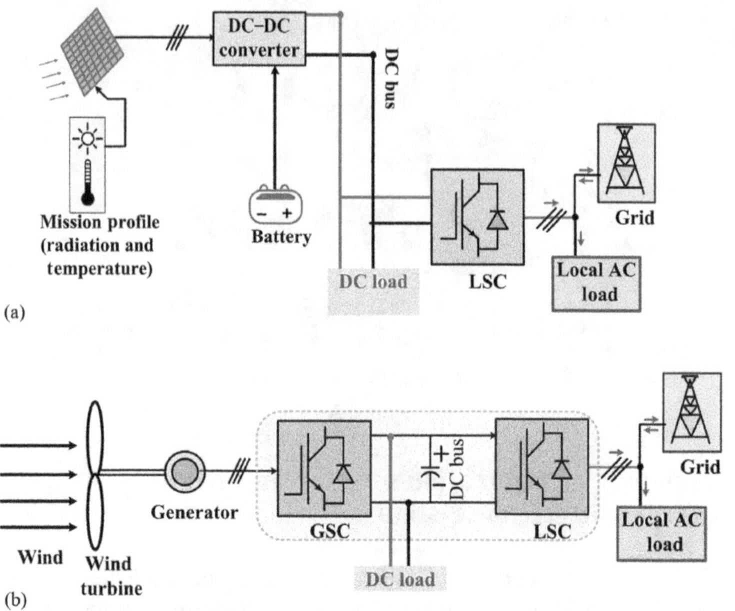

Figure 2.4 Energy harvesting configuration of (a) PV and (b) wind

The PV or WECS configuration can be designed to provide either DC or AC power service per the requirement. Therefore, this type of system is classified according to its component configuration and operational requirement broadly into two categories:

- Stand-alone system
- Grid-connected system

The isolated load objective is to keep load voltage magnitude and frequency constant fulfilled by regulating load voltage. Similarly, during grid connection, the current controller is acting to hold the grid current according to a reference given by the DC-link voltage controller. This control objective is achieved by generating a switching signal for the load side converter (LSC). In the grid-tied system, the utility is directly fed from these systems, and the presence of a battery energy storage system (BESS) is related to the level of grid reliability. If the grid is weak for supplying local loads, storage devices would be necessary, and the level of reliability will reduce. At the same time, BESS is not needed with a strong grid, so system reliability will enhance.

2.3 Reliability evaluation of SECS and WECS

It is reported in the earlier literature [2–5] that the temperature (thermal cycling) is the most significant factor for reducing the reliability of PV modules. The effect of temperature and humidity on the components is considered to estimate the reliability of the inverter.

In a solar energy conversion system (SECS), the inverter reliability and its life span are strongly affected by the operating condition, i.e., solar irradiance and ambient temperature. Both the operating conditions are known as mission profile of the PV system. As this profile is dependable upon the site and time, the inverter performance and its life span can vary considerably.

Additionally, component wears out, and failure can be achieved even under different mission profiles. The overall maintenance cost is increased with the same objective.

2.3.1 Mission profile-based reliability evaluation

The mission-profile [8] based reliability estimation process of PV and wind energy conversion systems is presented in Figure 2.5. The scheme is used to initially translate the environmental stress profiles to the system, electrical (converter) level, then to the component-level stress, and, finally, to estimate the component-level and the system-level reliability metrics. This procedure involves electrical, thermal, lifetime, and statistical reliability analysis for both solar and wind energy conversion systems, respectively.

2.3.1.1 Translation of mission profile into thermal stress profile for both solar and WECS

The mission profile (i.e., the solar irradiance and ambient temperature in case of PV and wind speed) is important in the reliability assessment and lifetime prediction of PV inverters and wind turbine power converters. From the mission profile, the components' power losses are determined from the solar panel model and the control strategy together with the loss model of the parts. Then, the power losses are applied to the thermal models of the elements to find the thermal loading during the operation, which is required for the lifetime model [8,10].

The mission profile is first translated into the electrical equivalent parameter of solar and wind at maximum power points. Then the converter power loss and junction temperature is obtained with an electro-thermal model. The accumulative damage can be obtained with the component lifetime and damage accumulation models. Subsequently, the system wear-out failure probability can be derived with the Weibull analysis and system reliability model as shown in Figure 2.6.

The mission profiles recorded in New Delhi are used in this article, as shown in Figure 2.4. It can be seen from the mission profiles that the average solar irradiance level is low from November to February. The same trend can also be seen in the ambient temperature profile. When translating the mission profile into the inverter loading (following Figure 2.7), it can be seen from the thermal stress profiles of the arrays and other components installed to experience a thermal loading during the operation. In that case, the reliability of power devices will be subjected to higher thermal stresses, as shown in Figure 2.8.

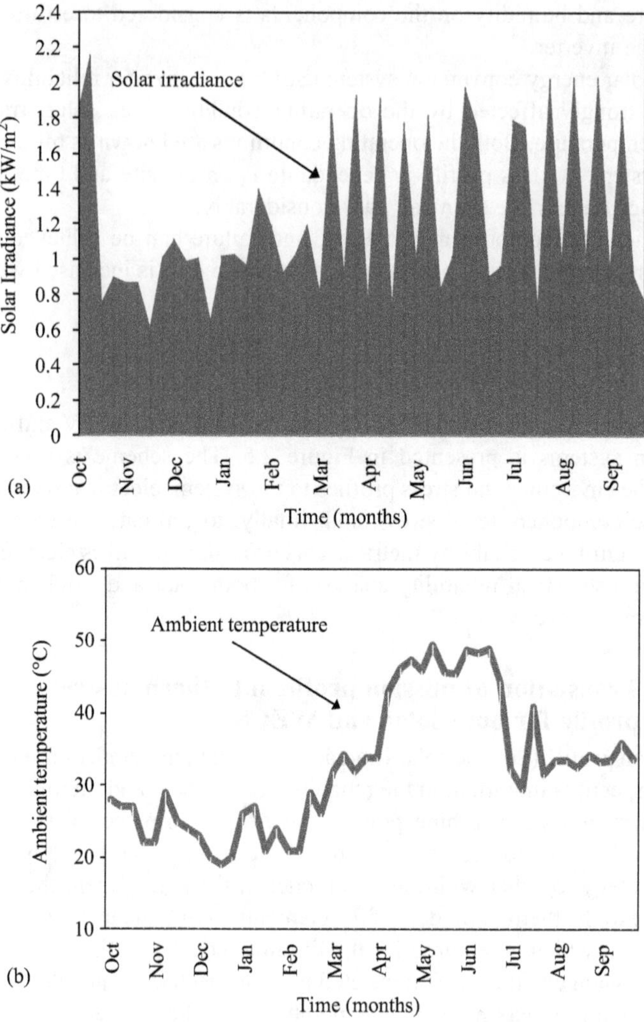

Figure 2.5 Yearly mission profiles: (a) solar irradiance and (b) ambient temperature data of India

Similarly with the considered mission profile approach, the percentage of failure rate per year of both generator and load side converter involved in WECS is portrayed in Figure 2.9. It is observed that more power loss of the converter occurred at the rated value of wind speed (11–25m/s). Therefore the failure rate of the power converter gradually increases around rated operation. Also generator side converter of the WECS contributes more failure rate (nearly 60%) compared with in load side converter. From this observation, it is clear that the DC-link capacitor smoothens the power variation in the case of both isolated load and grid-connected mode.

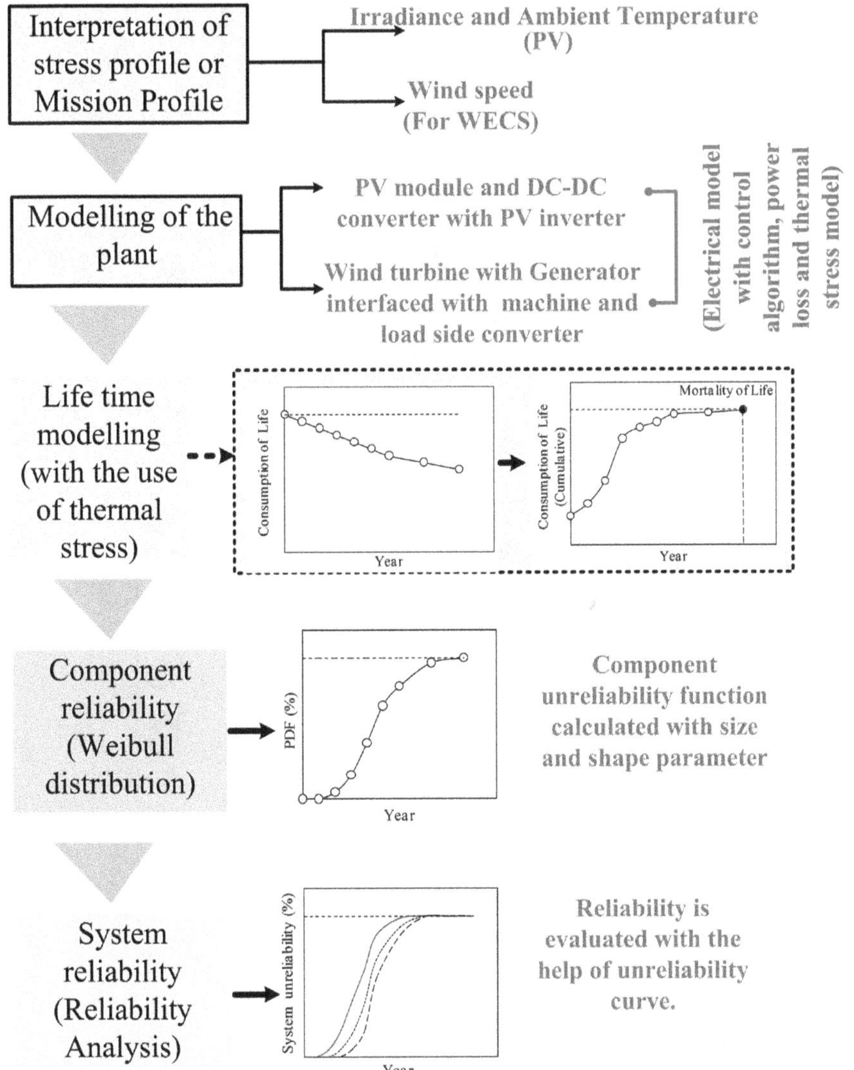

Figure 2.6 Flow diagram of the mission profile-based reliability evaluation of WECS and PV-based energy conversion system

Various types of generators are used in wind turbines, such as doubly-fed induction generators (DFIG), squirrel-cage induction generators (SCIG), and permanent magnet synchronous generators (PMSG) [9,10,13,14]. This work illustrates the wind turbine subassemblies failure rate [10] of DFIG with the use of Weibull parameter. The probability density distributions for MTBF of various wind turbine

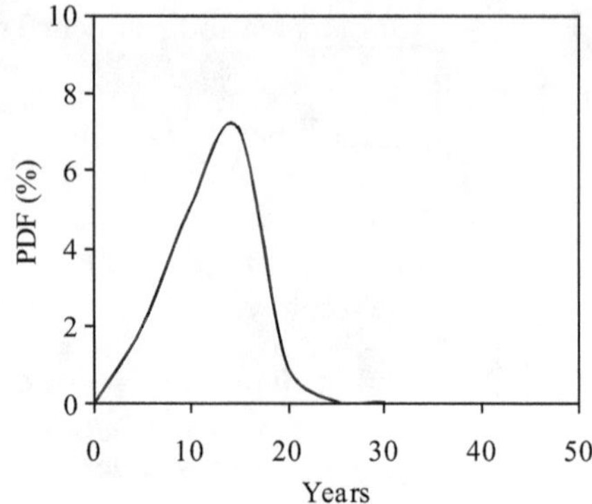

Figure 2.7 The survival rate of PV-based energy conversion system

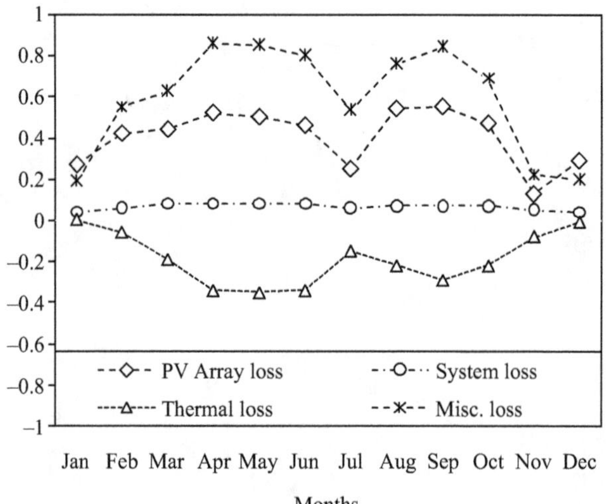

Figure 2.8 Various types of losses (in pu) involved in PV module

subassemblies are shown in Figure 2.10. It is clear from Figure 2.11 that the power converter records the higher failure probability among all electrical subassemblies in case of DFIG based WECS. Additionally, the survivability of the wind turbine subsystems of DFIG for 10 year is analyzed and presented in Figure 2.12. It is clear that due to installation and manufacturing problems, and climate issues, random failures can occur at any time in the WECS. However, mechanical subassemblies such as generator, gearbox, etc. require rigorous maintenance due to frequent failure.

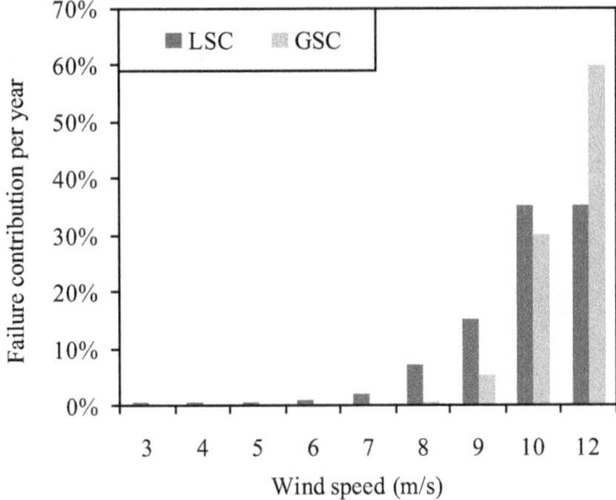

Figure 2.9 Contribution of GSC and LSC failure rate per year

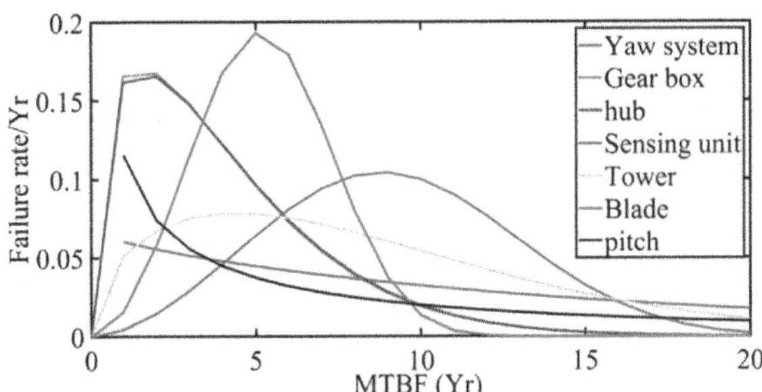

Figure 2.10 Failure rate/Yr for different assemblies of wind turbine

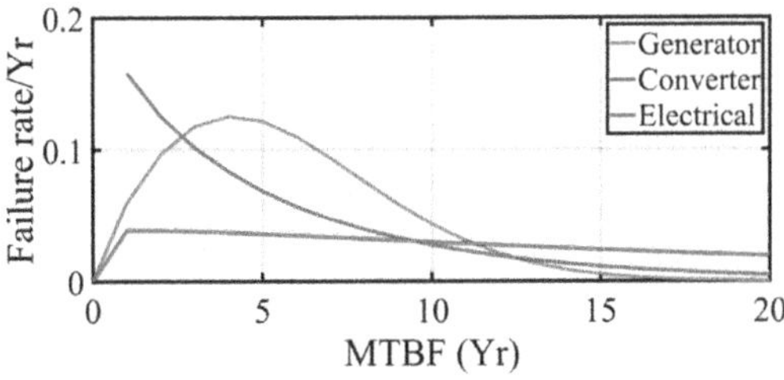

Figure 2.11 Failure rate/Yr for different section of WECS

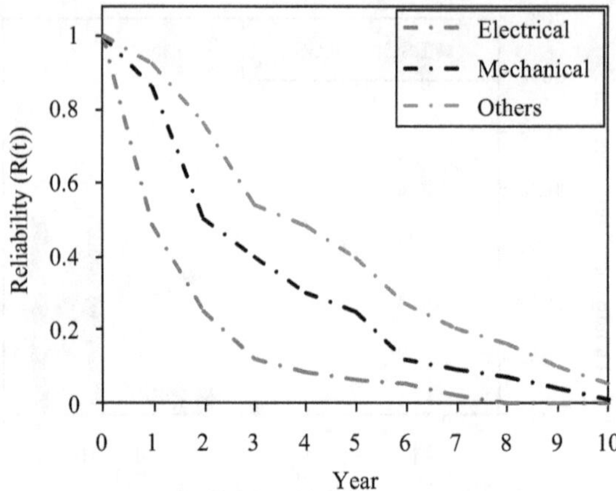

Figure 2.12 Survival rate of different parts of DFIG based WECS

2.4 Conclusion

In this study, a complete framework for evaluating the reliability of PV-based energy conversion systems and WECS has been carried out. The electrical and thermal stress in the power electronics converter integrated with the distributed generation system is investigated. With the failure rate evaluation, component and system-level reliability of the solar and wind energy conversion system is studied. A mission profile-based reliability evaluation for a case study with the data from the installation site is conducted. Using updated and more realistic data to model the impact of failures, MTTR on the energy yield allowed an assessment of the importance of the various failure modes.

Although a high-reliability rate of both the distributed generation system could be obtained under frequent maintenance, maintenance strategies can be optimized to reduce the economic burden. It will help to choose an appropriate maintenance strategy, which future work will help improve the overall system reliability.

References

[1] J. C. Balda and A. Mantooth, "Power-semiconductor devices and components for new power converter developments: a key enabler for ultrahigh efficiency power electronics," *IEEE Power Electron. Mag.*, vol. 3, no. 2, pp. 53–56, 2016, doi:10.1109/MPEL.2016.2551801.

[2] K. Chaiamarit and S. Nuchprayoon, "Modeling of renewable energy resources for generation reliability evaluation," *Renew. Sustain. Energy Rev.*, vol. 26, pp. 34–41, 2013.

[3] P. Georgilakis, P. Karafotis, and V. A. Evangelopoulos, "Reliability-oriented reconfiguration of power distribution systems considering load and RES

production scenarios," *IEEE Trans. Power Delivery*, vol. 87, pp. 1–1, 2022 doi:10.1109/TPWRD.2022.3153552.

[4] T. Wang, Z. Liu, M. Liao, N. Mrad, and G. Lu, "Probabilistic analysis for remaining useful life prediction and reliability assessment," *IEEE Trans. Reliab.*, vol. 71, no. 3, pp. 1207–1218, 2022, doi:10.1109/TR.2020.3032157.

[5] M. V. Kjaer, H. Wang, Y. Yang, and F. Blaabjerg, "Reliability analysis of power electronic-based power systems," in *2019 International Conference on Smart Energy Systems and Technologies* (*SEST*), 2019, pp. 1–6, doi:10.1109/SEST.2019.8849090.

[6] S. B. Q. Naqvi, S. Kumar, and B. Singh, "Weak grid integration of a single-stage solar energy conversion system with power quality improvement features under varied operating conditions," *IEEE Trans. Ind. Appl.*, vol. 57, no. 2, pp. 1303–1313, 2021.

[7] V. N. Lal and S. N. Singh, *"Control and performance analysis of a single-stage utility-scale grid-connected PV system," IEEE Syst. J.*, vol. 11, no. 3, pp. 1601–1611, 2017.

[8] T.-F. Wu, C.-H. Chang, L.-C. Lin, and C.-L. Kuo, *"Power loss comparison of single- and two-stage grid-connected photovoltaic systems," IEEE Trans. Energy Conv.*, vol. 26, no. 2, pp. 707–715, 2011.

[9] F. Spinato, P. J. Tavner, G. J. W. van Bussel, and E. Koutoulakos, "Reliability of wind turbine subassemblies," *IET Renew. Power Gener.*, vol. 3, no. 4, pp. 1–15, 2009.

[10] S. Wu, L. Qi, Z. He, W. Jia, and X. Zhang, "A voltage-boosting submodules based modular multilevel converter with temporary energy storage ability for fault ride through of offshore wind VSC-HVDC system," *IEEE Trans. Sustain. Energy*, vol. 13, no. 4, pp. 2172–2183, 2022, doi:10.1109/TSTE.2022.3188356.

[11] Renewable Energy Project Monitoring Division. Plant Wise Details of all India Renewable Energy Projects. 6/3/2019/1410. New Delhi, India: Central electricity authority, Ministry of Power, Government of India, 2020.

[12] F. Blaabjerg, M. Liserre, and K. Ma, "Power electronics converters for wind turbine systems," *IEEE Trans. Ind. Appl.*, vol. 48, no. 2, pp. 708–719, 2012.

[13] S. Golestan, J. M. Guerrero, F. Musavi, and J. C. Vasquez, "Single-phase frequency-locked loops: a comprehensive review," *IEEE Trans. Power Electron.*, vol. 34, no. 12, pp. 11791–11812, 2019.

[14] A. Ahmedi, M. Barnes, V. Levi, J. Carmona Sanchez, C. Ng, and P. Mckeever, "Modelling of wind turbine operation for enhanced power electronics reliability," *IEEE Trans. Energy Convers.*, vol. 37, no. 3, pp. 1764–1776, 2022, doi:10.1109/TEC.2022.3144499.

Chapter 3

Next-generation electrification of transportation systems: EV, ship, and rail transport

Carlos Reusser[1], Hector Young[2] and Marcelo A. Perez[3]

The ever-increasing global environmental concerns, the introduction of new energy sources to the power grid, added to the rapid technological development of power electronics, and control of electric drives, have recently provided new opportunities to reduce CO_2 and other polluting material emissions, toward a next generation of electrified transport systems.

The integration of power electronics has experienced a massive penetration in modern transportation systems such as electric vehicles (EV), ships, and rail applications, thereby playing a major role in the reduction of polluting emissions, either by decreasing the dependency on fossil fuels or by helping to increase the efficiency in hybrid systems.

In this context, the electrification of transport systems has resulted in the development of several configurations for hybrid and full-electric power-train arrangements, both introducing the use of high-performance electrical machines and power electronics drives.

Due to the critical importance of transportation systems in the society and economy, aspects such as safety, availability, sustainable operation, and efficiency constitute a central part in the design process. Therefore, electronic power converters must be analyzed in terms of voltage levels, power density, and system redundancy, among other criteria. Moreover, the infrastructure required to operate the new large-scale electric transportation systems must consider energy storage, charging capabilities, and accomplishment with renewable energies policies. These features introduce several challenges for the next generation of electrification transportation systems, which are examined within this chapter.

3.1 Power electronics for traction inverters

A general approach to classify power inverters is to separate the existing topologies according to the energy-storage element employed in the DC-link stage. Therefore,

[1]Pontificia Universidad Católica de Valparaíso, Valparaíso, Chile
[2]Universidad de La Frontera, Temuco, Chile
[3]Universidad Técnica Federico Santa María, Valparaíso, Chile

voltage–source inverters (VSIs) employ a capacitive filter at their DC-link, in order to feed a switched voltage waveform to the load. On the other hand, current–source inverters (CSIs) make use of an inductive DC-link in order to generate a switched current waveform to the load. The third type of power inverter does not feature a DC-link with large energy storage, which makes possible to increase the power density at the expense of and increased number of semiconductor switches and the requirement of more complex control strategies.

3.1.1 Two level voltage-source inverters (2L-VSI)

In the present, the prevalent power converter employed in the traction stage of light EVs is the two-level VSI (2L-VSI) based on IGBTs. This preference is explained by the simplicity of this topology along with its high efficiency and low cost [1]. Also, the most commonly employed inverter for electrical traction of urban and high-speed railways is the 2L-VSI [2,3]. Figure 3.1 contains the schematic of one leg of the 2L-VSI topology, where the power is provided by a battery pack or rectifier in the DC side, with a capacitive filter to achieve low ripple levels in the DC-link voltage. A decisive advantage of this configuration is its maturity level, achieved after several decades of successful implementation in industrial applications, and a wide availability of manufacturers.

3.1.2 Multilevel inverters

The use of higher DC-link voltages in electrified transportation systems pursues objectives such as enabling faster charging rates of battery-powered vehicles, as well as a decrease in the current rating of motors and cables thereby reducing weight and size while increasing efficiency. Multilevel inverters are well suited for operation at higher DC-link levels, due to lower voltage blocking requirements of individual semiconductor switches and reduced dv/dt [4,5]. Another advantage of multilevel converters over conventional 2L-VSIs is the ability to continue operating under fault conditions, due to their inherent redundancy. Therefore, methods for the detection and localization of faults in multilevel inverters have been proposed, with the aim of enabling the affected vehicle to arrive by its own means to a service station [6].

The most successful three-level inverter topologies in traction applications are the neutral point clamped (NPC) and the T-type inverter. In comparison with the 2L-VSI, these inverter topologies offer lower switching losses as well as lower harmonic distortion [7]. However, the capacitor banks of three-level inverters occupy a larger volume than those in 2L-VSIs and require special control strategies for capacitor voltage balancing [1].

3.1.2.1 NPC

The three-level neutral-point clamped (NPC) inverter topology, depicted in Figure 3.1, features clamping diodes and cascaded capacitors to generate the output voltage waveform. Albeit it is possible to employ the same operating principle to produce more than three-voltage levels, this implies an increased number of capacitors and semiconductor switches that have prevented a widespread use of these topologies in practical applications [8].

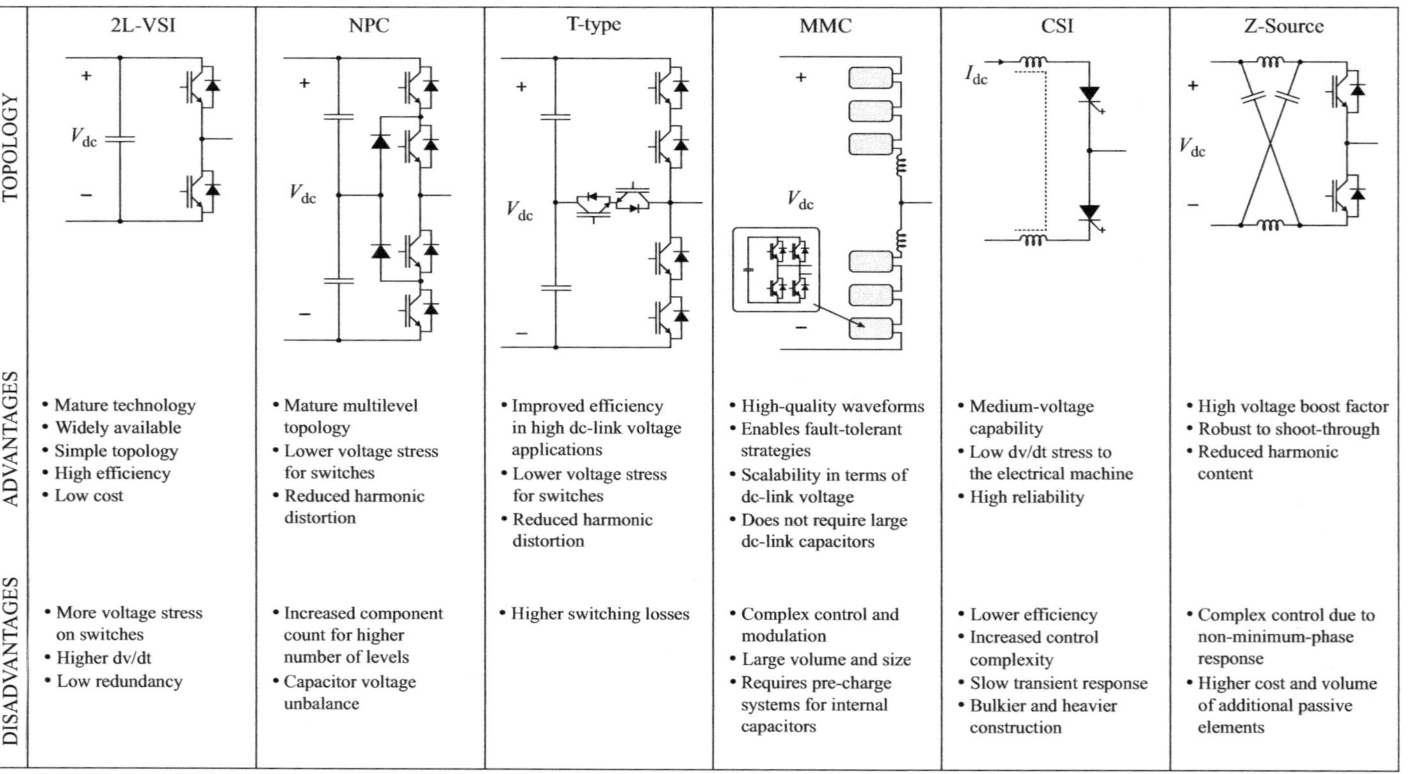

TOPOLOGY
ADVANTAGES
DISADVANTAGES
2L-VSI
+
V_{dc}
−
• Mature technology
• Widely available
• Simple topology
• High efficiency
• Low cost
• More voltage stress on switches
• Higher dv/dt
• Low redundancy
NPC
+
V_{dc}
−
• Mature multilevel topology
• Lower voltage stress for switches
• Reduced harmonic distortion
• Increased component count for higher number of levels
• Capacitor voltage unbalance
T-type
+
V_{dc}
−
• Improved efficiency in high dc-link voltage applications
• Lower voltage stress for switches
• Reduced harmonic distortion
• Higher switching losses
MMC
+
V_{dc}
−
• High-quality waveforms
• Enables fault-tolerant strategies
• Scalability in terms of dc-link voltage
• Does not require large dc-link capacitors
• Complex control and modulation
• Large volume and size
• Requires pre-charge systems for internal capacitors
CSI
I_{dc}
• Medium-voltage capability
• Low dv/dt stress to the electrical machine
• High reliability
• Lower efficiency
• Increased control complexity
• Slow transient response
• Bulkier and heavier construction
Z-Source
+
V_{dc}
−
• High voltage boost factor
• Robust to shoot-through
• Reduced harmonic content
• Complex control due to non-minimum-phase response
• Higher cost and volume of additional passive elements

Figure 3.1 Inverter topologies in traction applications

The NPC topology has been recently proposed to integrate multi-source inverters for rail traction systems fed by multiple independent electrical sources [9]. Recent research considering the NPC topology in electrical propulsion systems is directed to subjects such as minimization of electromagnetic interference [10], reduction of inverter losses [11], neutral point voltage balancing [12], and fault diagnosis [13].

3.1.2.2 T-type inverter

The three-level (3L) T-type inverter can be understood as an extension of the 2L-VSI topology, obtained by connecting the midpoints of the DC-link and each of the inverter legs using a bidirectional switch. The basic topology of the 3L T-type inverter is shown in Figure 3.1. The 3L T-type inverter has demonstrated to achieve higher efficiency than the 2L-VSI in applications with an 800 V DC-link [14] as well as lower conduction losses [15]. Topologies derived from the 3L T-type inverter have been applied as traction inverters in DC-electrified railways [16].

3.1.2.3 Modular multilevel converter (MMC)

Each phase of an MMC inverter is composed of a number of submodules in cascaded connection, distributed in two arms per inverter phase, as depicted in Figure 3.1. Among the advantages MMC may offer in traction applications, there is a high quality of current and voltage waveforms, the scalability of the system voltage by the addition of more submodules and no need for large DC-link capacitors [17]. Moreover, the modular structure of MMC inverters makes possible to integrate fault–tolerant strategies which increase the system reliability. The downside of MMC in traction applications is the complexity of control and modulation strategies, a large volume and size which makes it less attractive for lower power systems, and the requirement of pre-charge systems to regulate the DC-voltage in the converter cells.

3.1.3 CSIs

The CSI employs an inductive energy storage element at the DC-link instead of the capacitive one required by VSIs. One immediate advantage of this configuration is the replacement of DC-link capacitors by a more robust inductor, which leads to lower maintenance cost [18] and a reliable short-circuit protection. Additionally, the CSI produces output voltage waveforms with reduced dv/dt, which avoids stress on the electrical machine insulation [8]. One phase of the CSI topology is shown in Figure 3.1. A capacitive filter (not drawn in the diagram) needs to be included in the motor side of the inverter to enable the forced commutation of currents under inductive loads. Some examples of high-power traction drive fed by CSI are the French TGV Atlantique, and DC-fed railways in Germany, Switzerland, and South Africa [19].

Some drawbacks of CSIs for traction applications are a lower efficiency, increased control complexity, bulkier, and heavier construction in high-power applications and slow transient response [1,18].

3.1.4 Z-Source inverters (ZSI)

The ZSI is a topology that employs an impedance network instead of a capacitive or inductive DC-link energy storage found in VSI and CSI, respectively [20]. The topology of the ZSI is shown in Figure 3.1, with a distinctive capacitor–inductor network in its input stage. The ZSI has been proposed in electric transportation applications due to its high boost factor which enables the inverter output to deliver higher voltages, increased robustness to shoot-through, and reduced harmonic content [21]. Some downsides of the ZSI in comparison to the conventional 2L-VSI are the non-minimum-phase response due to a right-hand-plane zero in the ZSI network [22], and higher cost and volume due to additional passive components [1].

3.1.5 Multiphase traction inverters

In the last decades, the implementation of variable speed drives employing multiphase electric machines has received increasing attention. Multiphase drives are attractive for their inherent fault tolerance, low torque pulsations, and lower per-phase current [18]. Moreover, the additional degrees of freedom associated with multiphase machines can be used for fault–tolerant operation, torque improvement by means of harmonic current injection, multimotor drives with single inverter supply, capacitor voltage balancing, and integrated onboard battery chargers for EVs [23,24]. Recent applications of multiphase drives include ship propulsion systems [25], fault–tolerant power-trains for EVs [26], and heavy-duty EVs such as trucks and buses [27].

Two-power inverter topologies commonly employed in multiphase drives for traction applications are the n-phase 2L-VSI and the nine-switch inverter (NSI), depicted in Figure 3.2(a) and (b), respectively [18]. The multiphase 2L-VSI topology is obtained by adding the required number of legs to the conventional

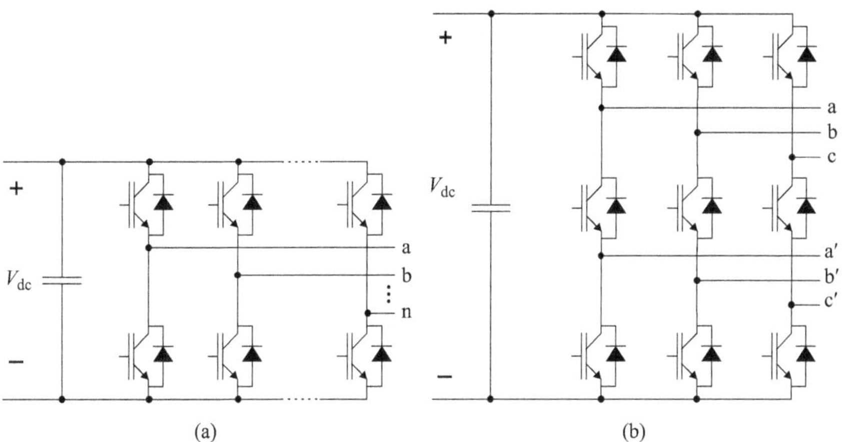

Figure 3.2 Multiphase inverter topologies: (a) multiphase 2L-VSI and (b) nine-switch inverter

three-phase 2L-VSI, according to the number of phases of the electrical machine. In the case of the NSI, its topology is able to feed one six-phase or two three-phase electrical machines using only three inverter legs.

3.2 Machines and control

Electric motors used in electric transportation are typically high torque and medium to low speed. Among the main types of equipment used in this application are brushed DC machines, induction machines, synchronous machines, wound-rotor machines, and permanent-magnet machines. Recent research has investigated homopolar and high-temperature superconducting machines [14]. Synchronous reluctance machines (SynRMs) also appear to be an attractive alternative for marine electric propulsion applications.

3.2.1 Induction machines

The induction machine is basically an AC polyphase machine connected to an AC power source, either in the stator or in the rotor, as in the case of the doubly fed induction machine.

The winding arrangement on the stator produces a rotating field in the machine airgap, which induce circulating currents, hence an induced rotating field in the rotor. In the case of the doubly-fed induction machine (DFIM), the rotor field is due to the independent excitation of the rotor winding, which is achieved using an external source connected to the terminal of each rotor winding using slip rings.

Induction machines are very attractive for traction applications because of their simplicity of their construction and working principle, robustness, and wide speed range operation, including field weakening operation. Implementation of control strategies, like field-oriented control (FOC) or direct torque control (DTC), which are today's industrial standard control schemes, enables the IM to develop maximum torque per Ampere ratio. However, due to their induction working principle, rotor losses increase, when going into higher power ratings, making its size bigger when compared with a PMSM of the same speed and power ratings.

3.2.1.1 Low-speed induction machines

The speed of a motor-driven system is typically determined by the application needs, for which electric motors are designed accordingly to run at the required speed. For motors working directly connected to the grid, the selection of the number of poles for a given application is fairly obvious; e.g., if the application requires operation at around 3,000–3,600 r/min, a 2-pole motor is designed and, if a 1,500–1,800 r/min is desired, a 4-pole motor is designed. The primary reason for the preselected number of poles is that once the r/min is known and the line frequency is specified, the number of poles is automatically determined. Specific pole numbers as they apply to the design of an induction machine do come with certain advantages and disadvantages but application considerations may override the opportunity to utilize any such advantages. However, the influence of pole number on performance is critical in machine design for applications involving variable frequency drives.

3.2.1.2 DFIMs

The DFIM is based on a WRIM, with the rotor windings brought out through slip rings. This type of induction machine is widely used as wind generators for small- and medium-power applications [29, 30] as well as for traction [31]. For marine applications, the DFIM is connected to the main busbar via two power converters, which supply stator and rotor windings, respectively, as described in Figure 3.3. DFIMs are similar to SCIMs in terms of their torque capability and power density. However, the speed range of the DFIM can be extended because the stator and the rotor windings can interact with different voltage frequencies to control the induced back electromotive forces. Moreover, active power can be supplied by both the stator and the rotor windings, depending on operating conditions and regardless of speed. However, controlling both the stator and the rotor winding converted power flow adds extra complexity, and it does not substantially contribute to the propulsion system redundancy.

3.2.2 Synchronous machines

The unique features of these machines, such as high efficiency, higher power density, and simple speed control, compared to early developed DC drives, made the devices the ideal choice for replacing DC machines as a ship propulsion motor. Depending on how the rotor field is produced, synchronous machine rotors are classified either as wound-rotor or permanent-magnet machines.

3.2.2.1 Wound-rotor synchronous machines (WRSMs)

In early developments of WRSMs, speed control was achieved by a variable-speed AC generator combined with the ability to change the number of poles on each winding. Later developments use load-commutated inverters to increase performance. The main disadvantages of WRSMs result from the need for slip rings and brushes to excite the rotor winding and the requirement for an external source to feed the field circuit.

3.2.2.2 Permanent magnet synchronous machines (PMSMs)

PMSMs have the advantage of a higher power density compared to wound-rotor, and induction machines, due to the high flux remanence of permanent magnets. Moreover, PMSM efficiency reaches values of nearly 98%, while the efficiency of induction machines is limited to 94–96% at full loads. This unique characteristic of high torque and high-power density makes them a very attractive alternative for traction applications. However, they have a narrow constant power operation range, due their limited capability in field weakening operation, because of the permanent magnet field, which can only be weakening through the stator flux linkage, which is limited by the power converter.

Type	Switched Reluctance (SRM)	Induction	Wound Rotor Synchronous (WRSM)	Permanent Magnet Synchronous (PMSM)	Open-end Multi-winding	Doubly-fed Induction (DFIM)	Synchronous Reluctance (SynRM)
Symbol	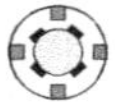				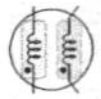	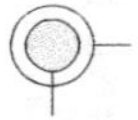	
Advantages	Robust and simple construction Easy control strategy. Wide speed operation range.	Robust and simple construction. Almost maintenance free. Wide range of load operation	Adjustable power factor. Larger air gap compared to induction machines, becoming more shock resistant.	High Power density. Maximum torque at zero speed.	Improved torque quality Multiple degrees of redundancy and assurance of power segmentation	Operation in sub and super synchronous regions. Allows to down size the power converter.	Does not require permanent magnets or separate excitation. Simple rotor design.
Disadvantages	Not widely used in electric drives. High costs.	Low efficiency at partial load. Variable slip depending on load. Low power factor at partial load.	Need separate excitation winding. Brushes and slip rings add complexity. Significant maintenance costs.	Thermally sensitive. High costs of magnets.	Additional power converters. Possibility of induce zero sequence currents. Complex inter-winding insulation and layout.	Slip rings and brushes add complexity. Low efficiency due to rotor currents.	Power factor depends on rotor design. Only reluctance torque component, leading to larger frames in high power applications

Figure 3.3 Electric machines for traction drives

3.2.3 Switched reluctance machines

Switched reluctance machine have been gaining a lot of research interest as a candidate of electric propulsion for EVs in the last years, because of its simple construction, control principle, and ability to operate under extremely high-speed ranges. These advantages become more attractive for traction application than SCIM or PMSM. However, since SRM are not yet widely produced, their cost can be higher than other alternatives.

3.2.4 Multiphase machines

Multiphase variable-speed drives, based on multiphase AC machines, are nowadays the most natural solution for high demanding industrial, traction and power generation and distribution applications.

Multiphase machines for variable-speed applications are in principle the same as their three-phase counterparts. These are synchronous machines, which depending on the excitation can be sub-classified into the following: wound rotor, permanent magnet or reluctance machines, and induction machines.

In a multiphase winding, the stator winding distribution becomes more concentrated, rather than distributed, as in the case of three-phase windings. This fact and the particularity of using quasi-sinusoidal, rather than sinusoidal voltages because of the inverting process in the power conversion stage, has several advantages that can be summarized as lower field harmonics content, better fault tolerance, because of extra degrees of freedom and less susceptibility to pulsation torque pulsations, due to an excitation field with less harmonic content.

Within the previous context, multiphase induction machines have become very popular for applications where high redundancy and power density are required. In particular, the use of multiple three-phase windings (six and nine phases) in naval propulsion systems has gained and aroused much interest and encouraged the research of new multiphase converter topologies and control schemes [4,5]. Some of their main advantages are the following:

- The total power can be divided into lower power converters.
- Due that each converter is insulated from each other, there is no circulating current between converters, which leads to no power derating for the converters.
- The increase of the number of phases in the generator voltages is phase shifted so that low order harmonics are reduced and consequently smaller filters can be used.
- Reduced torque pulsations.
- Fault–tolerant redundancy under winding fault conditions.
- Additional degrees of freedom which can be used to improve the machine performance.

3.2.5 Drive control strategies

Achieving fast, accurate, and robust control of speed and torque in electric machines have been a motivation for research and implementation mainly for traction applications. To this end, there are two control strategies that rise as today's standards, being FOC and DTC.

3.2.5.1 FOC

The characteristics of FOC have made this control strategy the most widely used for high demanding industrial applications. FOC is based on the decoupling of the current space vector, into a flux producing component and a torque producing component. This is achieved by rotating the current space vector from $\alpha\beta$ subspace, into the synchronous rotating reference frame dq, oriented respect the rotor flux linkage space vector. In this way, the magnetizing and torque producing currents can be controlled independently, thus ensuring the machine operating with optimum torque per Ampere ratio. Figure 3.4 shows a FOC control scheme.

The main drawbacks of FOC are the need of rotor flux linkage estimation and the rotation operations of the state-space variables into this synchronous reference frame, becoming a process, that requires much processing space of microcontrollers. Also, the dynamic response is limited by the maximum bandwidth achievable for the PI controllers.

3.2.5.2 DTC

DTC is based on the required torque and flux references are follows as close as possible, by applying a suitable voltage vector, depending on the estimation of torque and flux directly from the machine state variables.

The required voltage vector is chosen via a switching table depending on the output signal of the torque and flux hysteresis controllers. Implementation of the DTC scheme is shown in Figure 3.5.

The main characteristics of DTC are its simple implementation and a fast dynamic response achieved by using hysteresis controllers. Furthermore, the required switching states are directly assigned from the switching table algorithm, so no modulator is needed. However, the main problems with DTC are the following: variable switching frequency and sampling period dependence of hysteresis controllers.

3.3 EVs

The continuous use of fossil fuels, especially in recent decades, has led to various environmental issues, such as global warming and air pollution. In addition, energy crisis has affected the world economy to a great extent [28]. Considering that vehicles consume the overwhelming majority of fossil fuels used in the world, an effort has been made over the last few years to change the scene so that vehicles are as least-polluting as possible. This can be performed by the use of vehicle electrification technologies, including EVs and hybrid EVs (HEVs), on the basis that they use electricity produced from renewable energy sources [29–31]. However, EVs are a major technological challenge for power grids, since passive elements constitute a new kind of cargo. Therefore, a large number of EVs can appreciably burden the grid and adversely affect its smooth operation [32,33].

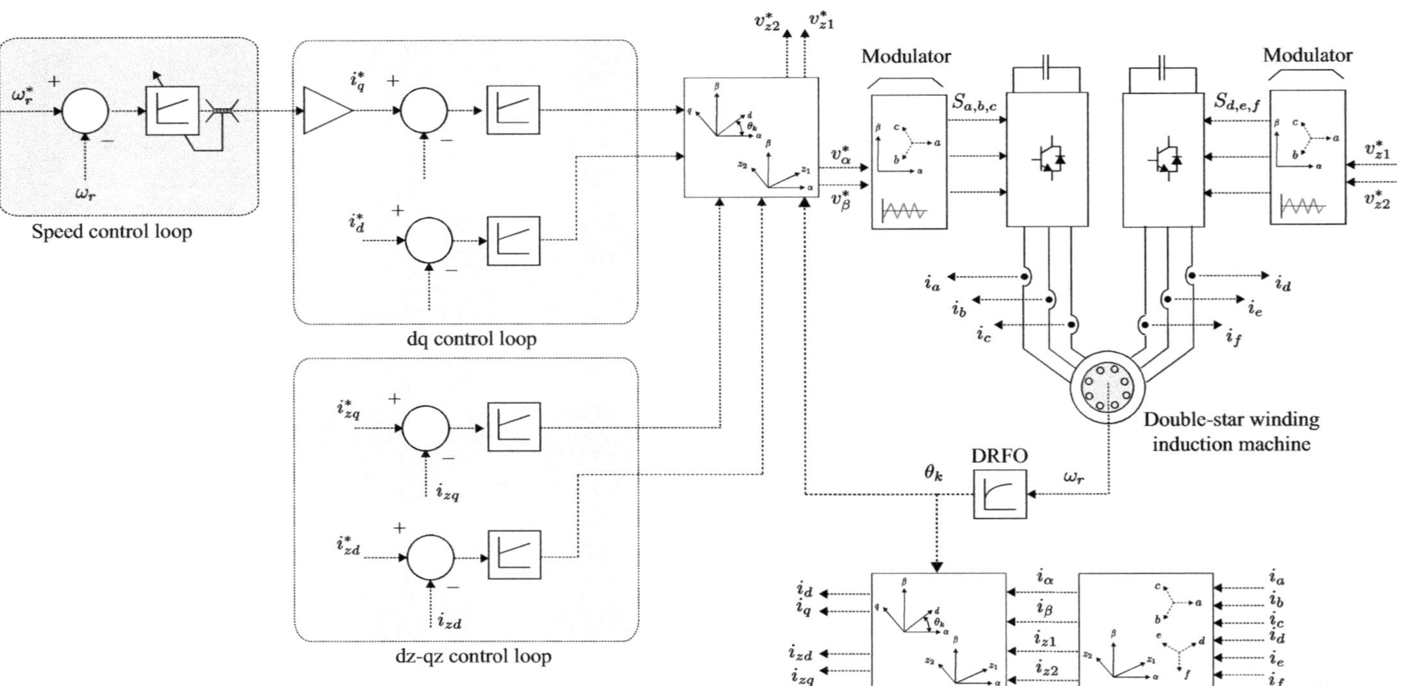

Figure 3.4 FOC scheme of a multiphase drive

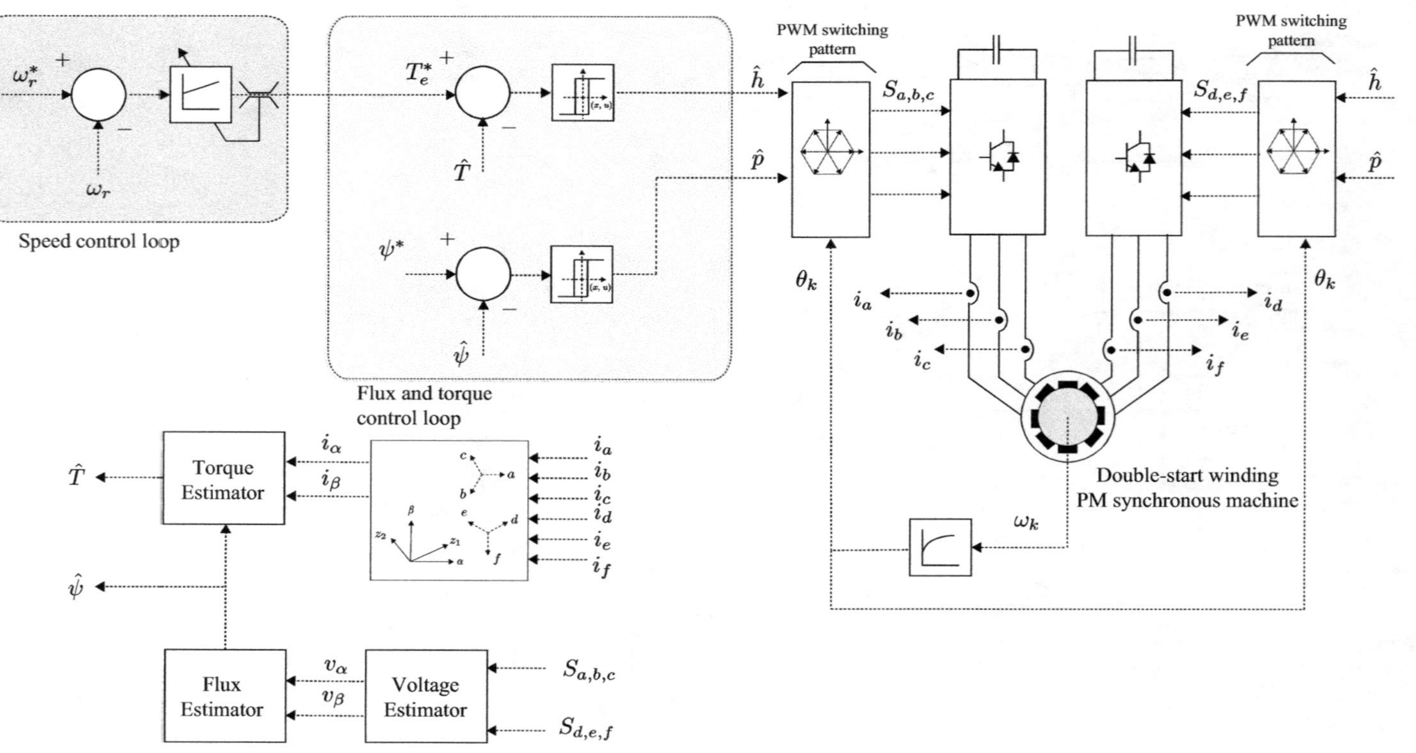

Figure 3.5 DTC scheme of a multiphase drive

In general, EVs are classified into three major categories according to the way and the place of production of electricity [34]: (a) vehicles using continuous power supply from an external power source, such as an overhead supply line. Unfortunately, these vehicles have a major limitation of having to move on specific routes in order to maintain continuous external electricity power supply for their operation. (b) Vehicles based on the storage of electricity supplied from an external source. In order to save energy, these vehicles use batteries or supercapacitors. (c) Vehicles that produce electricity within the vessel itself to meet their needs. These include electric hybrid cars that use thermal motors in series or parallel to electric motors, as well as EVs with fuel cells. Another separation of EVs is based on the source of energy type [35].

In this context, two major categories can be classified: (a) battery EVs (BEV) and (b) HEV. BEVs use batteries as a source of energy and they are also called green vehicles, or clean vehicles, or eco-friendly vehicles because they have zero emissions. In order to cover a travel distance, they are equipped with larger storage batteries than HEVs. However, the limited travelling distance of BEVs is an important drawback because it is often necessary to recharge the battery by connecting to an external power source (in city cars, autonomy starts from 100 to 120 km and reaches 500 km or more in high power cars Tesla Model). A HEV is classified as a car that uses two or more different technologies to achieve its movement. These technologies usually include the classic internal combustion engine and a more mild environmentally friendly technology, usually an electric motor. However, the electric motor is used as a supplementary power source in cases where the HEV requires more power.

It is apparent from the above that proper energy management is of vital importance for the smooth operation of EVs. A challenging research field includes the design and implementation of efficient charging schemes that ensure fast and reliable EV charging in order to increase vehicle autonomy. In this concept, the vehicle-to-grid (V2G) approach aims to optimize the way we transport, use, and produce electricity by turning electric cars into virtual power plants [36]. V2G technology refers to a bi-directional flow system operation, in which plug-in battery EVs communicate with a recipient and allow the reciprocal flow between the EV and an electric grid [37,38]. Under this relatively new concept, electric cars would store and dispatch electrical energy stored in networked vehicle batteries which together act as one collective battery fleet for peak shaving (sending power back to the grid when demand is high) and valley filling (charging at night when demand is low) [39]. V2G technology also improves stability and reliability of the grid, regulates the active power, and provides load balancing by valley fillings. These features enable better ancillary services, voltage control, frequency regulation, maintained peak power, and lead overall to a reduction of electric costs. In addition, owing to the inherent high mobility of EVs, flexible and timely on-demand response services against EV mobility in the V2G system must be provided [40,41]. To this end, several solutions have been proposed for integrating V2G technologies in fifth generation (5G) emerging wireless infrastructures, in order for the mobile user to experience a unified approach on application management (e.g., real-time navigation with traffic update and potential alarms regarding the energy autonomy of the EV) [42]. Additional research areas in EVs also include the design and deployment of self-driving objects, where efficient wireless coverage and zero latency are of utmost importance [43].

3.3.1 EV car configurations

3.3.1.1 HEVs

HEVs are propelled by an internal combustion engine (ICE) and an electric motor/generator (EM) in series, parallel or hybrid configurations. The ICE provides the vehicle an extended driving range, while the EM increases efficiency and fuel economy by regenerating energy during braking and storing excess energy from the ICE during coasting. Design and control of such power-trains involve modeling and simulation of intelligent control algorithms and power management strategies, which aim to optimize the operating parameters to any given driving condition.

HEVs can be graded according to their degree of hybridization, which is defined as the ratio resulting from dividing the power of the electric motor (or motors) into the power of the internal combustion engine. Nowadays, major automakers, such as Toyota, Honda, Ford, Saturn, Volkswagen, Peugeot, and others,

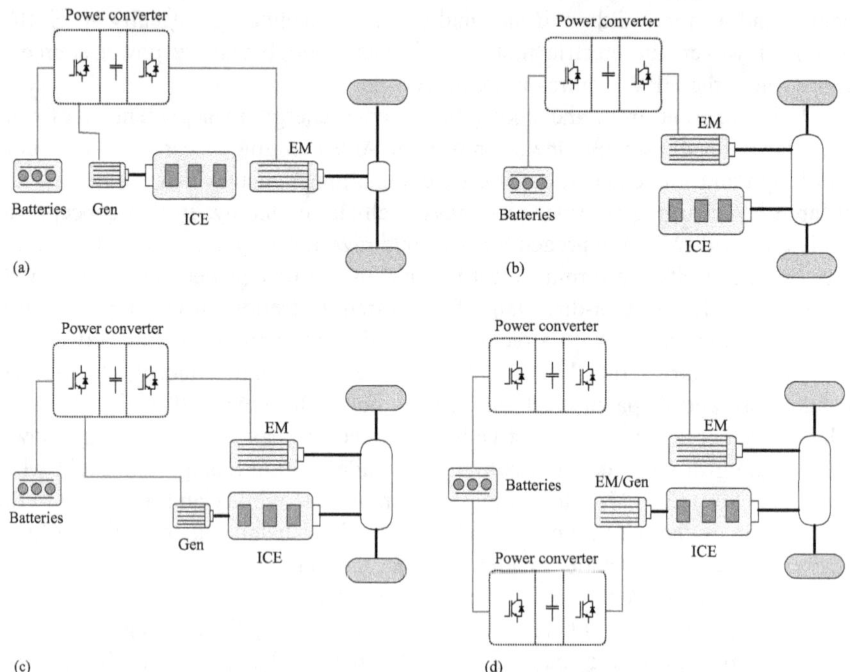

Figure 3.6 HEVs configurations: (a) series hybrid (SH), (b) parallel hybrid (PH), (c) series-parallel hybrid (SPH), and (d) complex hybrid (CH)

have developed many hybrid vehicles. The available models encompass passenger vehicles. According to the level of electric power and the function of the electric motor, HEVs can be classified into following categories [44]:

1. *Micro-hybrid*: The typical EM power for a micro-hybrid is about 2.5 kW at 12 V. The main function of the electric motor is for start and stop actions only, hence the energy saving gained is mainly due to the use of the EM for start and stop operations. In city driving where there are frequent starts and stops, the energy saving may reach about 5–10%.
2. *Mild-hybrid*: In this configuration, the power for the EM is in the range of 10–20 kW at 100–200 V. In city driving, energy savings reach about 20–30%. The drive-train configuration is designed in parallel.
3. *Full-hybrid*: Typical EM power in a full-hybrid configuration is around 50 kW at 200–300 V. In this configuration, the electric drive operates as motor/generator, depending on the direction of power flow, which enables the car to charge the batteries, during operation. To deal with this, the drivetrain is a parallel-series or complex-hybrid configuration. Typically, a full hybrid car in city driving can save energy about 30–50%.

Plug-in HEVs (PHEVs) are considered a subcategory of hybrid vehicles which are powered by a conventional combustible engine and an electric engine charged by a pluggable external electric source. PHEVs can store enough electricity from the grid to significantly reduce their fuel consumption in regular driving conditions. The Mitsubishi Out-lander PHEV [45] provides a 12 kWh battery, which allows it to drive around 50 km just with the electric engine.

3.3.1.2 Full EVs (FEVs)

FEVs can be categorized into three different types, depending on their energy storage system:

(a) BEVs: vehicles which are 100% are propelled by electric power. BEVs do not have an internal combustion engine and they do not use any kind of liquid fuel. They normally use large batteries stacks in order to give the vehicle the sufficient autonomy. A typical BEV will reach from 160 to 250 km, although some of them can travel as far as 500 km with just one charge. An example of this type of vehicle is the Nissan Leaf [46], which is 100% electric and it currently provides a 62-kWh battery energy, that allows users to have an autonomy of 360 km.
(b) Fuel cell electric vehicles (FCEVs): these vehicles are powered by a fuel cell that uses compressed hydrogen and oxygen obtained from the air in order to produce an electrochemical reaction, producing electric power and heat, having water as the only waste resulting chemical recombination this process. Although these kinds of vehicles are considered to present "zero emissions," it is worth highlighting that although there is green hydrogen, most of the used hydrogen is extracted from natural gas. The Hyundai Nexo FCEV [47] is an example of this type of vehicles, being able to travel 650 km without refueling.

(c) Extended-range EVs (ER-EVs): these vehicles are very similar to those previously mentioned in the BEV category. However, the ER-EVs are also provided with a supplementary combustion engine, which is only used to charge the batteries of the vehicle if needed; so unlike those the PHEVs and HEVs, the combustion engine is only used for charging, thus it is not mechanically connected to the vehicle power train. An example of this type of vehicles is the BMW i3 [48], which has a 42.2-kWh battery that results in a 260-km autonomy in electric mode, and users can benefit an additional 130 km from the extended-range mode.

3.3.2 Electric drives for HEVs–FEVs

Electric motor drives have a key role in the development of FEVs and HEVs. There are three major types of electric motors that are currently used for its application on EVs traction system: Permanent Magnet synchronous Machines (PMSM), Squirrel-Cage Induction Machines (SCIM), and Switched Reluctance Machines (SRM), which have been covered in Section 3.2. Typical requirements for motor and drive technology include the following: high torque under a large load profile, high-power density, high efficiency over wide speed range, reliability, and robustness.

The use of multilevel inverters in EV applications has been a topic of wide research. In this field of applications, the MMC previously referenced in Section 3.1.2 eliminates the need of a battery management system (BMS), due the fact that each battery is connected to each module, as a result, voltage balancing is ensured without a BMS. The main drawback of this configuration is the large number of modules required [49]. Another commonly used topology is the neutral point piloted (NPP) or T-Type converter, referenced in Section 3.1.2, which presents the main benefits of the three-level NPC topology, but having a more symmetrical distributed switching and conduction losses, thus leading to higher efficiency. The cascaded H-bridge topology has also risen as a possible solution; however, high-frequency transformer is required to provide isolated voltage sources [50]. In [51], a nested neutral point clamped (NNPC) topology is proposed using fewer diodes and floating capacitors compared to NPC and FC typologies, respectively, becoming an interesting option for its application EVs. Multi-drive converter configurations have also gain great interest as a solution for EV applications, based on the use on multiple two-level voltage source inverters [44] and the use of a nine switch inverter, presented in Section 3.1.5 [52].

3.3.3 Charging infrastructure

Modern plug-in vehicles have the same power-train configuration, consisting of the battery, typically operating in the high-voltage range to keep low currents; a battery management system (BMS) designed to protect the battery from operating outside its safe operating range (charge level, temperature, etc.); onboard charger (OBC); DC–DC power converters to provide the required power at a controlled voltage level and auxiliary systems (cooling system, heaters, chillers, inverter drivers, etc.).

In particular, the battery is the focus of charger design and control; it is highly sensitive and takes priority over all charging processes and units. In order to protect the battery, the connector unit incorporates a contactor and a fuse for protection against over-currents due to faulty operation. The BMS is the unit in charge to controlling the charging process, in order to control the required current and voltage levels during the charging process. Charging options can be classified into two main groups: AC and DC charging solutions. AC charging solutions cover the lower end of the spectrum, up to approximately 25 kW, while DC solutions handle up to roughly 400 kW, with projections as high as 900 kW. Wireless charging and battery swapping complete the charging infrastructure.

Regarding on charging standards, these can be divided depending on their maximum rated power and charging level and speed, as follows:

(a) American Standards

Today the Society of Automotive Engineering (SAE) J1772 defines EV charging system architecture used in North America. In function of the rated power, voltage, and current, the charging systems for EV in North America are classified into three categories, which are AC Level 1, AC Level 2, and DC Level 3, as follows:

- *Level I*: on-board charger providing AC voltage at 120 or 240 V with a maximum current of 15 A and a maximum power of 3.3 kW.
- *Level II*: on-board charger providing AC voltage at 240 V with a maximum current of 60 A and a maximum power of 14.4 kW.
- *Level III*: the off-board charger; the charging station provides DC voltage directly to the battery via a DC connector, with a maximum power of 240 kW.

(b) European Standards

Today the only standards available at European level, dealing with the charging system, plugs, and sockets, are contained in the IEC 61851. The actual standards provide a first classification of the type of charger in function of its rated power and so of the time of recharge, defining three categories:

- *Normal power or slow charging*: considers a rated power less than 3.7 kW, which is normally used for domestic application or for long-time EV parking.
- *Medium power or quick charging*: having a rated power from 3.7 to 22 kW, used for private and public EV.
- *High power or fast charging*: with a rated power higher than 22 kW, used for public EV.

Within this field of applications, the IEC 61851-1 Committee on "Electric vehicle conductive charging system" has then defined four modes of charging considering: the type of power received by the EV (DC, single-phase or three-phase AC), the voltage rating (110 V for single-phase AC to 480 V for three-phase), and the presence or absence of grounding and of control lines.

These standards and operating modes introduce operating constrains to be solved when defining the suitable converter topology to be implemented within this charging infrastructure. Converter topologies should also provide the necessary flexibility in order to complain with the different worldwide standards.

In this field of applications, most of the commercially available EV chargers are based on a two-stage topology, consisting of a grid-side passive rectifier feeding a boost converter operated as a power factor corrector. The power is supplied with a DC link, which stores energy and decouples the input stage from the output stage. The latter consists of a DC–DC converter, which is usually galvanically isolated [53].

3.4 Marine transportation

Research interest in electric ship propulsion systems was raised during the past years because of the improvements in power electronics devices, as well as in drive control schemes. In this context, different electrical propulsion systems have been developed based on two-level or multilevel converters depending on the power and arrangements with series-connected machines and with multi-winding machines. Several configurations have also been used either integrating all the electrical network in only one power system or separating in several power systems [54].

3.4.1 Electric propulsion configurations

A general classification of different propulsion system configurations according to the transmission system and the energy source is presented in Figure 3.7. The propulsion systems can be classified primarily as (a) geared in which an amplifying

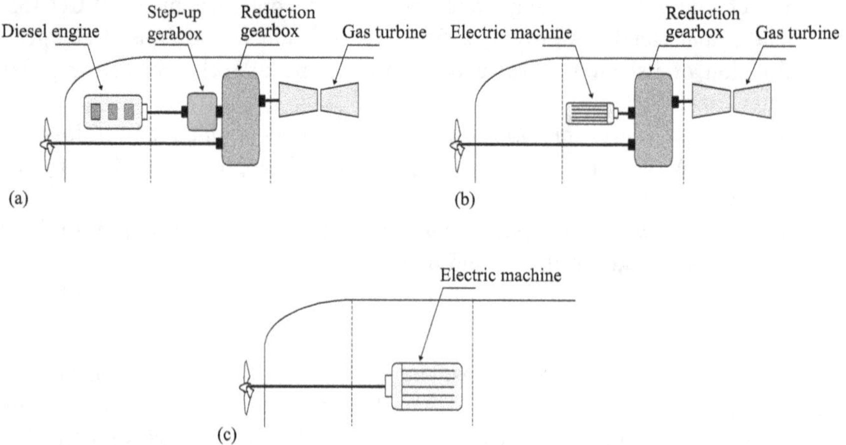

Figure 3.7 Types of ship propulsion systems: (a) thermal combined, (b) gas electric, and (c) full electric

or reductive gearbox is used to interconnect the gas turbine, diesel engine, or electrical motor to the main propeller; or (b) gear-less in which the machine is directly connected to the propeller. According to the energy source, the propulsion systems can be classified as thermal (i.e. diesel engines, gas turbines), electric, or hybrid. It can be noticed that the thermal and hybrid configurations commonly incorporate a gearbox due to the relative operating speed differences between prime mover and the propeller (e.g. propellers are typically low rpms efficient less than 200 rpm, whereas a diesel will be more like 700–1,000 rpm, and a gas turbine more like 3,000 rpm). A gearbox is therefore essential to "match" the speeds and to provide a mechanical coupling in combined thermal and hybrid configurations. On the other hand, electric propulsion systems are exclusively gear-less, because electrical machines can be designed for low-speed operation, optimized to suit the requirements of the propeller. Hence, the development of full electric propulsion systems would not only optimize efficiency and minimize emissions but also reduce the required space.

(a) *Thermal propulsion systems*: Thermal-based propulsion systems are commonly based on a combination of different prime movers (i.e. diesel engines, gas turbines/COGOG: combined gas or gas/CODOG: combined diesel or gas/COGAG: combined gas and gas/CODAD: combined diesel and diesel), with different nominal speeds, which are selected to work alone or combined by a synchronous clutching system, depending on the operating requirements of the vessel.

(b) *Hybrid propulsion systems*: this propulsion system consists in a combination of thermal-based engines and electric motors, providing a wide range of propeller speed. One of the main advantages of this configuration is the increasing efficiency at low loads, hence less fuel consumption, less emissions, and a small footprint is achieved. Recently, the concept of advanced hybrid drive system (AHDS) has been introduced for fuel savings at low ship speed [55]. A classical hybrid propulsion system is shown in Figure 3.7b where a gas turbine is used for high speed and an electric machine is used for low speed. In this configuration, the electric machine is fed by a bidirectional back-to-back power converter which can operate the machine either as a motor or generator. This configuration allows the ship to operate completely in electric drive mode, taking energy from the main busbar (power take-in). If large gas turbines are installed, the electric machines could operate in generator mode, supplying electrical power to main busbar (power take-off), supporting the power delivered by the main generators, so fewer generator sets are required in operation as shown in Figure 3.8(a) and (b). The AHDS system offers several benefits over classical thermal-based combined propulsion systems including: efficient operation of gas turbines at low ship speeds, additional power for sensor and weapons systems, and an efficient power dispatch for the main ship generators.

(c) *Full electric propulsion systems*: they consist in a configuration in which the main prime mover is an electric motor directly connected to the propulsion

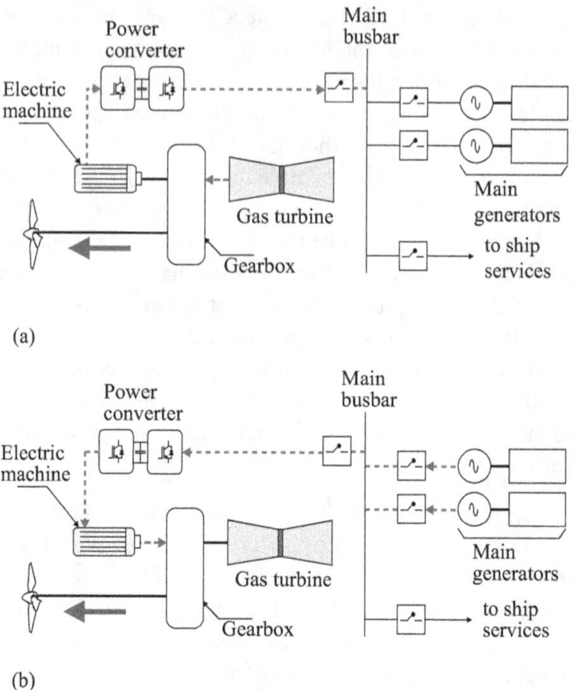

Figure 3.8 Advanced hybrid drive system: (a) power take-off mode and (b) power take-in mode

shaft, without gearbox, as shown in Figure 3.7c. The electric power is provided by ship's generators. The main advantage this configuration is that the on board generators can be operated within their fuel optimum point. Electric propulsion systems can be classified as segregated or integrated configurations:

1. *Segregated power system*: this is the natural evolution to electric propelled vessels, resulting in two separated electric systems, one dedicated to provide power for propulsion and another dedicated to feed auxiliaries loads as presented in Figure 3.9(a). As a consequence of this particular arrangement, the generator sets dedicated to each system do not have the ability to interact with other systems having their own voltage and frequency. This configuration very attractive for applications where high level of maneuverability is required such as icebreakers and towing ships [56].
2. *Integrated power system*: in this configuration, the total power required by the propulsion system and by the service loads is supplied by several generators connected to the medium voltage bus, and then distributed to the low-voltage bus through transformers. The propulsion motors are fed from power converters which are connected directly to the main busbar,

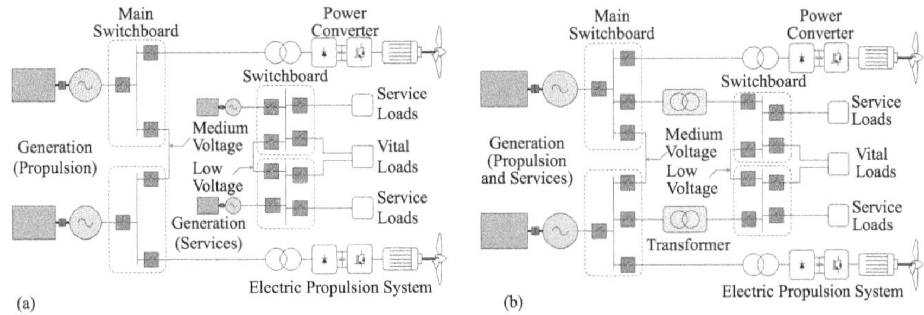

Figure 3.9 Types of full electric ship propulsion systems: (a) segregated power system and (b) integrated power system

and the service loads are fed from a low-voltage busbar, using step-down transformers as shown in Figure 3.9(b). The main features of an integrated grid architecture are its flexibility, availability, and system redundancy.

3.4.2 Future challenges

As a result of the harmful environmental impact of fossil fuels, more strict pollution policies, and the availability of high-performance machines and power electronics drives, electric propulsion has become an exciting option for the marine transportation industry. However, to enable its widespread application in the future, there are technological challenges that need to be resolved.

3.4.2.1 Wind-assisted ship propulsion (WASP)

In the last decade, the concept of WASP has shown an increasing research interest within electrified ships, by implementing wind turbine-assisted propulsion devices, such as Magnus effect or Flettner rotor turbines. Such devices can provide direct thrust, or as co-generation in an electrified ship propulsion system.

3.4.2.2 Solar-powered ship propulsion

Solar-powered propulsion systems are limited to the available surface required by the solar cells and weather conditions. Because of these limitations, solar-powered propulsion systems are currently being developed mostly for recreational marine crafts, small tourist boats, and patrol vessels. Future projects based on solar power ships are the Eoseas Cruise Ship (Builder: STH Europe), the Zero Emission Cruise Vessel (Builder: Delta-marine), the Super Eco Ship (Builder: Nippon Yusen HH), and the Aquarius Eco Ship (Builder: Eco Marine Power).

3.4.2.3 Energy storage devices (ESDs)

The use of ESD such as batteries and hybrid battery/supercapacitor configurations is meant to be used as an alternative to the on board generators, to support the

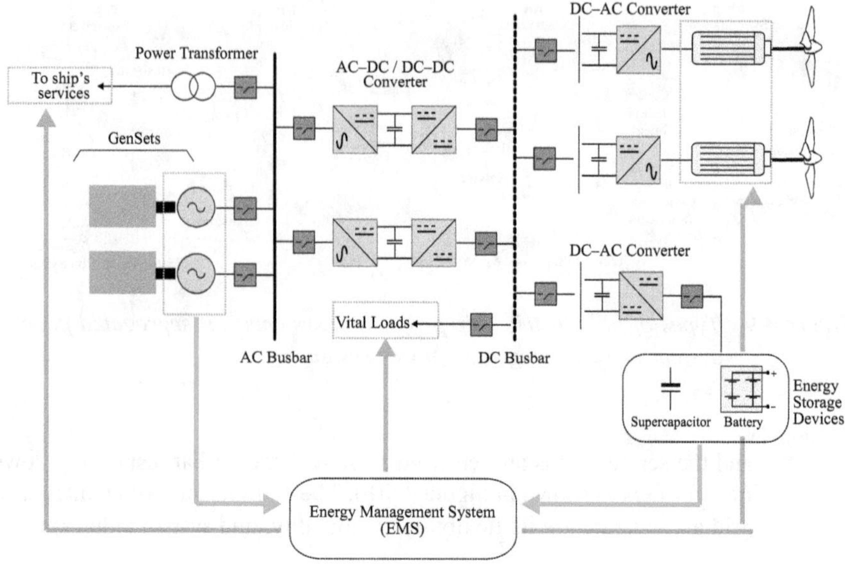

Figure 3.10 DC grid integration

distribution of electrical power. ESDs have been improved during the last few years, in terms of their capacity, allowing them to be used as a spinning energy reserve.

3.4.2.4 DC grid integration

Migration from AC- to DC-grid is an alternative of integrating ESDs to the ships main busbar, reducing the size and costs of power electronic devices and ensuring a stabilized voltage operation. The electric power distribution system consists of both AC and DC busbars. The last one provides the interface for the incorporation of the ESDs into the ship's power grid. The electric propulsion motors are fed directly from the DC busbar, thus reducing the complexity of the requited propulsion power converters. An AC–DC conversion stage linking both AC and DC busbars ensures full control decoupling and voltage stability in the DC busbar. Power flow is supervised by an EMS, ensuring optimize operation of the on board generators. Figure 3.10 shows a general arrangement of a DC-grid integration architecture.

3.5 Rail transport

The electrical traction of railways has its origins as early as the 1880s, evolving from urban tramways and later into long-distance interurban routes following technical advances in power supply and electromechanic conversion [57]. A major breakthrough in electrified traction of railways such as trains, metros, and trams, was the adoption, in the 1960s, of electronic power inverters which made possible to switch from DC to AC motors in a more efficient and controlled way. This technology change made possible to take advantage of the higher power density and reliability of AC electrical machines [58]. Inverters first based on thyristors were employed in early AC traction systems for railways in Finland in the decade of 1970 and replaced by controlled GTO switches near 1980 [59]. It is foreseen that in the following years, older configurations such as thyristor rectifiers will disappear, thereby reducing the concern about low-power factor and harmonic distortion of the line current [60]. The adoption of new highly controllable switching devices such as IGBT and IGCT allows to further increase efficiency, reliability, and power density [61].

3.5.1 Power-train configurations of railways

Electrical systems currently employed in Europe for rail transport systems can be AC with 15 kV/16.7 Hz or 25 kV/50 Hz, or medium voltage dc with 1.5/3 kV DC [62]. These configurations are depicted in Figure 3.11. The predominant electrical machine used for traction is the induction machine, due to its inherent advantages described in Section 3.2. The power-trains based on AC systems employ a rectifying stage in order to feed VSIs for variable-speed electrical drives. Conventionally, low-frequency step-down transformers are installed to adapt the line voltage to suitable levels for the rectifiers and existing electrical motors. However, by using AC/AC forced commutation converters, it is possible to increase the frequency of the AC voltage which allows the use of medium-frequency transformers with reduced weight and volume [63]. Modern railway propulsion systems include active-front-end rectifiers, which enable sinusoidal current input, unity power factor operation, and bidirectional power flow. This last feature have become important to increase the efficiency of transport systems avoiding the losses during breaking [60].

Configurations fed by a DC power line do neither require transformers nor power converters at the input side of the VSI, thereby reducing the overall complexity and weight of the propulsion system. However, a limitation of today's DC systems is their lower voltage, compared with AC, which imposes the need of higher currents, increasing power losses, and infrastructure size. Studies have determined that it would require to increase the DC voltage to 9 kV DC in order to reach a performance equivalent to the existing 25 kV AC systems [62].

3.5.2 Railway power stations

Many modern high-speed railways are powered by 25 kV single-phase AC-lines, with dedicated substations fed by one line-to-line voltage of the transmission system, creating a current imbalance which is a major concern for power quality of the electrical system [60,64]. A solution for this problem is the implementation of electronic power quality compensators (PQC), which have been deployed since the late 1990s [65,66]. The purpose of the PQC is to balance the loads between

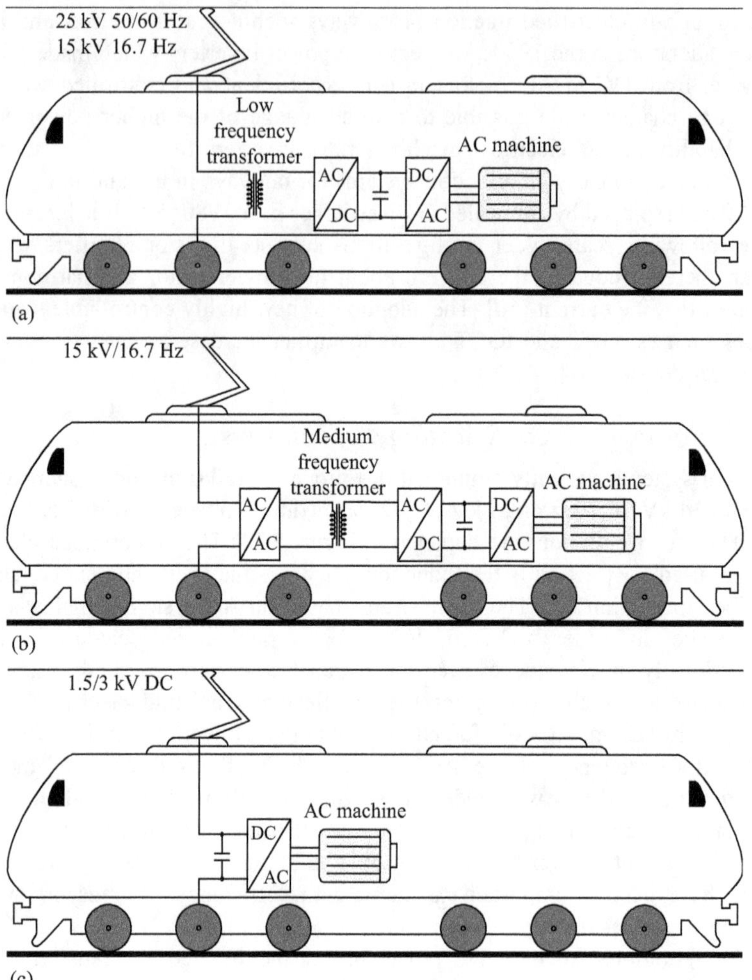

Figure 3.11 Power configurations of electrified railways. (a) Low-frequency transformer with indirect AC–AC drive. (b) Medium-frequency transformer with indirect AC–AC drive. (c) DC line and three-phase inverter

single-phase systems by means of a controlled active and reactive power exchange using a back-to-back configuration of single-phase power converters. Some recent advances in this regard are given by the implementation of PQC systems based on multilevel converters such as MMC [67,68] and cascaded H-bridge (CHB) [69]. A detailed comparison between modular power converter topologies for PQC is presented in [70].

Newest trains are equipped with active-front-end which provides bidirectional power flow. This feature allows to store or send the energy during the breaking back to the electrical grid. Therefore, to take full advantage of this feature, the power station must be able to store the energy or also have a bidirectional power flow [71].

3.5.3 Future challenges

Hybrid power-trains: Hybrid configurations have been implemented in order to foster the transition from carbon based on electric traction systems. In 2022, a battery-hybrid passenger train started its operation in the UK, operated by Chiltern Railways. Among the advantages of this configuration are the reduction of acoustic noise in stations, and zero emissions when operating from battery power, as well improvements in travel times due to enhanced acceleration performance. Overall, it is expected a reduction of 20% in fuel consumption and around 70% reduction in nitrogen oxide emissions. [72]. Figure 3.12 shows a general diagram of the supply system configuration of hybrid power-trains.

Battery-powered systems: In recent years, there has been significant progress on the practical applications of technologies related to the use of high-capacity lithium-ion batteries on traction power for rolling stock. In particular, the use of this kind of hybrid configurations to run on non-electrified and electrified sections of track can save energy and reduce maintenance requirements compared with conventional diesel-powered trains. In this field, Hitachi has successfully

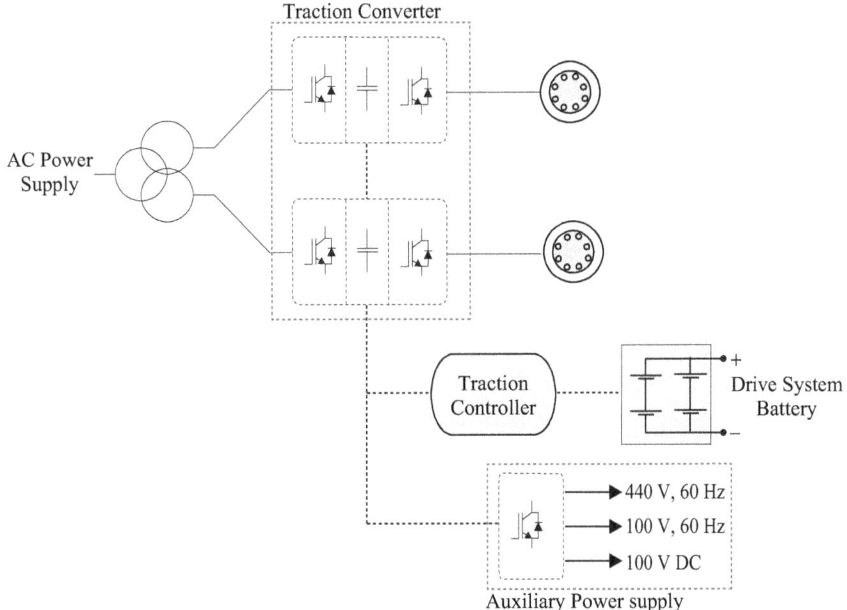

Figure 3.12 Hybrid power-train configuration

commercialized the JR Kyushu Series BEC819, a battery-powered train, that runs on non-electrified sections of track, using energy stored in lithium-ion batteries that are charged from the overhead lines or directly from the diesel engines for non-conventional diesel powered rolling stocks as the JR East Series HB-E210 and Series HB-E300 trains.

Flow batteries: In conventional batteries (e.g. lead-acid), ions diffuse from the electrolyte to the electrode surface, thus the limiting factor for the power outtake is a combination of diffusion and reaction rate at the electrode. In a flow battery, liquid is pumped through the cell which adds a convection term to the diffusion, and thus reduces the limitation of taking ions from the electrolyte into the electrode surface. The limiting factor then becomes the reaction rate at the electrode surface. This is a significant difference, which makes it easier to take out or charge the battery with higher current rates. Another aspect is the reduced heat production compared to conventional batteries.

Finally, as the electrolyte is a liquid, a flow battery can be recharged just by pumping in charged electrolyte while pumping out the used one.

Implementation of flow batteries can be considered as a replacement for existing diesel powered trains, by introducing a partial or hybrid configuration (diesel–flow batteries), and a full implementation consisting only on flow batteries. Both concepts have been implemented by East Japan Railway Company [73]. The use of this devices may also be included in full electrified trains, where flow batteries are used as storage device during regenerative breaking, to overcome the short-term energy storage capability of super-capacitors [74,75].

High-voltage DC systems: Currently, classical DC electrification systems have reached their power limits and it is not possible to operate locomotives at their nominal ratings with the increased demand of service. In this context, new medium voltage DC (MVDC) railway electrification appears as alternative to become the next evolution in railway traction power supply. This power system offers identical performances to the conventional 25 kV AC system while reducing the power installed in substations. For this application, high efficiency and reduced volume are essential requirements. High-power, isolated DC–DC converters using 3.3-kV SiC-MOSFET power modules with high efficiency appear as a suitable solution for this applications [76,77].

Inductive power transfer: Wireless power transfer (WPT) techniques for inductive charging of EV have gained an increased interest. WPT technology has been applied to various low-power applications (laptops, mobile phones, etc.) very successfully using self-resonance system and inductive coupling system. However, for high-power application, this technology has important issues to solve such as economic feasibility, and covering areas. In order to deal with these constraints ensuring constant power supply, it is necessary to change the impedance, depending on the output power demand, so if the output power is high, the impedance of the power supply line is reduced, otherwise, if output power is low, the impedance of the power supply should be increased. This can be achieved by using a resonance compensation method as proposed in [78].

Linear motor trains: Linear motors are applied to railway traction for special purposes, e.g., space saving mini subways with substantially flexible vertical or horizontal track profiles, advanced quiet and rapid rail transits for urban and sub-urban areas, and super-high speed Maglev ground transport. The commercial operations have been gradually implemented in recent decades with a successful start of the commercial operation of Chuo-Linear Express, Japanese super-high speed Maglev. However, linear motors for its application on railway systems have several disadvantages compared to rotatory motors such as relatively low-power factor and efficiency, no possibility to reduce speed by mechanical gears: request of large direct thrust, and the requirement to deal and compensate three-dimensional acting forces.

References

[1] J. Reimers, L. Dorn-Gomba, C. Mak, and A. Emadi, "Automotive Traction Inverters: Current Status and Future Trends," *IEEE Trans. Veh. Technol.*, vol. 68, no. 4, pp. 3337–3350, 2019.

[2] J. Hu, W. Liu, and J. Yang, "Application of Power Electronic Devices in Rail Transportation Traction System," in *Proc. Int. Symp. Power Semicond. Devices ICs*, vol. June 2015, pp. 7–12, 2015.

[3] B. Gou, X. Ge, S. Wang, X. Feng, J. B. Kuo, and T. G. Habetler, "An Open-Switch Fault Diagnosis Method for Single-Phase PWM Rectifier Using a Model-Based Approach in High-Speed Railway Electrical Traction Drive System," *IEEE Trans. Power Electron.*, vol. 31, no. 5, pp. 3816–3826, 2016.

[4] A. Poorfakhraei, M. Narimani, and A. Emadi, "A Review of Multilevel Inverter Topologies in Electric Vehicles: Current Status and Future Trends," *IEEE Open J. Power Electron.*, vol. 2, pp. 155–170, 2021.

[5] A. Choudhury, P. Pillay, and S. S. Williamson, "Performance Comparison Study of Space-Vector and Modified-Carrier-Based PWM Techniques for a Three-Level Neutral-Point-Clamped Traction Inverter Drive," *IEEE J. Emerg. Sel. Top. Power Electron.*, vol. 4, no. 3, pp. 1064–1076, 2016.

[6] A. Kersten, K. Oberdieck, A. Bubert, *et al.*, "Fault Detection and Localization for Limp Home Functionality of Three-Level NPC Inverters with Connected Neutral Point for Electric Vehicles," *IEEE Trans. Transp. Electrif.*, vol. 5, no. 2, pp. 416–432, 2019.

[7] S. Bhattacharya, D. Mascarella, G. Joós, J. M. Cyr, and J. Xu, "A Dual Three-Level T-NPC Inverter for High-Power Traction Applications," *IEEE J. Emerg. Sel. Top. Power Electron.*, vol. 4, no. 2, pp. 668–678, 2016.

[8] B. Wu, *High Power Converters and AC Drives*, 1st ed., New York, NY: Wiley-IEEE Press, 2006.

[9] E. Fedele, D. Iannuzzi, P. Tricoli, and A. Del Pizzo, "NPC-Based Multi-Source Inverters for Multimode DC Rail Traction Systems," *IEEE Trans. Transp. Electrif.*, pp. 1–1, 2022. Available: https://ieeexplore.ieee.org/document/9774380/

[10] A. Kersten, K. Oberdieck, J. Gossmann, *et al.*, "Measuring and Separating Conducted Three-Wire Emissions from a Fault-Tolerant, NPC Propulsion Inverter with a Split-Battery Using Hardware Separators Based on HF Transformers," *IEEE Trans. Power Electron.*, vol. 36, no. 1, pp. 378–390, 2021.

[11] A. Choudhury, P. Pillay, and S. S. Williamson, "Discontinuous Hybrid-PWM-Based DC-Link Voltage Balancing Algorithm for a Three-Level Neutral-Point-Clamped (NPC) Traction Inverter Drive," *IEEE Trans. Ind. Appl.*, vol. 52, no. 4, pp. 3071–3082, 2016.

[12] A. Choudhury, P. Pillay, and S. S. Williamson, "DC-Bus Voltage Balancing Algorithm for Three-Level Neutral-Point-Clamped (NPC) Traction Inverter Drive with Modified Virtual Space Vector," *IEEE Trans. Ind. Appl.*, vol. 52, no. 5, pp. 3958–3967, 2016.

[13] Y. Si, R. Wang, S. Zhang, W. Zhou, A. Lin, and Y. Wang, "Fault Diagnosis Based on Attention Collaborative LSTM Networks for NPC Three-Level Inverters," *IEEE Trans. Instrum. Meas.*, vol. 71, pp. 1–16, 2022. Available: https://ieeexplore.ieee.org/document/9761886/

[14] W. Taha, B. Nahid-Mobarakeh, and J. Bauman, "Efficiency Evaluation of 2L and 3L SiC-Based Traction Inverters for 400 V and 800 V Electric Vehicle Powertrains," in *2021 IEEE Transp. Electrif. Conf. Expo, ITEC 2021*, pp. 625–632, 2021.

[15] M. Schweizer and J. W. Kolar, "Design and Implementation of a Highly Efficient Three-Level T-Type Converter for Low-Voltage Applications," *IEEE Trans. Power Electron.*, vol. 28, no. 2, pp. 899–907, 2013.

[16] T. Mizobuchi, K. Kondo, Y. Dairaku, T. Shinomiya, and K. Ishikawa, "Energy-Management Method to Reduce the Capacity of Lithium-Ion Batteries in Hybrid-Voltage-Source Three-Level Inverter for DC-Electrified Railway Vehicles," *IEEE Open J. Ind. Appl.*, vol. 3, pp. 41–55, 2022.

[17] X. He, J. Peng, P. Han, Z. Liu, S. Gao, and P. Wang, "A Novel Advanced Traction Power Supply System Based on Modular Multilevel Converter," *IEEE Access*, vol. 7, pp. 165 018–165 028, 2019.

[18] W. Taha, P. Azer, A. D. Callegaro, and A. Emadi, "Multiphase Traction Inverters: State-of-the-Art Review and Future Trends," *IEEE Access*, vol. 10, pp. 4580–4599, 2022.

[19] R. J. Hill, "Electric Railway Traction Part 2: Traction Drives with Three-Phase Induction Motors," *Power Eng. J.*, vol. 8, no. 3, pp. 143–152, 1994.

[20] F. Z. Peng, "Z-Source Inverter," *IEEE Trans. Ind. Appl.*, vol. 39, no. 2, pp. 504–510, 2003.

[21] D. Mande, J. P. Trovão, and M. C. Ta, "Comprehensive Review on Main Topologies of Impedance Source Inverter Used in Electric Vehicle Applications," *World Electr. Veh. J., 2020*, vol. 11, no. 2, pp. 37, 2020. Available: https://www.mdpi.com/2032-6653/11/2/37/htm

[22] A. Florescu, O. Stocklosa, M. Teodorescu, C. Radoi, D. A. Stoichescu, and S. Rosu, "The Advantages, Limitations and Disadvantages of Z-Source Inverter," in *Proc. Int. Semicond. Conf. CAS*, vol. 2, pp. 483–486, 2010.
[23] E. Levi, "Multiphase Electric Machines for Variable-Speed Applications," *IEEE Trans. Ind. Electron.*, vol. 55, no. 5, pp. 1893–1909, 2008.
[24] E. Levi, "Advances in Converter Control and Innovative Exploitation of Additional Degrees of Freedom for Multiphase Machines," *IEEE Trans. Ind. Electron.*, vol. 63, no. 1, pp. 433–448, 2016.
[25] C. Alosa, G. Migliazza, F. Immovilli, and E. Lorenzani, "Reconfigurable Multi-Three-Phase Drive for Naval Rim-Driven Propulsion System," *IEEE Trans. Ind. Appl.*, vol. 58, no. 2, pp. 2075–2087, 2022.
[26] M. Gleissner, J. Haring, M. M. Bakran, W. Wondrak, M. Hepp, and M. B. Ag, "Advantageous Fault-Tolerant Multilevel and Multiphase Inverter Systems for Automotive Electric Powertrains," in *2020 15th Int. Conf. Ecol. Veh. Renew. Energies, EVER 2020*, September 2020.
[27] A. Salem and M. Narimani, "A Review on Multiphase Drives for Automotive Traction Applications," *IEEE Trans. Transp. Electrif.*, vol. 5, no. 4, pp. 1329–1348, 2019.
[28] M. Sthel, J. Tostes, and J. Tavares, "Current Energy Crisis and its Economic and Environmental Consequences: Intense Human Cooperation." *Nat. Sci.*, vol. 5, no. 2A, pp. 244–252, 2013.
[29] J. Y. Yong, V. K. Ramachandaramurthy, K. M. Tan, and N. Mithulananthan, "A Review on the State-of-the-Art Technologies of Electric Vehicle, its Impacts and Prospects," *Renew. Sust. Energy Rev.*, vol. 49, pp. 365–385, 2015. Available: http://www.sciencedirect.com/science/article/pii/S1364032115004001
[30] J. Du and M. Ouyang, "Review of Electric Vehicle Technologies Progress and Development Prospect in China." *World Electr. Veh. J.*, vol. 6, no. 4, pp. 1086–1093, 2013.
[31] F. Un-Noor, S. Padmanaban, L. Mihet-Popa, M. Mollah, and E. Hossain, "A Comprehensive Study of Key Electric Vehicle (EV) Components, Technologies, Challenges, Impacts, and Future Direction of Development." *Energies*, vol. 10, no. 8, pp. 1086–1093, 2017.
[32] J. A. P. Lopes, F. J. Soares, and P. M. R. Almeida, "Integration of Electric Vehicles in the Electric Power System," *Proc. IEEE*, vol. 99, no. 1, pp. 168–183, 2011.
[33] K. J. Dyke, N. Schofield, and M. Barnes, "The Impact of Transport Electrification on Electrical Networks," *IEEE Trans. Ind. Electron.*, vol. 57, no. 12, pp. 3917–3926, 2010.
[34] K.-T. Chau, C. Jiang, W. Han, and C. H. T. Lee, "State-of-the-Art Electromagnetics Research in Electric and Hybrid Vehicles (Invited Paper)," *Prog. Electromagn. Res.*, vol. 159, pp. 139–157, 2017.
[35] A. Khaligh and Z. Li, "Battery, Ultracapacitor, Fuel Cell, and Hybrid Energy Storage Systems for Electric, Hybrid Electric, Fuel Cell, and Plug-in Hybrid

Electric Vehicles: State of the Art," *IEEE Trans. Veh. Technol.*, vol. 59, no. 6, pp. 2806–2814, 2010.

[36] B. Wang, P. Dehghanian, S. Wang, and M. Mitolo, "Electrical Safety Considerations in Large-Scale Electric Vehicle Charging Stations," *IEEE Trans. Ind. Appl.*, vol. 55, no. 6, pp. 6603–6612, 2019.

[37] M. Etezadi-Amoli, K. Choma, and J. Stefani, "Rapid-Charge Electric-Vehicle Stations," *IEEE Trans. Power Deliv.*, vol. 25, no. 3, pp. 1883–1887, 2010.

[38] P. Kong and G. K. Karagiannidis, "Charging Schemes for Plug-in Hybrid Electric Vehicles in Smart Grid: A Survey," *IEEE Access*, vol. 4, pp. 6846–6875, 2016.

[39] E. S. Rigas, S. D. Ramchurn, and N. Bassiliades, "Managing Electric Vehicles in the Smart Grid using Artificial Intelligence: A Survey," *IEEE Trans. Intell. Trans. Syst.*, vol. 16, no. 4, pp. 1619–1635, 2015.

[40] K. Ginigeme and Z. Wang, "Distributed Optimal Vehicle-to-Grid approaches with Consideration of Battery Degradation Cost under Real-Time Pricing," *IEEE Access*, vol. 8, pp. 5225–5235, 2020.

[41] H. M. Khalid and J. C. . H. Peng, "Bidirectional Charging in v2g Systems: An In-Cell Variation Analysis of Vehicle Batteries," *IEEE Syst. J.*, vol. 14, no. 3, pp. 3665–3675, 2020.

[42] Y. Li, G. Yu, J. Liu, and F. Deng, "Design of v2g Auxiliary Service System based on 5g Technology," in *2017 IEEE Conf. Energy Internet Energy Syst. Integr. (EI2)*, pp. 1–6, 2017.

[43] R. Bonetto, I. Sychev, O. Zhdanenko, A. Abdelkader, and F. H. P. Fitzek, "Smart Grids for Smarter Cities," in *2020 IEEE 17th Annu. Consumer Commun. Netw. Conf. (CCNC)*, pp. 1–2, 2020.

[44] C. C. Chan, "The State of the Art of Electric, Hybrid, and Fuel Cell Vehicles," *Proc. IEEE*, vol. 95, no. 4, pp. 704–718, 2007.

[45] Mitsubishi-Motors, Mitsubishi outlander phev 2018, 2019. Available: https://www.mitsubishicars.com/outlander-phe v/2018/specifications

[46] InsideEVs, Nissan reveals leaf e-plus: 62 kwh battery, 226-mile range, 2019. Available: https://insideevs.com/nissan-reveals-leaf-e-plus-ces/

[47] Hyundai, All-new Hyundai Nexo—technical specifications, 2019. Available: https://www.hyundai.news/eu/press-kits/all-new-hyundai-nexo-technical-specifications/

[48] InsideEVs, BMW i3, i3 rex, i3s & i3s rex: Full specs, 2019. Available: https://insideevs.com/2019-bmw-i3-rex-i3s-rex-full-spec/

[49] I. Husain, B. Ozpineci, M. S. Islam, *et al.*, "Electric Drive Technology Trends, Challenges, and Opportunities for Future Electric Vehicles," *Proc. IEEE*, vol. 109, no. 6, pp. 1039–1059, 2021.

[50] A. Emadi, Y. J. Lee, and K. Rajashekara, "Power Electronics and Motor Drives in Electric, Hybrid Electric, and Plug-In Hybrid Electric Vehicles," *IEEE Trans. Ind. Electron.*, vol. 55, no. 6, pp. 2237–2245, 2008.

[51] M. Narimani, B. Wu, Z. Cheng, and N. R. Zargari, "A New Nested Neutral Point-Clamped (nnpc) Converter for Medium-Voltage (MV) Power Conversion," *IEEE Trans. Power Electron.*, vol. 29, no. 12, pp. 6375–6382, 2014.
[52] M. Wang and K. Tian, "A Nine-Switch Three-Level Inverter for Electric Vehicle Applications," in *2008 IEEE Veh. Power Propulsion Conf.*, pp. 1–5, 2008.
[53] S. Rivera, S. Kouro, S. Vazquez, S. M. Goetz, R. Lizana, and E. Romero-Cadaval, "Electric Vehicle Charging Infrastructure: From Grid to Battery," *IEEE Ind. Electron. Mag.*, vol. 15, no. 2, pp. 37–51, 2021.
[54] C. A. Reusser, H. A. Young, J. R. Perez Osses, M. A. Perez, and O. J. Simmonds, "Power Electronics and Drives: Applications to Modern Ship Propulsion Systems," *IEEE Ind. Electron. Mag.*, vol. 14, no. 4, pp. 106–122, 2020.
[55] J. Langston, M. Andrus, M. Steurer, *et al.*, "System Studies for a Bi-Directional Advanced Hybrid Drive System (ahds) for Application on a Future Surface Combatant," in *2013 IEEE Elect. Ship Technol. Symp. (ESTS)*, pp. 509–513, 2013.
[56] S.-Y. Kim, S. Choe, S. Ko, and S.-K. Sul, "A Naval Integrated Power System with a Battery Energy Storage System: Fuel Efficiency, Reliability, and Quality of Power." *IEEE Electrif. Mag.*, vol. 3, no. 2, pp. 22–33, 2015.
[57] M. Guarnieri, "Challenging Steam: Early Electric Railways [Historical]," *IEEE Ind. Electron. Mag.*, vol. 15, no. 3, pp. 49–53, 2021.
[58] E. Spooner, N. Rash, M. Lilley, and M. Lockwood, "Novel Traction System for Railway Applications," *Electron Power*, vol. 24, no. 10, pp. 737–740, 1978.
[59] J. A. Taufiq, "AC Traction in Finland," *IEE Rev.*, vol. 34, no. 10, pp. 381–383, 1988.
[60] P. Ladoux, G. Raimondo, H. Caron, and P. Marino, "Chopper-Controlled Steinmetz Circuit for Voltage Balancing in Railway Substations," *IEEE Trans. Power Electron.*, vol. 28, no. 12, pp. 5813–5822, 2013.
[61] H. Li, Y. Sun, S. Xie, W. Xiong, and M. Su, "Improved Branch Energy Balance Control for Three-Phase to Single-Phase Modular Multilevel Converter for Railway Traction Power Supply," *IEEE Trans. Trans. Electrif.*, pp. 1–1, 2022.
[62] A. Verdicchio, P. Ladoux, H. Caron, and C. Courtois, "New Medium-Voltage DC Railway Electrification System," *IEEE Trans. Transp. Electrif.*, vol. 4, no. 2, pp. 591–604, 2018.
[63] D. Ronanki and S. S. Williamson, "Evolution of Power Converter Topologies and Technical Considerations of Power Electronic Transformer-Based Rolling Stock Architectures," *IEEE Trans. Trans. Electrif.*, vol. 4, no. 1, pp. 211–219, 2017.

[64] S. M. Mousavi Gazafrudi, A. Tabakhpour Langerudy, E. F. Fuchs, and K. Al-Haddad, "Power Quality Issues in Railway Electrification: A Comprehensive Perspective," *IEEE Trans. Ind. Electron.*, vol. 62, no. 5, pp. 3081–3090, 2015.

[65] I. Perin, G. R. Walker, and G. Ledwich, "Load Sharing and Wayside Battery Storage for Improving AC Railway Network Performance, with Generic Model for Capacity Estimation, Part 1," *IEEE Trans. Ind. Electron.*, vol. 66, no. 3, pp. 1791–1798, 2019.

[66] Z. Li, X. Li, Y. Lin, *et al.*, "Active Disturbance Rejection Control for Static Power Converters in Flexible AC Traction Power Supply Systems," *IEEE Trans. Energy Convers.*, vol. 37, no. 4, pp. 1–12, 2022.

[67] M. Lei, Y. Li, Z. Li, *et al.*, "A Single-Phase Five-Branch Direct AC-AC Modular Multilevel Converter for Railway Power Conditioning," *IEEE Trans. Ind. Electron.*, vol. 67, no. 6, pp. 4292–4304, 2020.

[68] P. Guo, Q. Xu, Y. Yue, *et al.*, "Analysis and Control of Modular Multilevel Converter with Split Energy Storage for Railway Traction Power Conditioner," *IEEE Trans. Power Electron.*, vol. 35, no. 2, pp. 1239–1255, 2020.

[69] M. Lei and Y. Wang, "A Transformer-less Railway Power Quality Compensator Based on Cascaded H-Bridge Featuring Reduced Branch Capacity Requirement," *IEEE Trans. Power Deliv.*, vol. 37, no. 6, pp. 1–1, 2022.

[70] Q. Xu, F. Ma, Z. He, *et al.*, "Analysis and Comparison of Modular Railway Power Conditioner for High-Speed Railway Traction System," *IEEE Trans. Power Electron.*, vol. 32, no. 8, pp. 6031–6048, 2017.

[71] Q. Zhang, Y. Zhang, K. Huang, *et al.*, "Modeling of Regenerative Braking Energy for Electric Multiple Units Passing Long Downhill Section," *IEEE Trans. Transp. Electrif.*, vol. 8, no. 3, pp. 3742–3758, 2022.

[72] Chiltern Railways, "Chiltern Railways puts Britain's first hybrid-powered train to the test," 2022. Available: https://www.chilternrailways.co.uk/news/chiltern-tests-britains-first-hybrid-powered-train

[73] Y. Kono, N. Shiraki, H. Yokoyama, and R. Furuta, "Catenary and Storage Battery Hybrid System for Electric Railcar Series ev-e301," in *2014 Int. Power Electron. Conf.* (*IPEC-Hiroshima 2014 – ECCE ASIA*), pp. 2120–2125, 2014.

[74] H. Hirose, K. Yoshida, and K. Shibanuma, "Development of Catenary and Storage Battery Hybrid Train System," in *2012 Electrical Systems for Aircraft, Railway and Ship Propulsion*, pp. 1–4, 2012.

[75] M. Brenna, F. Foiadelli, E. Tironi, and D. Zaninelli, "Ultracapacitors application for energy saving in subway transportation systems," in *2007 Int. Conf. Clean Electr. Power*, pp. 69–73, 2007.

[76] A. Verdicchio, P. Ladoux, H. Caron, and S. Sanchez, "Future dc Railway Electrification System – go for 9 kv," in *2018 IEEE International Conference on Electrical Systems for Aircraft, Railway, Ship Propulsion and Road Vehicles & International Transportation Electrification Conference* (*ESARS-ITEC*), pp. 1–5, 2018.

[77] J. Fabre, P. Ladoux, H. Caron, *et al.*, "Characterization and Implementation of Resonant Isolated dc/dc Converters for Future mvdc Railway Electrification Systems," *IEEE Trans. Transp. Electrif.*, vol. 7, no. 2, pp. 854–869, 2021.

[78] J. H. Kim, B.-S. Lee, J.-H. Lee, *et al.*, "Development of 1-mw Inductive Power Transfer System for a High-Speed Train," *IEEE Trans. Ind. Electron.*, vol. 62, no. 10, pp. 6242–6250, 2015.

Chapter 4

Multilevel inverter topologies and their applications

Faramarz Faraji[1], *Amir Abbas Aghajani*[2], *Mojtaba Eldoromi*[2], *Ali Akbar Moti Birjandi*[2], *Amer M.Y.M. Ghias*[3] *and Honnyong Cha*[1]

Multilevel inverter (MLI) topologies have gained more attention in many applications. These MLIs can classify into standard/traditional and hybrid/advanced structures. The conventional MLIs, including neutral-point-clamped (NPC), flying capacitor (FC), and cascaded H-bridge (CHB), have some structural/topological and control/modulation drawbacks which limit their applications. Many novel topologies have been reported to overcome structural constraints associated with these topologies. Likewise, several new control/modulation methods have been introduced to improve the performance of classical converters under various operating conditions. This chapter aims to review and shed light upon the merits and demerits of these recent contributions. To this end, at first, the limitations of well-established standard multilevel converters are highlighted. Then new topologies introduced in the recent 5 years are discussed along with their applications. Recent advances made in the control and modulation methods of MLIs are also addressed. A comparative case study is conducted between the conventional three-level NPC and one of the recently proposed topologies (3L-NPCI^2) by simulation and experimental results. Finally, some challenges and future trends in the development of this technology are highlighted.

Nomenclature

ANNPC	active NNPC
APOD	alternative phase opposite disposition
CEC	California energy commission

[1]School of Energy Engineering, Kyungpook National University, Korea
[2]Electrical Faculty, Shahid Rajaee Teacher Training University, Iran
[3]School of Electrical and Electronic Engineering, Nanyang Technological University, Singapore

CHB	Cascaded H-bridge
CM	common-mode
CMV	common-mode voltage
DSP	digital signal processors
DF	distortion factor
EU	European Union
FLBI	five-level boost inverter
FC	flying capacitor
HG	high gain
HERIC	highly efficient and reliable inverter concept
HANPC	hybrid active neutral point clamped
HNPC	hybrid neutral point clamped
H-PWM	hybrid-PWM
IPUC	improved packed U-cell
LV	large vector
MV	medium vector
MANPC	modified active-neutral-point-clamped
MLI	multilevel inverter
NLC	nearest level control
NLM	nearest level modulation
NNPC	nested neutral point clamped
NT2	nested T-type
NPC	neutral-point-clamped
NUPF	none-unity power factor
UPC	pack U-cell
POD	phase opposite disposition
PD-PWM	phase-disposition pulse-width modulation
PS-PWM	phase-shifted pulse width modulation
PV	photovoltaic
PI	proportional–integral
PID	proportional–integral–derivative
qCHB	quasi cascaded H-bridge
SHE	selective harmonic elimination
SV	short vector
SPWM	sinusoidal pulse width modulation
SVD	space vector diagram
SVM	space vector modulation
SI-NPCI2	split-inductor neutral point clamped inverter-improved

SBD2T	switched-boost dual t-type
THD	total harmonic distortion
TRL	transformer-less
TPUC	T-type packed U-cell
UPF	unity power factor
ZV	zero vector
ZVS	zero-voltage switching

4.1 Introduction

Multilevel inverter (MLI) topologies have gained significant attention in various applications [1–7]. This is because they have lower THD, lower voltage transients *dv/dt*, lower common-mode (CM) voltage, higher output voltage, high efficiency, and the small size of filter elements compared to their two-level counterparts [1]. There are three well-known classical MLIs, namely neutral-point-clamped (NPC), also called diode clamped [1–6], cascaded H-bridge (CHB) [3–5], and the flying capacitor (FC) [3–5]. However, these traditional topologies have several shortcomings. For example, the NPC topology does not possess sufficient redundant switching states to maintain the dc-link voltage in an acceptable balancing value using the traditional sinusoidal pulse width modulation (SPWM) method. Furthermore, this converter requires many series connection clamping diodes, especially in more than five levels, to block the higher voltage, making the inverter implementation and maintenance process more challenging. The uneven distribution of losses among inner and outer power devices is another drawback that makes NPC less attractive for high-power applications [5]. Moreover, during date-time in the commutation process, the unbalanced voltage distribution occurs on the power device in conventional NPC, particularly in medium-voltage applications [5]. The CHB is used in different industrial applications to function at more than 6.6 kV [3–5]. Although CHB topology has a modular structure and fault–tolerant capability, it requires a bulky, heavy and costly phase-shift transformer(s). Flying capacitor (FC) is another attractive conventional converter that does not have such limitations mentioned above [3–5]. Moreover, in this MLI, the balancing of FC voltage is obtained without generating fundamental frequency ripples of voltage, even if the converter has many output levels. Nonetheless, the number of FCs increases in this structure by soaring the voltage level and challenging to balance FC voltages. In light of this drawback, FC topology is unsuitable, especially for high-power MV applications, affecting the inverter efficiency.

For MLI's modulations usually alternative phase opposite disposition (APOD) [5, 6], in-phase disposition (IPD) [5, 6], phase-opposite disposition (POD) [5, 6], phase-shifted (PS) [5, 6], double sinusoidal (DS) [6], and space vector modulation (SVM) [5] are used in industrial applications. Even so, these conventional modulation methods have several drawbacks. For instance, standard IPD, POD, and APOD cannot balance the FCs' voltage in a flying capacitor structure.

The PS-PWM offers a natural balancing of the capacitors' voltage. However, the dynamic of the voltage balance is slow especially under different operation conditions. Likewise, it might be prolonged in some real/practical applications. On the flip side, conventional PS-PWM fails to operate NPC safely.

Recently, many state-of-the-art MLIs have been reported to overcome topological deficiencies connected with standard NPC, CHB, and FC converters. Likewise, several new control/modulation strategies have been suggested to enhance the performance of classical MLIs such as improved PD-PWM [8–10], optimized PS-PWM [11–14], model predictive control (MPC) [15, 16], selective harmonic elimination (SHE) [17–20], and nearest level control (NLC) [21, 22]. By considering earlier explanations, recent contributions are reviewed in this chapter to shed light on the merits and demerits of these novel proposals. Besides, some challenges and promising solutions are discussed to enhance this technology further to stimulate upcoming contributions addressing open problems and exploring new solutions/possibilities. The rest of this chapter is organized as follows: Section 4.2 reviews new and hybrid MLIs. The modulation methods are surveyed in Section 4.3. Section 4.4 includes a case study. The advancement, challenges, and future trends are discussed and highlighted in Section 4.5, followed by the conclusion in Section 4.6.

4.2 Hybrid/advanced MLIs

With a combination of traditional multilevel topologies, advanced/hybrid MLIs are extracted that can tackle some constraints as mentioned earlier in the standard converters. More than 20 advanced and new/modified topologies introduced in recent years are opted and reviewed deeply in this section from the critical point of view as follows.

The shoot-through (leg short circuit) occurs in the standard NPC topology when the complementary power switches are turned on simultaneously. In [6], a novel three-level (3L) topology, so-called split-inductor NPC inverter-improved (SI-NPCI^2), is suggested for transformer-less (TRL) photovoltaic (PV) application that does not need to insert dead-time among complementary switches in PWM [see Figure 4.1(a)]. Therefore, the shoot-through issue is eliminated. Removing the dead-time intervals in PWM leads to higher voltage/current gain and deduction in total harmonic distortion (THD) than conventional MLIs. This novel topology resolves the unbalanced voltage distribution problem caused by dead-time intervals on the power devices in conventional NPC. Moreover, suppose classical NPC and SI-NPCI^2 topologies are employed for variable frequency/speed motor drive applications. In that case, the electrical machine's adverse heating effects increase because of NPC's higher THD. This phenomenon decreases the reliability of the NPC construction compared to the SI-NPCI^2. As reported in [23], removing/reducing dead-time or overcoming the shoot-through problem decreases the common-mode voltage (CMV) considerably. Thus, the three-phase configuration of the SI-NPCI^2 can be an exciting option in this regard. Besides, the SI-NPCI^2

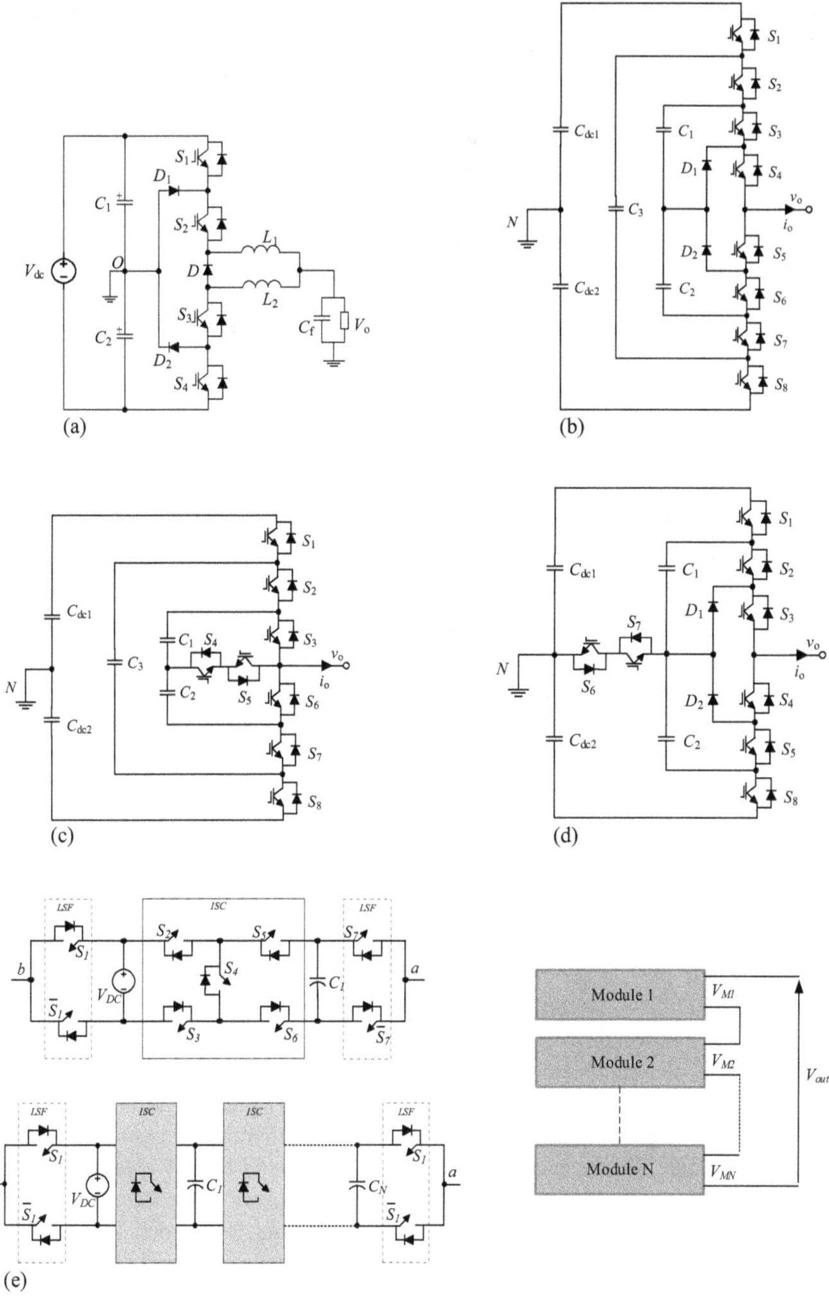

Figure 4.1 New and advanced MLIs: (a) 3L-SI-NPCI[2] [6]. (b) 5L-NNPC [26].(c) 5L-NT[2] [7, 27]. (d) 5L-ANNPC [28]. (e) 5L-Boosting MLI [29]

suppresses the leakage current even with asymmetrical output filter inductors in TRL PV application; this issue is one of the most critical challenges in the mentioned application. Likewise, in [6], a comprehensive comparison has been made between the SI-NPCI^2 and several different SI topologies. This study demonstrated that the SI-NPCI^2 could meet the power quality standard in many loads and satisfy the California Energy Commission (CEC) efficiency requirement [1]. However, uneven distribution of losses on power devices still exists in the SI-NPCI^2 like traditional NPC inverter. Also, requiring an additional low-voltage diode is another demerit of the SI-NPCI^2 compared to conventional NPC structure.

The nested neutral point clamped (NNPC) topology is among the promising advanced/hybrid MLIs introduced in [24, 25]. This topology has fewer clamping diodes over conventional NPC. Besides, the voltage balancing of the FCs can be obtained very easily compared to the NPC inverter. The same stress voltage on power devices is another merit of NNPC topology.

The nested T-type (NT^2) structure extracted from the NNPC converter is introduced in [7] that not only can decrease the number of devices further (regardless of different voltage stress on power switches) but also has almost acceptable output current waveforms quality during the open-circuit fault on power switches [7]. The 5L version of these NNPC [see Figure 4.1(b)] and NT^2 [Figure 4.1(c)] inverters are introduced in [26, 27]. In recent two topologies, the FCs' voltage balancing cannot be achieved by traditional PWM and SVM control strategies when these converters operate at low frequency and high-power factors [27]. In addition, the FCs of the NNPC and NT^2 inverters suffer from inherently low-frequency voltage ripple. This ripple might be large under some certain operating conditions. It may impact the power quality, performance, semiconductor life span, and safe operation of the converter.

Another advanced topology is introduced in [28], the so-called active NNPC (ANNPC) [see Figure 4.1(d)]. The ANNPC topology can tackle the shortcoming mentioned for NNPC and NT^2 inverters and reduce THD value. On top of that, ANNPC removes one highly voltage stressed capacitor [C_3 in Figure 4.1(b) and (c)] in NNPC and NT^2 inverters and offers lower cost, reducing complexity with improved reliability. Nevertheless, the voltage stress distribution is not the same among all the power semiconductors. It is worth noting that unbalanced voltage distribution is also associated with NNPC, ANNPC, and NT^2 inverters.

In [29], a hybrid MLI is introduced to boost the front-end booster in the conventional two-stage configuration without soaring structural and control complexity. Figure 4.1(e) illustrates the circuit diagram of two different extensions of this single-stage boosting MLI. This topology has nine power switches and only one capacitor in its structure to generate 5L output voltage. Therefore, the mentioned architecture has intrinsic self-balancing ability and does not need extra balancing circuitry. However, all switches must withstand dc voltage stress in the single-stage boosting MLI. This major issue limits its application for medium-voltage applications. On the other hand, it does not have two split dc-link capacitors and needs only half the dc source voltage compared to conventional FC and NPC inverters. Therefore, it is suitable for low-voltage TRL PV grid-connected applications. A comparative study is also conducted in [29] to highlight the merits of the suggested topology in comparison with several state-of-the-art counterparts. In single-stage boosting MLI, the load current through many power devices leads to more conduction losses. Likewise, the power losses or efficiency is not reported for this MLI to validate that this topology can meet efficiency standards.

In [30], a hybrid 5L-ANPC full-bridge inverter with a combination of Si and SiC MOSFETs is proposed to achieve the tradeoff between performance and cost [see Figure 4.2(a)]. In the mentioned research, the authors also proposed a new

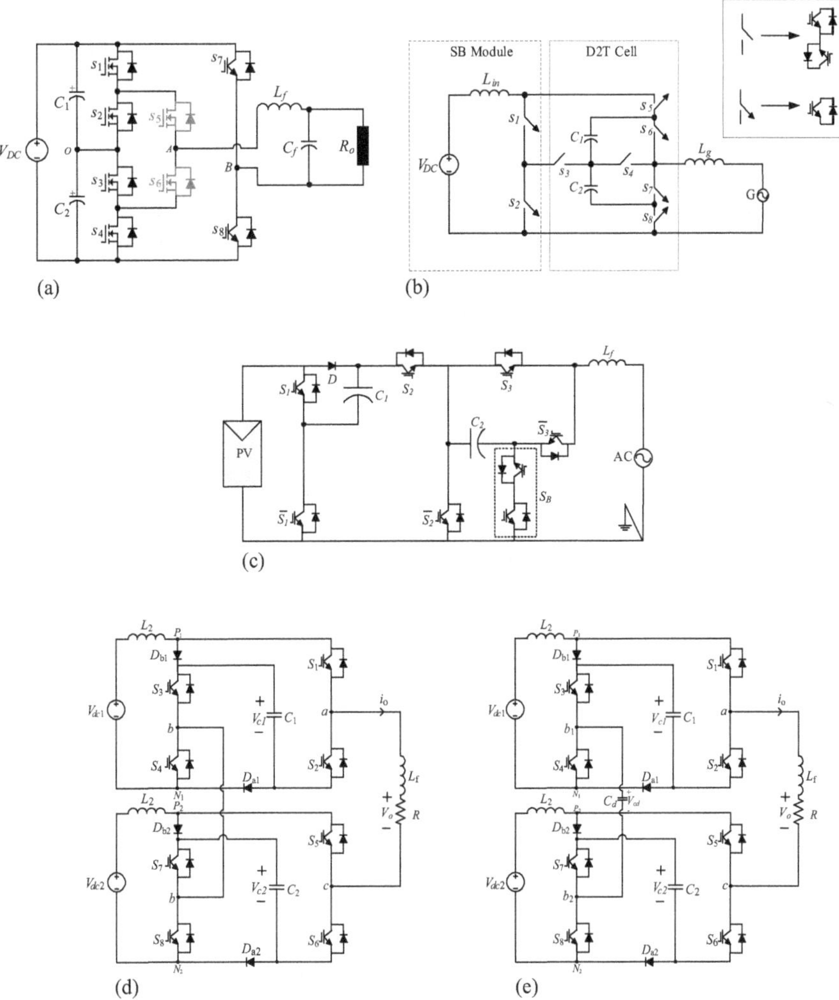

Figure 4.2 New and advanced MLIs: (a) 5L-hybrid-ANPC [30]. (b) 5L-SBD2T-TL inverter [31]. (c) 5L-8 switch inverter [32]. (d) 5L-qCHB-FLBI without filter capacitor [33]. (e) 5L-qCHB-FLBI with filter capacitor [33]

modulation method to reduce the conduction losses. The proposed modulation method is compared with the conventional one. Moreover, the comparison is demonstrated that the proposed modulation strategy exhibits higher efficiencies under high-power conditions. Nonetheless, the traditional technique of modulation illustrates superior efficiency under low-power conditions. Consequently, a hybrid modulation algorithm is also proposed to achieve high efficiency. Furthermore, the proposed hybrid modulation methods can provide the soft-switching ability for the hybrid converter. However, in this hybrid MLI, the switches S_7 and S_8 [see Figure 4.2(a)] must withstand stress voltage equal to dc source voltage. This limits this topology's application for medium-voltage applications.

MLIs with a common-ground (CG) and step-up voltage capability can benefit TRL PV grid-connected applications. In the light of this concept, the hybrid 5L topology [31] is proposed for grid-connected application [see Figure 4.2(b)]. The proposed topology is called the 5L-SBD2T-TL inverter. An inductor in the integrated switched-boost module boosts the voltage across two involved capacitor of the dual T-type cell. Thus, a desirable ac output voltage for the grid-connected application can be obtained for a wide input voltage range. Moreover, the current stress is maintained within a permissible input current range by the inductor's soft-charging function of the involved capacitors. A corresponding dead-beat continuous-current-control-set controller with a sinusoidal PWM modulator is implemented for controlling the reactive and active/real powers of the 5L-SBD2T-TL inverter. By utilizing the mentioned control method, constant switching frequency for power switches is achieved. The proposed topology is compared with more than 15 different topologies. This study shows that the proposed inverter can suppress the leakage current considerably and reduce it to zero. Besides, the 5L-SBD2T-TL inverter can meet the CEC or European Union (EU) efficiency standards and satisfy the power quality requirement of output current with iTHD= 2.2%. However, the higher number of components is the main disadvantage of this topology, which makes it complex, costly, less reliable, and bulky.

Figure 4.2(c) depicts a hybrid 5L-8 switch inverter introduced for TRL PV grid-tie application [32]. The boost ability without needing an additional dc-dc converter, natural balancing of FCs' voltage, working at a wide range of modulation indices, simplicity of control method, requiring 50% of input dc voltage, and reducing leakage current to zero are considered the main merits of the hybrid 5L-8 switch inverter. The proposed converter is compared with 16 different TRL PV inverters. Additionally, the performance of the proposed MLI is tested by an inductive load in steady-state and dynamic conditions. If the purely resistive load is used at the ac side, the THD of output current would be beyond the defined standard.

In [33], a two-hybrid MLIs with and without filter capacitor [C_d in Figure 4.2(e)] are introduced. The proposed inverter cuts down the number of components and improves the output current THD over the CHB quasi-Z-source inverter. Likewise, this topology can function in the shoot-through state like Z-source topologies. A low-voltage rating capacitor is incorporated into the 5L-qCHB-FLBI to eliminate the offset voltage at the output during the unbalanced input voltages of the two modules.

This ability makes these topologies an appropriate choice for TRL PV grid-connected applications, especially in residential areas where the shadowing phenomenon caused by separated dc-source, i.e., PV panels, is one of the main concerns. Therefore, to control the voltage of each module's capacitor at the desired level, a simple PID controller and PS-PWM are employed. Although these structures reduce the number of components compared to the CHB quasi-Z-source inverter, their power devices are still high. Suppressing the leakage current produced between PV panels and ground [1] in TRL PV grid-tie application may be a significant issue for these configurations.

An enhanced step-up 5L inverter is proposed for PV systems in [34], and the proposed MLI circuit diagram is shown in Figure 4.3(a). The proposed 5L-MLI has fewer power devices than the conventional 5L-NPC, 5L-FC, and 5L-CHB topologies. Likewise, it has a boost ability compared to classical MLIs, suitable for low-voltage renewable applications. Using the level shift multicarrier-based PWM method, the switches S_1 and S_2 are working at high frequency and the remainder at line/lower frequency. Therefore, switching losses are significantly reduced. The enhanced step-up 5L inverter is compared with nine other topologies in terms of devices and boost capability. However, the dynamic performance of the proposed structure is not tested. Also, this topology can only operate at the unity power factor (UPF), which limits its application.

Figure 4.3(b) shows a hybrid seven-level (7L) MLI called modified UPC [35]. This MLI can produce 7L output ac voltage with fewer switches than the conventional CHB and NPC MLIs. This topology also has a boosting feature, making it a competitive topology for TRL PV applications. The IPD-PWM is used for the 7L-modified UPC topology along with a new switching technique to decrease the

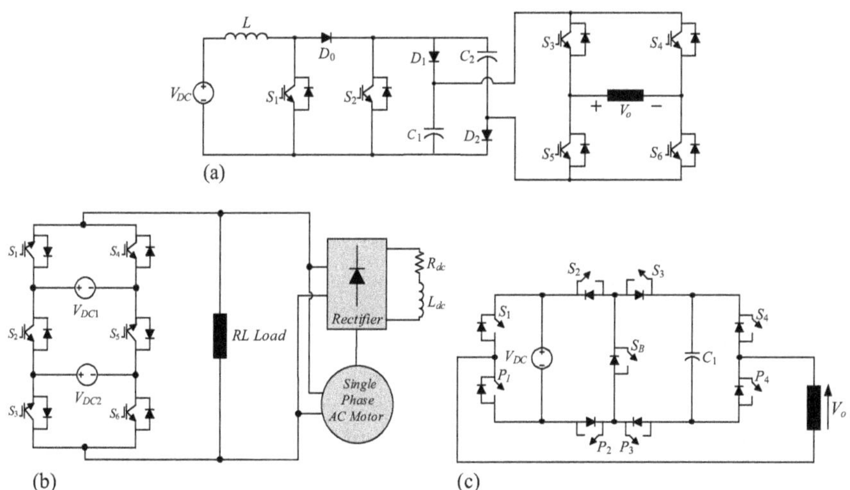

Figure 4.3 New and advanced MLIs: (a) enhanced step-up 5L inverter [34]. (b) 7L-modified PUC [35]. (c) 7L-IPUC inverter [36]

switching frequency. The proposed method has two redundant zero-zone switching states that significantly reduce the switching frequency. The proposed work tested for PV application with UPF. Although the output THD is within the defined PV codes and standards (vTHD = 3.1%, iTHD = 2.2%), the proposed topology cannot meet the power quality standard in light load operating conditions. The dynamic performance and non-unity power factor (NUPF) operation of the 7L-modified UPC are not addressed. This inverter does not have the inherent feature of mitigating leakage current defined by PV codes and standards. Hence, the 7L-modified UPC needs to be equipped with additional software or hardware (like ref [33]) to overcome the mentioned problem.

Another 7L-UPC inverter is introduced in [36] the so-called 7L improved packed U-cell (7L-IPUC) inverter [Figure 4.3(c)]. The advantages of boosting ability and needing fewer power devices like the previous inverter still exist for this converter. Hence, a novel balancing method uses logic form equations for balancing the capacitors' voltage. The proposed inverter is compared with 12 existing new/modified 7L inverters in respect of the number of devices, stress voltage on power devices and capacitors, and voltage gain. On top of that, the load voltage THD and RMS are provided by different PWM (IPD, POD, and APOD) methods at three other modulation indices (0.5, 0.7, and 0.9). Compared to the previous converter [Figure 4.3(b)], this inverter has three additional switches that reduce the reliability and increase complexity and cost. The effectiveness of the proposed inverter is evaluated under unity and NUPF factor loads in conjunction with dynamic performance. The voltage stress on power devices is the dc voltage, limiting this inverter's application for a low-voltage system. Like the former inverter, the 7L-IPUC needs an extra control strategy to suppress the leakage current if it is considered for TRL PV application. The efficiency/losses are not reported to this inverter.

A high gain seven-level (7L) active neutral point clamped (HG-7L-ANPC) self-balancing inverter [see Figure 4.4(a) below] is proposed for grid-tie renewable applications [37]. This topology decreases the dc-link voltage requirement by 50% in comparison with several advanced converters. The performance of the proposed structure is investigated under various operating situations in terms of modulation indices and the load. The nearest level modulation method generates the gating pulses for the switches. As a result, the HG-7L-ANPC has 95% efficiency. Although the HG-7L-ANPC does not need any sensor to keep the voltage of FC in balancing conditions, decreasing the control design's cost and complexity, needing a higher number of power devices makes it less reliable. Another major flaw of this inverter is that the discharging period is higher than the charging time, which increases the size of FC.

In [38], a new single-stage 7L boost type topology called 7L-TG inverter is proposed that can generate a peak value for output voltage three times higher than the input dc voltage [see Figure 4.4(b)]. The proposed topology is suitable for renewable-based distributed systems and electric vehicle applications. In addition, the 7L-TG inverter has the self-balancing capability of FCs' voltage. The proposed inverter is compared with 14 different advanced MLIs with regard to the number of

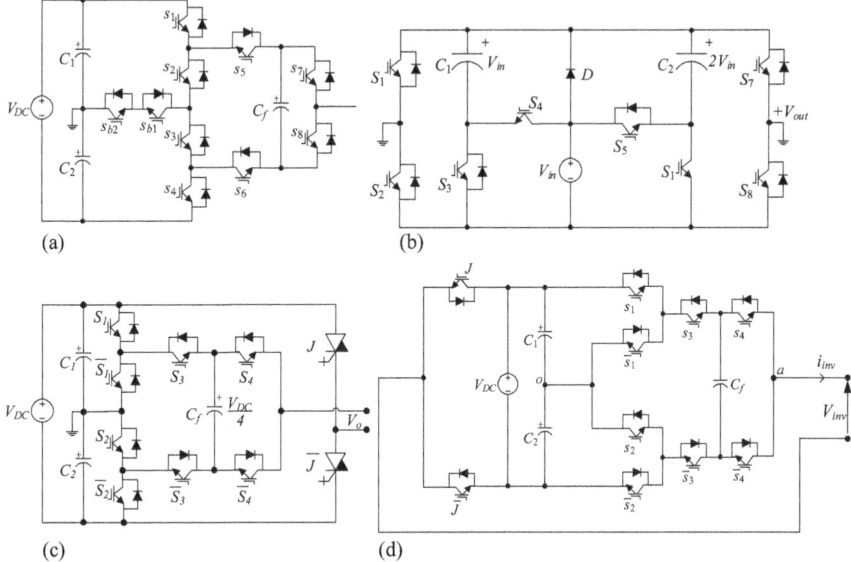

Figure 4.4 New and advanced MLIs: (a) HG-7L-ANPC [37]. (b) 7L-TG Inverter [38]. (c) 9L-MANPC inverter [39]. (d) 9L-RDC ANPC inverter [40]

gate driver circuits, power devices, capacitors, and voltage gain. The comparison attests benefits of the proposed topology. Likewise, the POD-PWM produces the gating signals for power switches. There are no sensors for balancing the FC voltage, operating in both lagging/leading power factors, an equal number of charging and discharging times in each fundamental cycle, smaller capacitor size because of high volt-amp per sec, and fewer number of power devices are other benefits of the proposed structure. Furthermore, the behavior of the 7L-TG inverter under the sudden change in load and modulation indices is measured experimentally. As a result, 98.2% maximum efficiency at UPF is reported for this MLI. However, from the commercial prospect, the proposed topology cannot be a strong competitor for the conventional HERIC topology [1] because of possessing a large number of power devices, one more capacitor, and lower efficiency.

Figure 4.4(c) shows a hybrid MLI topology named 9L-MANPC inverter [39]. In this study, an *ad hoc* switching state redundancy-based modulation method is exerted to ensure that the voltage across the FC is tightly balanced. The appropriate operating features of the introduced inverter are: the design procedure complexity of the voltage balancing controller is less and is accomplished by utilizing look-up tables and a few comparators; the number of input dc sources and its magnitude is halved compared to traditional dc-link midpoint-based converters for the same peak-to-peak output voltage; the amalgamation of the 2L leg with 5L-ANPC needs minimum topological disruption while extending the latter to output a 9L voltage. Therefore, to further clarify the superiority of the 9L-MANPC inverter, a comparative assessment is also conducted between the 9L-MANPC and eight existing

MLIs. Additionally, the performance of the proposed inverter is tested by considering a three-phase grid-tied case study with satisfactory results. The system response to the transient and steady-state condition are examined as well. The efficiency of 91.1% is reported when the inverter works at a modulation index higher than 0.9. However, the function of this inverter at NUPF is not tested. The control system of the 9L-MANPC inverter would be slightly complex due to split dc-link capacitors compared to topologies that need only one capacitor [1]. In addition, the unbalanced voltage distribution among power switches exists for this topology like conventional NPC.

The same authors have introduced another 9L topology by modifying the previous converter so-called 9L-RDC ANPC inverter [40]. Figure 4.4(d) represents the circuit diagram of this topology, which is built by including two switches that function at line/low frequency to the classical 5L-ANPC inverter. Also, a logic form equation-based voltage balancing method is developed to balance the FC voltage. Like a previous study, a three-phase grid-connected medium-voltage system is implemented in MATLAB®/Simulink® environment. The practicability of the 5L-ANPC inverter is justified by experimental results extracted from a low-scale low-voltage prototype. Moreover, a comprehensive comparison is made between the proposed and several MLIs by targeting all topologies for medium-voltage applications. As a result, the reported maximum efficiency is more than 98%. Moreover, the balancing method is robust against the change of modulation index from 0.2 to 0.98. The major shortcoming of this topology is inequality of the voltage stress on semiconductor devices.

By incorporating two split capacitors instead of the upper dc source in Figure 4.3(b) [35] and connecting two bidirectional switches in the middle of the mentioned capacitors, the 9L version of PUC inverters is derived [41]. Figure 4.5(a) shows the implantation of this topology called the 9L-TPUC inverter. The single carrier modulation method produces gaiting pulses for power switches. Half of the semiconductors are switched in high frequency in this topology, and the rest are at line/lower frequency. This leads to uneven distribution of losses, which is a significant problem. The proposed inverter is compared with various topologies from different aspects. Therefore, the combination of Si and SiC power semiconductors is used to increase the converter's efficiency, and approximately 98% maximum efficiency is achieved. The functioning of the 9L-TPUC inverter is evaluated with purely resistive load with linear and nonlinear loads. The feasibility of the inverter for NUPF is not addressed. Like the previous topology, the stress voltage on the switches is not equal.

In [42] a new 11L-IANPC topology is introduced for ground power unit application [see Figure 4.5(b)]. The proposed MLI possesses only one FC and reduced numbers of high-frequency power switches. Thus, stored energy, voltage diversity and ratings, and the number of expensive and bulky FCs are remarkably decreased. Furthermore, the augmented submodule's functioning at the low-voltage/power results in doubling the output voltage levels. In contrast, the low additional cost and size are imposed on the 11L-IANPC converter. Additionally, the suggested sensor-less switching strategy leads to self-balancing of FC and

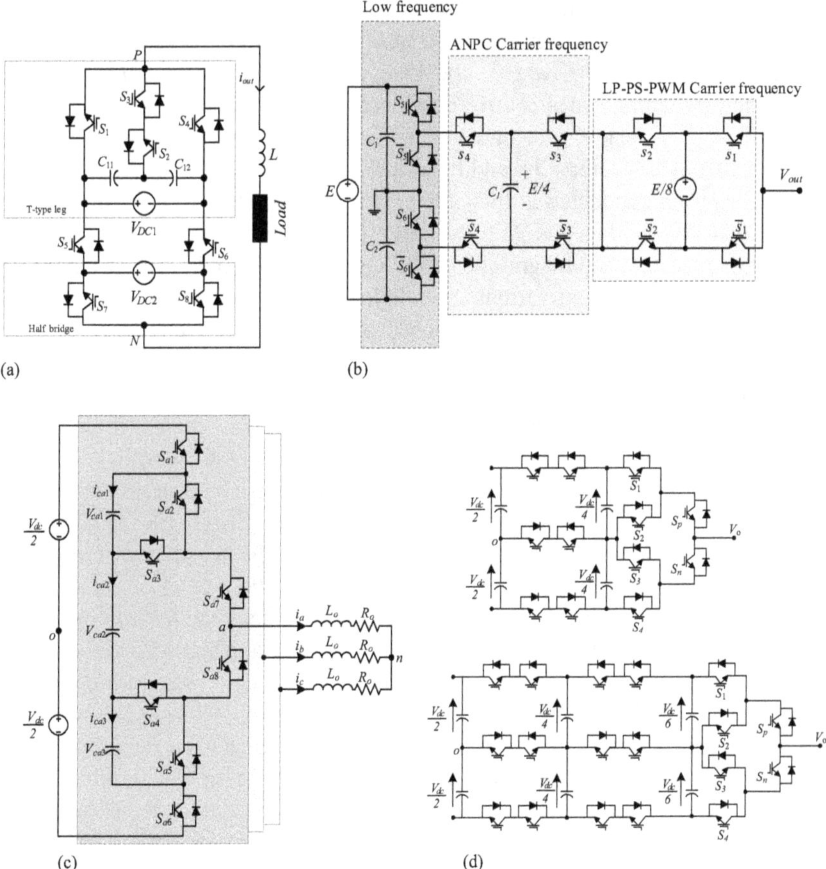

Figure 4.5 New and advanced MLIs: (a) 9L-TPUC inverter [41]. (b) 11L-IANPC topology [42]. (c) 5L-VSI [43]. (d) 5L-TRLMLI (top) and 7L-TRLMLI (bottom) [44]

considerably amends the output voltage frequency spectrum. These enhancements remarkably reduce control complexity and the size of the output LC filter, thus, enhancing the dynamic and steady-state operation of the proposed static GPU. The authors have compared the proposed topology with several conventional converters from different perspectives. The performance of the suggested inverter is tested with inductive load and dynamic change in loads. The efficiency of the proposed MLI is 97.33 %. However, the functionality of the inverter is not analyzed under a lower modulation index and higher PF. Likewise, uneven stress voltage on power semiconductors is a significant problem for this converter.

A hybrid 5L topology [see Figure 4.5(c)] is introduced in [43]. Mentioned topology is competitive in terms of control complexity, performance, and component count over the classical topologies. As this topology does not require any dc-link neutral points, unlike other hybrids' MLIs, its structure is simple. Moreover, the proposed inverter can be installed back-to-back because of a common dc-link. Besides, to balance the FC voltages, a simple method using the redundancy switching states is also introduced. The performance of the inverter is validated against steady-state and transient conditions with various modulation indices and power factors. In addition, the design procedure of FC and losses analysis has been conducted in this paper. The reported efficiency of this structure is 85.92% at PF= 0.3 lag, and 93.92% at PF= 0.7lag, while the modulation index is one. In this research, the magnitude of FC voltage ripple and THD of output voltage/current are offered in various modulation indices with PF=0.3 lag and PF= 0.7 lag. The performance of the proposed converter is not verified at a critical operating condition such as a modulation index lower than 0.5 and higher PF, e.g., PF=0.9–1. The higher number of FCs with different voltage stress is the main drawback for this inverter over existing advanced/state-of-the-art hybrid MLIs. Another deficiency is the lopsided stress voltage on power devices.

Figure 4.5(d) shows an interesting commercialized TRL multilevel topology introduced Nidec ASI for the medium voltage drive SILCOVERTFH [44]. The proposed MLI amalgamates two Macro-Cells and has the possibility of expanding output levels. The number of semiconductors consolidated in series does not depend on the dc bus/input voltage. Similar power devices can be employed in 5L and 7L configurations to address the 4.16 and 6.6 kV applications, which is substantial merit from the industrialization perspective. Also, the zero-crossing is performed naturally at zero-voltage switching (ZVS) on the selected cell because of its intrinsic characteristic. A comparison study is made between the proposed and several commercialized converters to shed light on the merits. Furthermore, a balancing method can maintain the FC voltage at the desired value. The design of FC is represented for the proposed converter as well. The experimental results are derived from the real application with the following parameters: rated power= 1.5 MVA, dc-link voltage= 5.8 kV, dc capacitance= 300 μF, FC capacitance= 900 μF, and switching frequency= 976 Hz. The experimental test bench is implemented with two configurations: (a) inverter with RL load; (b) back-to-back configuration for the motor drive. Additionally, during the start-up procedure, an auxiliary pre-charge circuit connected with the dc bus was used to charge the FC voltages to their rated values. As the rated dc voltage is not equal to the rated FCs' voltages, a proper start-up procedure is essential for charging whole the FC installed in the converter. The output voltage THD is reported with and without the third harmonic injection. Moreover, the power loss analysis is carried out, and the efficiency of the proposed technology is measured at different modulation indices and various power factors. For example, at PF=1, 98.8% is obtained as the maximum efficiency. However, the switching losses are not equal among switches, and voltage stress distribution on power devices was not addressed.

A modified configuration called 3L-10S-3P is proposed in [45] to reduce the number of power devices in the conventional NPC structure. Figure 4.6(a) depicts the circuit structure of the mentioned architecture. In this study, a comparison is made between 10S-3P-3L and several existing inverters regarding stress voltage/

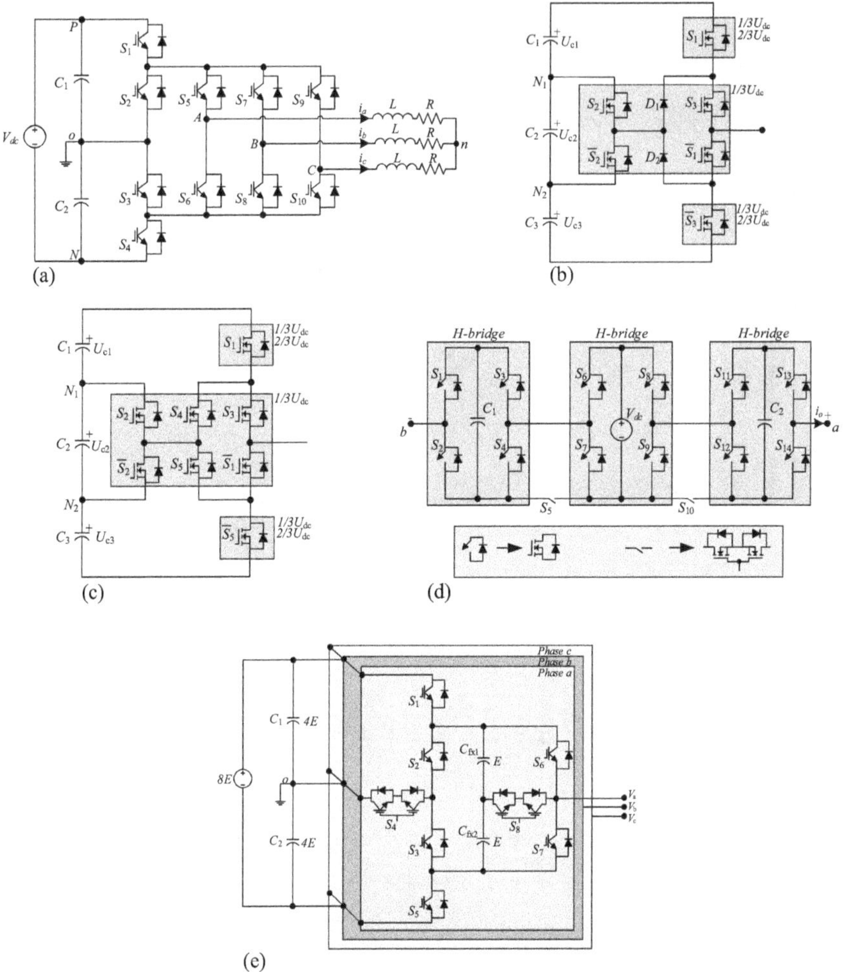

Figure 4.6 New and advanced MLIs: (a) 3L-10S-3P [45]. (b) 4L-HNPC [46]. (c) 4L-HANPC [46]. (d) 7L-TVG [47]. (e) 9L-T^2 [48]

current on power devices and switching and conduction losses. Improved SVM operates the inverter in a wide range of modulation indices. The critical demerit of this topology is that more than 50% of power devices have to withstand the dc voltage, which makes the proposed topology impractical in MV applications.

Figure 4.6(b) and (c) shows the circuit diagram of two 4L hybrid inverters (4L-HNPC and 4L-HANPC) [46]. The inverter provides higher efficiency and almost equal distribution of losses among power devices. Moreover, since these inverters have a smaller number of flying capacitors when compared with 5L counterparts, light-weight and higher power density are expected. Therefore, variable-reference voltage-balancing control method and a proportional-integral (PI) regulator are

Table 4.1 Comparison among MLIs

Topology name	Number of levels	Reference	Figure	V_{in} (V)	(%)	f_{sw} (kHz)	Number of switches	Number of diodes	Number of capacitors
3L-SI-NPCI2	3	[6]	4.1(a)	250	97.82	20	4	3	2
5L-NNPC	5	[26]	4.1(b)	300	–	0.5	8	2	5
5L-NT^2	5	[27]	4.1(c)	300	–	5	8	0	5
5L-ANNPC	5	[28]	4.1(d)	400	–	2.5	5	2	4
5L-boosting MLI	5	[29]	4.1(e)	100	97.1	5	9	0	1
5L-hybrid-ANPC	5	[30]	4.2(a)	400–700	98.1	40	8	0	2
5L-SBD2T-TL	5	[31]	4.2(b)	50	98.2	20	10	0	2
5L-8 switch	5	[32]	4.2(c)	50	97.8	1	8	1	2
5L-qCHB-FLBI (1)	5	[33]	4.2(d)	50–60	96	10	8	4	2
5L-qCHB-FLBI (1)	5	[33]	4.2(e)	50–60	96	10	8	4	3
Enhanced 5L	5	[34]	4.3(a)	60	–	15	6	3	2
7L-modified PUC	7	[35]	4.3(b)	50	–	2	6	0	0
7L-IPUC	7	[36]	4.3(c)	100	–	–	9	0	1
HG-7L-ANPC	7	[37]	4.4(a)	200	97	2.5	10	0	3
7L-TG	7	[38]	4.4(b)	100	98.2	2.5	8	1	2
9L-MANPC	9	[39]	4.4(c)	20	99	2.5	10	0	3
9L-RDC ANPC	9	[40]	4.4(d)	50	–	2.5	10	0	3
9L-TPUC	9	[41]	4.5(a)	200	98	20	8	0	2
11L-IANPC	11	[42]	4.5(b)	400	–	12	12	0	3
5L-VSI	5	[43]	4.5(c)	200	93.92	2	8	0	5
5L-TRLMLI	5	[44]	4.5(d)	5800	99.27	0.976	12	0	4
3L-10S-3P	3	[45]	4.6(a)	400	97.2	6	10	0	2
4L-HNPC	4	[46]	4.6(b)	45	–	4	6	2	3
4L-HANPC	4	[46]	4.6(c)	45	–	4	8	0	3
7L-TVG	7	[47]	4.6(d)	30	90	1	16	0	2
9L-T^2	9	[48]	4.6(e)	400	98.91	3	10	0	4

proposed to balance the voltage of flying capacitors in one-third of the dc-link voltage.

In [47], a novel 7L inverter was proposed that needs only one dc voltage source [see Figure 4.6(d)]. This topology is capable of boosting voltage gain to three times. Three H-bridges are connected via two bidirectional switches with voltage blocking ability to generate seven output levels. Natural voltage balancing of the capacitors also is achieved with this structure. However, the dynamic performance of this topology is not tested.

Figure 4.6(e) demonstrates the novel 9L-T^{2}C extracted from the conventional 3L-T2C [48]. The proposed inverter has a lower number of power switches and capacitors compared to conventional 9L inverters. The PD-PWM and the dead-beat model predictive control method are used for the proposed inverter. The proposed converter was comprehensively compared with more than ten different structures regarding the number of power devices, capacitors, cost, power loss distribution among components, and efficiency. Furthermore, the feasibility of the proposed architecture is validated under various operating conditions. Table 4.1 represents a comparison among the reviewed MLIs.

4.3 Modulation/control methods

The modulation techniques play a substantial role in the implementation of any MLI's topology. One of the most popular modulation methods among multilevel inverters is sinusoidal pulse width modulation. The simple SPWM method for a traditional two-level inverter is divided into unipolar and bipolar PWM methods. In the bipolar method, a reference voltage (V_{ref}) is compared with a carrier voltage (V_C), in order to produce the required gate signals of switches. Two reference voltages that differ in phases must be compared with the same carrier voltage for the unipolar strategy to have the three-level voltage. These SPWM techniques can be improved and extended for MLIs using multi-carrier signals and one sinusoidal reference signal [5]. The multi-carrier or carrier-based modulation strategies for multilevel inverters are classified into level-shifted and phase-shifted modulations.

4.3.1 Level-shifted (LS) and phase-shifted (PS) PWM

In m-level MLI structure, the (m-1) carrier signals are required, all having the same frequency and amplitude. These carrier signals are placed vertically to each other. The reference voltage is constantly compared with the corresponding saw-tooth carrier voltages to create the switching signals [5, 6, 8–10]. The level-shifted sinusoidal PWM (LS-PWM) techniques are categorized as phase opposition disposition PWM (POD-PWM), phase disposition or in phase PWM (PD-PWM), and alternative phase opposition disposition PWM (APOD-PWM).

4.3.1.1 PD-PWM

In this strategy, the sinusoidal reference voltage is continuously compared with $(m-1)$ triangular carrier voltages of the same frequency, phase, and amplitude to

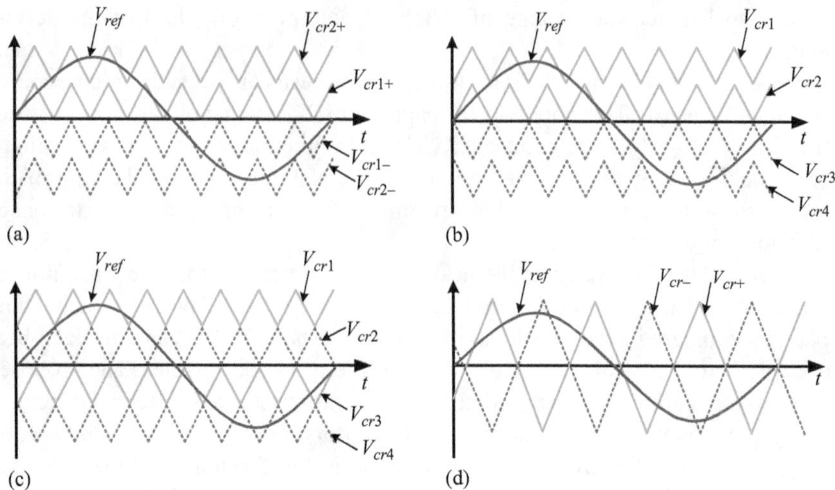

Figure 4.7 LS-PWM and PS-PWM techniques for MLIs [5], [6], [8–10]:(a) PD-PWM. (b) POD-PWM. (c) APOD-PWM. (d) PS-PWM

build the proper switching signals for the m-level inverter. Where these carriers are level shifted with a dc offset. Figure 4.7(a) illustrates the PD-PWM for a 5L-MLI.

4.3.1.2 POD-PWM

In this method, carrier signals which contain the same frequency and amplitude are different in the phase. For example, in the m-level MLI, half of the triangular carrier voltages which are located above the time axis are 180° phase-shifted, while the other half is below the zero references. Then the sinusoidal reference signal will then be compared with these carriers individually to generate switching signals. Figure 4.7(b) shows this method for 5L-MLI.

4.3.1.3 APOD-PWM

Considering m-level topology, (m-1) identical triangular carrier voltages must be situated in a phase opposition form and will be compared alternately with the desired voltage (V_{ref}), to produce the required switching signals. The APOD-PWM for 5L-MLI is depicted in Figure 4.7(c).

4.3.1.4 PS-PWM

In the PS-PWM, all the carrier voltages have the same peak-to-peak amplitude and frequency as the previous method. In contrast, there is a phase difference between any two adjacent triangular carrier signals, presented by $360°/m-1$. The modulating signal is generally a three-phase sinusoidal wave with adjustable magnitude and frequency. The gating pulses/signals are produced by comparing the carrier waves with the modulating wave. Figure 4.7(d) demonstrates the principle of the PS-PWM for a 3L inverter, where two carrier signals are required with a 180° phase difference between any two related carriers. The conventional PD-PWM, PO-DPWM, and APOD-PWM stated above cannot balance the voltage of FC(s) in the flying capacitor converter. Although PS-PWM offers natural/self-balancing of FCs, it might not be adequately robust to keep the existing FC voltages at the desired quantity, notably when the inverter works under critical transient conditions. Likewise, it might be prolonged in critical operating applications. Thus, the modified version of this modulation method should be considered [8–10, 49].

4.3.2 SV-PWM technique

In this modulation method, a complex vector must be compared with the desired vector to reach the output voltage. The mentioned desired vector (reference vector) of the first-order frequency is variable in magnitude but rotates in space with a defined angular frequency ($\omega = 2\pi f$). The dramatic rules of this method are: the closest fundamental vectors to the desired vector must be detected, the time intervals of each fundamental vector must be calculated, and finally, the appropriate switching sequence is produced by an inverter. In the basic SV-PWM technique, the three-phase voltage vector system i.e., *a–b–c* frame is converted to the simple two-phase voltage vector system i.e., α–β frame, by Clarke transformation using Eq. (4.1). At any sampling time, these vectors are located in one of the six main sectors of the (α, β) plan [5]:

$$\begin{bmatrix} V_{\alpha\_ref} \\ V_{\beta\_ref} \end{bmatrix} = \frac{2}{3} \begin{bmatrix} 1 & -\frac{1}{2} & -\frac{1}{2} \\ 0 & \frac{\sqrt{3}}{2} & -\frac{\sqrt{3}}{2} \end{bmatrix} \begin{bmatrix} V_1 \\ V_2 \\ V_3 \end{bmatrix} \tag{4.1}$$

Each *m*-level MLI consists of m^3 switching states. For example, the 3L-NPC inverter has 27 switching states (see Figure 4.8). The desired vector due to the three-phase voltages is determined by:

$$v(t) = \frac{2}{3}\left[v_a(t) + a.v_b(t) + a^2.v_c(t)\right] \tag{4.2}$$

where

$$a = e^{j\frac{2\pi}{3}} = -\frac{1}{2} + j\frac{\sqrt{3}}{2} \tag{4.3}$$

By using Clarke's theory, the three-phase coordinate system is converted into a two-dimensional frame. By replacing all the output voltage quantities generated by switching states in (4.2), both the space vectors of the MLI and the space vector diagram (SVD) are determined. The SVD of the 3L-NPC inverter is demonstrated in Figure 4.8. In this case, SVD is divided into six separated sectors, and each sector must be divided into four triangles. The switching states for any level inverters are classified into zero-vector (ZV), the vectors concluded (0, 0, 0), (−1, −1, −1) and (1, 1, 1) switching states called zero vectors, SV, MV, LV

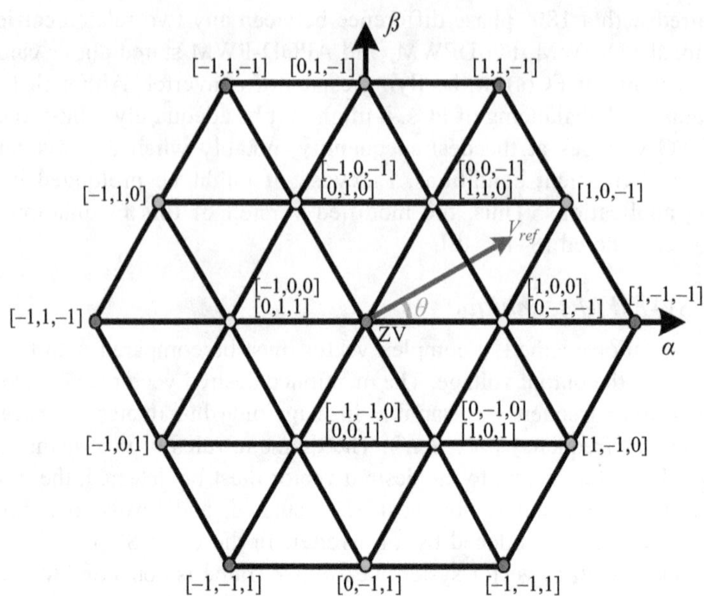

Figure 4.8 Three-level MLI SVD

Table 4.2 3L-MLI switching sates [51]

Switching vector	Switching state redundancy	Number of redundancies
ZV (blue)	[000], [111], [−1−1−1]	3
MV (green)	[10–1], [01−1], [−110], [−101], [0−11], [1−10]	6
SV (yellow)	[100], [0−1−1], [110], [00−1], [010], [101],[011], [100], [001], [110], [101], [0−10]	12
LV (red)	[1−1−1], [11−1], [−11−1], [−111], [−1−11], [1−11]	6

which refer to small vector, medium vector, and large vector, respectively. Here the ZV, MV, and LV are located on the boundaries of the SVD, while SV is situated in the central part of the SVD. Table 4.2 summarizes the space vector diagram switching states for the 3L-NPC inverter [50, 51]. After arranging the SVD, the nearest vectors to the reference vector must be selected. Note that in the 3L space vector diagram, as illustrated in Figure 4.8, the vector V_0 and V_1–V_6 have redundancy-switching states. As observed with the increasing number of levels in MLIs, the number of redundancy-switching states is increased. After selecting the closest vectors to the desired vector, the selected vectors' time sequence has to be determined. For example, the time equation based on sector-1 in Figure 4.8 is:

$$\begin{aligned} V^* T_s &= V_1 T_1 + V_2 T_2 + V_3 T_0 \\ T_s &= T_1 + T_2 + T_0 \end{aligned} \tag{4.4}$$

In which T_s is the corresponding sampling time, T_1, T_2, and T_0 are the switching on-times for V_1, V_2, and V_0, respectively. By replacing with the Clarke transformation formula, the transferred system can be expressed by:

$$\begin{aligned} V_d{}^* T_s &= V_{1d} T_1 + V_{2d} T_2 + V_{3d} T_0 \\ V_q{}^* T_s &= V_{1q} T_1 + V_{2q} T_2 + V_{3q} T_0 \end{aligned} \tag{4.5}$$

By solving this equation for the switching on-times, the time duration of the selected nearest vectors is calculated as:

$$\begin{bmatrix} T_1 \\ T_2 \\ T_3 \end{bmatrix} = \begin{pmatrix} V_{1d} & V_{2d} & V_{3d} \\ V_{1q} & V_{2q} & V_{3q} \\ 1 & 1 & 1 \end{pmatrix} \begin{bmatrix} V_d^* . T_s \\ V_q^* . T_s \\ T_s \end{bmatrix} \tag{4.6}$$

After determining the closest vectors to the reference vector and their corresponding time sequence, the proper switching sequence must be applied to the MLI. There are several types of switching sequences in the literature. The types of these switching sequences are determined due to the target and application. For example, in Figure 4.9, the target is to make sequences symmetrical. This figure selects the sequence based on the two switches in the on-state when passing from one switching state to another. This sequence is based on minimizing the

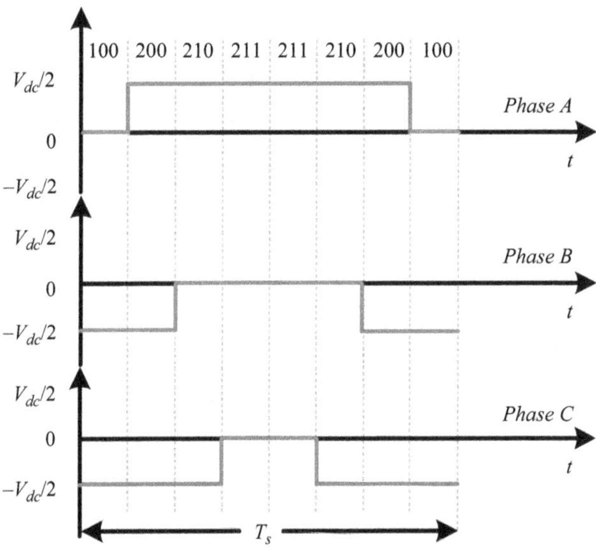

Figure 4.9 Example of switching sequence for region 2

switching losses and improving the output voltage quality. The switching sequence of this approach is summarized in Table 4.3 [52].

4.3.3 SV-PWM based on the gh coordinate system

In this method, the two-dimension (α, β) coordinate system is converted to the 60° (gh) coordinate system. The g-axis co-occurred with the α-axis and the h-axis, is 60° away from the g-axis in the counterclockwise direction, as illustrated in Figure 4.10. As it is observed, all the vectors in the (gh) system have only integer values. Thus, the real quantities of the coordinates can be rounded down to the integer quantities by considering the equation below [52]:

$$g = \mathrm{int}(V_g)$$
$$h = \mathrm{int}(V_h) \tag{4.7}$$

Table 4.3 Switching states of region 2

Voltage vector		[100]	[200]	[210]	[211]	[211]	[210]	[200]	[100]
Switching state	Phase A	0	$\frac{V_{dc}}{2}$	$\frac{V_{dc}}{2}$	$\frac{V_{dc}}{2}$	$\frac{V_{dc}}{2}$	$\frac{V_{dc}}{2}$	$\frac{V_{dc}}{2}$	0
	Phase B	$-\frac{V_{dc}}{2}$	$-\frac{V_{dc}}{2}$	0	0	0	0	$-\frac{V_{dc}}{2}$	$-\frac{V_{dc}}{2}$
	Phase C	$-\frac{V_{dc}}{2}$	$-\frac{V_{dc}}{2}$	$-\frac{V_{dc}}{2}$	0	0	$-\frac{V_{dc}}{2}$	$-\frac{V_{dc}}{2}$	$-\frac{V_{dc}}{2}$

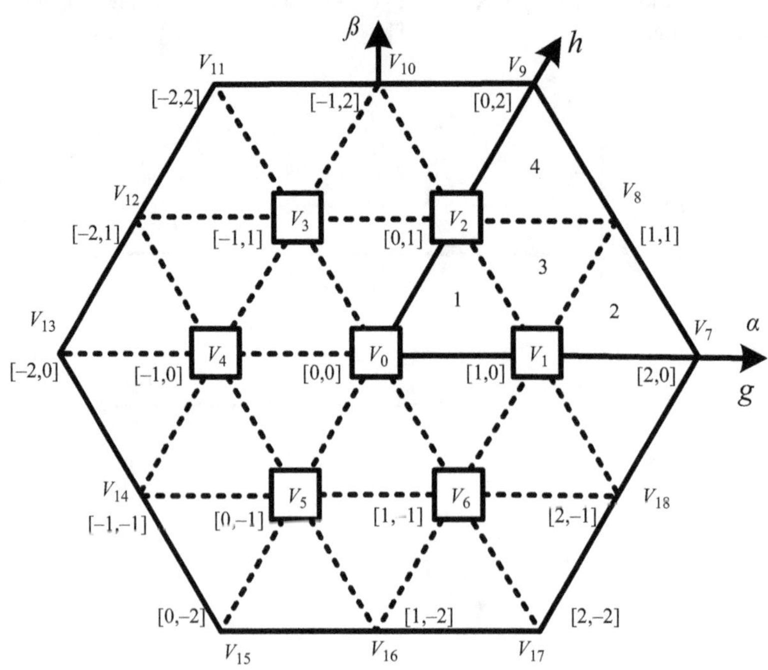

Figure 4.10 SVD of an inverter in (gh) frame [52]

The (α, β) coordinate system can transfer to the (gh) coordinate system by the equation below:

$$\begin{bmatrix} V_g \\ V_h \end{bmatrix} = \begin{bmatrix} 1 & -\dfrac{1}{\sqrt{3}} \\ 0 & \dfrac{2}{\sqrt{3}} \end{bmatrix} \begin{bmatrix} V_\alpha \\ V_\beta \end{bmatrix} \tag{4.8}$$

where V_g and V_h are the elements of the desired vector in the (gh) coordinate system. Considering (4.2) the transformed coordinate system yield as:

$$\begin{bmatrix} V_g \\ V_h \end{bmatrix} = \begin{bmatrix} 1 & -1 & 0 \\ 0 & 1 & -1 \end{bmatrix} \begin{bmatrix} V_a \\ V_b \\ V_c \end{bmatrix} \tag{4.9}$$

The closest four-vectors include a rhombus contained triangle I and triangle II as illustrated in Figure 4.11. All the vectors will be replaced in the formula. The desired vector must be situated in the triangle I or triangle II; therefore, the choice should be made by:

$$V_{ref} \text{ in the first triangle} \quad V_{ca} + 1 \geq -(g + h) \tag{4.10}$$

$$V_{ref} \text{ in the second triangle} \quad V_{ca} + 1 \leq -(g + h) \tag{4.11}$$

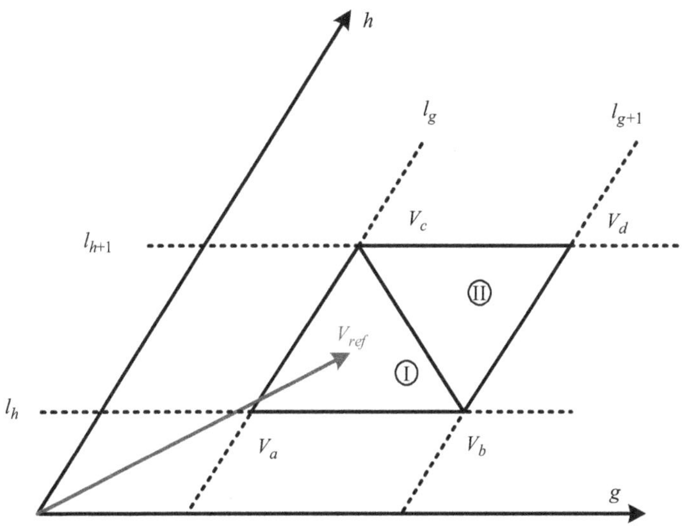

Figure 4.11 The composition of space vector in (gh) frame

Considering the desired vector is located in triangle I, the closest three vectors are V_a, V_b, and V_c. The duty cycle of the related vectors can be estimated by:

$$\begin{cases} V_{ref}T_s = V_a.t_a + V_b.t_b + V_c.t_c \\ T_s = t_a + t_b + t_c \end{cases} \tag{4.12}$$

Using *gh* transformation, the equation below is written:

$$\begin{cases} V_{rg}T_s = V_{ag}.t_{gh} + V_{bg}.t_{g+1.h} + V_{cg}.t_{g.h+1} \\ V_{rh}T_s = V_{ah}.t_{g.h} + V_{bh}.t_{g+1.h} + V_{ch}.t_{g.h+1} \\ T_s = t_{g.h} + t_{g+1.h} + t_{g.h+1} \end{cases} \tag{4.13}$$

Then, the on-time duration of the selected vectors can be calculated as:

$$\begin{cases} t_{g.h} = T_s\left[\left(\frac{V_{ca}}{V_{dc}}\right) + g + h + 1\right] \\ t_{g+1.h} = T_s\left[\left(\frac{V_{ab}}{V_{dc}}\right) - g\right] \\ t_{g.h+1} = T_s\left[\left(\frac{V_{bc}}{V_{dc}}\right) - h\right] \end{cases} \tag{4.14}$$

The procedure for the situation where the desired vector is situated in triangle II is similar, and the final result is yielded:

$$\begin{cases} t_{g+1.h} = T_s\left[-\left(\frac{V_{bc}}{V_{dc}}\right) + g + 1\right] \\ t_{g.h+1} = T_s\left[-\left(\frac{V_{ab}}{V_{dc}}\right) + g + 1\right] \\ t_{g+1.h+1} = T_s\left[-\left(\frac{V_{ca}}{V_{dc}}\right) - g - h - 1\right] \end{cases} \tag{4.15}$$

The two-dimensional coordinate system is converted back to a three-dimensional coordinate system as:

$$\begin{bmatrix} S_A \\ S_B \\ S_C \end{bmatrix} = \begin{bmatrix} i \\ i - g \\ i - h \end{bmatrix} \in (0.1.2). \quad \in (0.1.2) \tag{4.16}$$

The (*gh*) coordinate method is more beneficial for implementation in microcontrollers such as digital signal processors (DSP) [53].

4.3.4 Fundamental frequency modulation technique

Some methods based on fundamental frequency modulation have been proposed for eliminating the lower order harmonics [17–20]. The selective harmonic elimination (SHE) method was introduced to decrease either the distortion factor (DF) or THD of the output voltage waveform of the MLI. First, the output voltage must be expressed in the Fourier series. Then, the selected harmonics will be eliminated by calculating several unknown switching angles from the equations and adding them to the output voltage. The SHE-PWM technique is more suitable for high-power applications [17–20]. Likewise, this method can reduce the size of the electromagnetic interference (EMI) filters. Compared to SPWM and SV-PWM, the SHE-PWM technique has lower switching losses because this modulation technique makes MLI operate at a low switching frequency [17–20]. Calculating several nonlinear transcendental equations containing trigonometric terms and setting look-up tables are the problems of the SHE-PWM technique. Therefore, this modulation method is mainly used for MLIs with low levels [17–20]. SHE-PWM is based on the Fourier series generated from the periodic output voltage of MLI. Considering the Fourier series formula of the simple square waveform in Figure 4.12, it can be written as:

$$v(\omega t) = \sum_{n=1.3....}^{\infty} \frac{4E}{n\pi} \cos(n\alpha).\sin(n\omega t) \tag{4.17}$$

With the above concept, the Fourier series of the MLIs can be introduced. Here, the Fourier series formula of the output voltage waveform of the three-level NPC inverter (Figure 4.13) is expressed as:

$$v_{aN}(\omega t) = \sum_{n=1,3,...}^{\infty} \frac{4E}{n\pi} [\cos(n\alpha_1) - \cos(n\alpha_2) + \cos(n\alpha_3)] \sin(n\omega t) \tag{4.18}$$

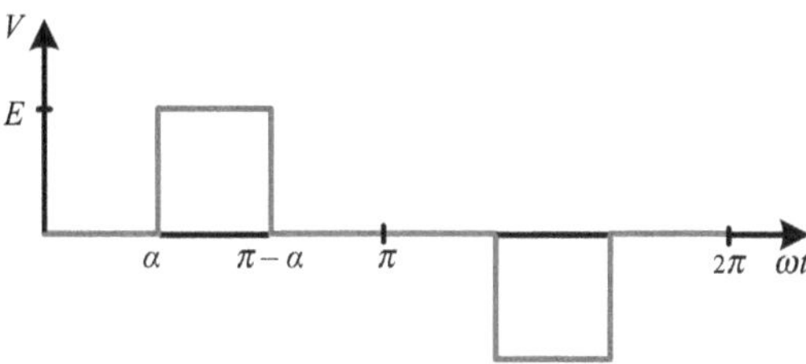

Figure 4.12 A simple square waveform

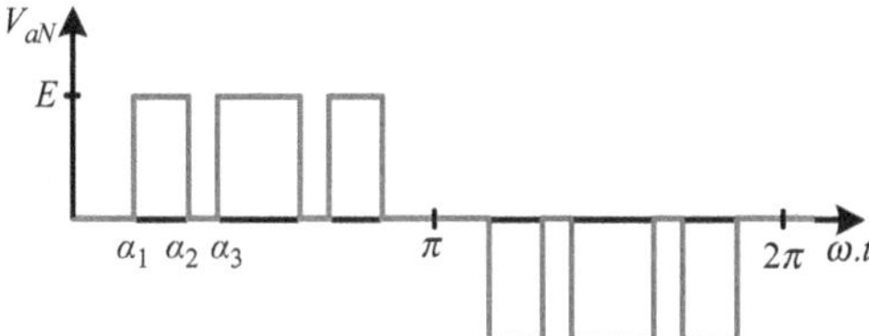

Figure 4.13 Three-level square waveform

Considering the elimination of fifth- and seventh-order harmonics, (4.19) can be yield by (4.18) [54]:

$$\begin{cases} V_1 = \frac{4E}{\pi}[\cos(\alpha_1) - \cos(\alpha_2) + \cos(\alpha_3)] \\ 0 = \cos(5\alpha_1) - \cos(5\alpha_2) + \cos(5\alpha_3) \\ 0 = \cos(7\alpha_1) - \cos(7\alpha_2) + \cos(7\alpha_3) \end{cases} \tag{4.19}$$

There are several approaches to solving the above equations. These approaches have been deeply reviewed in [55]. After solving the equations, the achieved angles are applied to the MLI to eliminate the selected harmonics' order.

4.3.5 Low-frequency modulation techniques

Nearest level control (NLC) and nearest level modulation (NLM) methods are two examples of low-frequency switching techniques and are famous for high-power applications in industries.

4.3.5.1 NLC

The NLC method compares the sinusoidal reference voltage with the possible output voltage that the MLI can produce to determine the nearest voltage level. The operating principle of the NLC method is illustrated in Figure 4.14, where V_{aN} is the closest voltage level to the reference voltage. For selecting the proper voltage level, the equation below can be used [56]:

$$v = V_{dc}.round\left\{\frac{V_{ref}}{V_{dc}}\right\} \tag{4.20}$$

In which, V_{dc} is the difference of voltages between two levels. The round function ($round\{V\}$) is determined as the closest quantity to the V. If V is in fraction and the decimal fraction of V is greater than 0.5, e.g., 1.6 it is round up to

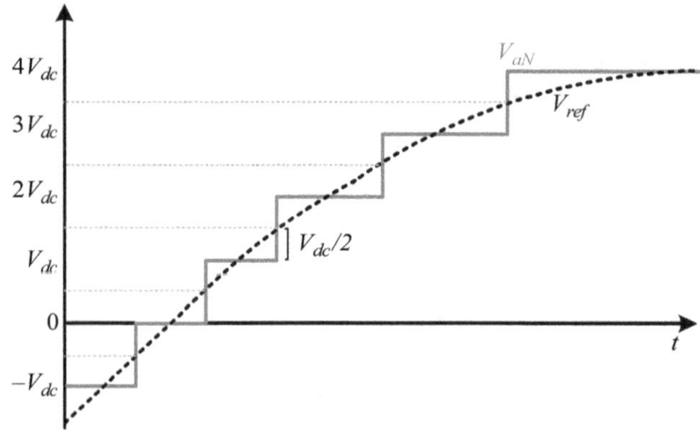

Figure 4.14 Space vector diagram of an inverter in (gh) frame [54]

the next whole number 2 and if it is lower than 1.5, it is round down to 1. The block diagram of the NLC strategy is illustrated in Figure 4.15 [54]. The NLC method is very suitable for high-level MLIs because of the convenience of calculation and minor output current ripple [54].

4.3.5.2 NLM

The NLM technique controls the phase voltage of the MLI instantly. First, the desired voltage is estimated utilizing any pair of voltage levels. Then the estimation is expanded to all the current pairs of voltage levels. Finally, the average of all the reference levels can be placed, which is then the exact quantity of the reference.

Figure 4.16 shows the basic approach of the NLM technique. In every cycle, there are two parameters in the switching state S_h of phase h (where h=A, B, or C) i.e., K_h and K_{h+1}, then, the related duty ratios are $1 - D_h$ and D_h, respectively. The desired voltage of each phase is expressed by [55]:

$$V_{h_{ref}} = (1 - D_h).\frac{K_h V_{dc}}{n-1} + D_h.\frac{(K_h + 1)V_{dc}}{n-1} = \frac{(K_h + D_h)V_{dc}}{n-1} \tag{4.21}$$

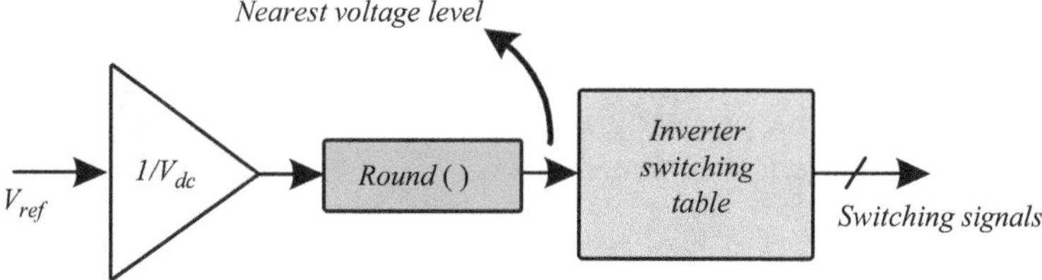

Figure 4.15 The composition of space vector in (gh) frame [54]

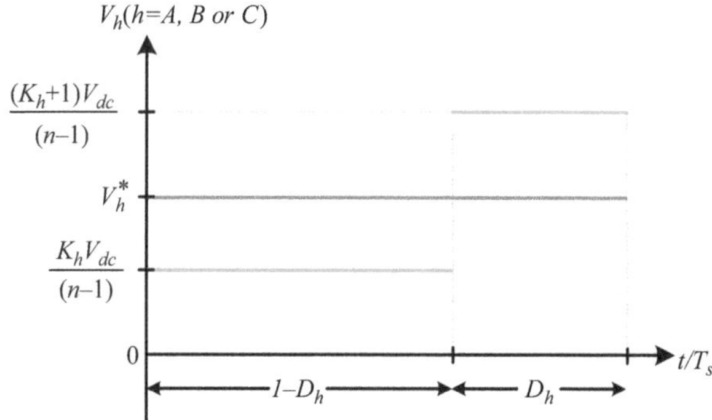

Figure 4.16 The reference voltage and the related output phase voltage during the switching cycle for the NLM method [56]

where the switching state and the duty ratio are written as [56]:

$$\begin{cases} K_h = \text{int}\left(\dfrac{\dfrac{V_{h_{ref}}}{V_{dc}}}{(n-1)}\right) \\ D_h = \dfrac{\dfrac{V_{h_{ref}}}{V_{dc}}}{(n-1)} - K_h \end{cases} \tag{4.22}$$

If the common-mode voltage is a constant value, the modulation index value must be smaller than 0.866. However, this restriction does not exist in the SVM method [56]:

$$0 \le V_{h_{ref}} \le V_{dc} \tag{4.23}$$

The equations of each phase's voltages are express by:

$$\begin{cases} V_{a_{ref}} = m\dfrac{V_{dc}}{\sqrt{3}}\cos(\theta) + V_{com} \\ V_{b_{ref}} = m\dfrac{V_{dc}}{\sqrt{3}}\cos\left(\theta - \dfrac{2\pi}{3}\right) + V_{com} \\ V_{c_{ref}} = m\dfrac{V_{dc}}{\sqrt{3}}\cos\left(\theta + \dfrac{2\pi}{3}\right) + V_{com} \end{cases} \tag{4.24}$$

By substituting the two early equations into each other, the region of the common-mode voltage is determined:

$$-\min(V_{a1}.V_{b1}.V_{c1}) \le V_{com} \le V_{dc} - \max(V_{a1}.V_{b1}.V_{c1}) \tag{4.25}$$

Therefore, the following equations can be expressed for the phase's voltages:

$$\begin{cases} V_{a1} = m\dfrac{V_{dc}}{\sqrt{3}}\cos(\theta) \\ V_{b1} = m\dfrac{V_{dc}}{\sqrt{3}}\cos\left(\theta - \dfrac{2\pi}{3}\right) \\ V_{c1} = m\dfrac{V_{dc}}{\sqrt{3}}\cos\left(\theta + \dfrac{2\pi}{3}\right) \end{cases} \tag{4.26}$$

4.3.6 Hybrid-PWM (H-PWM) technique

A hybrid-PWM (H-PWM) scheme is more suitable for the asymmetrical CHB inverters. In this modulation method, only the lowest power rate cell operates with a conventional PWM scheme. Other cells with higher voltage rates operate at low-frequency modulation. The configuration of this strategy is illustrated in Figure 4.17. It is observed from this figure that the reference voltage of the MLI is the reference signal for *m*th cell. This signal is compared to the given number of constant levels to determine the desired output voltage of the high-power cells.

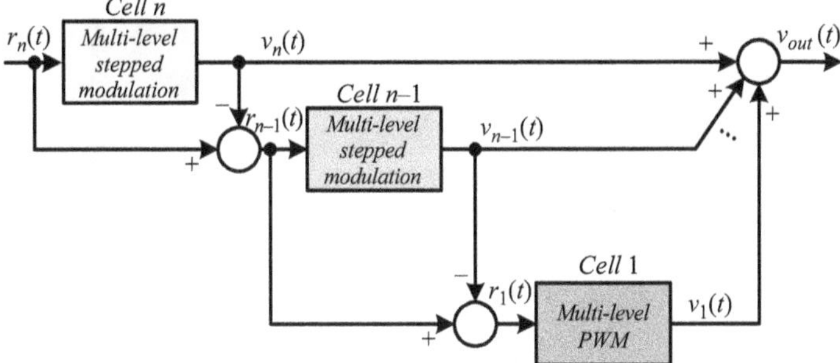

Figure 4.17 Block diagram of H-PWM

Therefore, the reference voltage signal of the mth cell is achieved by subtracting the output voltage signal of the $m + 1$ cell from its respective reference voltage signal. This reference is also compared to several constant levels. Finally, the last block is the conventional PWM method, which compares the simple high-frequency triangular carrier signal with the reference signal of the lowest power cell. It is so beneficial to compare the different aspects of modulation techniques to address their characteristics and applications for MLIs. Table 4.4 outlines the merits, demerits, and application of reviewed various modulation methods.

4.4 Case study

As stated earlier, new/modified topologies and novel/improved control/modulation methods are two main approaches for the performance enhancement of MLIs. Besides, there are several essential aspects for improvement of MLIs such as power quality issues, loss reduction, equal distribution of losses, reliability, reducing/removing dead-time, and shoot-through elimination, to name but a few. The priority and importance of the mentioned enhancement views are defined based on the application. For example, using a high switching frequency for switches in the low-voltage application is not as constrainer as for MV. In light of facets, as mentioned earlier, the 3L-NPC inverter [Figure 4.18(a)] that is one of the most popular MLIs for MV class application in the industry, and one of the recently proposed topologies, the so-called 3L-SI-NPCI^2 [6], are selected to study in this section. The feasibility of the 3L-SI-NPCI^2 shown in Figure 4.18(b) for improving/overcoming some limitations for the conventional 3L-NPC inverter, including THD issue, shoot-through problem, adverse effect dead-time, and static voltage sharing challenge among power devices are investigated and addressed as a topological solution. Also, various tests are conducted, like the abrupt changes in load, modulation index, carrier frequency, power factor, and output frequency, to validate the usefulness of the proposed strategy. In addition, both topologies are simulated in MATLAB/Simulink with the exact parameters selected from [57] represented in Table 4.5. In all the following simulation waveforms Y-axis

Table 4.4 Comparison among different modulation methods

Main modulation type	Modulation technique	Sub-modulation technique	Advantages	Disadvantages	Application
High switching frequency	LS-PWM	PD-PWM POD-PWM APOD-PWM	Excellent line voltage performance	The harmonic sidebands do not fully get canceled between the three-phase legs	Small scale application that does not need to control well
		PS-PWM	Distortion reduction Equal Power distribution between cells	Poor Reliability	Small scale application that does not need to control well
		SV-PWM	1. Reduces the common mode voltage 2. Reduces switching losses 3. Controlling the DC link voltages 4. Optimal switching sequence	1. Not suitable for large number of levels 2. Common-mode voltage variation	1. Capacitor voltage balancing 2. High cell CHBs
Low switching frequency		NLC	1. High quality output voltage level 2. Inverse relation between active power and frequency	1. No accurate mean value tracking of the reference 2. No average value calculation	High power converters
		NVC	1. Minimum space error 2. NVC control is simple	1. Low switching frequency 2. Low order harmonics generation 3. High THD and load current ripple	Higher number of output voltage levels applications

	SHE	1. Low order harmonic elimination 2. Maintain the fundamental component of the waveform 3. Reduction of switching losses	1. Low switching frequency operation 2. Maximum value of individual low order harmonics component 3. Low accuracy in case of higher number of levels	Low number of MLIs
	NLM	1. The common-mode voltage is fixed 2. Easy to implement when in case of High-level MLIs	Increasing the probity of fault	1. MMC converters 2. Multi-phase MLIs
Hybrid modulation	H-PWM	loss reduction of the inverter	Reducing the switching losses of the converter by decreasing the switching frequency of the higher power cells	1. Systems with unequal dc sources 2. Photovoltaic panels with different voltages.

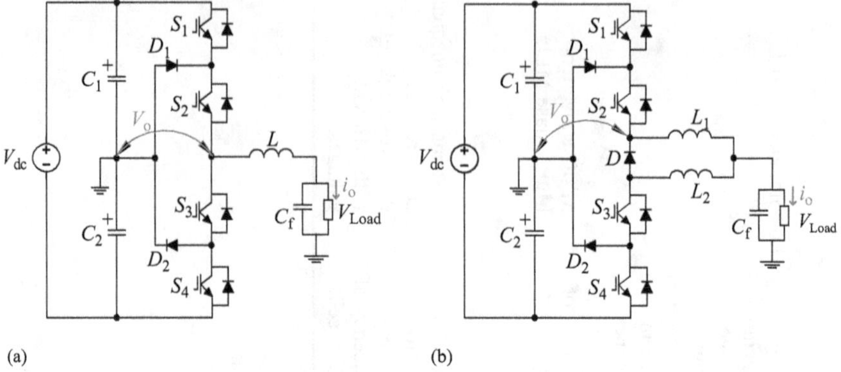

Figure 4.18 MLIs: (a) 3L-NPC. (b) 3L-SI-NPCI² [6]

Table 4.5 Simulation parameters [57]

Variable	Description	Values
V_{dc}	Input dc voltage (kV)	7
f_c	Carrier frequency (Hz)	1,430
f_o	Output frequency (Hz)	50
m	Modulation index	0.85
$C_1 = C_2$	Dc-link capacitors (μF)	2,200
L_1=L_2=L	Output filter inductor (mH)	1.2
C_f	Output filter capacitor (μF)	187
δ	Dead-time (μs)	2

illustrates the voltage in volt and current in ampere; meanwhile, X-axis is an indicator of the time in seconds.

In practice, it should be pointed out that a 6.5 kV Si-IGBT incorporating an antiparallel SiC-JBS diode can be used [58]. In general, the current and the voltage rating of the diodes and power switches are advised to be 1.5 times higher than their theoretically computed values. Because in practice, there are some small spikes created by stray inductance capacitances [59].

Several studies [58, 60] illustrated that the turn-on voltage of the SiC-JBS diode is 1V, while the turn-on voltage of the SiC body diode is near 3V, and hence, the Si body diode was effectively bypassed. Therefore, taking these two aforementioned recommendations into account, an IGBT together with an external diode are installed in parallel as a power module. Then, to meet the current safety margin on semiconductor devices, two of these power modules are paralleled in the simulation [58]. For diodes D_1 and D_2 in both inverters, two SiC-JBS diodes are parallel. Finally, DO-9 (DO-205AB) has opted for diode D in the 3L-SI-NPCI^2 structure.

4.4.1 Simulation results

Figure 4.19(a) indicates the voltage of S_1 in the classical 3L-NPC inverter without inserting dead-time among complementary power switches. Ideally, the switches of a converter can function without employing dead-time. However, in practice, it can cause short-circuit (shoot-through) problems. Therefore, to avoid this issue in the 3L-NPC topology, it is necessary to include proper dead-time among complementary power switches in PWM. By incorporating 2 μs dead-time, the voltage distribution on power switches is unbalanced. In other words, some notches are noticeable during the commutation/dead time intervals [Figure 4.19(b)]. The static voltage sharing resistor (SVSHR) [5] is used in parallel with all switches in the traditional 3L-NPC converter to lessen the adverse effect of dead-time, and the static voltage equalization is obtained [Figure 4.19(c)]. Even so, employing SVSHRs may decrease the inverter's efficiency and results in higher weight, volume, and cost. Figure 4.20 shows the voltage of S_1 in the 3L-SI-NPCI^2. In the 3L-SI-NPCI^2, inductors L_1 and L_2 will damp the shoot-through problem among all complementary switches [see Figure 4.18(b)]; thus, no dead-time interval is required in PWM. Consequently, the unbalance voltage distribution among all the switches can be eliminated (Figure 4.20). Although some small notches/spikes are observed from Figure 4.20 caused by diode D in the 3L-SI-NPCI^2, they are ignorable compared to Figure 4.19(b) and (c). Figures 4.21(a), 4.22(a), and 4.23(a)

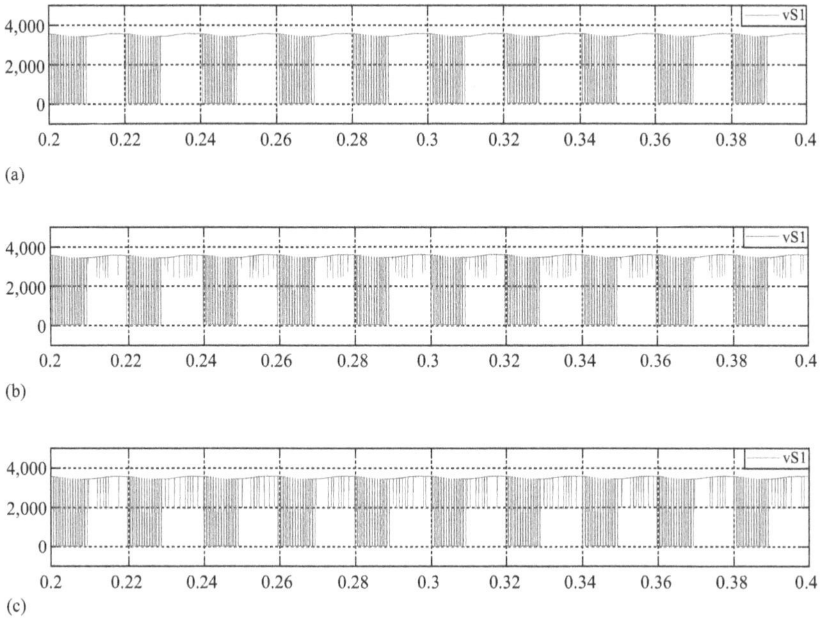

Figure 4.19 Simulation results: (a) the voltage of S_1 in 3L-NPC without introducing dead-time. (b) The voltage of S_1 in 3L-NPC by introducing 2 μs dead-time. (c) The voltage of S_1 in 3L-NPC by introducing 2 μs dead-time and utilizing SVSHRs

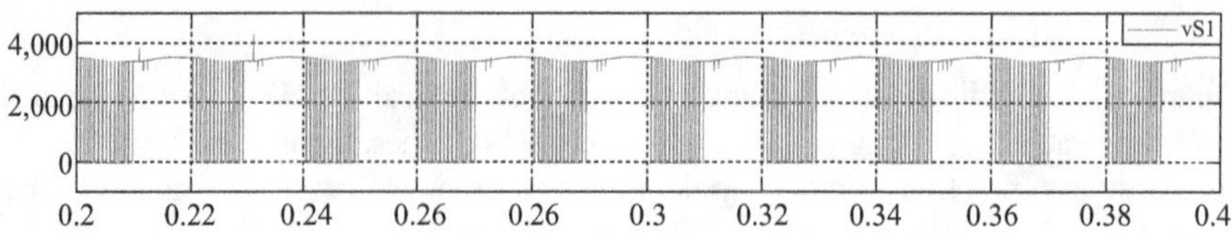

Figure 4.20 Simulation results: the voltage of S_1 in 3L-SI-NPCI²

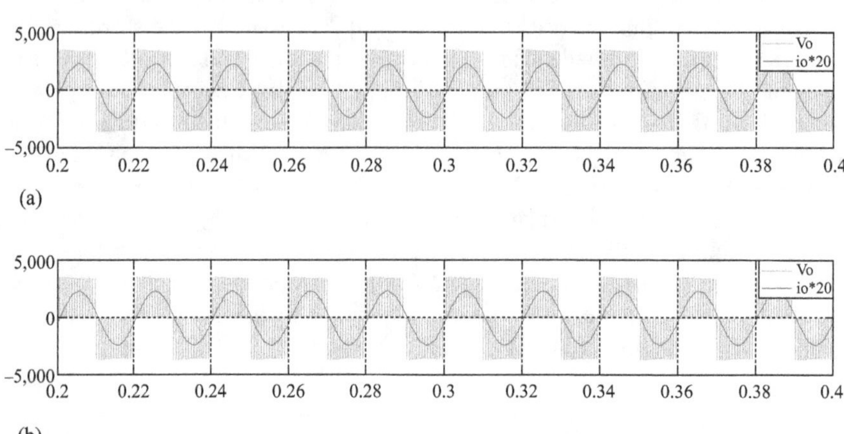

Figure 4.21 Simulation results: (a) the output voltage (V_o) and load current (i_o) of the 3L-NPC at PF= 0.98. (b) the output voltage (V_o) and load current (i_o) of the 3L-SI-NPCI² at PF= 0.98

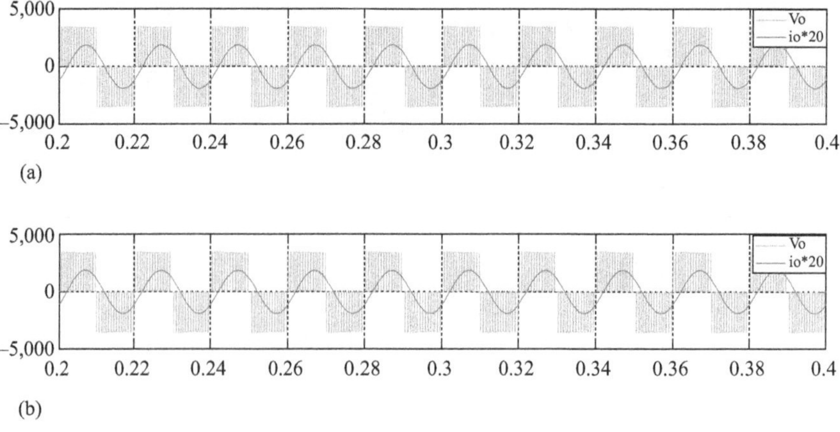

Figure 4.22 Simulation results: (a) the output voltage (V_o) and load current (i_o) of the 3L-NPC at PF= 0.8. (b) the output voltage (V_o) and load current (i_o) of the 3L-SI-NPCI² at PF= 0.8

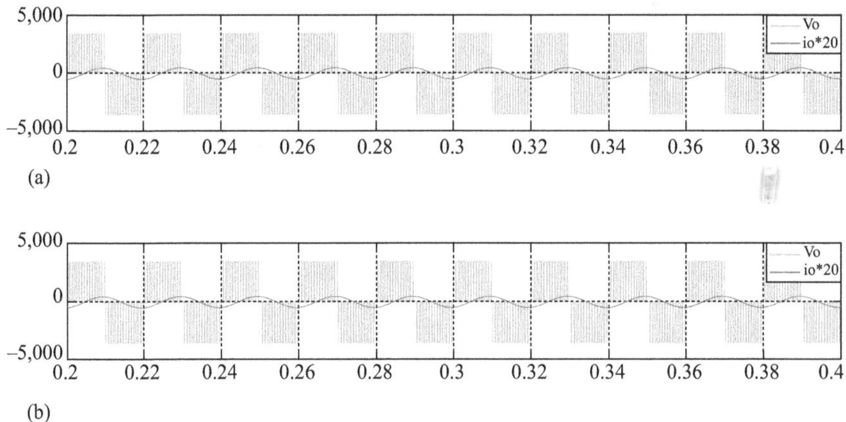

Figure 4.23 Simulation results: (a) the output voltage (V_o) and load current (i_o) of the 3L-NPC at PF= 0.2. (b) The output voltage (V_o) and load current (i_o) of the 3L-SI-NPCI^2 at PF= 0.2

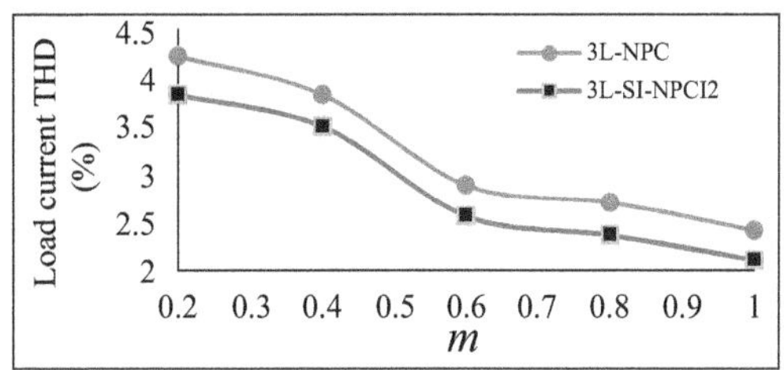

Figure 4.24 iTHD comparison between 3L-NPC and 3L-SI-NPCI^2 at PF= 0.98 and under various m

demonstrate the 3L-NPC inverter's output voltage and load current at PF= 0.98, PF= 0.8, and PF= 0.2, respectively. Likewise, the simulation waveforms of load current and voltage at PF= 0.98, PF= 0.8, and PF= 0.2 for the 3L-SI-NPCI^2 topology are shown in Figures 4.21(b), 4.22(b), and 4.23(b), respectively. From these figures and comparing the simulation waveforms of the load current between the conventional 3L-NPC and 3L-SI-NPCI^2 topologies, the better iTHD for the latter inverter can be noticed.

Figures 4.24–4.26 compare the iTHD value between the traditional 3L-NPC and 3L-SI-NPCI^2 at different PF and modulation indices (*m*). Observe that the 3L-SI-NPCI^2 has a high-quality load current over the 3L-NPC inverter.

Stress caused by transient/dynamic conditions on power converters dramatically influences the system's reliability. Thus, only the steady-state test is not enough for the applicability of the power converter. Consequently, the proposed

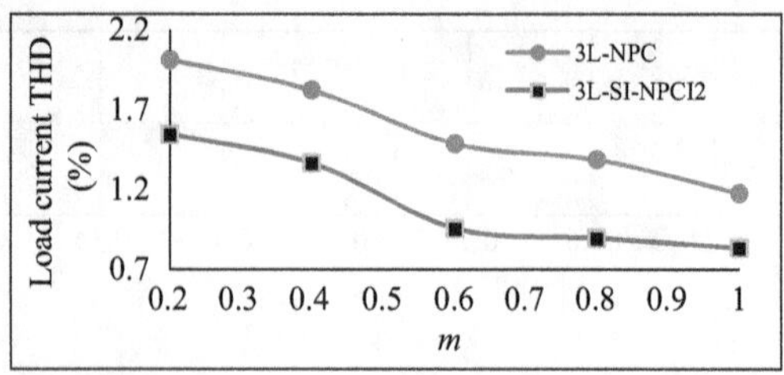

Figure 4.25 iTHD comparison between 3L-NPC and 3L-SI-NPCI² at PF= 0.8 and under various m

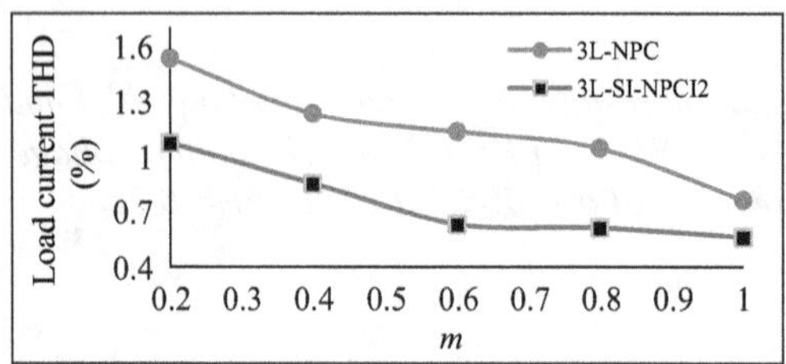

Figure 4.26 iTHD comparison between 3L-NPC and 3L-SI-NPCI² at PF= 0.2 and under various m

converters should handle a wide range of dynamic conditions. Accordingly, the dynamic performance of the 3L-SI-NPCI^2 is tested in various operating conditions where V_o is the inverter's output voltage before the filter, and i_o depicts the load current [see Figure 4.18(b)]. Figure 4.27(a) illustrates the sudden change in the output load (PF) from PF= 0.2 to PF= 0.98. Initially, the inverter functions under a lower power factor (PF= 0.2), and at time t= 665 ms, the PF changes from 0.2 to 0.98. An excellent smooth and rapid response can be observed during the load change.

The performance of the 3L-SI-NPCI^2 is also examined under modulation index (m) change. The proper operation of the power converters under lower modulation indices, notably lower than 0.5, is crucial. To this end, the dynamic behavior of the 3L-SI-NPCI^2 is measured under harsh conditions by changing m from 0.2 to 0.85. Figure 4.27(b) and (c) shows the simulation results of these dynamic tests with two different power factors (PF= 0.98 and PF= 0.8). Although a slight distortion in load current is observable 1.5 cycles after the change (t= 305 ms), it is acceptable as it is very severe.

In motor drive applications, the electrical motor needs to runs from low speed to high/rated speed. Hence, the proper operation of inverters at a lower output frequency (f_o) is essential. Moreover, to operate the power converters at lower than 20 Hz adequately, a robust control/modulation strategy is required. Figure 4.27(d)

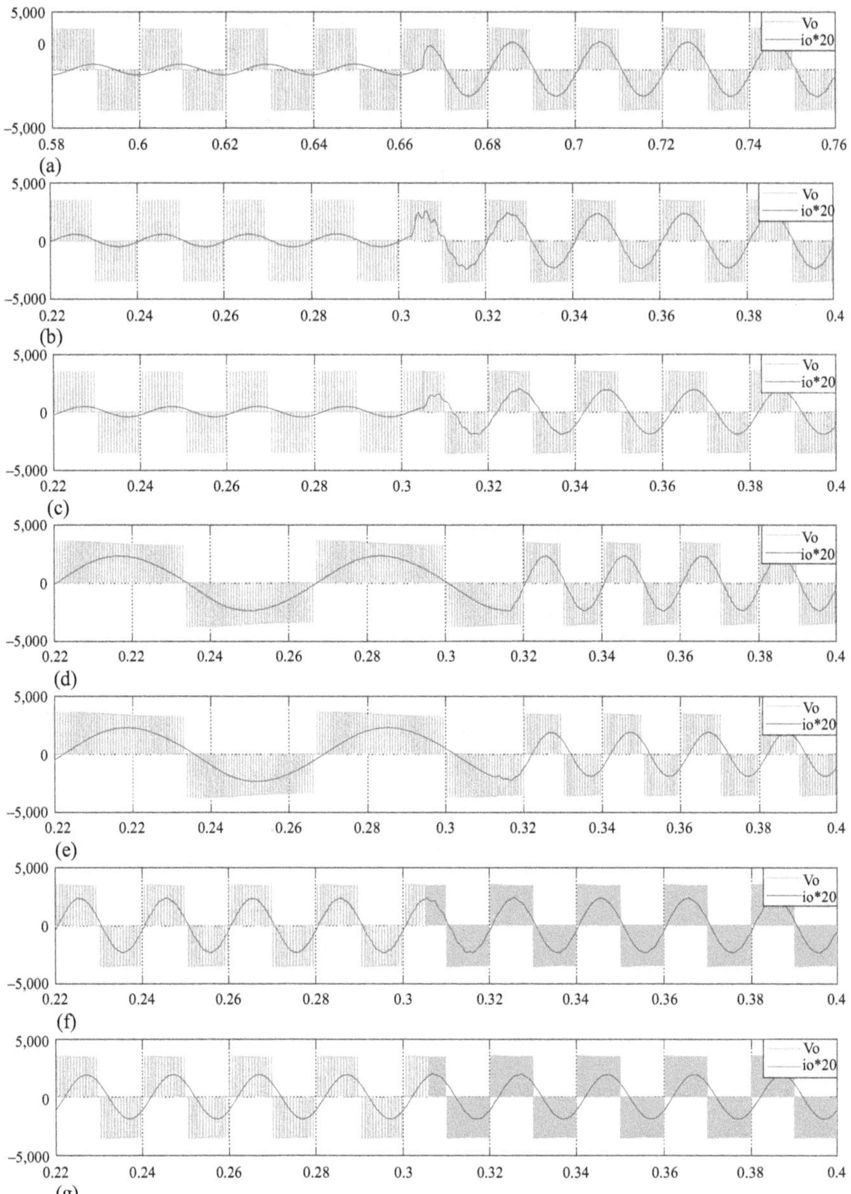

Figure 4.27 Simulation results: (a) sudden change in load (PF) from PF= 0.2 to PF= 0.98. (b) Sudden change in modulation index (m) from m= 0.2 to m= 0.85 at PF= 0.98. (c) Sudden change in modulation index (m) from m= 0.2 to m= 0.85 at PF= 0.8. (d) Sudden change in output frequency (f_o) from f_o= 15 Hz to f_o= 50 Hz at PF= 0.98. (e) Sudden change in output frequency (f_o) from f_o= 15 Hz to f_o= 50 Hz at PF= 0.8. (f) Sudden change in carrier frequency (f_c) from f_c= 1,430 Hz to f_c= 5 kHz at PF= 0.98. (g) Sudden change in carrier frequency (f_c) from f_c= 1,430 Hz to f_c= 5 kHz at PF= 0.8

and (e) shows the performance of the 3L-SI-NPCI^2 against the sudden change in output frequency, f_o = 15 Hz to 50 Hz at about t = 315 ms. In brief, a resilience performance of the inverter is observed under this test at different PF.

The behavior of the 3L-SI-NPCI^2 is investigated against the sudden change in the carrier frequency (f_c). From Figure 4.27(f), in the beginning, the 3L-SI-NPCI^2 operates in normal conditions (f_c= 1,430 Hz) with a load power factor, PF = 0.98. At time t = 305 ms, the carrier frequency changes to f_c= 5 kHz. Notwithstanding a minor distribution in load current after a change that takes about one and a half cycles, the response is acceptable since the difference is extreme. The same test is done at PF= 0.8 and the appropriate reaction of the 3L-SI-NPCI^2 can be noticed in Figure 4.27(g).

4.4.2 Experimental results [6]

Figure 4.28 illustrates the experimental waveforms of the 3L-SI-NPCI^2 that have been obtained for a low-power prototype. The experimental parameters are listed in Table 4.6. The input dc source voltage and dc-link capacitors are shown in Figure 4.28(a). In this figure, channels #1, #3, and #4 represent the voltages of the C_1, C_2, and the input dc source, respectively. Figure 4.28(b) shows the inverter and load voltage. The output/load current and the inductor currents are depicted in Figure 4.28(c). The high-quality waveform for output current is visible in this figure. From Figure 4.28(c), the inductor current's low ripple can be observed. Figure 4.28(d)–(g) depicts the voltage stress on power devices, including (S_1, S_2), (S_3, S_4), (D_1, D_2), and (D), respectively.

4.4.3 Case study discussion

According to the following information/recommendation/benchmark for MV application and addressing some shortcomings/limitations associated with the conventional 3L-NPC converter, one of the recently proposed topologies (3L-SI-NPCI^2) has been opted and studied by the simulation and experimental results:

- The split-inductor concept [6, 61] can reduce/eliminate the following adverse effects of the dead-time:

1. Reduction in the voltage/current's gain [61–65].
2. Even employing a tiny dead time (e.g., less than 1%) can dramatically deteriorate quality of waveforms. In MV applications, the dead time value is larger [66, 67]. Furthermore, high harmonics generated by converters cause extra losses and reduce the motor's efficiency [5, 68]. Furthermore, THD contributes to the motor's higher temperature, damaging the insulation material [5, 68].
3. Degrading the converter's performance.
4. Decreasing the input DC of the inverter [69].
5. Detrimental effect on common-mode voltage [23].
6. Unbalanced/uneven voltage distribution in conventional NPC structure [5]. Additional components are required to tackle the issue [5, 61].

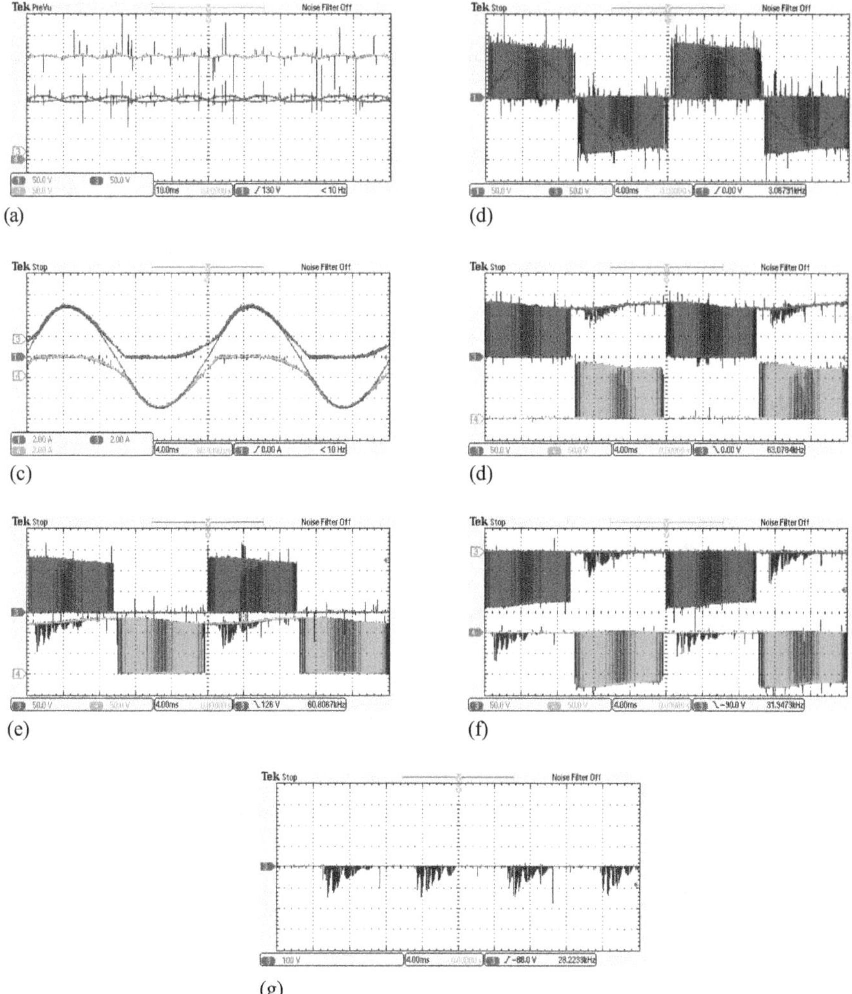

Figure 4.28 Experimental results. (a) Input and dc-link capacitors voltage. (b) V_{out} and V_{inv}. (c) i_{out} and $i_{L1\&L2}$. (d) V_{S1} and V_{S2}. (e) V_{S3} and V_{S4}. (f) V_{D1} and V_{D2}. (g) V_D

- Transient conditions such as a sudden change in load, power factor, and modulation index can cause stress in MLIs, which severely influence the reliability of the converter. However, from the simulation results, it can be observed that the proposed method can handle these changes appropriately (see Figure 4.27).
- In the variable speed drive, an electrical motor operates at a wide range of speeds. Hence, the inverter must function from low to high frequencies [5, 61]. The simulation study investigates this requirement.

Table 4.6 Experimental parameters [6]

Variable	Description	Values
V_{dc}	Input dc voltage (V)	250
f_c	Carrier frequency (kHz)	20
f_o	Output frequency (Hz)	50
m	Modulation index	0.85
$C_1 = C_2$	Dc-link capacitors (μF)	940 μF (in parallel with $R_1=R_2= 20$ kΩ)
$L_1=L_2=L$	Output filter inductor (mH)	5
C_f	Output filter capacitor (μF)	9.4
R	Load resistance (Ω)	20

- Efficiency is also another essential aspect of MV converter topologies. Since in the 3L-SI-NPCI^2, employing voltage equalization techniques/components are not needed, the 3L-SI-NPCI^2 may have competitive efficiency in MV application. This topic is comprehensively elaborated in [61] for different MLIs.

4.5 Advancements, challenges, and future trends

It is essential to mention that standards and codes regarding TRL PV grid-connected systems are becoming more and more strict [1, 70–73]. For example, the operation of the next generation of TRL PV inverters should not be limited to PF= 1. Also, it must function appropriately at low-voltage ride-through (LVRT) under grid faults condition. Furthermore, the study in [74] confirmed that even though the HERIC inverter is one of the best candidates in terms of efficiency, it is not very feasible in the case of voltage sag. Therefore, in deriving new TRL PV topologies, mentioned recommendations should be considered.

The extensive use of high-frequency switching devices in power converters, especially in MV applications, is agreeable for rapid voltages and current transitions [4]. However, using a high switching frequency led to higher power losses and the creation of common-mode (CM) currents, EMI, and degradation of motor and transformer insulation, to name but a few. Thus, the distribution of power device losses is a crucial issue, and this aspect can be considered in extracting new topologies or modulation methods [8–14].

An isolation transformer in some applications, e.g., MV drive, is responsible for 30–50% of total drive size and 50–70% of the system's weight and increasing raw material costs, cooling, etc., [4]. Hence, the TRL MLIs are an exciting solution to save energy dramatically. Likewise, TRL MLIs are predicted to alleviate CMV and generate substantially sinusoidal output waveform and restricted *dv/dt* [4]. Furthermore, this phenomenon can be reduced further by removing/reducing dead-time [6, 23, 61].

Notably, most contributions related to high or MV multilevel inverter topologies are obtained by low-power prototypes, especially in academia.

Besides, it is admissible to use a power device with high average switching frequencies (e.g., several kilohertz) in these low-power prototypes. However, in contrast, the lower switching frequency is recommended in actual application due to limitations, e.g., cooling system, and power loss issues [4]. Therefore, from the real/commercial point of view, the experimental result obtained from these low-power prototypes may not be satisfactory enough. Furthermore, although in most of these contributions, simulation studies/results are provided with realistic specifications, many of these simulation results are obtained in ideal conditions (e.g., the dead-time is ignored in the simulation). Therefore, an extensive effort (at least by simulation study) should be made to provide results with more realistic conditions.

Other issue that should be considered is the stress on converters induced by a transient/dynamic condition (e.g., a sudden change in load). The transient condition has a remarkable impact on the reliability of the power converters [75]. Thus, a wide range of dynamic tests such as a sudden change in loads at different modulation indices (m= 0.1–1), PF form 0.1–1 at m= 0.1–1, and other switching and output frequency should be conducted to verify the effectiveness and feasibility of the new topologies.

Diodes and bidirectional power switches can plausibly decrease the converter efficiency [76]. For example, during turnoff of diodes, they create high reverse recovery currents. This phenomenon can contribute high switching losses in other power devices. The methodology represented in [58, 61] can be helpful in this regard when extracting new/modified MLI topologies.

The size and the weight reduction are other essential aspects that can stimulate further developments in MLIs [76]. Although weight and space limitations are not critical in some industrial applications (e.g., compressors, pumps, and fans), there are several applications in which mentioned aspects are crucial (e.g., marine propulsion and train traction). Extracting topologies like NNPC, NT^2, and ANNPC MLIs can be helpful in this regard [7, 26–28].

4.6 Conclusion

Nowadays, multilevel inverter topologies have gained a maturity level that confirms their industrial existence and practical application. Nonetheless, the considerable number of current publications on the subject reveals that there is still plenty of room for further enhancement. Therefore, in this chapter, after discussing the advantages and disadvantages of traditional MLIs, more than 20 state-of-the-art topologies have been reviewed from a critical point of view. Although many of these contributions have not been commercialized yet, reviewing these structures may give a clear picture of their merits and demerits to the potential readers to develop new solutions. Moreover, several modulation methods have been surveyed as well. In addition, some challenges associated with MLIs have been highlighted to encourage future works that tackle open limitations and explore novel possibilities.

References

[1] F. Faraji, S. M. Mousavi, G. A. Hajirayat, A. A. Motie Birjandi, and K. Al-Haddad, "Single-stage single-phase three-level neutral-point-clamped transformerless grid-connected photovoltaic inverters: topology review," *ELSE., Renew. Sustain. Energy Rev.*, vol. 80, pp. 197–214, 2017.

[2] S. M. Mousavi, G., F. Faraji, A. Majazi, and K. Al-Haddad, "A comprehensive review of flywheel energy storage system technology," *ELSE. Renew. Sustain. Energy Rev.*, vol. 67, pp. 477–490, 2017.

[3] J. Rodríguez, S. Bernet, B. Wu, J. O. Pontt, and S. Kouro, "Multilevel voltage-source-converter topologies for industrial medium-voltage drives," *IEEE Trans. Ind. Electron.*, vol. 54, no. 6, pp. 2930–2945, 2007.

[4] H. Abu-Rub, J. Holtz, J. Rodriguez, and G. Baoming, "Medium-voltage multilevel converters-state of the art, challenges, and requirements in industrial applications," *IEEE Trans. Ind. Electron.*, vol. 57, no. 8, pp. 2581–2596, 2010.

[5] B. Wu and M. Narimani, *High-Power Converters and AC Drives*. John Wiley & Sons, New York, NY, 2017.

[6] F. Faraji, A. A. M. Birjandi, S. M. Mousavi G, J. Zhang, B. Wang, and X. Guo, "An improved multilevel inverter for single-phase transformerless PV system" *IEEE Trans. Energy. Convers.*, vol. 36, no. 1, pp. 281–290, 2021.

[7] F. Faraji, A. Hajirayat, and M. Narimani, "Fault tolerant T-type nested neutral point clamped (T-NNPC) converter," in *IEEE Southern Power Electronics Conference (SPEC)*. Puerto Varas, Chile, December 2017.

[8] A. M. Y. M. Ghias, J. Pou, V. G. Agelidis, and M. Ciobotaru, "Voltage balancing method for a flying capacitor multilevel converter using phase disposition PWM," *IEEE Trans. Ind. Electron.*, vol. 61, no. 12, 2014.

[9] A. M. Y. M. Ghias, J. Pou, V. G. Agelidis, and M. Ciobotaru, "Optimal switching transitions-based voltage balancing method for flying capacitor multilevel converters," *IEEE Trans. Power. Electron.*, vol. 30, no. 4, pp. 4521–4531, 2015.

[10] A. M. Y. M. Ghias, J. Pou, G. J. Capella, V. G. Agelidis, R. P. Aguilera, and T. Meynard, "Single-carrier phase-disposition PWM implementation for multilevel flying capacitor converters," *IEEE Trans. Power. Electron.*, vol. 30, no. 10, 2015.

[11] A. K. Sadigh, V. Dargahi, and K. A. Corzine, "New active capacitor voltage balancing method for flying capacitor multicell converter based on logic-form-equations," *IEEE Trans. Ind. Electron.*, vol. 64, no. 5, 2017.

[12] A. M. Y. M. Ghias, J. Pou, G. J. Capella, P. Acuna, and V. G. Agelidis, "On improving phase-shifted PWM for flying capacitor multilevel converters," *IEEE Trans. Power. Electron.*, vol. 31, no. 8, 2016.

[13] G. Farivar, A. M. Y. M. Ghias, B. Hredzak, J. Pou, and V. G. Agelidis, "Capacitor voltages measurement and balancing in flying capacitor multilevel converters utilizing a single voltage sensor," *IEEE Trans. Power. Electron.*, vol. 32, no. 10, pp. 8115–8123, 2014.

[14] A. M. Y. M. Ghias, J. Pou, M. Ciobotaru, and V. G. Agelidis, "Voltage-balancing method using phase-shifted PWM for the flying capacitor multilevel converter," *IEEE Trans. Power. Electron.*, vol. 29, no. 9, 2014.

[15] P. Acuna, R. P. Aguilera, A. M.Y.M. Ghias, M. Rivera, C. R. Baier, and V. G. Agelidis, "Cascade-free model predictive control for single-phase grid-connected power converters," *IEEE Trans. Ind. Electron.*, vol. 64, no. 1, pp. 285–294, 2017.

[16] A. Merabet, L. Labib, and A. M.Y.M. Ghias, "Robust model predictive control for photovoltaic inverter system with grid fault ride-through capability," *IEEE Trans. Smart. Grid.*, vol. 9, no. 6, pp. 5699–5709, 2017.

[17] H. Zhao, S. Wang, and A. Moeini, "Critical parameter design for a cascaded H-bridge with selective harmonic elimination/compensation based on harmonic envelope analysis for single-phase systems," *IEEE Trans. Ind. Electron.*, vol. 66, no. 4, pp. 2914–2925, 2018.

[18] M. Sharifzadeh, H. Vahedi, R. Portillo, L. Garcia Franquelo, and k. Al-Haddad, "Selective harmonic mitigation based self-elimination of triplen harmonics for single-phase five-level inverters," *IEEE Trans. Power. Electron.*, vol. 34, no. 1, pp. 86–96, 2018.

[19] M. Wu, Y. Wei Li, and G. Konstantinou, "A comprehensive review of capacitor voltage balancing strategies for multilevel converters under selective harmonic elimination PWM," *IEEE Trans. Power. Electron.*, vol. 36, no. 3, pp. 2748–2767, 2020.

[20] M. Hossein Etesami, D. Mahinda Vilathgamuwa, N. Ghasemi, and D. P. Jovanovic, "Enhanced metaheuristic methods for selective harmonic elimination technique," *IEEE Trans. Ind. Electron.*, vol. 14, no. 12, pp. 5210–5220, 2018.

[21] M. Hoang Nguyen, and S. Kwak, "Nearest-level control method with improved output quality for modular multilevel converters," *IEEE Access.*, vol. 8, 2020.

[22] S. Jiang, K. Ma, Y. Yang, and X. Cai, "Mission profile emulator for individual submodule in modular multilevel converter with nearest level control," *IEEE Trans. Power. Electron.*, vol. 36, no. 9, pp. 9722–9730, 2021.

[23] Z. Shen and D. Jiang, "Dead-time effect compensation method based on current ripple prediction for voltage source inverters," *IEEE Trans. Power. Electron.*, vol. 34, no. 1, pp. 971–983, 2019.

[24] M. Narimani, B. Wu, Z. Cheng, and N. R. Zargari, "Multilevel voltage source converters and systems," US Patent 9083230, July 14, 2015.

[25] M. Narimani, B. Wu, Z. Cheng, and N. R. Zargari, "A new nested neutral point-clamped (NNPC) converter for medium-voltage (MV) power conversion," *IEEE Trans. Power. Electron.*, vol. 29, no. 12, pp. 6375–6382, 2014.

[26] M. Narimani, B. Wu, and N. R. Zargari, "A novel five-level voltage source inverter with sinusoidal pulse width modulator for medium-voltage applications," *IEEE Trans. Power. Electron.*, vol. 31, no. 3, 2016.

[27] A. Bahrami and M. Narimani, "A new five-level t-type nested neutral point clamped (T-NNPC) *converter," IEEE Trans. Power Electron.*, vol. 34, no. 11, 2019.

[28] A. M. Y. M. Ghias, P. Acuna, J. Lu, and A. Merabet, "Finite control set model predictive control of an active nested neutral-point-clamped converter," in *IEEE 4th Southern Power Electronics Conference (SPEC)*, 2018.

[29] N. Sandeep, J. Sathik, U. Yaragatti, and V. Krishnasamy, "A self-balancing five-level boosting inverter with reduced components," *IEEE Trans. Power Electron.*, vol. 34, no. 5, pp. 6020–6024, 2018.

[30] L. Zhang, Z. Zheng, and C. Li, "A Si / SiC hybrid five-level active NPC inverter," *IEEE Trans. Power. Electron.*, vol. 35, no. 5, pp. 4835–4846, 2020.

[31] R. Barzegarkhoo, S. Sing Lee, Y. P. Siwakoti, S. Ahamed Khan, and F. Blaabjerg, "Design, control, and analysis of a novel grid-interfaced switched-boost dual t-type five-level inverter with common-ground concept," *IEEE Ind. Power. Electron.*, vol. 68, no. 9, pp. 8193–8206, 2020.

[32] S. Kumari, H. R. Pota, A. K. Verma, N. Sandeep, and U. R. Yaragatti, "An eight-switch five-level inverter with zero leakage current," *IET Power. Electron.*, vol. 14, no. 3, pp. 590–601, 2021.

[33] M. Nguyen and T. Tran, "Quasi cascaded H-bridge five-level boost inverter," *IEEE Trans. Ind. electron.*, vol. 64, no. 11, pp. 8525–8533, 2017.

[34] F. Gao, "An enhanced single phase step-up five-level inverter," *IEEE Trans. Power Electron.*, vol. 31, no. 12, pp. 8024–8030, 2016.

[35] H. Vahedi, M. Sharifzadeh, and K. Al-haddad, "Modified seven-level pack U-cell inverter for photovoltaic applications," *IEEE J. Emerg. Selected Top. Power Electron.*, vol. 6, no. 3, pp. 1508–1516, 2018.

[36] M. J. Sathik, K. Bhatnagar, N. Sandeep, and F. Blaabjerg, "An improved seven-level PUC inverter topology with voltage boosting," *IEEE Trans. Circuits Syst. II Express Briefs*, vol. 67, no. 1, pp. 127–131, 2019.

[37] M. J. Sathik, N. Sandeep, and F. Blaabjerg, "High gain active neutral point clamped seven-level self-voltage balancing inverter," *IEEE Trans. Circuits Syst. II Express Briefs.*, vol. 67, no. 11, pp. 2567–3571, 2019.

[38] M. J. Sathik, N. Sandeep, M. Daula Siddique, D. Almakhles, and S. Mekhilef, "Compact seven-level boost type inverter topology," *IEEE Trans. Circuits Syst. II Express Briefs*, vol. 68, no. 4, pp. 1358–1362, 2020.

[39] N. Sandeep and U. R. Yaragatti, "Operation and control of a nine-level modified ANPC inverter topology with reduced part count for grid-connected applications," *IEEE Trans. Ind. Electron.*, vol. 65, no. 6, pp. 4810–4818, 2018.

[40] N. Sandeep and U. R. Yaragatti, "Design and implementation of active neutral-point-clamped nine-level reduced device count inverter: an application to grid integrated renewable energy sources," *IET Power Electron.*, vol. 11, no. 1, 2017.

[41] D. Niu, G. Feng, W. Panrui, Z. Kangjia, Q. Futian, and M. Zhan. "A nine-level T-type packed U-cell inverter," *IEEE Trans. Power Electron.*, vol. 35, no. 2, pp. 1171–1175, 2019.

[42] M. Abarzadeh and K. Al-haddad, "An improved active-neutral-point-clamped converter with new modulation method for ground power unit application," *IEEE Trans. Ind. Electron.*, vol. 66, no. 1, pp. 203–214, 2018.

[43] A. Dekka, A. Ramezani, S. Ounie, and M. Narimani, "A new five-level voltage source inverter: modulation and control," *IEEE Trans. Ind. Appl.*, vol. 56, no. 5, 2020.

[44] J. W. Zapata, G. Postiglione, D. Falchi, G. Borghetti, T. A. Meynard, and G. Gateau, "Multilevel converter for 4.16 and 6.6 kV variable speed drives," *IEEE Trans. Power. Electron.*, vol. 36, no. 3, pp. 3172–3180, 2021.

[45] X. Zhu, H. Wang, X. Deng, W. Zhang, H. Wang, X. Yue, "Coupled three-phase converter concept and an example: a coupled ten-switch three-phase three-level inverter," *IEEE Trans. Power Electron.*, vol. 36, no. 6, pp. 6475–6468, 2021.

[46] J. Chen, Y. Zhong, C. Wang, and Y. Fu, "Four-level hybrid neutral point clamped converters," *IEEE J. Emerg. Selected Top. Power Electron.*, vol. 9, no. 4, 2021.

[47] S. S. Lee, "A single-phase single-source 7-level inverter with triple voltage boosting gain," *IEEE Access*, vol. 6, pp. 30005–30011, 2018.

[48] I. Harbi, M. Ahmed, J. Rodr´ıguez, R. Kennel, and M. Abdelrahem, "A nine-level t-type converter for grid-connected distributed generation," *IEEE J. Emerg. Selected Top. Power Electron.*, Apr. 2022.

[49] J. C. Kartickl, B. K. Sujit, and K.C. Suparna, "Dual reference phase shifted pulse width modulation technique for a N-level inverter based grid connected solar photovoltaic system," *IET Power Electron.*, vol. 10, no. 7, pp. 928–935, 2016.

[50] S. K. Mondal, B. K. Bose, V. Oleschuk, and J. O. P. Pinto, "Space vector pulse width modulation of three-level inverter extending operation into overmodulation region," *IEEE Trans. Power Electron.*, vol. 18, no. 2, pp. 604–611, 2003.

[51] P. Madasamy, R. K. Pongiannan, S. Ravichandran, *et al*, "A simple multilevel space vector modulation technique and MATLAB system generator built FPGA implementation for three-level neutral-point clamped inverter". *Energies*, vol. 12, no. 22, pp. 4332, 2019.

[52] C. Surajit, M. Mitra, and S. Sengupta, "Clarke and park transform," *Electric Power Quality, Springer, New York, NY*, 2011.

[53] Q. M. Attique, Y. Li, and K. Wang, "A survey on space-vector pulse width modulation for multilevel inverters," *CPSS Trans. Power Electron. Appl.*, vol. 2, no. 3, pp. 226–236, 2017.

[54] Kamaldeep and J. Kumar, "Notice of removal: performance analysis of H-bridge multilevel inverter using selective harmonic elimination and nearest level control technique," in *IEEE 2015 International Conference on Electrical, Electronics, Signals, Communication and Optimization (EESCO)*, Jan. 2015, pp. 24–25.

[55] M. S. A. Dahidah, G. Konstantinou, and V. G. Agelidis, "A review of multilevel selective harmonic elimination PWM: formulations, solving algorithms, implementation and applications," *IEEE Trans. Power Electron.*, vol. 30, no. 8, pp. 4091–4106, 2015.

[56] Y. Deng and R. G. Harley, "Space-vector versus nearest-level pulse width modulation for multilevel converters," *IEEE Trans. Power Electron.*, vol. 30, no. 6, pp. 2962–2974, 2015.

[57] S. S. Fazel, S. Bernet, D. Krug, and K. Jalili, "Design and comparison of 4-kV neutral-point-clamped, flying-capacitor, and series-connected H-bridge multilevel converters," *IEEE Trans. Ind. Appl.*, vol. 43, no. 4, 2007.

[58] H. Mirzaee, A. De, A. Tripathi, and S. Bhattacharya, "Design comparison of high-power medium-voltage converters based on a 6.5-kV Si-IGBT/Si-PiN Diode, a 6.5-kV Si-IGBT/SiC-JBS Diode, and a 10-kV SiCMOSFET/SiC-JBS Diode," *IEEE Trans. Ind. Appl.*, vol. 50, no. 4, pp. 2728–2740, 2017.

[59] Y. P. Siwakoti and F. Blaabjerg, "Common-ground-type transformerless inverters for single-phase solar photovoltaic systems," *IEEE Trans. Ind. Electron.*, vol. 65, no. 3, 2018.

[60] A. Agarwal, H. Fatima, S. Haney, and S.-H. Ryu, "A new degradation mechanism in high-voltage SiC power MOSFETs," *IEEE Electron Device Lett.*, vol. 28, no. 7, pp. 587–589, 2007.

[61] F. Faraji, A. M. Y. M. Ghias, X. Guo, Z. Chen, Z. Lu, and C. Hua, "A split-inductor flying capacitor converter for medium-voltage application," *IEEE Journal of Emerging and Selected Topics in Power Electronics*, early access, 2022.

[62] F. Akbar, H. Cha, H. F. Ahmed, and A. A. Khan, "A family of single-stage high-gain dual-buck split-source inverters," *IEEE J. Emerg. Selected Top. Power Electron.*, vol. 8, no. 3, 2020.

[63] A. Ali Khan, H. Cha, and J.-S. Lai, "Cascaded dual-buck inverter with reduced number of inductors," *IEEE Trans. Power. Electron.*, vol. 33, no. 4, 2018.

[64] A. Ali Khan, Y. W. Lu, W. Eberle, *et al*, "Single-stage bidirectional buck–boost inverters using a single inductor and eliminating the common-mode leakage current," *IEEE Trans. Power Electron.*, vol. 35, no. 2, 2020.

[65] A. Ali Khan, H. Cha, H. F. Ahmed, J. Kim, and J. Cho, "A highly reliable and high-efficiency quasi single-stage buck–boost inverter," *IEEE Trans. Power Electron.*, vol. 32, no. 6, 2017.

[66] F. Chierchie, L. Stefanazzi, E. E. Paolini, and A. R. Oliva, "Frequency analysis of PWM inverters with dead-time for arbitrary modulating signals," *IEEE Trans. Power. Electron.*, vol. 29, no. 6, pp. 2850–2860, 2014.

[67] L. Chen and F. Zheng Peng, "Dead-time elimination for voltage source inverters," *IEEE Trans. Power. Electron.*, vol. 23, no. 2, pp. 574–580, 2008.

[68] The next generation of variable frequency drives efficiency without draw-backs (https://smartd.tech/), 2020.

[69] A. Guha and G. Narayanan, "Impact of dead-time on inverter input current, dc-link dynamics and light-load instability in rectifier-inverter-fed induction motor drives," *IEEE Trans. Ind. Appl.*, vol. 99, p. 1, 2017.

[70] W. Li, Y. Gu, H. Luo, W. Cui, X. He, and C. Xia, "Topology review and derivation methodology of single phase transformerless photovoltaic inverters for leakage current suppression," *IEEE Trans. Ind. Electron.*, vol. 62, no. 7, pp. 4537–4551, 2015.

[71] D. Meneses, F. Blaabjerg, O. García, and José A. Cobos, "Review and comparison of step-up transformerless topologies for photovoltaic ac-module

application," *IEEE Trans. Power. Electron.*, vol. 28, no. 6, pp. 2649–2663, 2013.

[72] J. M. Carrasco, L. G. Franquelo, J. T. Bialasiewicz, *et al.*, "Power-electronic systems for the grid integration of renewable energy sources: a survey," *IEEE Trans. Ind. Electron.*, vol. 53, no. 4, pp. 1002–1016, 2006.

[73] H. Patel and V. Agarwal, "A single-stage single-phase transformerless doubly grounded grid-connected PV interface," *IEEE Trans. Energy Convers.*, vol. 24, no. 1, pp. 93–101, 2009.

[74] Y. Yang, F. Blaabjerg, and H. Wang, "Low-voltage ride-through of single-phase transformerless photovoltaic inverters," *IEEE Trans. Ind. Appl.*, vol. 50, no. 3, pp. 1942–1952, 2014.

[75] S. Yang, A. Bryant, P. Mawby, D. Xiang, L. Ran, and P. Tavner, "An industry-based survey of reliability in power electronic converters," *IEEE Trans. Ind. Appl.*, vol. 47, no. 3, pp. 1441–1451, 2011.

[76] S. Couro, M. Malinowski, K. Gopakumar, *et al.*, "Recent advances and industrial applications of multilevel converters," *IEEE Trans. Ind. Electron.*, vol. 57, no. 8, pp. 2553–2580, 2010.

Chapter 5

Multilevel inverters: topologies and optimization

Ebrahim Babaei[1] *and Mohammadamin Aalami*[1]

Generally, there are four sorts of power electronic converters based on their input and output sides, which can be either AC or DC. The DC/DC converters are mainly utilized as a magnitude changer which can either increase or decrease the magnitude of the input DC voltage on the output side. If the output voltage magnitude exceeds the input side, this converter is called a boost one. Otherwise, it is called a buck converter. An AC/AC converter can be used to change the magnitude and frequency of an AC voltage. Furthermore, AC/DC rectifiers can be used in order to extract a DC voltage from an AC input. Another type of power electronic converters used to generate an AC output from a DC input is called an inverter. These converters generate a sinusoidal-like output using the input source and an appropriate control method. The similarity of the generated waveform with an ideal sinusoidal shape is measured by a parameter called total harmonic distortion (THD). The less THD means that the output waveform is more similar to a sinusoidal waveform. The inverters have many applications, such as connecting renewable power sources such as photovoltaic panels to the power grid, driving AC motors, and FACTS devices.

The simplest topology, which can be regarded as an inverter, is the two-level or half-bridge configuration. This configuration is demonstrated in Figure 5.1. As this figure shows, this topology has two switches, where if S_1 or S_2 are connected, the output voltage will be equal to $0.5V_{DC}$ and $-0.5V_{DC}$, respectively.

Regarding these output voltages, both switches must endure a voltage of V_{DC}, which is high voltage stress for the switches. Also, regarding the switches' states, their voltages change from V_{DC} to zero between the different switching states, which can increase the switching losses for this configuration. Besides, the output voltage has a high THD because of having only two levels. The mentioned problems for the half-bridge topology show that this topology is not an excellent multilevel inverter topology, and it is better to increase the number of output voltage levels. Nevertheless, it should be noted that a high number of output voltage

[1]Faculty of Electrical and Computer Engineering, University of Tabriz, Iran

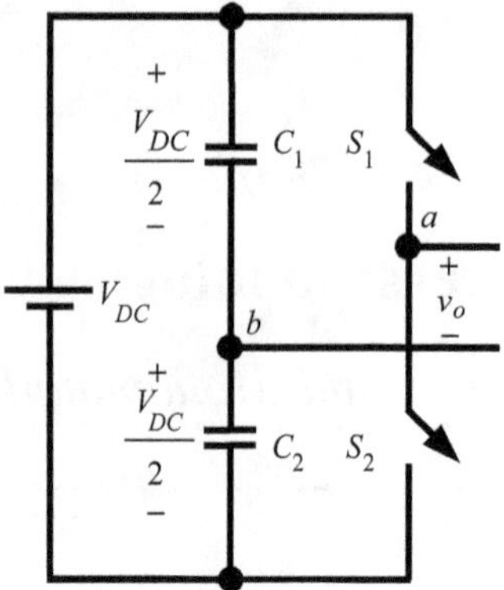

Figure 5.1 Half-bridge inverter

levels should not lead to a high number of switches and DC voltage sources. Thus, suggesting a topology that generates a high number of voltage levels with the least number of devices is so significant.

In this chapter, the presented multilevel inverters are surveyed to find which topology is better. Also, different comparisons are shown to expand this study. Furthermore, the optimization of multilevel inverters, a means to minimize the number of the required DC voltage sources and switches, is described.

5.1 Types of multilevel inverters

Three-level inverters, with a zero voltage added to the past positive and negative voltages, were the first multilevel inverter topologies suggested for maintaining better harmonic characteristics. Considering the configuration of these topologies, they can be categorized into three main groups:

- Neutral point clamped (NPC) or diode clamped topology
- Flying capacitor (FC) or capacitor clamped topology
- H-bridge topology

These topologies are described in the following subsections, and the merits and disadvantages of each of these topologies are mentioned in summary.

5.1.1 Three-level neutral-point-clamped topology

This structure was presented by Nabae *et al.* in 1981 [1]. Figure 5.2 depicts the configuration of this topology. As this figure shows, this topology includes two capacitors of C_1 and C_2, two diodes of D_1 and D_2, and four switches of S_1, S_2, S_1', and S_2'. In this topology, the two capacitors divide the input voltage into two equal parts, so the middle point has a neutral voltage. Also, the switching states of

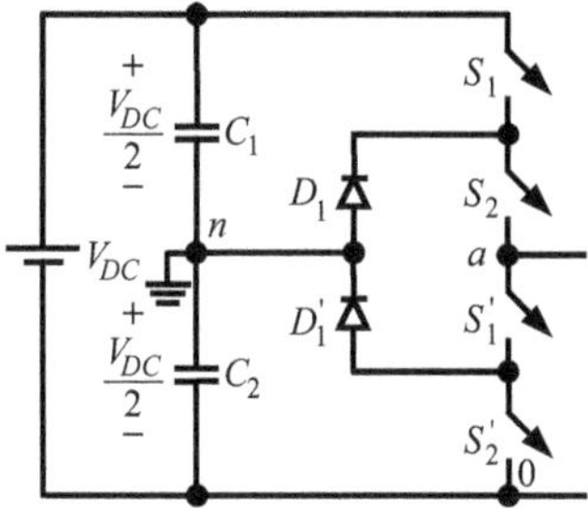

Figure 5.2 Three-level NPC configuration

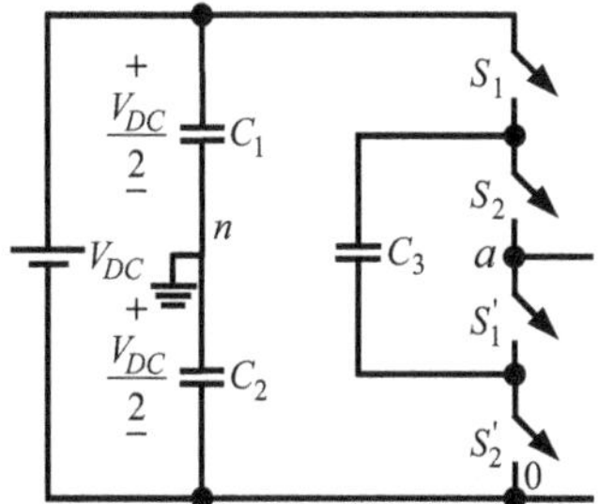

Figure 5.3 Three-level FC structure

switches in pairs of (S_1 and S_1') and (S_2 and S_2') are supposed to be complementary to prevent any short circuits on the input side and open circuits on the output side.

Regarding Figure 5.2, the voltage level of $0.5V_{dc}$ can be obtained when two switches of S_1 and S_2 are switched on. Also, the voltage level of $-0.5V_{dc}$ can be obtained when two switches of S_1' and S_2' are switched on. Moreover, the zero voltage level can be obtained by switching the switches of S_2 and S_1' on.

From the limitations of this topology, some problems can be mentioned. First, the voltage of the middle point must be maintained at the neutral voltage and achieving this aim may sometimes be challenging. Thus, the control method of this topology will have limitations. Also, the number of required diodes will be dramatically high in the high-voltage level numbers.

5.1.2 Three-level capacitor clamped topology

This topology was presented by Meynard in 1990 [2]. This topology is so similar to its NPC peer one. In the three-level configuration, diodes were used to fix the voltage of the middle point in a neutral voltage. However, this voltage is maintained using the capacitors in the flying capacitor topology. This structure is shown in Figure 5.3.

The switching states of this topology are identical to the three-level NPC structure.

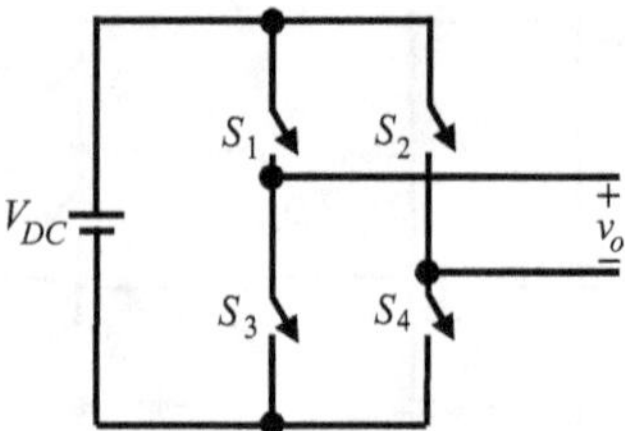

Figure 5.4 Three-level H-bridge structure

Similar to the NPC structure, the number of the required flying capacitors will be dramatically high in the high-voltage level numbers, which can increase the size and volume of the converter.

5.1.3 Three-level H-bridge topology

The idea of this topology was presented by Peng *et al.* in 1996 [3]. The configuration of a three-level H-bridge topology is depicted in Figure 5.4. As shown in this figure, this topology includes a DC voltage source along with four switches of S_1, S_2, S_3, and S_4. As it can be seen, this topology does not have any diodes or capacitors, which shows that it has a simpler topology than the past two groups.

In this topology, the voltage level of V_{DC} can be obtained by switching S_1 and S_4 on. Also, the voltage levels of $-V_{DC}$ can be obtained by switching S_2 and S_3 on. To generate the zero voltage level, two different combinations of switches can be used, where if two switches of S_1 and S_2 or S_3 and S_4 are switched on, the output voltage will be zero.

The simple topology and an easy control method are the main advantages of this three-level inverter topology over the others. Some cells have been introduced in [4], and all multilevel inverter topologies can be driven using them. In the next subsection, these cells are presented, and the procedure of creating a configuration based on them is described.

5.2 Fundamentals of multilevel inverter topologies

Ref. [4] presents five different cells where every multilevel inverter topology can be obtained using them. Table 5.1 illustrates these essential cells along with their switching states.

- Type-IV cell, named the T-type cell, has four output ports and the potential for three sorts of connections. This cell uses two unidirectional switches, one bidirectional switch, and two DC voltage sources.
- Type-V cell has four output ports and the potential for six types of connections. This topology includes two DC voltage sources, four unidirectional switches, and one bidirectional switch. The unidirectional switch and the bidirectional one block the voltages of $2V_{DC}$ and V_{DC}, respectively.

Table 5.1 Basic cells for multilevel inverters and their switching table

Basic cells	On-state switches	V_{ba}	V_{ca}	V_{da}
I	S_2	V_{DC}	0	–
	S_1	0	$-V_{DC}$	–
II	S_1	–	$-V_{DC}$	–
	S_2	–	–	0
	S_3	–	–	–
	S_1, S_2	0	$-V_{DC}$	0
	S_1, S_3	0	$-V_{DC}$	$-V_{DC}$
	S_2, S_3	$-V_{DC}$	0	0
III	S_1, S_3	–	–	0
	S_2, S_4	–	–	0
	S_2, S_3	–	–	V_{DC}
	S_1, S_4	–	–	$-V_{DC}$
IV	S_1	0	$-V_{DC}$	$-2V_{DC}$
	S_2	$2V_{DC}$	V_{DC}	0
	S_3	V_{DC}	0	$-V_{DC}$
V	S_1, S_5	0	0	$-2V_{DC}$
	S_2, S_4	0	0	$2V_{DC}$
	S_1, S_4	0	0	0
	S_2, S_5	0	0	0
	S_3, S_5	0	0	$-V_{DC}$
	S_3, S_4	0	0	V_{DC}

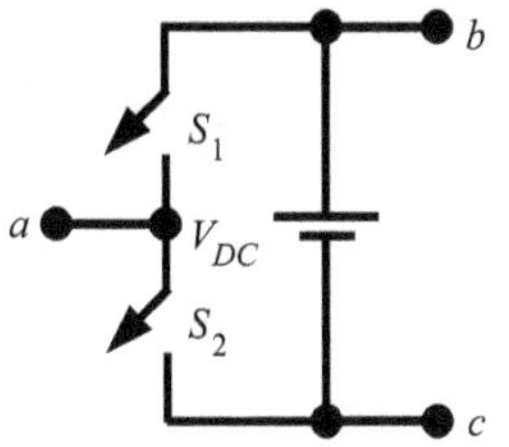

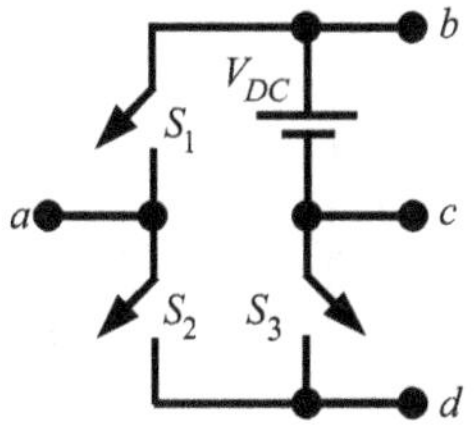

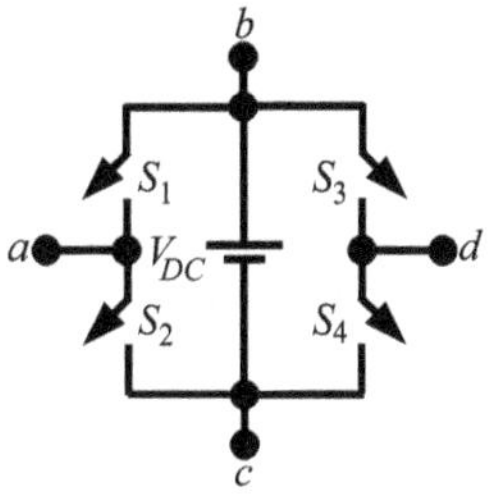

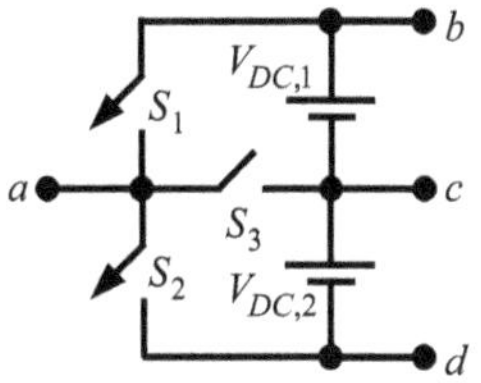

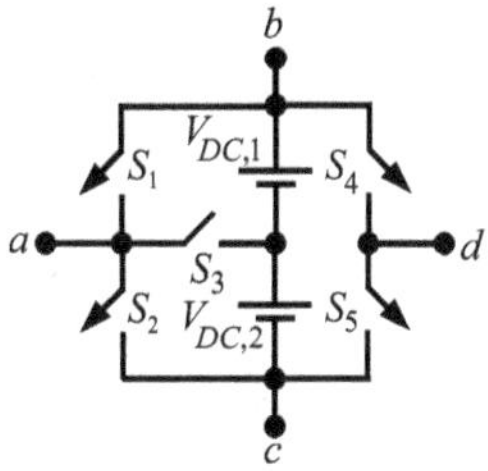

It is important to note that the shown cells in Table 5.1 can be considered modified. Figure 5.5 shows the modified type of the cells of III and V.

The switches must be able to pass the current in both directions in every multilevel inverter topologies, which have an inductive-resistive load. Also, it

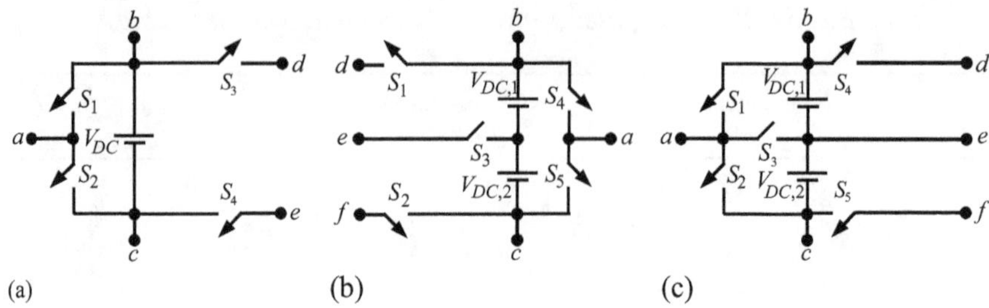

Figure 5.5 (a) Type-III cell, (b) and (c) type-V cell

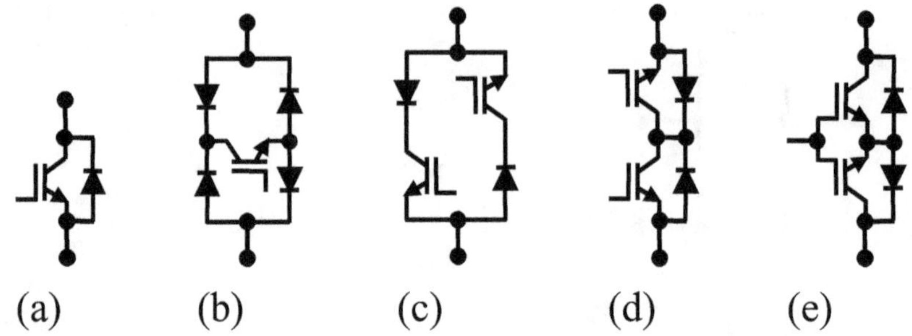

Figure 5.6 (a) A unidirectional voltage-bidirectional current switch and (b)–(e) a bidirectional voltage-bidirectional current switch

should be noticed that this sort of load is a typical load for every multilevel inverter. Besides, the sign of the blocked voltage on a switch is the other parameter that identifies its kind. Some switches block a positive voltage during a switching period, meaning they only block a positive voltage. In contrast, some other switches block both positive and negative voltages in a switching table, meaning they must be bidirectional from the voltage point of view. Thus, two kinds of unidirectional voltage-bidirectional current and bidirectional voltage-bidirectional current switches are used in multilevel inverter configurations, typically. These types of switches are illustrated in Figure 5.6. A bidirectional voltage-bidirectional current switch can be made in different forms, as shown in Figure 5.6.

Four different configurations can be considered for a bidirectional voltage-bidirectional current switch, regarding Figure 5.6. The depicted configuration in Figure 5.6b wants the highest count of the active devices, so it is not a good choice to consider in the multilevel inverter topologies. The other configurations have the same number of switches and diodes, but the shown configuration in Figure 5.6e is the best because it wants only one driver circuit. So, this configuration is considered in the other topologies that use bidirectional switches. It is important to note that the unidirectional voltage-bidirectional current and bidirectional voltage-bidirectional current switches contract as unidirectional and bidirectional switches in the other parts of this chapter.

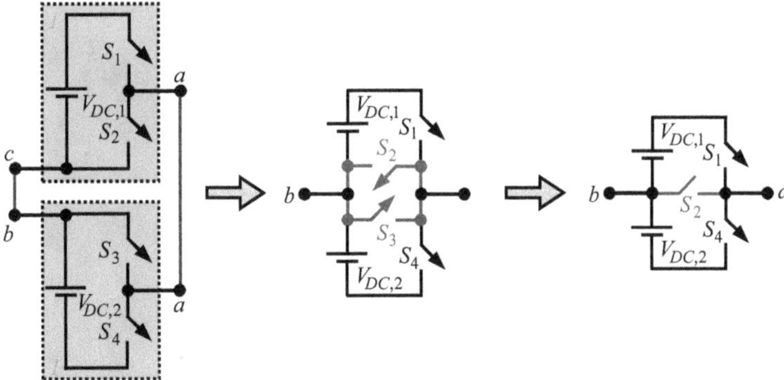

Figure 5.7 (a) Procedure of making the T-type inverter using the type-I cell

As an example, the procedure of making the T-type inverter using the depicted cells in Table 5.1 is shown in Figure 5.7. In this procedure, two unidirectional switches are connected in parallel, which leads to a bidirectional switch.

In the following, the existing cascaded multilevel inverter topologies are described, and the procedure for making some of them is given in detail. First, it should be stated that the existing multilevel inverter topologies can be grouped into two main categories regarding their ability to generate negative voltage levels. Some topologies can generate both positive and negative voltage levels inherently, which can be named as bipolar configurations. In contrast, some other topologies cannot generate negative voltage levels alone, requiring additional parts such as H-bridge cells to generate negative voltage levels, and they name unipolar configurations. Using these descriptions, the presented cascaded multilevel inverter topologies are described in the following. It is important to note that, the presented cells in Table 5.1 can be connected in either series, parallel, or cross forms to construct the different multilevel inverter topologies.

5.3 Bipolar multilevel inverter configurations

In this section, the bipolar multilevel inverter topologies that have been presented in the literature are described in detail. As introduced earlier, the bipolar topologies can inherently generate positive and negative voltage levels. Thus, they do not need other cells to generate the negative cells. Besides, the procedure of making some of them regarding the basic cells is described.

5.3.1 Cascaded H-bridge topology

The cascaded H-bridge configuration, which can be considered the simplest bipolar topology, is depicted in Figure 5.8. This topology can work well under faulty conditions regarding its modular configuration, which is an advantage of this structure. This topology can be obtained from the series connection of type-III cell, as shown in Figure 5.9.

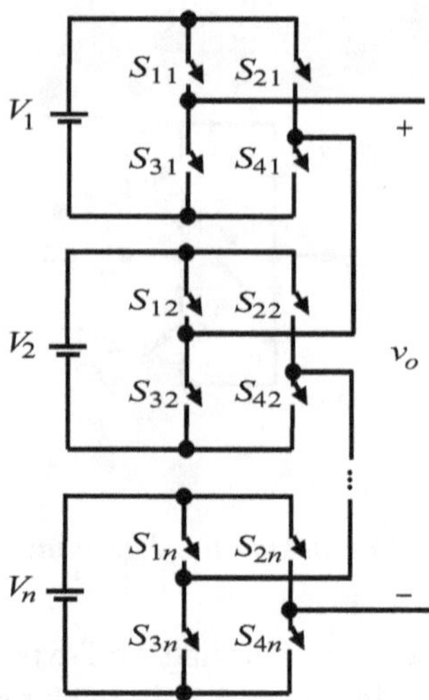

Figure 5.8 Cascaded H-bridge configuration

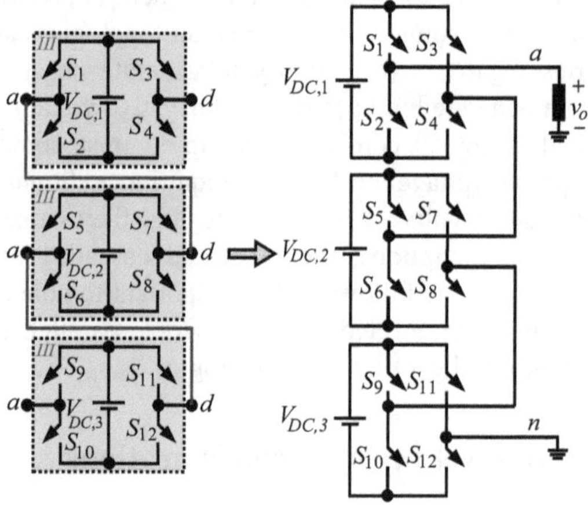

Figure 5.9 Construction of the cascaded H-bridge configuration using type-III cell

All switches block a positive voltage, as seen in Figure 5.9. Thus, every switch that is used in this topology will be unidirectional. Regarding that this topology has n DC voltage sources, their magnitudes can be defined as some patterns. Besides, for a multilevel inverter topology, some indexes such as the count of output voltage levels (N_{level}), the number of required switched (N_{switch}), required IGBTs (N_{IGBT}), required DC voltage sources (N_{source}), variety in $V_j = V_{dc}$ magnitudes of the DC voltage sources ($N_{variety}$), the total blocking voltage (V_{block}), and the maximum number of on-state switches to generate different voltage levels ($N_{\text{on,state}}$) can be defined. Regarding the given descriptions, the characteristics of the shown topology in Figure 5.8 can be listed in Table 5.2.

Table 5.2 Characteristics of cascaded H-bridge topology (j = 1,2, …, n)

Pattern	N_{level}	N_{switch}	N_{IGBT}	N_{source}	$N_{variety}$	V_{block}/V_{DC}	$N_{on,state}$
$V_j = V_{DC}$	$2n + 1$	$4n$	$4n$	n	1	$2N_{level} - 2$	$2n$
$V_j = 2^{j-1}V_{DC}$	$2^{n+1} - 1$				n		
$V_j = 3^{j-1}V_{DC}$	3^n				n		

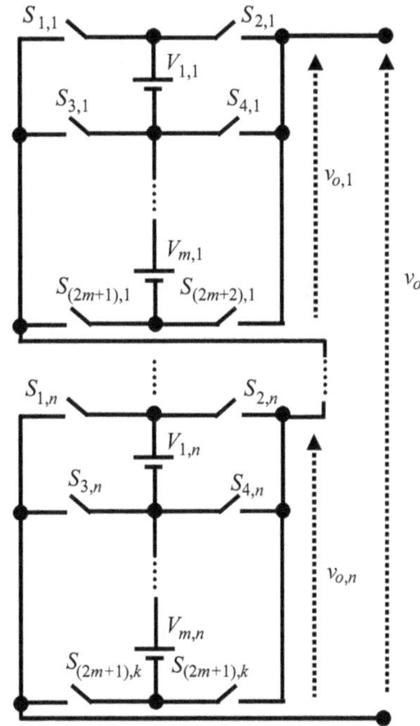

Figure 5.10 Suggested structure in [5]

Regarding Table 5.2, the cascaded H-bridge topology is able to generate a higher number of voltage levels when the third pattern is used.

5.3.2 Suggested structure in [5]

The presented topology in [5] is demonstrated in Figure 5.10. This topology includes several basic cells, where each of these cells can be obtained using several type-I cells. The procedure of making a basic cell of the presented topology in [5] using these cells is shown in Figure 5.11. Also, it is seen that each cell has $2n+2$ bidirectional switches, where each of these switches includes two unidirectional ones.

Regarding Figure 5.11, the connection of six type-I cells leads to a basic cell of the suggested structure in [5], where m is four.

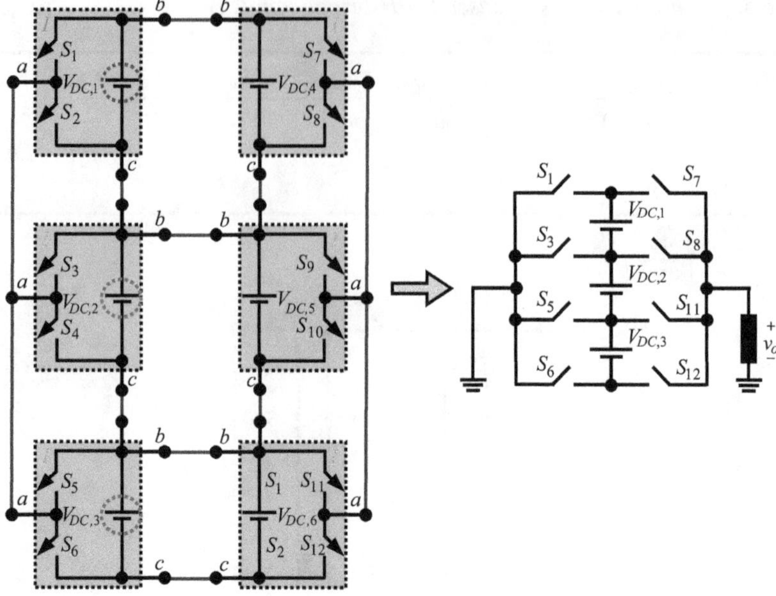

Figure 5.11 Construction of a cell of the suggested structure in [5]

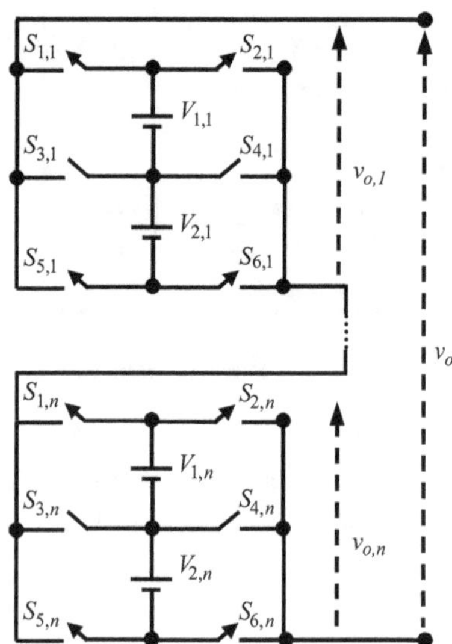

Figure 5.12 Existing topologies in [6]

5.3.3 Suggested structure in [6]

This topology can be obtained by the presented topology in [5] if m is supposed to be three. This topology is shown in Figure 5.12.

As previously described, this topology can be obtained from the connection of several type-I cells. In [6], three different patterns are presented to set the magnitudes of the DC voltage sources. Table 5.3 shows each of these algorithms along with the other key parameters of this structure.

From Table 5.3, all of the patterns are asymmetric.

5.3.4 Suggested structures in [7,8]

A topology based on the half-bridge cells is presented in [7]. This topology is shown in Figure 5.13. As seen in this figure, each of the basic cells includes two half-bridge cells, or type-I cells, that are connected. Regarding that this topology is able to generate all positive and negative voltage levels, it does not need an H-bridge cell on its output stage.

Regarding this topology and its patterns, its key characteristics can be listed in Table 5.4.

Another topology in [8] is shaped by the series connection of cascaded H-bridge cells with the presented topology in [7]. This topology is illustrated in Figure 5.14.

5.3.5 Existing topology in [9]

Based on the half-bridge cells, another multilevel inverter topology is presented in [9]. This topology is shown in Figure 5.15. As seen in this figure, each basic cell includes three type-I cells, where each cell can generate all of the negative and positive voltage levels.

Regarding this topology, each switch blocks only positive voltages, so they are unidirectional. Regarding this topology and the patterns to choose its DC voltage sources, its characteristics can be listed in Table 5.5.

Regarding this table, the first pattern is symmetric, where the number of generated voltage levels is low. However, the second pattern is an asymmetric type, and p can be chosen as any number.

Table 5.3 Characteristics of the suggested structure in [6] ($j = 1,2, \ldots, n$)

Pattern	N_{level}	N_{switch}	N_{IGBT}	N_{source}	$N_{variety}$	V_{block}/V_{DC}	$N_{on,state}$
$V_{1,1} = \frac{1}{2}V_{2,1} = V_{DC}$	7^n	$6n$	$8n$	$2n$	$2n$	$\frac{8}{3}(N_{level} - 1)$	$2n$
$V_{1,j} = \frac{1}{2}V_{2,j} = V_{DC} + \sum_{i=1}^{j-1}(V_{1,i} + V_{2,i})$	$(3 \times 2^{n+1}) - 5$						
$2V_{1,j} = V_{2,j} = 2^j V_{DC}$	$3^{n+1} - 4$						
$V_{1,1} = V_{2,1} = V_{DC}$ $2V_{1,j} = V_{2,j} = 2 \times 3^{j-1} V_{DC}$							

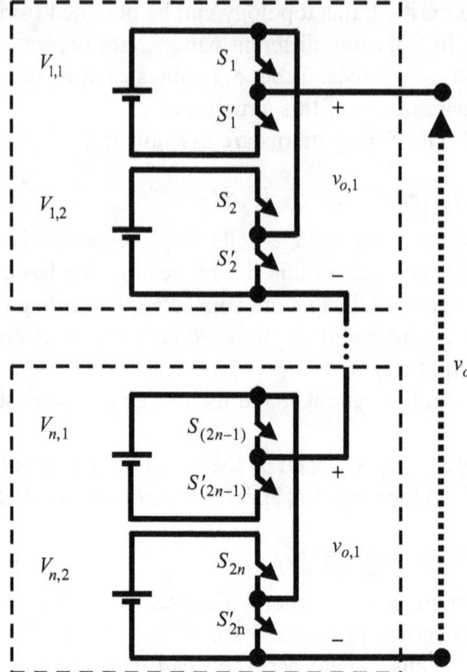

Figure 5.13 Presented topology in [7]

Table 5.4 Characteristics of the presented topology in [7] (j = 1,2, ..., n and k = 1 and 2)

Pattern	N_{level}	N_{switch}	N_{IGBT}	N_{source}	$N_{variety}$	V_{block}/V_{DC}	$N_{on,state}$
$V_{j,k} = 2^{j-1} V_{DC}$	$2^{n+1} - 1$	$4n$	$4n$	$2n$	n	$2N_{level} - 2$	$2n$
$V_{j,k} = 3^{j-1} V_{DC}$	3^n				n		
$V_{1,1} = V_{DC},\ V_{1,2} = 2V_{DC}$ $V_{2,1} = 5V_{DC},\ V_{2,2} = 4V_{DC}$ $V_{j,1} = V_{j,2} = 13 \times 3^{j-3} V_{DC}$	$13 \times 3^{n-2}\ n \geq 2$				n		

5.3.6 Existing topology in [10]

The topology in [5] can obtain the basic cell of the presented topology in [10] if m is supposed to be two. This topology is shown in Figure 5.16. As seen in this figure, two of the switches are bidirectional ones, which shows that these switches must be able to block both negative and positive voltage levels.

Regarding this topology, four different patterns are presented for it. These patterns, along with the number of voltage levels and the other parameters, are listed in Table 5.6.

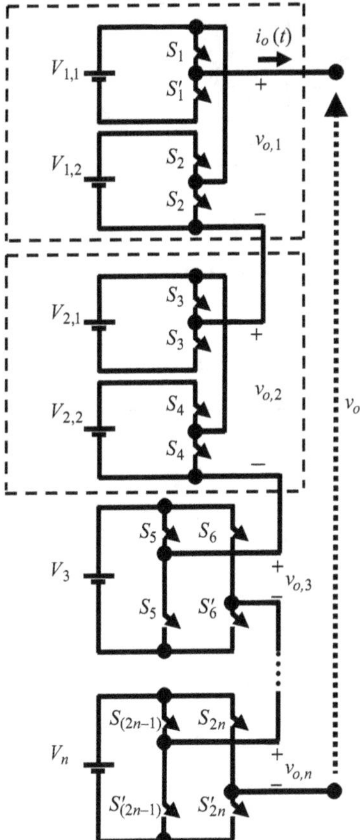

Figure 5.14 Existing topology in [8]

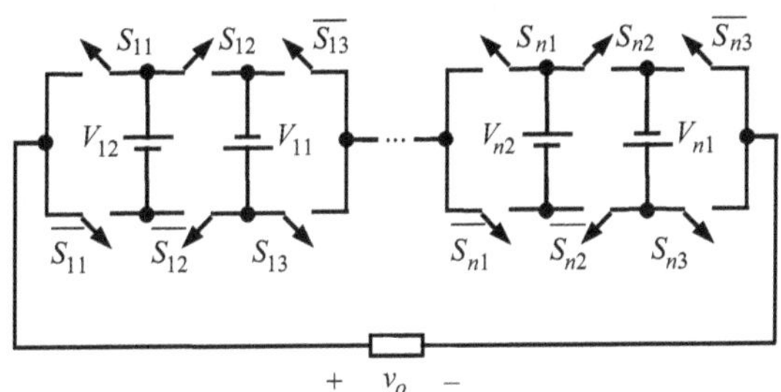

Figure 5.15 Presented topology in [9]

Table 5.5 Characteristics of the presented topology in [9] (j = 1,2, ..., n)

Pattern	N_{level}	N_{switch}	N_{IGBT}	N_{source}	$N_{variety}$	V_{block}/V_{DC}	$N_{\text{on,state}}$
$V_{j1} = V_{j2} = V_{DC}$	$4n + 1$	$6n$	$6n$	$2n$	1	$2N_{level} - 2$	$3n$
$V_{j1} = V_{j2} = p^{j-1}V_{DC}$	$4\frac{1-p^n}{1-p} + 1$				n		

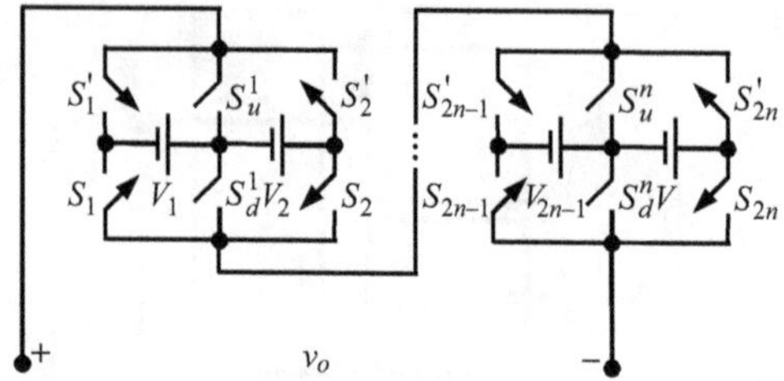

Figure 5.16 Presented topology in [10]

Table 5.6 Characteristics of the presented topology in [10] (j = 1,2, ..., n)

Pattern	N_{level}	N_{switch}	N_{IGBT}	N_{source}	$N_{variety}$	V_{block}/V_{DC}	$N_{\text{on,state}}$
$V_{2j-1} = V_{2j} = V_{DC}$	$4n+1$	$6n$	$8n$	$2n$	1	$2.5\ (N_{level} - 1)$	$2n$
$V_{2j-1} = V_{2j} = 5^{j-1} V_{DC}$	5^n				n		
$2\ V_{2j-1} = V_{2j} = 2^{2j-1} V_{DC}$	$2^{2n+1} - 1$				$2n$		
$2\ V_{2j-1} = V_{2j} = 2 \times 7^{j-1} V_{DC}$	7^n						

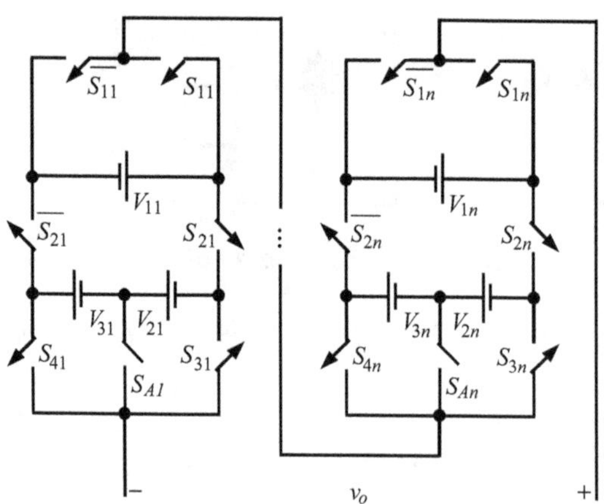

Figure 5.17 Suggested topology in [11]

Regarding Table 5.5, the first pattern is symmetric. However, the other ones are asymmetric. Also, it can be seen that the number of the generated voltage levels can be maximized by using the fourth pattern.

5.3.7 Existing topology in [11]

The presented topology in [11] is shown in Figure 5.17. This figure shows that each basic unit includes six unidirectional switches, one bidirectional switch, and three

DC voltage sources. Regarding this structure, its characteristics are listed in Table 5.7. A basic cell of this topology can be constructed using type-III and type-IV cells, as shown in Figure 5.18.

As seen in Table 5.7, the presented pattern is asymmetric. Also, each of the cells can generate five voltage levels.

5.3.8 Existing topology in [12]

The presented topology in [12] is shown in Figure 5.19. As seen in this figure, this topology has similar basic cells, each including six unidirectional switches and one bidirectional switch. Also, it has four DC voltage sources. One type-I and two type-IV cells can be used to make a basic cell of this topology as can be seen in Figure 5.20.

Regarding this topology and its pattern for choosing the DC voltage sources, its characteristics can be listed in Table 5.8.

Table 5.7 Characteristics of the presented topology in [11] (j = 1,2, ..., n)

Pattern	N_{level}	N_{switch}	N_{IGBT}	N_{source}	$N_{variety}$	V_{block}/V_{DC}	$N_{on,state}$
$V_{1,j} = 2\ V_{2,j} = 2\ V_{3,j} = 2\ V_{DC}$	$8n + 1$	$7n$	$8n$	$3n$	2	$2.5\ (N_{level} - 1)$	$2n$

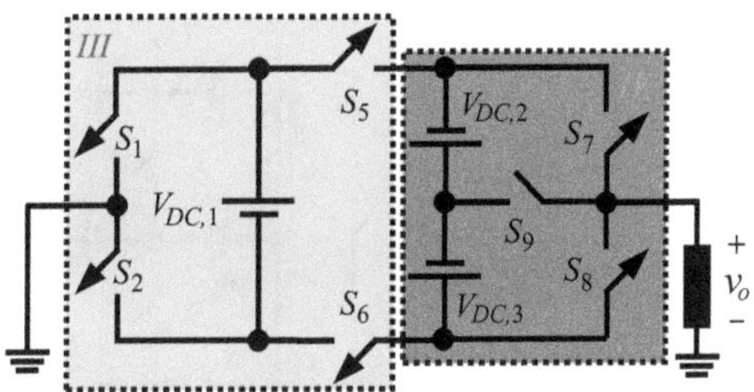

Figure 5.18 Construction of a cell of the suggested structure in [11]

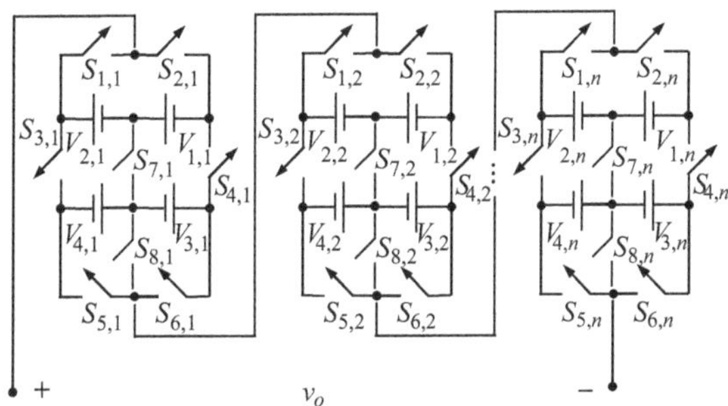

Figure 5.19 Suggested topology in [12]

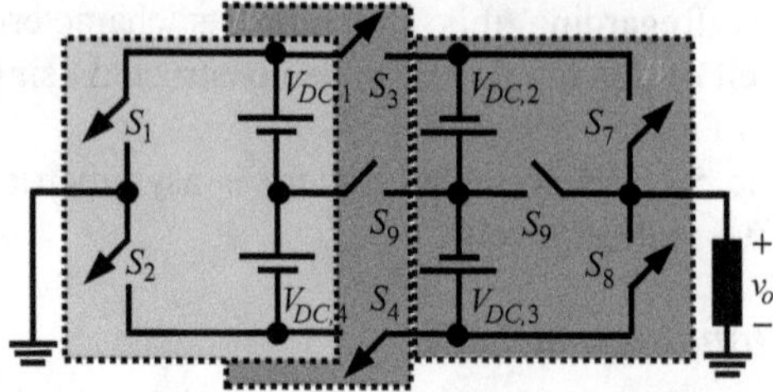

Figure 5.20 Construction of a cell of the suggested structure in [12]

Table 5.8 Characteristics of the presented topology in [12] (j = 1,2, ..., n)

Pattern	N_{level}	N_{switch}	N_{IGBT}	N_{source}	$N_{variety}$	V_{block}/V_{DC}	$N_{on, state}$
$V_{1,j} = V_{2,j} = 2\ V_{3,j} = 2\ V_{4,j} = 2\ V_{DC}$	$12n + 1$	$8n$	$10n$	$4n$	2	$\frac{5}{3}(N_{level} - 1)$	$3n$

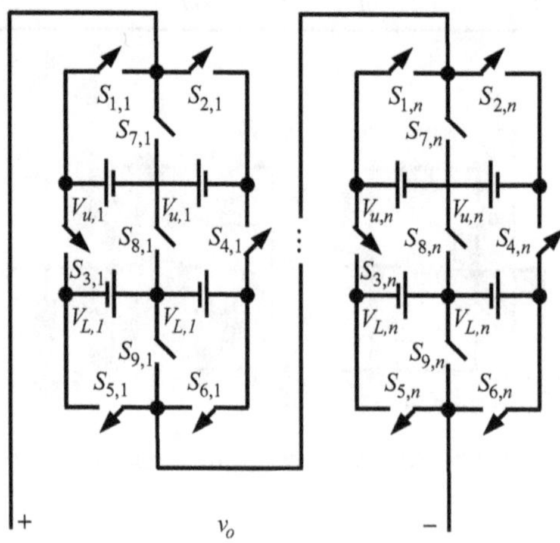

Figure 5.21 Suggested topology in [13]

Table 5.8 shows that this pattern uses two kinds of DC voltage sources, and each cell generates seven different voltage levels.

5.3.9 Existing topology in [13]

Similar to [12], another topology is presented in [13]. This topology is depicted in Figure 5.21. Regarding this figure, it can be seen that each cell has six unidirectional switches, three bidirectional switches, and four DC voltage sources. Three type-IV cells can be used to make a basic cell of this topology, as can be seen in Figure 5.22.

Regarding this topology, its key features can be listed in Table 5.9.

Table 5.9 shows that both patterns lead to the same result for n=1.

5.3.10 Existing topology in [14]

The existing topology in [14] is shown in Figure 5.23. As seen in this figure, each basic cell includes eight unidirectional switches, two bidirectional switches, and four DC voltage sources. Also, each of these cells can be made using the series connection of two type-V cells. Regarding a type-V cell, it has one bidirectional switch, so this topology will have two bidirectional switches per one basic unit.

Regarding this topology, three different patterns are presented to determine the magnitudes of the DC voltage sources. These algorithms and the other vital factors are listed in Table 5.10.

Regarding this table, the first pattern is symmetric, and the other algorithms are asymmetric.

5.3.11 Existing topology in [15]

Two different structures for the symmetric and asymmetric algorithms are presented in [15]. These structures are demonstrated in Figure 5.24.

As seen in these figures, the difference between these topologies is the number of switches. Regarding these topologies, their patterns and characteristics can be listed in Table 5.11.

As seen in Table 5.11, two different algorithms can be considered for the presented topologies in [15]. Moreover, they will lead to different results because

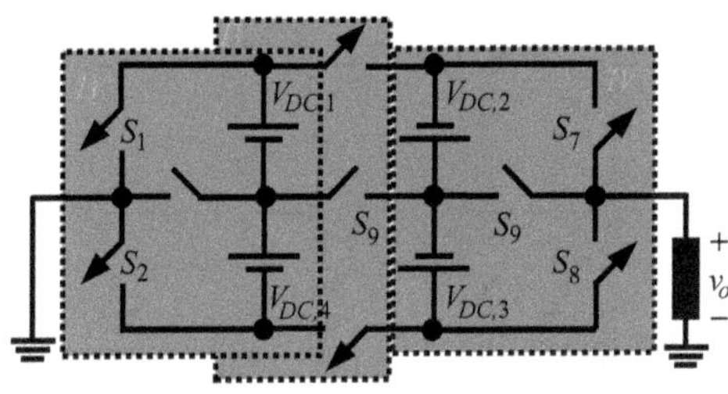

Figure 5.22 Construction of a cell of the suggested structure in [13]

Table 5.9 Characteristics of the presented topology in [12] (j = 1,2, ..., n)

Pattern	N_{level}	N_{switch}	N_{IGBT}	N_{source}	$N_{variety}$	V_{block}/V_{DC}	$N_{on,state}$
$3V_{u,j} = V_{L,j} = 3\ V_{DC}$	$16n+1$	$9n$	$12n$	$4n$	2	$\frac{5}{2}(N_{level}-1)$	$3n$
$3V_{u,j} = V_{L,j} = 3 \times 17^{j-1}\ V_{DC}$	17^n				$2n$		

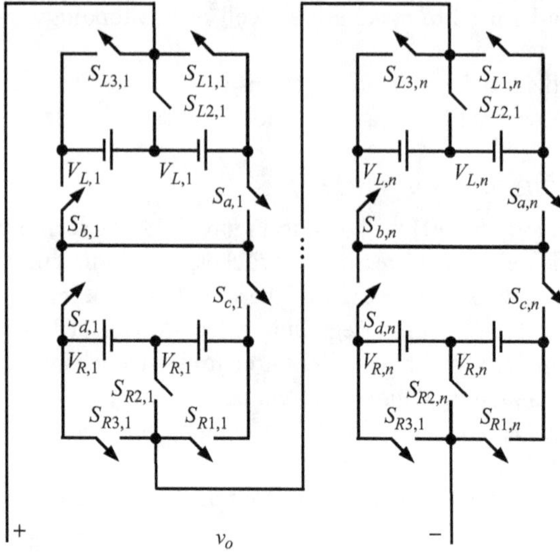

Figure 5.23 Suggested topology in [14]

Table 5.10 Characteristics of the presented topology in [14] (j = 1,2, ..., n)

Pattern	N_{level}	N_{switch}	N_{IGBT}	N_{source}	$N_{variety}$	V_{block}/V_{DC}	$N_{on,state}$
$V_{L,j} = V_{R,j} = V_{DC}$	$8n + 1$	$10n$	$12n$	$4n$	1	$\frac{9}{4}(N_{level} - 1)$	$4n$
$5\ V_{L,j} = V_{R,j} = 5\ V_{DC}$	$24n + 1$				2		
$5\ V_{L,j} = V_{R,j} = 5^{\,2j-1} V_{DC}$	25^n				$2n$		

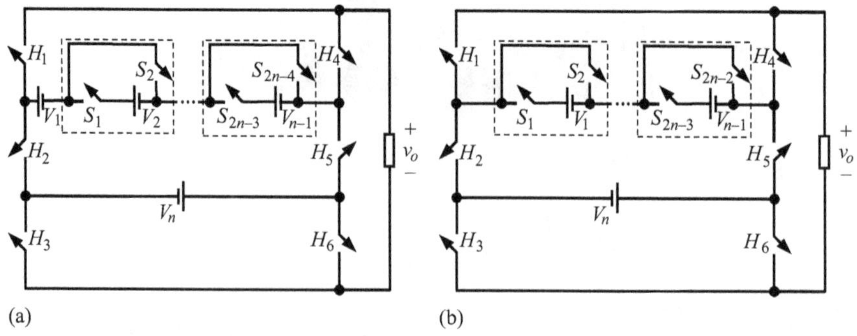

Figure 5.24 Suggested topologies in [15]: (a) for symmetric pattern and (b) for asymmetric pattern

Table 5.11 Characteristics of the presented topologies in [15] (j = 1,2, ..., n)

Pattern	N_{level}	N_{switch}	N_{IGBT}	N_{source}	$N_{variety}$	V_{block}/V_{DC}	$N_{\text{on,state}}$
Fig. $1-26a$ $V_j = V_{DC}$	$2n+3$	$2n+4$	$2n+4$	$n+1$	1	$3n-6$	$n+1$
Fig. $1-26b$ $V_j = p^{j-1}\, V_{DC}$	$2\frac{1-p^{n+1}}{1-p}+1$	$2n+6$	$2n+6$		$n+1$	$6\frac{1-p^{n+1}}{1-p}-2p^n$	$n+2$

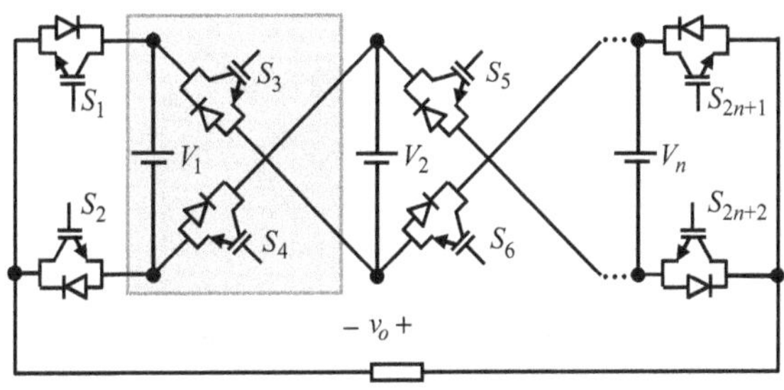

Figure 5.25 Suggested topology in [16]

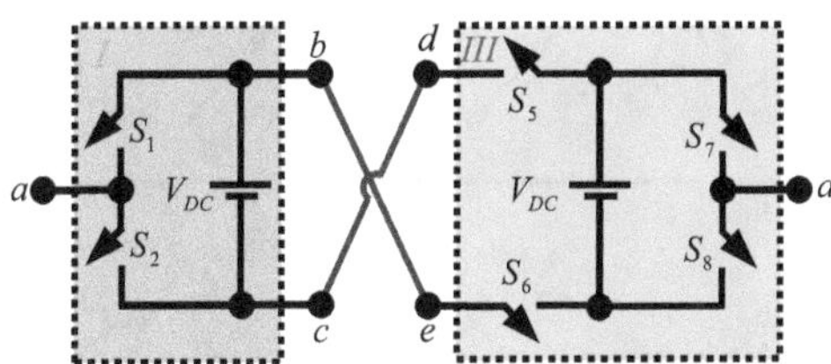

Figure 5.26 Construction of a cell of the suggested structure in [16]

there is a difference in the topologies used for the symmetric and asymmetric patterns.

5.3.12 Existing topologies in [16,17]

The presented topology in [16] is shown in Figure 5.25. As seen in this figure, this topology can generate all positive and negative voltage levels because it does not need additional parts such as an H-bridge cell to generate these voltage levels. Also, it is seen that each cell includes two unidirectional switches and one DC voltage source. Also, if n is supposed to equal one, the H-bridge structure can be obtained from this topology. The procedure of making a basic cell of this topology using type-I and type-III cells is shown in Figure 5.26.

As shown in Figure 5.26, each cell of this structure can be made through cross-connection of type-I and type-III cells. In [17], another topology is presented, which includes a series connections of several cells presented in [17]. This topology is shown in Figure 5.27.

Regarding this topology and the DC voltage sources, two different algorithms are presented in [17]. These patterns, along with the other important factors, are listed in Table 5.12.

As seen in Table 5.30, the first pattern is symmetric. Also, the other algorithm is an asymmetric pattern that has the potential to generate more voltage levels in the output stage.

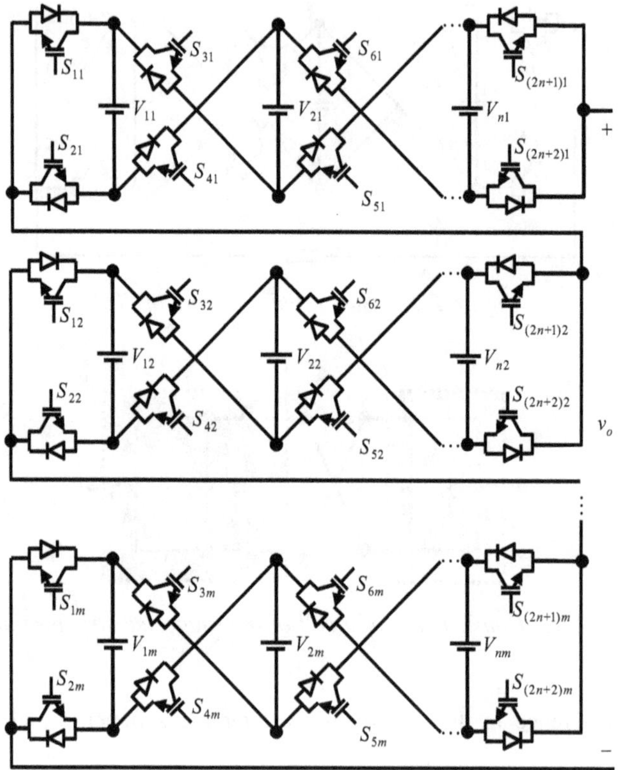

Figure 5.27 Presented topology in [17]

Table 5.12 Key factors of the presented topology in [17] (j = 1,2, . . . , n and k = 1,2, . . . , m)

Pattern	N_{level}	N_{switch}	N_{IGBT}	N_{source}	$N_{variety}$	V_{block}/V_{DC}	$N_{on,state}$
$V_{jk} = V_{DC}$	$2mn + 1$	$m(2n + 2)$	$m(2n + 2)$	mn	1	$2N_{level} - 2$	$m(n + 1)$
$2V_{1k} = V_{jk} = 2\,(4n - 1)^{k-1} V_{DC}$	$(4n - 1)^m$				mn		

5.3.13 Existing topology in [18]

The basic unit of the presented topology in [18] is shown in Figure 5.28. Regarding the similarity of this topology with the conventional H-bridge topology, it is cited as the developed H-bridge cell. An H-bridge cell includes four switches and a DC voltage source, which can generate three different voltage levels on its output stage. In comparison, the developed H-bridge cell includes six switches and two DC voltage sources and has the potential to generate seven voltage levels. A parallel connection of type-I and type-III cells can be used to construct this developed H-bridge cell, and this procedure is illustrated in Figure 5.29. A cascaded topology can be obtained by connecting *n* developed H-bridge cells in series. This topology is shown in Figure 5.30. Regarding this topology, several patterns are suggested to choose the magnitudes of the DC voltage sources, which are listed in Table 5.13. Table 5.13 shows all patterns, except the first one, are asymmetric algorithms.

5.3.14 Existing topologies in [19,20]

The presented topology in [19] is shown in Figure 5.31. This figure shows that this structure can be obtained from developing the developed H-bridge [18] topology.

The existing topology in [20] includes several of these cells connected in series. This topology can be seen in Figure 5.32. The characteristics of this topology and its patterns can be listed in Table 5.14.

5.3.15 Existing topology in [21]

The existing topology in [21] is shown in Figure 5.33. As seen in this figure, this topology includes several similar topologies in series, where each of these cells can

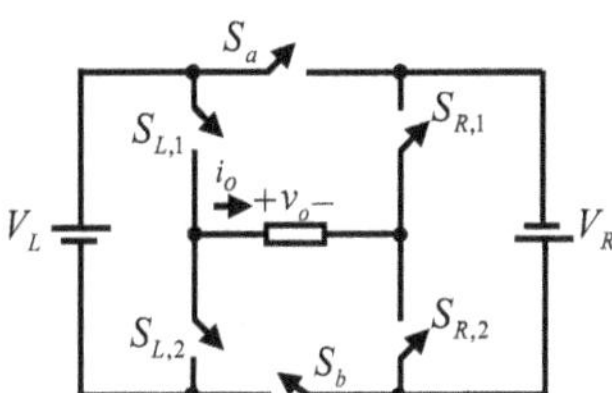

Figure 5.28 Developed H-bridge cell [18]

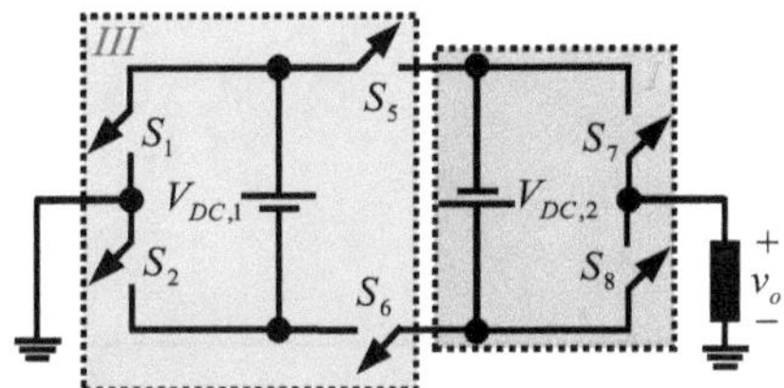

Figure 5.29 Construction of a cell of the suggested structure in [18]

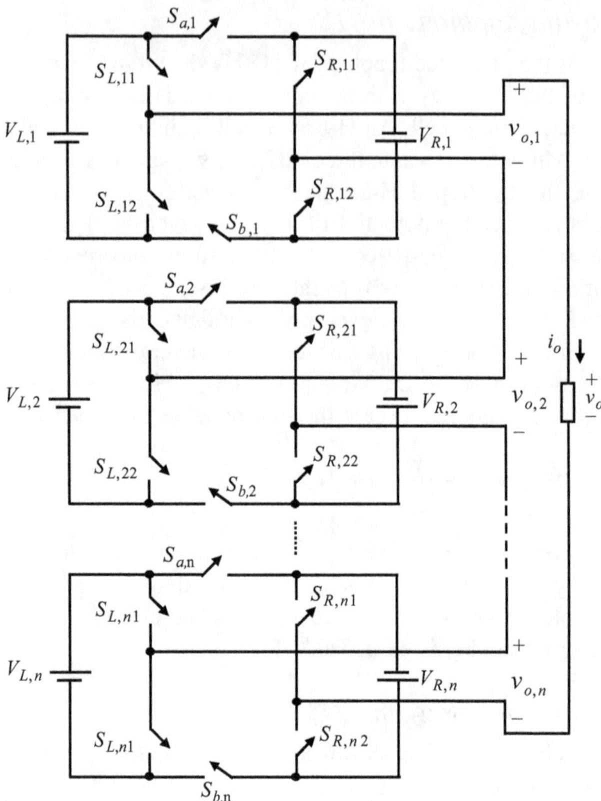

Figure 5.30 Cascaded topology in [18]

Table 5.13 Key factors of the presented topology in [18] (j = 1,2, …, m)

Pattern	N_{level}	N_{switch}	N_{IGBT}	N_{source}	$N_{variety}$	V_{block}/V_{DC}	$N_{on,state}$
$V_{R,j} = V_{L,j} = V_{DC}$	$4n+1$	$6n$	$6n$	$2n$	1	$2\,N_{level} - 2$	$3n$
$2V_{R,j} = V_{L,j} = 2V_{DC}$	$6n+1$				2		
$2V_{R,1} = V_{L,1} = 2\,V_{DC}$ $V_{R,j} = V_{L,j} = 2^{j-1}V_{DC}$	$2^{n+2} - 1$				n		
$V_{R,j} = V_{L,j} = 3^{j-1}V_{DC}$	$2 \times 3^n - 1$				n		
$V_{R,1} = V_{L,1} = V_{DC}$ $V_{R,j} = 0.5V_{L,j} = 2^{j-1}V_{DC}$	$3 \times 2^{n-1} - 7$				$\begin{cases} n+1 & n>1 \\ 1 & n=1 \end{cases}$		
$V_{R,1} = V_{L,1} = V_{DC}$ $V_{R,j} = 0.5V_{L,j} = 3^{j-1}V_{DC}$	$3^{n+1} - 4$				$2n-1$		
$V_{R,j} = V_{L,j} = 4^{j-1}V_{DC}$	$\frac{4^{n+1}-1}{3}$				n		
$V_{R,j} = V_{L,j} = 5^{j-1}V_{DC}$	5^n				n		
$V_{R,j} = 0.5V_{L,j} = 7^{j-1}V_{DC}$	7^n				$2n$		

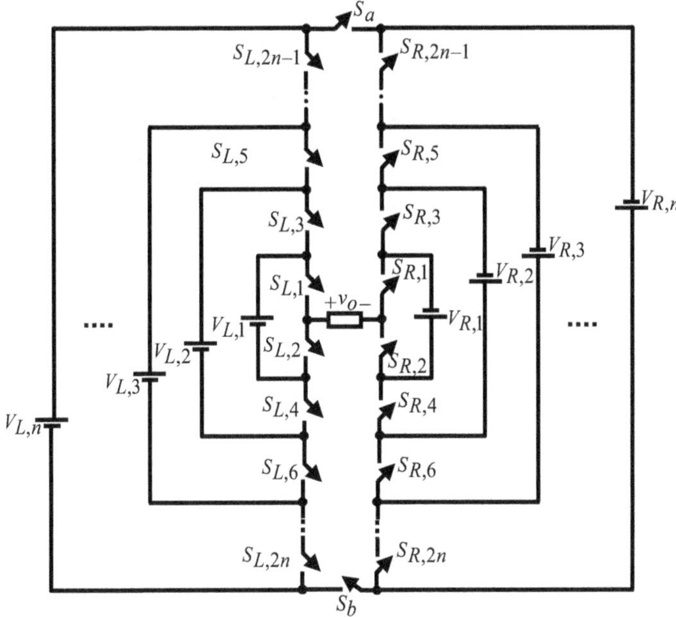

Figure 5.31 Presented topology in [19]

generally have different DC voltage sources. Nevertheless, it is assumed that all these cells have an equal number of DC voltage sources to simplify the calculations.

The procedure of making a cell of this structure using type-I and type-II cells is shown in Figure 5.34.

Regarding this topology, two patterns are suggested to obtain the magnitudes of the DC voltage sources. These algorithms, along with the other related equations, are listed in Table 5.15.

5.3.16 Existing topology in [22]

The suggested topology in [22] consists of a cascaded connection of several cells, as shown in Figure 5.35. Figure 5.36 demonstrates the procedure of making a cell of this topology with n=2 using two type-IV cells and two additional switches.

Two different patterns are suggested to determine this topology's magnitudes of the DC voltage sources. These patterns, along with the other related equations, are listed in Table 5.16.

5.3.17 Existing topology in [23]

Another topology based on the developed H-bridge topology is presented in [23]. This suggested topology is shown in Figure 5.37.

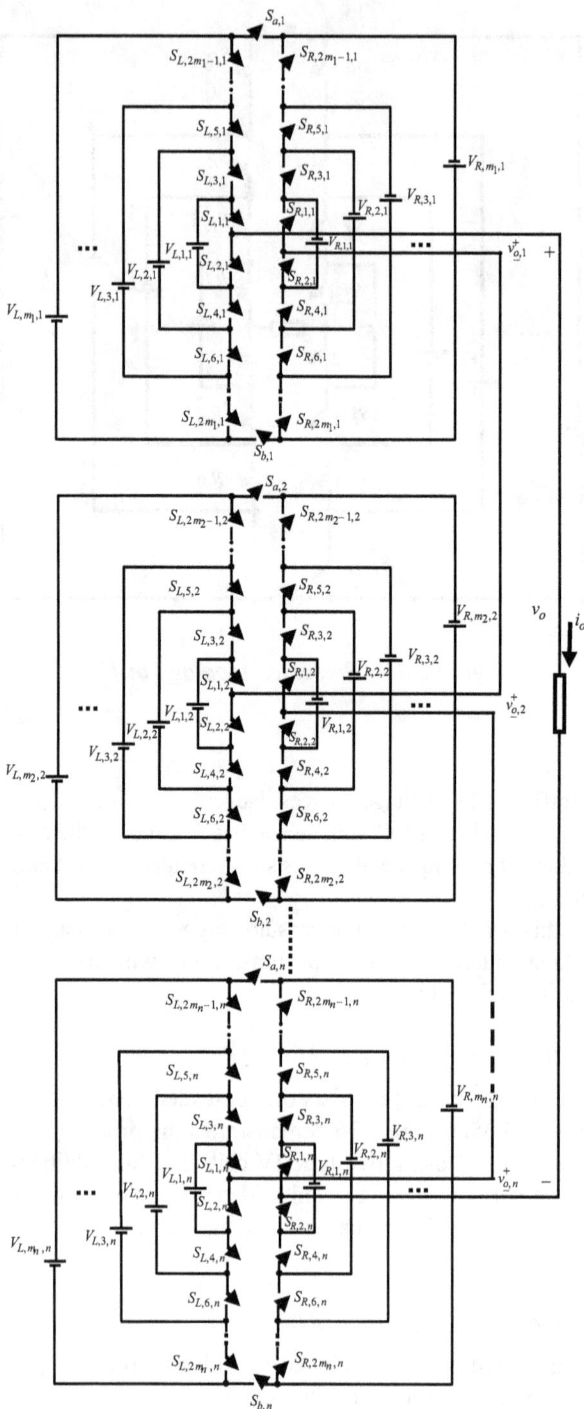

Figure 5.32 Presented topology in [20]

Table 5.14 Key factors of the presented topology in [20] (j = 1,2, ..., m and k = 1,2, ..., n)

Pattern	N_{level}	N_{switch}	N_{IGBT}	N_{source}	$N_{variety}$	V_{block}/V_{DC}	$N_{on,state}$
$2\,V_{L,j,k} = V_{R,j,k} = 2 \times 5^{k-1} V_{DC}$	$n \times 2^{2m+1} - 2n + 1$	$n\,(4m+2)$	$n\,(4m+2)$	$2mn$	$2m$	$2\,N_{level} - 2$	$n\,(2m+1)$
$V_{L,j,k} = 5^{k-1}(3 \times 5^{m-1} + 1)^{n-1} V_{DC}$ $V_{R,j,n} = (2 \times 5^{k-1}) V_{L,j,n}$	$2 \times (3 \times 5^{m-1} + 1)^{n} - 1$				$2n$		

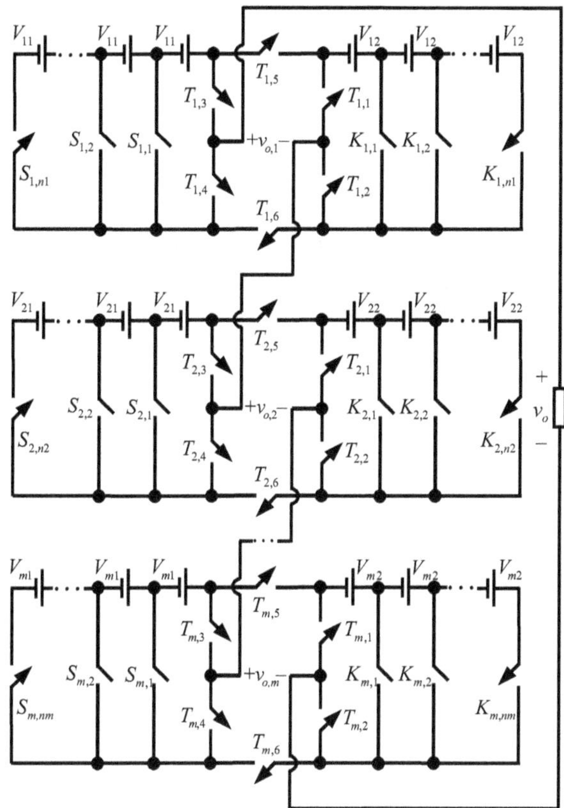

Figure 5.33 Presented topology in [21]

Two different algorithms are suggested in [23] for determining the magnitudes of the DC voltage sources. These two algorithms along with the other related equations are listed in Table 5.17.

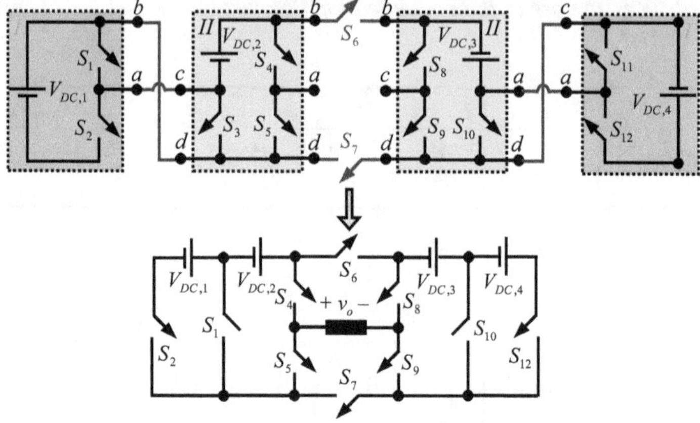

Figure 5.34 Construction of a cell of the suggested structure in [21]

Table 5.15 Key factors of the presented topology in [21] (j = 1,2, . . . , m)

Pattern	N_{level}	N_{switch}	N_{IGBT}	N_{source}	$N_{variety}$	$N_{on,state}$
$V_{j1} = V_{j2} = (4n+1)^{j-1}V_{DC}$	$(4n+1)^m$	$m(2n+6)$	$m(4n+6)$	$2mn$	m	$3m$
$V_{j1} = (2n^2+4n+1)^{j-1}V_{DC}$ $V_{j2} = (n+1)V_{j1}$	$(2n^2+4n+1)^m$				$2m$	

5.4 Unipolar multilevel inverter configurations

This chapter describes the other existing multilevel inverter topologies that require H-bridge cells in their output stages to generate negative levels.

5.4.1 Suggested topology in [24]

The presented topology in [24] is depicted in Figure 5.38. As this figure shows, the cascaded part includes several half-bridge cells, and a full-bridge topology is placed on the output side of these cells to generate both negative and positive voltage levels. In fact, this topology cannot generate negative voltage levels inherently, and to generate these voltage levels, a full-bridge topology is utilized. The procedure of making the cascaded part of the existing topology in [24] is shown in Figure 5.39. This part includes several type-I cells connected in series according to this figure.

Regarding the shown topology in Figure 5.38, its characteristics can be listed in Table 5.18.

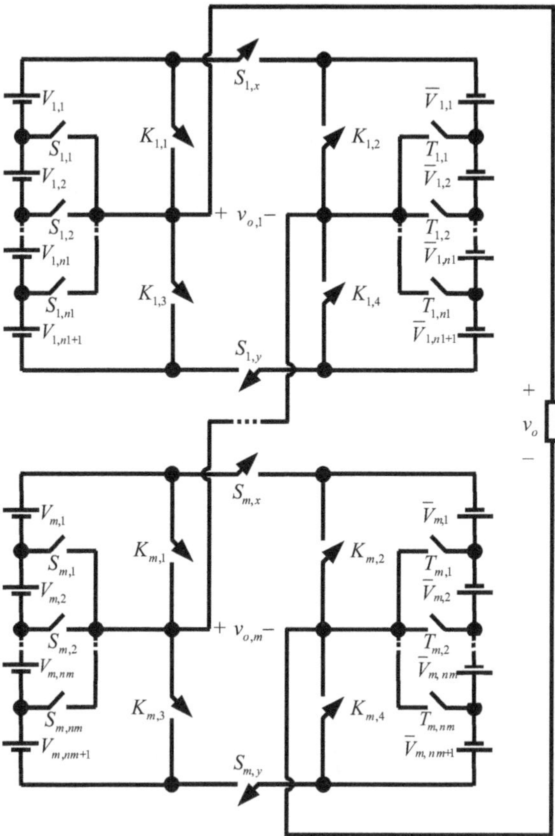

Figure 5.35 Presented topology in [22]

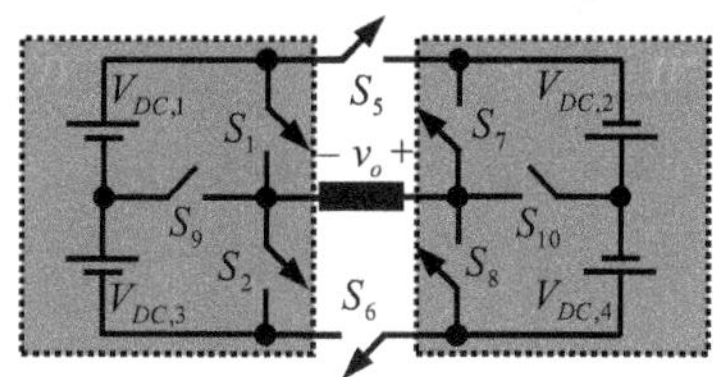

Figure 5.36 Construction of a cell of the suggested structure in [22]

Table 5.16 Key factors of the presented topology in [22] (j = 1,2, ..., m and k = 1,2, ..., n, n+1)

Pattern	N_{level}	N_{switch}	N_{IGBT}	N_{source}	$N_{variety}$	$N_{on,state}$
$V_{j,k} = (4n+5)^{j-1} V_{DC}$ $\overline{V}_{j,k} = V_{j,k}$	$(4n+5)^m$	$m\,(2n+6)$	$m\,(4n+6)$	$m\,(2n+2)$	m	$n\,(2m+1)$
$V_{j,k} = (2n^2+8n+7)^{j-1} V_{DC}$ $\overline{V}_{j,k} = (n+2)\, V_{j,k}$	$(2n^2+8n+7)^m$				$2m$	

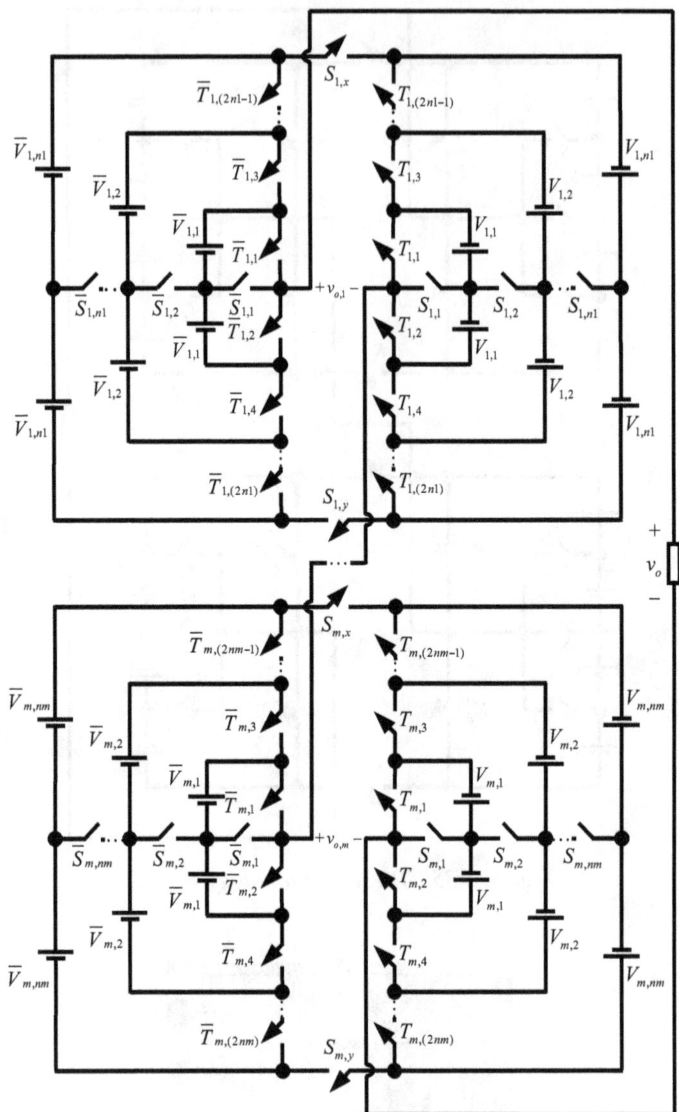

Figure 5.37 Presented topology in [23]

Table 5.17 Key factors of the presented topology in [23] (j = 1,2, …, m and k = 1,2, …, n)

Pattern	N_{level}	N_{switch}	N_{IGBT}	N_{source}	$N_{variety}$	V_{block}/V_{DC}	$N_{on,state}$
$\bar{V}_{jk} = 5^{k-1}(1+6\times 5^{n-1})^{j-1}V_{DC}$ $V_{jk} = 2\bar{V}_{jk}$	$(12\times 5^{n-1}+1)^m$	$m\,(6n+2)$	$m\,(8n+2)$	$4mn$	$2m$	$\frac{9}{4}(N_{level}-1)$	$m(2n+1)$
$\bar{V}_{jk} = 10^{k-1}(1+8\times 10^{n-1})^{j-1}V_{DC}$ $V_{jk} = 3\bar{V}_{jk}$	$(16\times 10^{n-1}+1)^m$				$2m$		

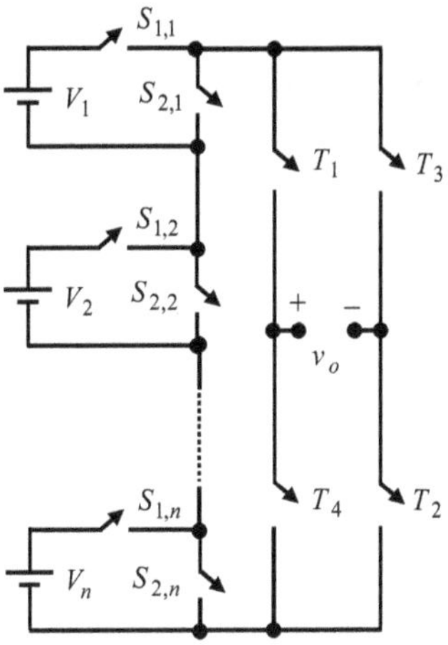

Figure 5.38 The presented topology in [24]

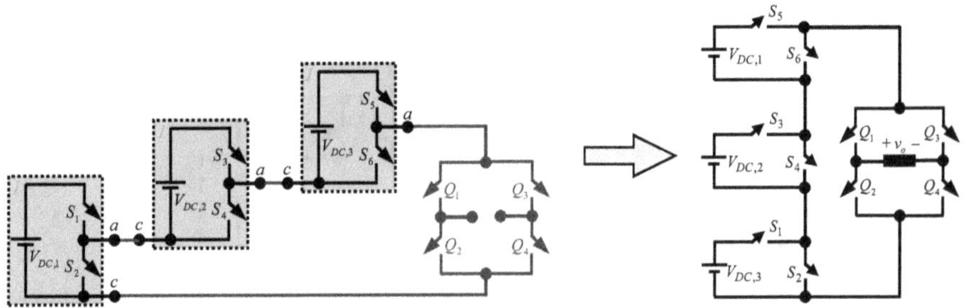

Figure 5.39 Construction of cascaded part of the suggested structure in [24]

Table 5.18 Characteristics of the presented topology in [24] (j = 1,2, . . . , n)

Pattern	N_{level}	N_{switch}	N_{IGBT}	N_{source}	$N_{variety}$	V_{block}	$N_{on,state}$
$V_j = V_{DC}$	$2n+1$	$2n+4$	$2n+4$	n	1	$3N_{level}-3$	$n+2$
$V_j = 2^{j-1}V_{DC}$	$2^{n+1}-1$				n		
$V_1 = V_{DC}$ $V_j = 2V_{DC}$	$4n-1$				2		

From Table 5.18, it is seen that the first pattern is a symmetric one and the other ones are asymmetric, whereas the third pattern only uses two sorts of magnitudes for the DC voltage sources.

5.4.2 Suggested topology in [25]

By connecting several numbers of presented cells in [24], the suggested structure in [25] can be obtained as shown in Figure 5.40.

Regarding the shown topology in Figure 5.40, its characteristics can be listed in Table 5.19.

It is seen that all of the presented patterns are asymmetric, whereas the third one has the widest variety of the DC voltage sources' algorithm.

5.4.3 Suggested topology in [26]

By using the presented structure in [24], the suggested structure in [26]. The proposed topology in [24] includes several half-bridge cells connected in series. Composing these cells with each other can lead to the presented topology in [26].

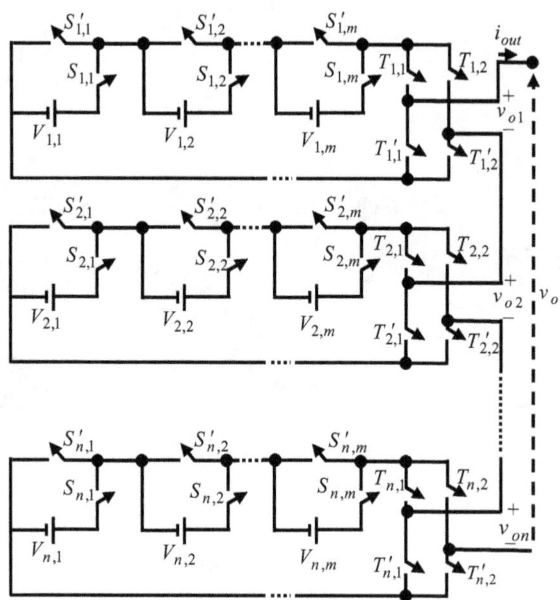

Figure 5.40 The presented topology in [25]

Table 5.19 Characteristics of the presented topology in [25] (j = 1,2, …, m and k = 1,2, …, n)

Pattern	N_{level}	N_{switch}	N_{IGBT}	N_{source}	$N_{variety}$	V_{block}/V_{DC}	$N_{on,state}$
$V_{k,j}=(2m+1)^{k-1}V_{DC}$	$(2m+1)^n$	$(2m+4)n$	$(2m+4)n$	mn	n	$3N_{level}-3$	$(m+2)n$
$V_{k,1}=(4m-1)^{n-1}V_{DC}V_{k,j}=2\,V_{k,1}$	$(4m-1)^n$				$2n$		
$V_{k,j}=2^{j-1}(2^{m+1}-1)^{n-1}V_{DC}$	$(2^{m+1}-1)^n$				mn		

As seen in Figure 5.41, this topology includes several similar cells, where each cell has an H-bridge inverter in its output cell.

Regarding the shown topology in Figure 5.41, its characteristics can be listed in Table 5.20.

5.4.4 Suggested topology in [27]

By having the obtained cells in [26] and connecting them in series, the suggested structure in [27] can be obtained. This topology is demonstrated in Figure 5.42. The construction of a cell of this topology using type-I and type-II cells is shown in Figure 5.43.

Regarding the shown topology in Figure 5.42, its characteristics can be listed in Table 5.21.

5.4.5 Suggested topology in [28]

The suggested structure in [28] is depicted in Figure 5.44. This figure shows that each cell includes a half-bridge cell in series with a DC voltage source, whereas a

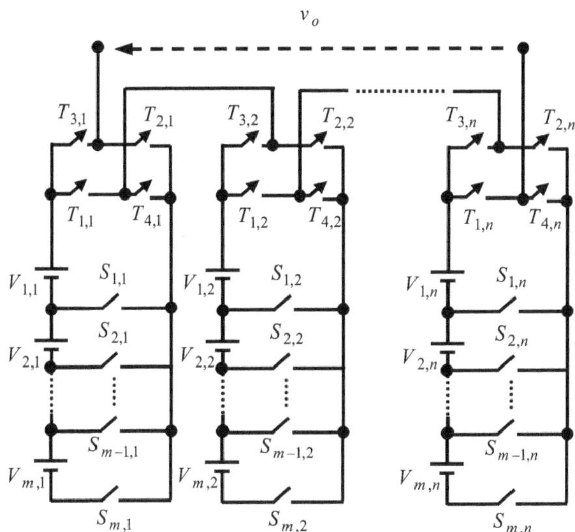

Figure 5.41 The suggested structure in [26]

Table 5.20 Characteristics of the presented topology in [26] (j = 1,2, ..., m and k = 1,2, ..., n)

Pattern	N_{level}	N_{switch}	N_{IGBT}	N_{source}	$N_{variety}$	V_{block}/V_{DC}	$N_{on,state}$
$V_{k,j} = (2m+1)^{k-1}V_{DC}$	$(2m+1)^n$	$(m+4)n$	$(2m+4)n$	mn	n	$2m(N_{level}-1)$	$3n$

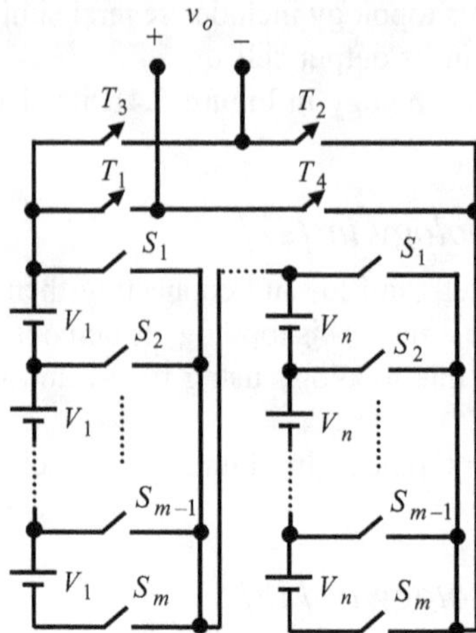

Figure 5.42 The suggested structure in [27]

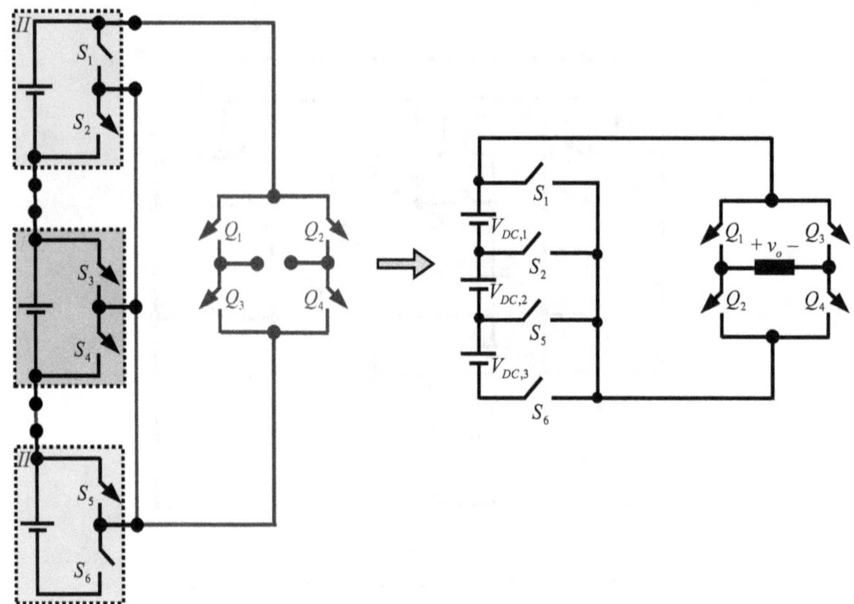

Figure 5.43 Construction of cascaded part of the suggested structure in [27]

Table 5.21 Characteristics of the presented topology in [27] (j = 1,2, …, n)

Pattern	N_{level}	N_{switch}	N_{IGBT}	N_{source}	$N_{variety}$	$N_{on,state}$
$V_j = (m^{j-1})V_{DC}$	$2m^n - 1$	$mn + 4$	$2mn + 4$	$mn - n$	n	$n + 2$

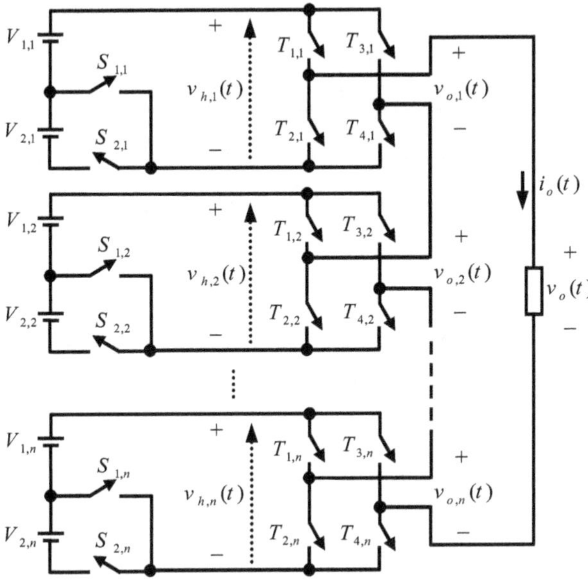

Figure 5.44 Suggested structure in [28]

Table 5.22 Characteristics of the suggested structure in [28] (j = 1,2, ..., n)

Pattern	N_{level}	N_{switch}	N_{IGBT}	N_{source}	$N_{variety}$	V_{block}/V_{DC}	$N_{on,state}$
$V_j = V_{DC}$	$4n+1$	$6n$	$6n$	$2n$	1	$2.5(N_{level}-1)$	$3n$
$V_j = 2^{j-1}V_{DC}$	$2^{n+2}-3$				n		
$V_j = 3^{j-1}V_{DC}$	$(2\times 3^n)-1$						
$V_j = 4^{j-1}V_{DC}$	$\frac{4^{n+1}-1}{3}$						
$V_j = 5^{j-1}V_{DC}$	5^n						

full-bridge cell generates all negative and positive voltage levels. Regarding the different patterns for selecting the magnitudes of the DC voltage sources, the characteristics of this structure can be listed in Table 5.22. As shown in Figure 5.44, the cells can be made using the type-I cells and an additional DC voltage source.

As seen in this table, the first pattern is a symmetric algorithm. In contrast, the other ones are asymmetric. Also, it is seen that the number of voltage levels can be maximized using the fifth pattern.

5.4.6 Suggested structure in [29]

Supposing the existing topology in [26], if the number of the DC voltage sources in each cell is supposed to equal two, the suggested topology in [29] can be obtained. This topology is shown in Figure 5.45. These topologies' characteristics can be listed in Table 5.23.

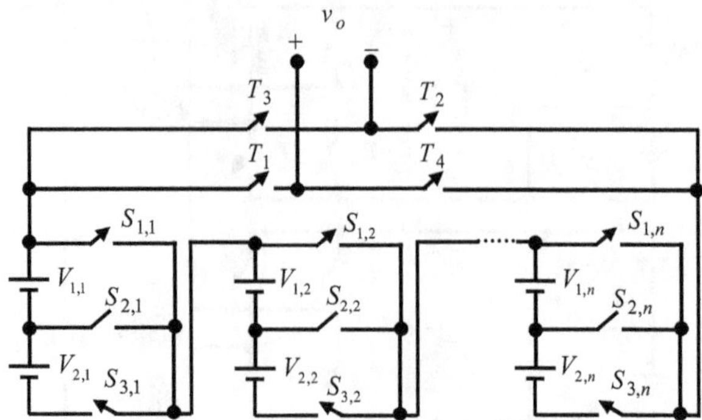

Figure 5.45 Existing topologies in [29]

Table 5.23 Characteristics of the suggested structure in [29] (j = 1,2, …, n and k=1,2)

Pattern	N_{level}	N_{switch}	N_{IGBT}	N_{source}	$N_{variety}$	V_{block}/V_{DC}	$N_{on,state}$
$V_{k,j} = V_{DC}$	$4n + 1$	$3n + 4$	$4n + 4$	$2n$	1	$\frac{11}{4}(N_{level} - 1)$	$n + 2$
$V_{k,1} = V_{DC}$ $V_{k,j} = 2V_{DC}$	$8n - 3$				2		
$V_{k,1} = V_{DC}$ $V_{k,j} = 3V_{DC}$	$12n - 7$				2		
$V_{k,j} = 2^{j-1}V_{DC}$	$2^{n+2} - 3$				n		
$V_j = 3^{j-1}V_{DC}$	$2 \times 3^n + 1$				n		

The first algorithm is symmetric in Table 5.23, whereas the others are asymmetric. Also, it is seen that the second and the third algorithm have two types of DC voltage sources.

5.4.7 Suggested structures in [30–32]

The existing structure in [30] is shown in Figure 5.46. As seen in this figure, it is obtained from the connection of some half-bridge structures, and a full-bridge cell is used on its output stage to generate both positive and negative voltage levels. This topology is depicted in Figure 5.46. Regarding this topology, its key equations can be listed in Table 5.24.

Using this topology, other structures were presented in [31] and [32]. These structures can be seen in Figure 5.47.

As was described before, the presented topologies in [30–32] are obtained from half-bridge or type-I cells. The procedure of making these topologies using these cells is shown in Figure 5.48.

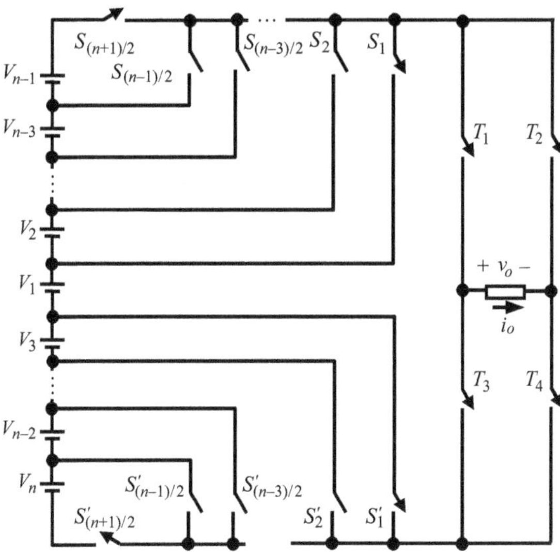

Figure 5.46 Suggested topology in [30]

Table 5.24 Characteristics of the presented topology in [30] (j = 1,2, . . . , n)

Pattern	N_{level}	N_{switch}	N_{IGBT}	N_{source}	$N_{variety}$	$N_{\text{on,state}}$
$V_j = V_{DC}$	$2n+1$	$n+5$	$2n+2$	n	1	4

5.4.8 Suggested structure in [33]

Another topology, which is similar to the existing topologies in [31] and [32], is suggested in [33]. This topology is shown in Figure 5.49. This figure shows that this structure can be obtained from the half-bridge cells, similar to the other topologies.

5.4.9 Suggested structures in [34]

The existing topology in [34] is depicted in Figure 5.50. This figure shows that this structure has a full-bride cell on its output stage to generate all positive and negative voltage levels. In other words, the cascaded part cannot inherently generate negative voltage levels. Regarding the topology of this structure, it can be seen that it can be obtained using the presented topology in [27].

To maximize the number of the generated voltage levels, an H-bridge cell can be connected in series with the presented topology in Figure 5.50. This structure is shown in Figure 5.51. Regarding Figure 5.51, its characteristics can be listed in Table 5.25.

The pattern which is presented for the cascaded part is a symmetric kind. In other words, the magnitudes of all DC voltage sources are assumed to be equal.

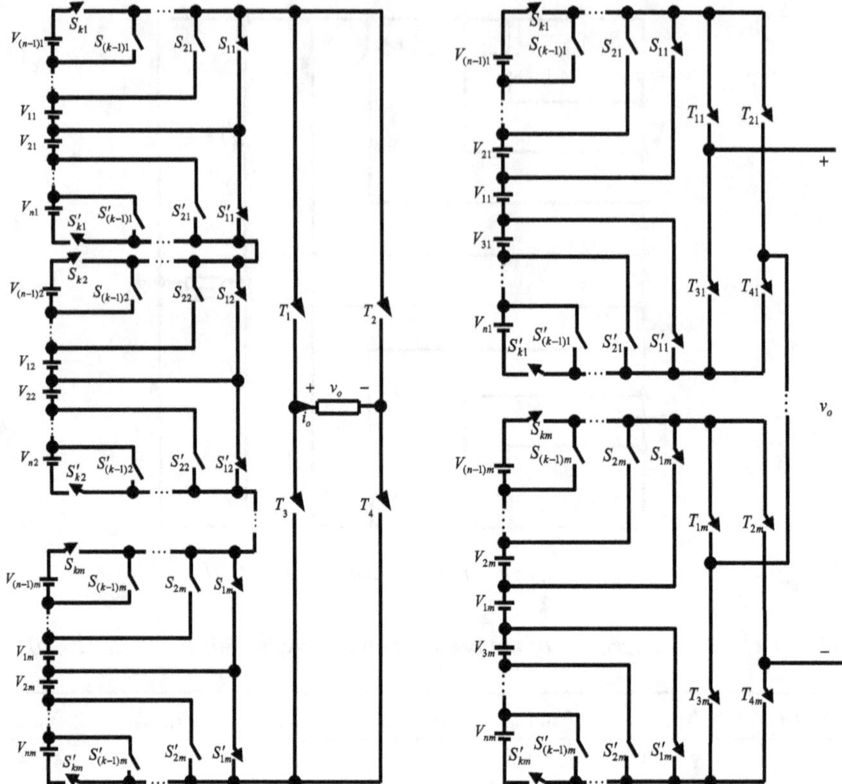

Figure 5.47 Presented topologies in [31] and [32]: (a) [31] and (b) [32]

5.4.10 Suggested structures in [35]

The presented topology in [35] can be obtained from a cascaded connection of the existing cell in [34]. This topology is shown in Figure 5.52.

Regarding the presented pattern for this topology, its characteristics can be listed in Table 5.26.

Regarding this pattern, the magnitudes of DC voltage sources are considered the same in each cell. However, each cell's DC voltage sources have different magnitudes than the adjacent cells.

5.4.11 Suggested structures in [36]

The presented topology in [36] is depicted in Figure 5.53. Regarding this topology, it can be seen that it is similar to the presented topology in [25].

As seen in Figure 5.53, the H-bridge cells are used for each multilevel cell to generate both negative and positive voltage levels. In other words, this topology cannot generate negative voltage levels by itself, similar to the topology presented in [25].

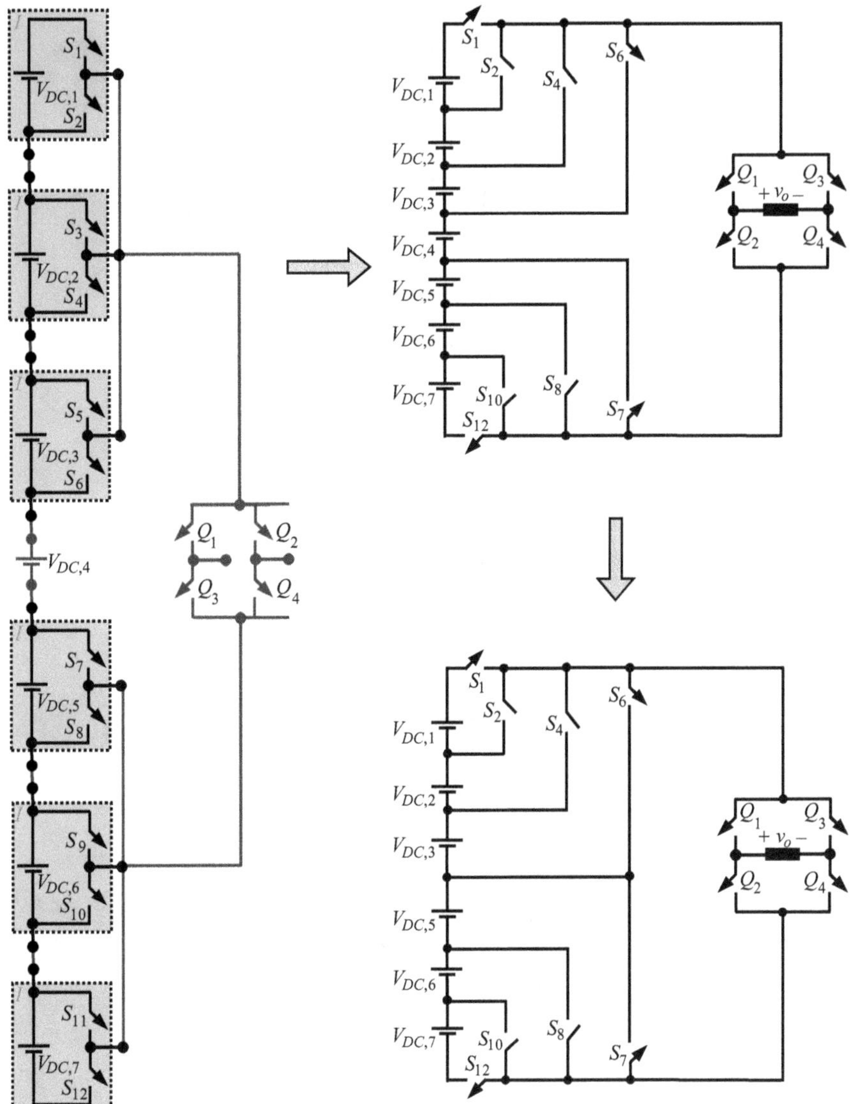

Figure 5.48 Construction of the presented topologies in [31] and [32] using type-I cells

5.4.12 Existing topology in [37]

Figure 5.54 shows the presented basic unit in [37]. This figure shows that this cell has five unidirectional switches and three DC voltage sources. This cell can generate three voltage levels, where all of which are positive. So, an H-bridge cell must be used in order to generate negative voltage levels.

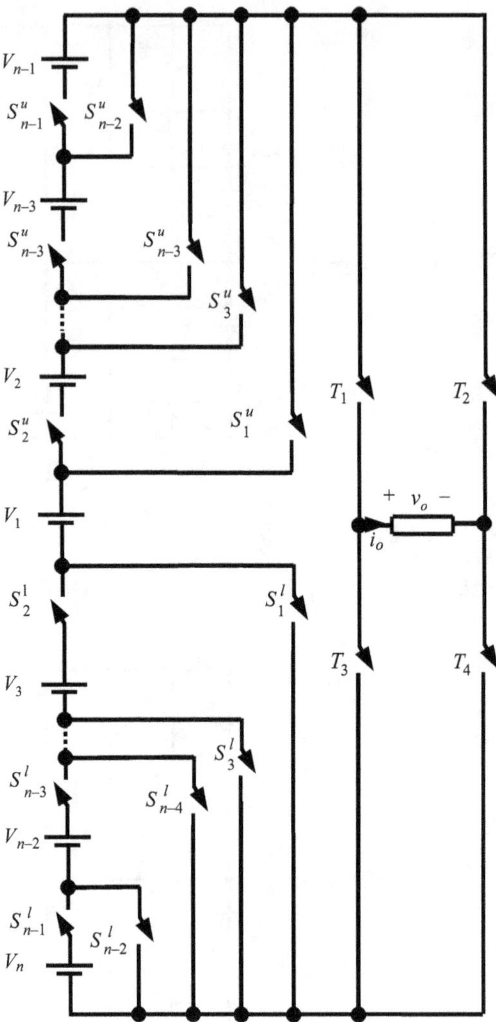

Figure 5.49 Presented topology in [33]

The cascaded topology, which includes a series of connections of several basic units, is depicted in Figure 5.55. As shown in this figure, an additive half-bridge cell is used to maximize the number of generated voltage levels on the output stage. Four different patterns are presented regarding this cascaded topology, which are listed in Table 5.27. As seen in this table, the first pattern is symmetric, and the others are asymmetric.

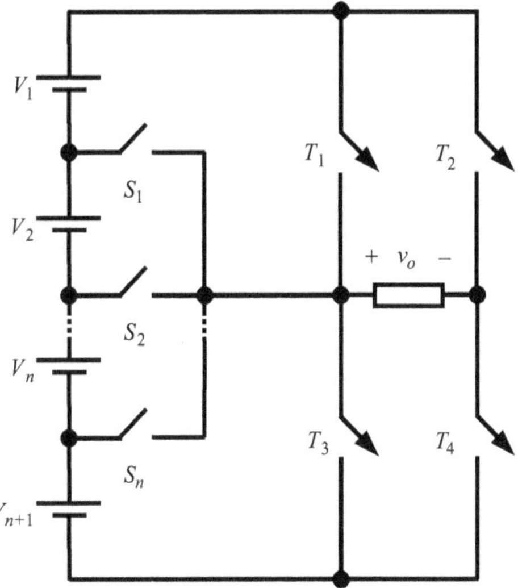

Figure 5.50 Presented topology in [34]

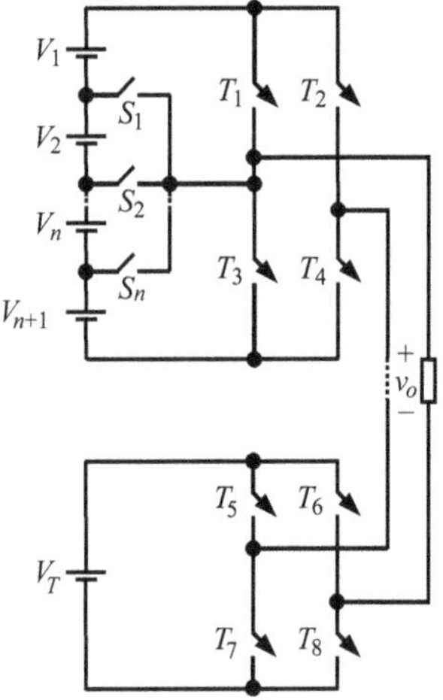

Figure 5.51 Suggested topology in [34] in series with an H-bridge cell

Table 5.25 Characteristics of the presented topology in [34] (j = 1,2, ..., n, n+1)

Pattern	N_{level}	N_{switch}	N_{IGBT}	N_{source}	$N_{variety}$	$N_{\text{on,state}}$
$V_j = V_{DC}$ $V_T = (2n+3)\ V_{DC}$	$6n+9$	$n+8$	$2n+8$	$n+2$	2	4
$V_j = V_{DC}$ $V_T = (n+2)\ V_{DC}$	$4n+7$				2	

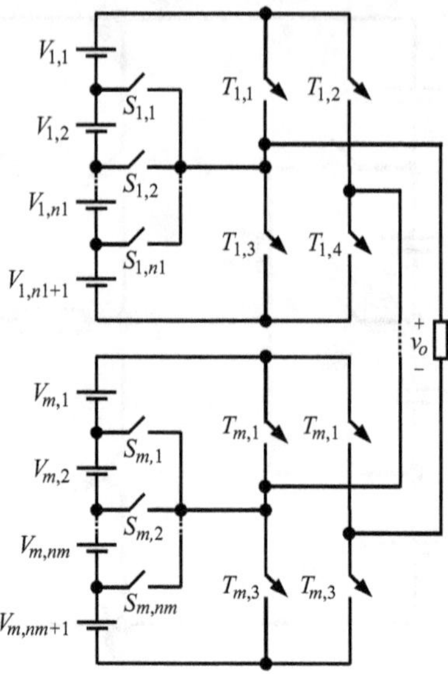

Figure 5.52 Suggested topology in [35]

Table 5.26 Characteristics of the presented topology in [35] (j = 1,2, ..., n, n+1, k = 1,2, ..., m)

Pattern	N_{level}	N_{switch}	N_{IGBT}	N_{source}	$N_{variety}$	$N_{on,state}$
$V_{j,k} = (2n+3)^{k-1} V_{DC}$	$(2n+3)^m$	$m\ (n+4)$	$m\ (2n+4)$	$m\ (n+1)$	m	$2m$

5.4.13 Existing topology in [38]

The presented topology in [38] is shown in Figure 5.56. As this figure shows, the cascaded unit includes several similar cells, where each cell includes four unidirectional power switches and three DC voltage sources. Also, an H-bridge cell is used on the output stage of this topology in order to generate the negative voltage levels. In other words, the cascaded unit cannot generate these voltage levels inherently. Regarding the basic cells of this topology, it can be seen that each can be obtained using a parallel connection of two type-I cells, where an additional DC voltage source is connected with them in series. Also, it is required to mention that every switch only has to block a positive voltage, so all of them are unidirectional.

The pattern which is used for this topology is a symmetric one. Using this pattern, the characteristics of this topology can be listed in Table 5.28.

5.4.14 Existing topology in [39]

Using the presented cells in [38] and substituting them in the existing topology in [15], the suggested topology in [39] can be obtained. This topology is shown in

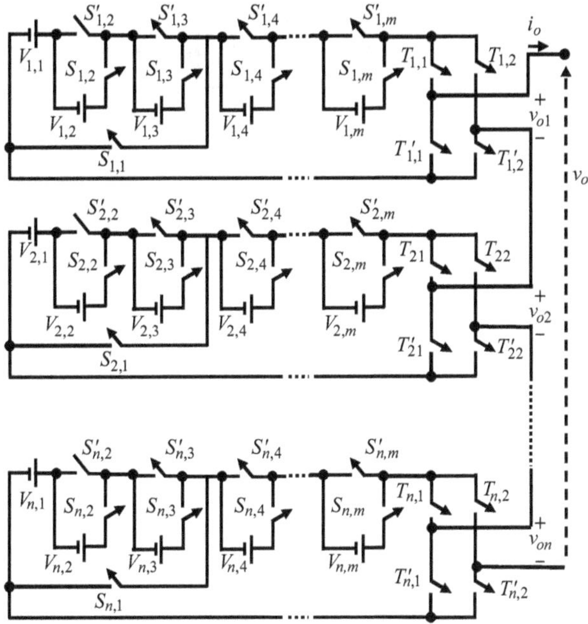

Figure 5.53 Suggested topology in [36]

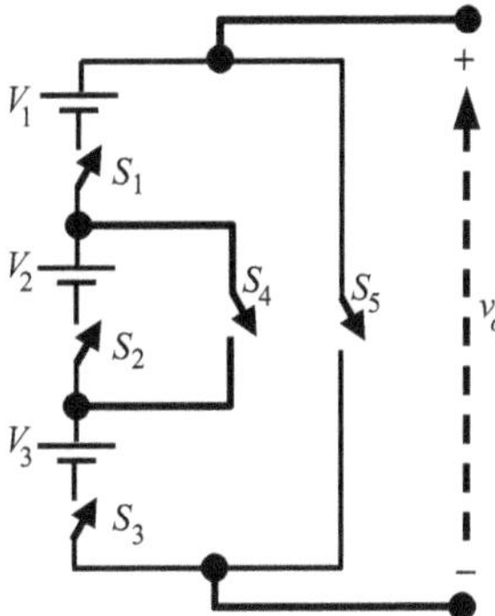

Figure 5.54 Proposed basic unit in [37]

Figure 5.57. As seen in this figure, an individual DC voltage source is used in this topology, which can be omitted or not.

This topology uses the symmetric algorithm to generate the output voltage levels, but two different conditions can be given regarding the additional DC voltage source. These conditions are explained in Table 5.29.

Regarding Table 5.29, it can be seen that the number of voltage levels can be increased by two by using the individual DC voltage source.

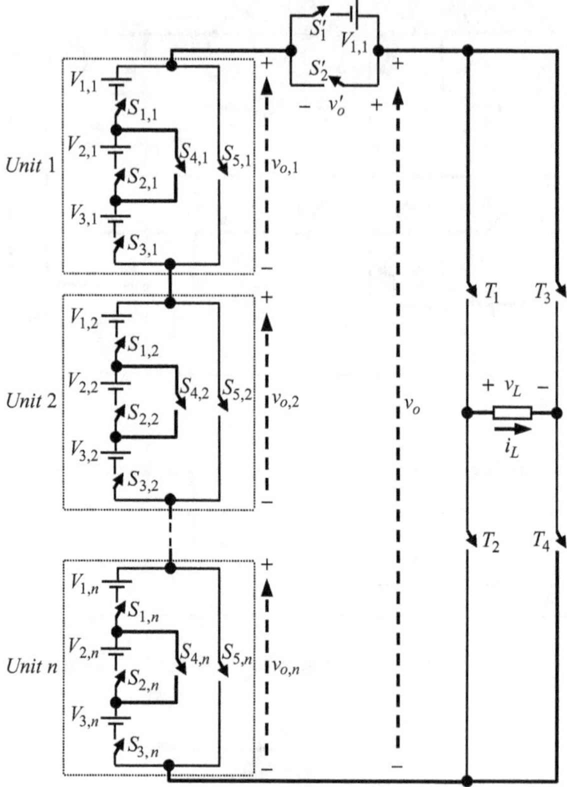

Figure 5.55 Proposed basic unit in [37]

Table 5.27 Characteristics of the presented topology in [37] (j = 1,2, …, n)

Pattern	N_{level}	N_{switch}	N_{IGBT}	N_{source}	$N_{variety}$	V_{block}/V_{DC}	$N_{on,state}$
$V_{1,j} = V_{2,j} = V_{3,j} = V_{DC}$	$6n+3$	$5n+6$	$5n+6$	$3n+1$	1	$21n+6$	$3n+3$
$V_{1,1} = V_{2,1} = V_{3,1} = V_{DC}$ $V_{1,j} = V_{2,j} = V_{3,j} = 2\ V_{DC}$	$12n-9$				2	$40n-13$	
$V_{1,1} = V_{2,1} = V_{3,1} = V_{DC}$ $3V_{1,j} = V_{2,j} = 3V_{3,j} = 3^{j-1}V_{DC}$	$5(3^{n-1})+4$				n	$82\ (3^{n-1})-7$	
$2V_{1,j} = V_{2,j} = 2V_{3,j} = 2^{j}V_{DC}$	$2^{n+3}-5$					$7\ (2^{n+2})-22$	

5.4.15 Existing topology in [40]

The suggested topology in [40] is shown in Figure 5.58. As seen in this topology, this structure includes several half-bridge cells that are connected. Also, this topology cannot generate negative voltage levels inherently and requires an

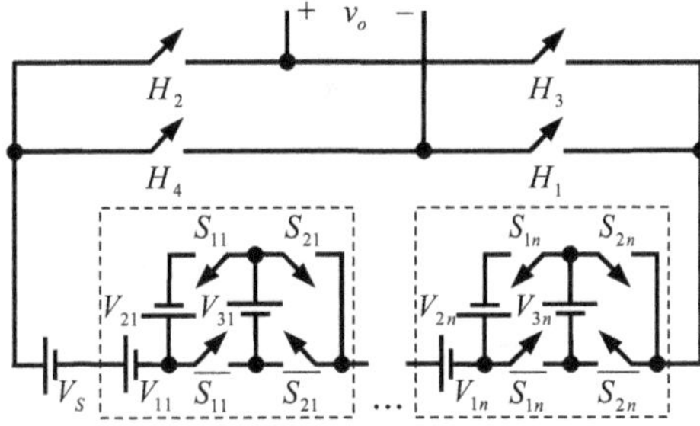

Figure 5.56 Suggested structure in [38]

Table 5.28 Key factors of the presented topology in [38] (j = 1,2, …, n)

Pattern	N_{level}	N_{switch}	N_{IGBT}	N_{source}	$N_{variety}$	V_{block}/V_{DC}	$N_{on,state}$
$V_{1j} = V_{2j} = V_{3j} = V_S = V_{DC}$	$6n+3$	$4n+4$	$4n+4$	$3n+1$	1	$3N_{level}-3$	$2n+2$

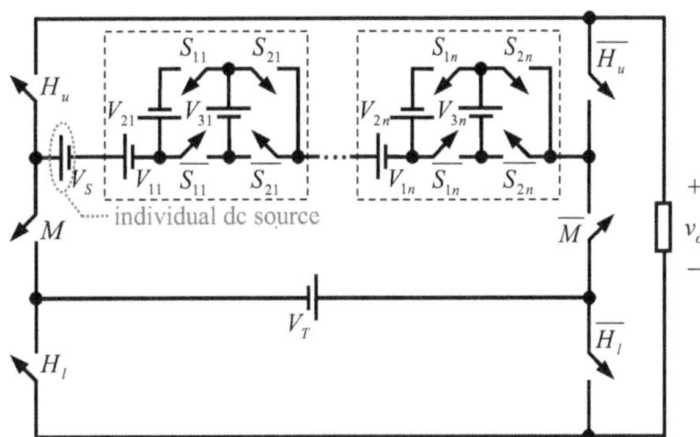

Figure 5.57 Presented topology in [39]

Table 5.29 Key factors of the presented topology in [39] (j = 1,2, …, n)

Pattern	N_{level}	N_{switch}	N_{IGBT}	N_{source}	$N_{variety}$	V_{block}/V_{DC}	$N_{on,state}$
$V_{j1} = V_{j2} = V_{j3} = V_{DC}$	$6n+3$	$4n+6$	$4n+6$	$3n+1$	1	$3N_{level}-5$	$2n+3$
$V_{j1} = V_{j2} = V_{j3} = V_S = V_{DC}$	$6n+5$			$3n+2$		$3N_{level}-7$	

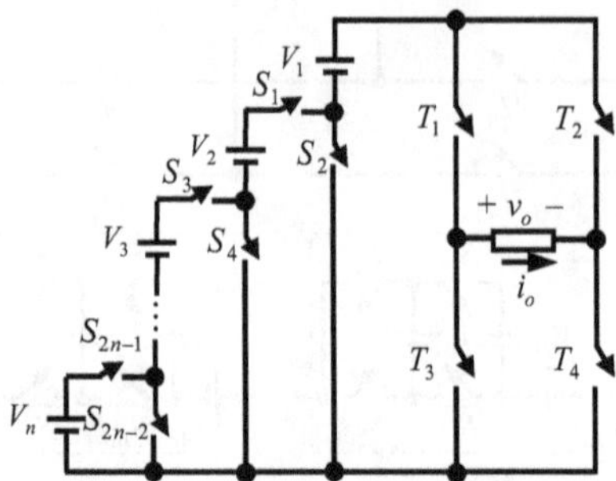

Figure 5.58 Suggested topology in [40]

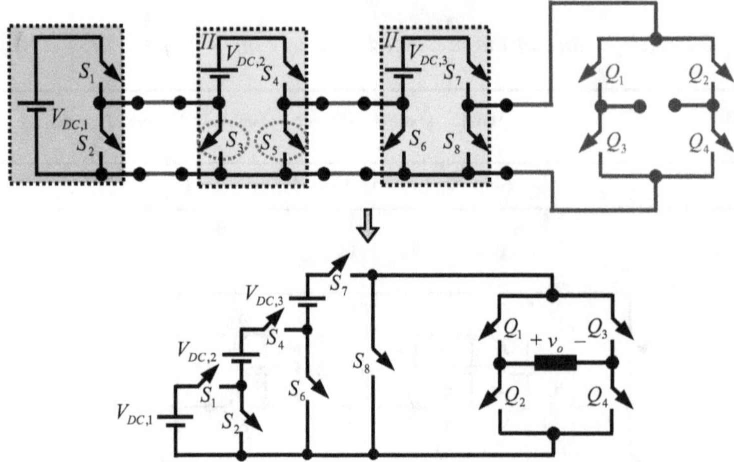

Figure 5.59 Construction of the presented topologies in [40] using type-I and type-II cells

Table 5.30 Key factors of the presented topology in [40] (j = 1,2, …, n)

Pattern	N_{level}	N_{switch}	N_{IGBT}	N_{source}	$N_{variety}$	$N_{\text{on,state}}$
$V_j = V_{DC}$	$2n + 1$	$2n + 2$	$2n + 2$	n	1	n

H-bridge cell to generate them. This topology can be obtained using the type-I and type-II cells; this procedure is shown in Figure 5.59. This figure shows this procedure for n equals three.

Regarding Figure 5.59, some switches are omitted in this procedure to make the presented topology in [40]. The used pattern for this topology is a symmetric one. Regarding this pattern, the key factors of this topology can be obtained. Table 5.30 shows the results of these calculations.

5.4.16 Existing topology in [41]

The suggested topology in [41] is depicted in Figure 5.60. As seen in this figure, this structure includes several similar cells that are connected in a cascaded form. Also, it can be observed that an H-bridge cell is used in the output stage of this structure, which shows that it cannot generate negative voltage levels by itself. Moreover, each cell includes two unidirectional switches, two bidirectional switches, and two DC voltage sources.

Two different patterns are presented in [41] to determine the magnitudes of the DC voltage sources, which are listed in Table 5.31.

As seen in Table 5.31, one of the patterns is symmetric, and the other is an asymmetric one. Also, it can be seen that more voltage levels can be generated using the asymmetric pattern.

5.4.17 Existing topology in [42]

The presented topology in [42] constructs from the series connection of an H-bridge cell with an additional part. This topology is shown in Figure 5.61. As seen in this figure, the additional part includes a series connection of several switched DC voltage sources. Also, this part has an H-bridge cell used to generate the negative voltage levels.

To determine the magnitudes of the DC voltage sources, two different patterns are suggested in [42]. These patterns and associated parameters are listed in Table 5.32.

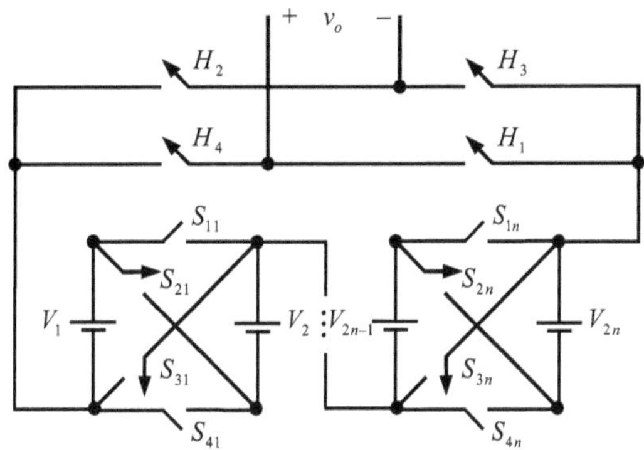

Figure 5.60 Existing topology in [41]

Table 5.31 Key factors of the presented topology in [41] (j = 1,2, …, n)

Pattern	N_{level}	N_{switch}	N_{IGBT}	N_{source}	$N_{variety}$	V_{block}/V_{DC}	$N_{on,state}$
$V_{2j-1} = V_{2j} = V_{DC}$	$4n+1$	$4n+4$	$6n+4$	$2n$	1	$14n$	$n+2$
$p\,V_{2j-1} = V_{2j} = p^{2j-1}\,V_{DC}$	$2\,\frac{1-p^{2n}}{1-p}+1$				$2n$	$(8p+6)\,\frac{1-p^{2n}}{1-p^2}$	

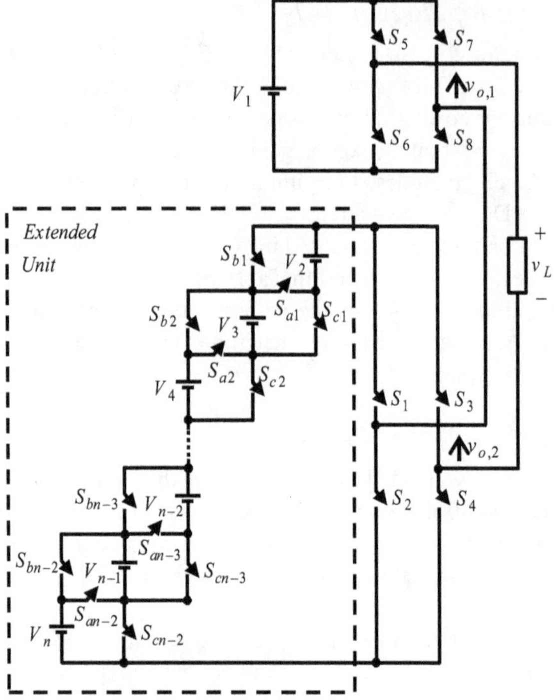

Figure 5.61 Suggested topology in [42]

Table 5.32 Key factors of the presented topology in [42] (j = 1,2, . . . , n)

Pattern	N_{level}	N_{switch}	N_{IGBT}	N_{source}	$N_{variety}$	V_{block}/V_{DC}	$N_{on,state}$
2 $V_1 = V_j = 2V_{DC}$	$4n - 1$	$3n + 2$	$3n + 2$	n	2	$3.5\ (N_{level} - 15)$	$n + 2$
3 $V_1 = V_j = 3V_{DC}$	$6n - 3$					$(8p + 6)\ \frac{1-p^{2n}}{1-p^2}$	

As seen in Table 5.32, both suggested patterns are asymmetric. However, the second one has the potential to generate more voltage levels. Moreover, the parallel connection of type-I and type-III cells can be used to make this cell. This procedure is shown for n=4 in Figure 5.62.

5.4.18 Existing topology in [43]

The suggested topology in [43] is shown in Figure 5.63. As seen in this figure, this topology includes a series of connections of the additional part presented in [42]. As seen in Figure 5.63, the magnitudes of the DC voltage sources are supposed the same in each of the stages. Considering this description, the characteristics of this topology can be listed in Table 5.33.

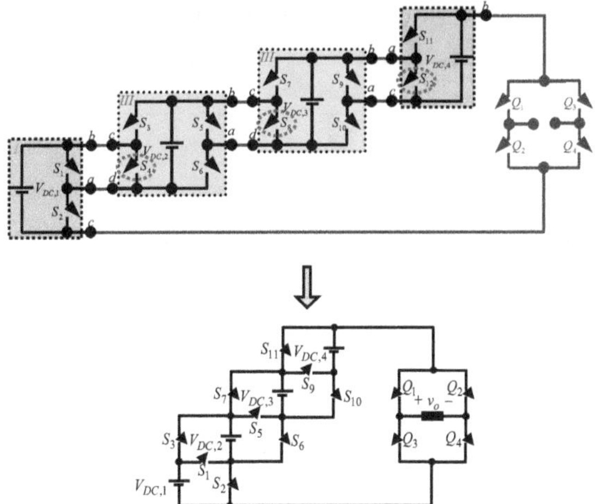

Figure 5.62 Construction of the presented topologies in [42] using type-I and type-III cells

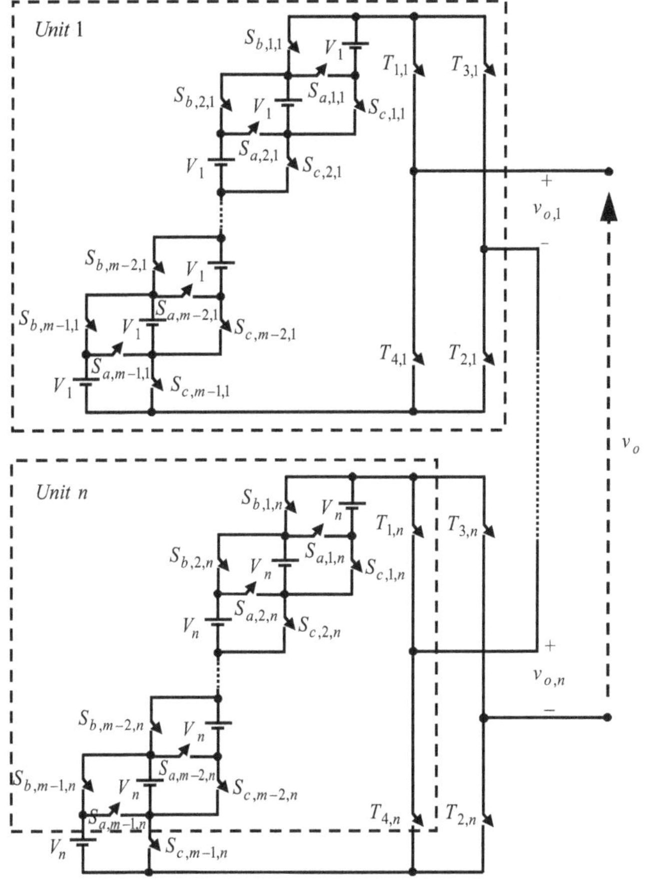

Figure 5.63 Suggested topology in [43]

Table 5.33 Key factors of the presented topology in [43] (j = 1,2, …, m)

Pattern	N_{level}	N_{switch}	N_{IGBT}	N_{source}	$N_{variety}$	$N_{on,state}$
$V_j = (2m+1)^{j-1} V_{DC}$	$(2m+1)^n$	$n(3m+1)$	$n(3m+1)$	mn	m	mn

5.5 Comparison of existing multilevel inverters

This section compares some of the existing topologies with the conventional cascaded H-bridge (CHB) configuration to study their merits and disadvantages. The topologies that are used in these comparisons are presented in [6, 9–11, 13, 15, 17, 18, 23, 24, 29, 37, 39, 41]. Moreover, these comparisons include several points of view, such as the count of the switches, IGBTs, DC voltage sources, the variety in the magnitudes of the DC voltage sources, the total blocked voltage by the switches, and the maximum count of switches that are switched on to pass the output current.

The binary algorithm is supposed for the CHB topology. For the other ones, their best patterns, which have the potential to generate the maximum count of the voltage levels, are considered. The parameter of p is considered equal to two for the existing topologies in [9, 15, 41], and the number of the submodules is supposed to equal to one in the suggested structure in [23].

Figure 5.64 shows the comparison of the count of the required switches. It can be seen that the presented topology in [23] generates a number of voltage levels with the minimum number of switches. Also, the existing topologies in [10, 17, 18] require the exact count of the power switches. Also, it can be seen that the presented topologies in [9, 12, 24, 37, 39] need a higher number of switches than the CHB topology, that is, a disadvantageous point for these topologies.

The number of required IGBTs in the different topologies is compared in Figure 5.65. By comparing Figures 4.42 and 4.43, it can be shown that the number of required switches can vary from the IGBT count of the IGBTs. As mentioned before, a bidirectional switch includes two unidirectional ones. The leading cause of this difference is the use of bidirectional switches in some structures. Figure 5.65 depicts that the suggested topologies in [17, 18] generate a higher count of the

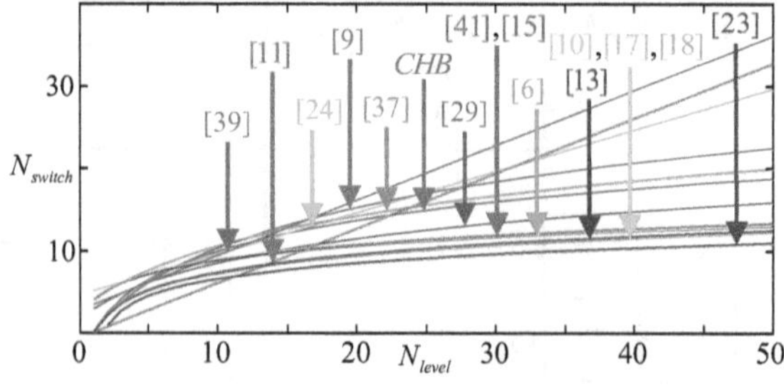

Figure 5.64 Switches' count comparison

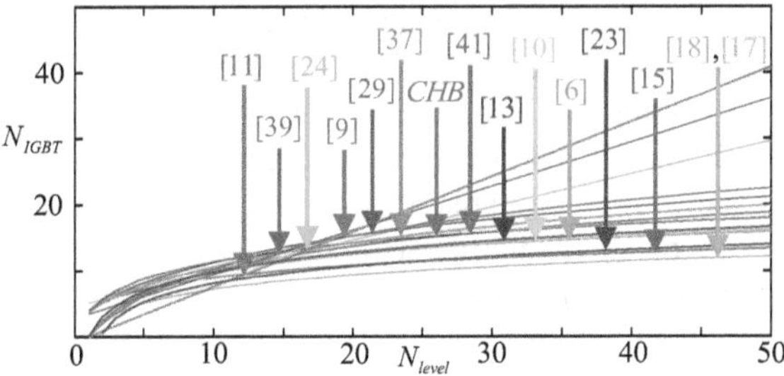

Figure 5.65 IGBTs' count comparison

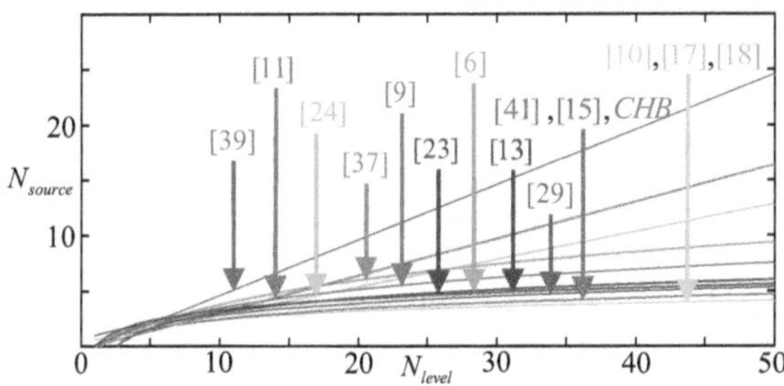

Figure 5.66 DC voltage sources' count comparison

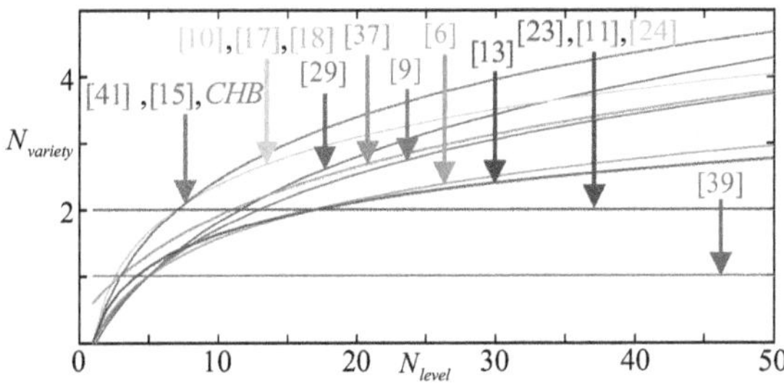

Figure 5.67 Variety in magnitudes of DC voltage sources comparison

voltage levels with the lower IGBTs. Moreover, it can be seen that the presented topologies in [9, 12, 24, 29, 37, 39] need a higher number of IGBTs than the CHB topology, that is a disadvantageous point for these topologies.

The comparison from the number of the required DC voltage sources is demonstrated in Figure 5.66. If a topology requires a high number of DC voltage sources, its volume and size will increase. The presented topologies in [10, 17, 18] generate a high number of voltage levels with a lower number of DC voltage sources regarding Figure 5.66.

The comparison from the variety in the magnitudes of the DC voltage sources point of view is illustrated in Figure 5.67. A high difference in the magnitudes of the DC voltage sources means the switches have to endure different voltages on them; in other words, they have different voltage ratings. This parameter can lead to higher cost and complexity of a topology. The presented topology in [39] has superiority in decreased cost and complexity regarding the given descriptions.

Figure 5.68 shows the comparison from the switches' total blocked voltage point of view. A topology's volume, size, and cost will increase if its switches block a high voltage on them. Some topologies can generate negative voltages by themselves, as described before. However, others require an H-bridge cell to generate these voltage levels that can be considered a parameter that can increase the total blocking voltage by the switches in a structure because each of the switches of this part must be able to block a voltage equal to the output voltage. The presented topologies in [6, 24, 37, 41] are some topologies that generate negative voltages by means of H-bridge cells. These topologies block a higher voltage on their switches regarding Figure 5.68, confirming the given descriptions. Also, the switches of the presented topology in [38] have the least blocking voltage.

The other important parameter is the number of switches that conduct the output current. It can be shown that the number of switches that are turned on directly relates to the power losses of a structure. The presented topology in [23] has the least number of turned-on switches regarding Figure 5.69, which means this topology has lower power losses.

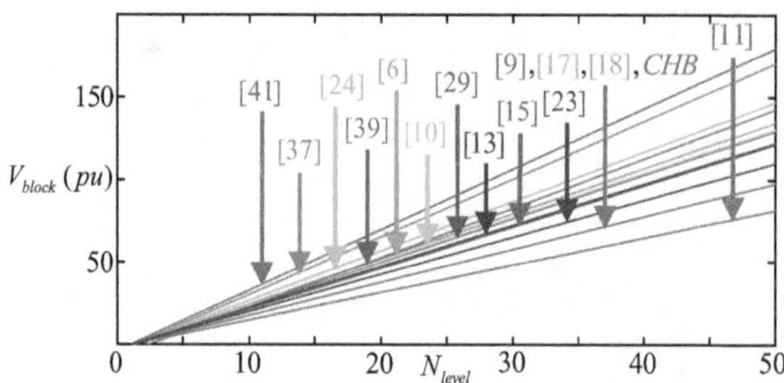

Figure 5.68 Comparison from total blocked voltage by switches point of view

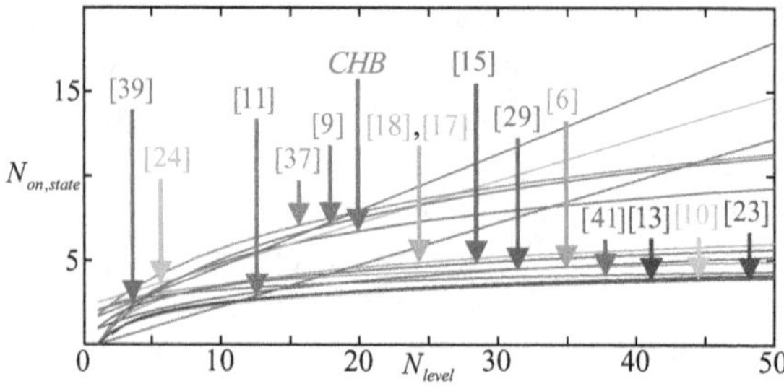

Figure 5.69 Comparison of the number of turned-on switches point of view

5.6 Optimization of cascaded multilevel inverters

Ref. [27] presents a topology depicted in Figure 5.70. As seen in this figure, in this configuration, the first unit, second unit, ..., and kth unit have $n_1, n_2, \cdots, n_k$ switches, respectively.

In Figure 5.70, the maximum number of values for v_o is obtained as follows:

$$N_{level} = \prod_{i=1}^{k} n_i = n_1 \times n_2 \times \cdots \times n_k \tag{5.1}$$

and the maximum output voltage $\left(V_{o,\,\max}\right)$ can be evaluated by:

$$V_{o,\,\max} = \sum_{j=1}^{k} (n_j - 1) V_i \tag{5.2}$$

The reason for using the term "maximum" in (5.1) is that it is possible to have an equal value for v_o over different states of the switches. In order to have unequal values for v_o in Figure 5.70 and produce linear steps, the values of the dc voltage sources must be chosen according to the following algorithm:

Unit 1:

$$V_1 = V_{dc} \tag{5.3}$$

Unit 2:

$$V_2 = (n_1 - 1)V_1 + V_1 = n_1 - V_{dc} \tag{5.4}$$

Unit 3:

$$V_3 = (n_1 - 1)V_1 + (n_2 - 1)V_2 + V_1 = n_1 \cdot n_2 \cdot V_{dc} \tag{5.5}$$

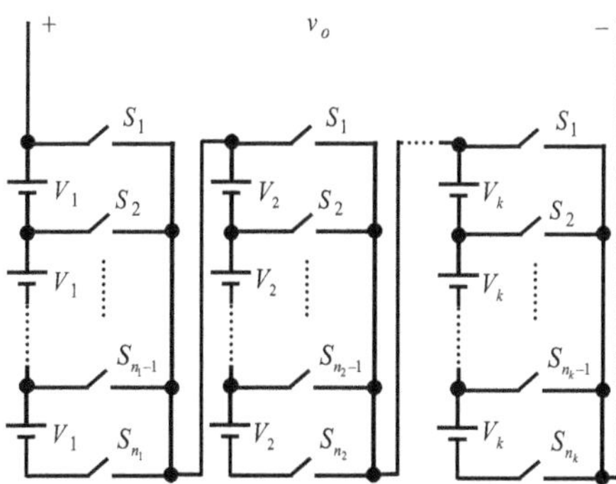

Figure 5.70 Suggested topology in [27]

In general, the dc voltage sources in jth unit are calculated using the following equation:

$$V_j = V_1 + \sum_{i=1}^{j-1} [(n_i - 1)V_i] = \prod_{i=1}^{j-1} (n_i \cdot V_{dc})\, j \geq 2 \tag{5.6}$$

5.6.1 Optimal structure for the maximum number of voltage steps with a constant number of switches

The desirable objective in a multilevel converter is obtaining the maximum step number for minimum switches. The question concerning the suggested structure is whether the number of switches is constant; which topology can provide a maximum number of output voltage steps?

Suppose the proposed topology consists of a series of k extended base units each with n_i switches ($i = 1, 2, \ldots, k$). This topology is shown in Fig. 5.70. Having this structure:

$$N_{Swich} = n_1 + n_2 + \cdots + n_k \tag{5.7}$$

In this case, the maximum number of voltage steps is given by (5.1). Considering (5.1) and (5.7) and that the product of numbers whose summation is constant will be maximum when all are equal:

$$n_1 = n_2 = \cdots = n_k = n \tag{5.8}$$

From (5.7) and (5.8), it is evident that:

$$k = \frac{N_{Swich}}{n} \tag{5.9}$$

The value of n must now be determined. Considering (5.1) and (5.8), the maximum number of voltage steps will be:

$$N_{Step} = n^k \tag{5.10}$$

Considering (9) and (10), it is clear that:

$$N_{level} = \left(n^{\frac{1}{n}}\right)^{N_{switch}} \tag{5.11}$$

Figure 5.71 shows the variation of $n^{1/n}$ versus n. It is clear that the maximum number of voltage steps is obtained for $n = 3$. Thus, a structure consisting of a series of the extended basic units with three switches can provide maximum step voltages for v_o.

5.6.2 Optimal structure for the maximum number of voltage steps with a constant number of DC voltage sources

An alternative question is expressed as follows: suppose the number of DC voltage sources is constant and equal to N_{source}, which topology provides the maximum number of voltage steps?

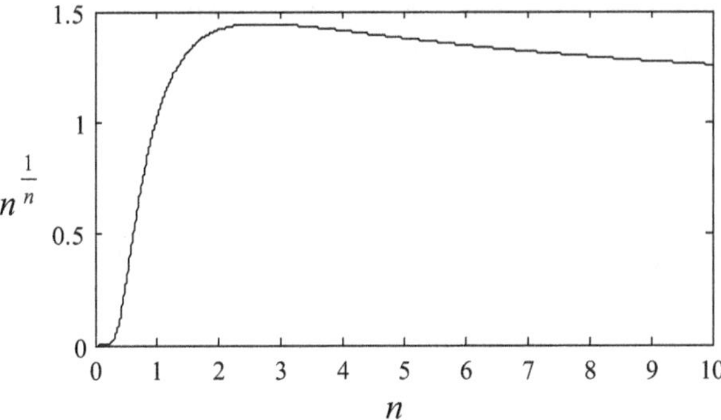

Figure 5.71 Variation of $n^{1/n}$ *versus n*

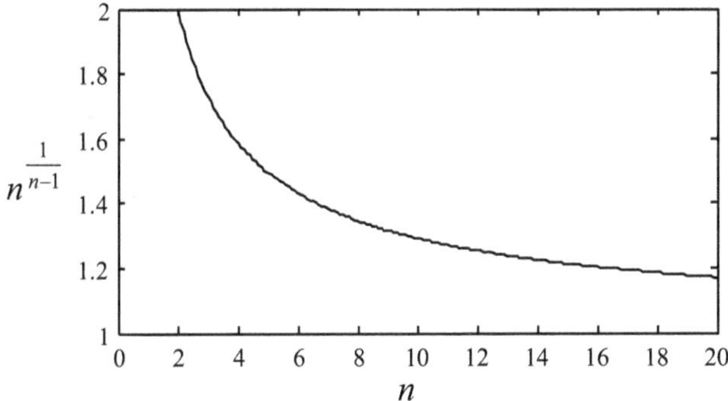

Figure 5.72 Variation of $n^{1/(n-1)}$ *versus* n

Suppose the proposed topology consists of a series of k extended base units as shown in Figure 5.70. Each unit with n_i switches and $n_i - 1$ DC voltage sources $(i = 1, 2, \ldots k)$. Thus:

$$N_{source} = \sum_{i=1}^{k}(n_i - 1) = n_1 + n_2 + \cdots + n_k - k \tag{5.12}$$

Considering (5.8), the number of DC voltage sources can be written as:

$$N_{source} = nk - k \tag{5.13}$$

From (5.10), the maximum number of voltage steps will be:

$$N_{level} = \left(n^{\frac{1}{n-1}}\right)^{N_{source}} \tag{5.14}$$

Figure 5.72 shows the variation of $n^{1/(n-1)}$ versus n. It is clear that the maximum number of voltage steps is obtained for $n = 2$. Thus, a structure consisting of a series of the extended basic units with two switches or one DC voltage source can provide maximum step voltages for v_o.

5.6.3 *Optimal structure for the minimum number of switches with a constant number of voltage steps*

The next question is that if N_{level} is the number of voltage steps considered for voltage v_o, which topology with a minimum number of switches can satisfy this need?

It can be proven that the maximum number of voltage steps may be obtained for equal switches. Thus, if the number of switches in each extended unit is assumed equal to n, considering (5.11), the total number of switches (N_{switch}) can be obtained as follows:

$$N_{switch} = n \times \log_n N_{level} = \ln(N_{level}) \times \frac{n}{\ln(n)} \tag{5.15}$$

Since N_{level} is constant, N_{switch} will be minimized when $n/\ln(n)$ tends to minimum. Figure 5.73 shows the corresponding figure, where $n = 3$ gives again the minimum number of switches to realize N_{level} values for voltage.

It is necessary to notice that the number of components is integers. Thus, if an integer number has not been obtained, the nearest integer number is certainly the proposed solution.

5.6.4 *Optimal structure for minimum standing voltage of switches with a constant number of voltage steps*

Voltage and current ratings of the switches in a multilevel converter play important roles on the cost and realization of the multilevel converter. In all topologies, currents of all switches are equal to the rated current of the load. This is, however, not the case for the voltage. The question is that if N_{level} voltages are proposed for v_o, which topology uses the switches with the minimum voltage?

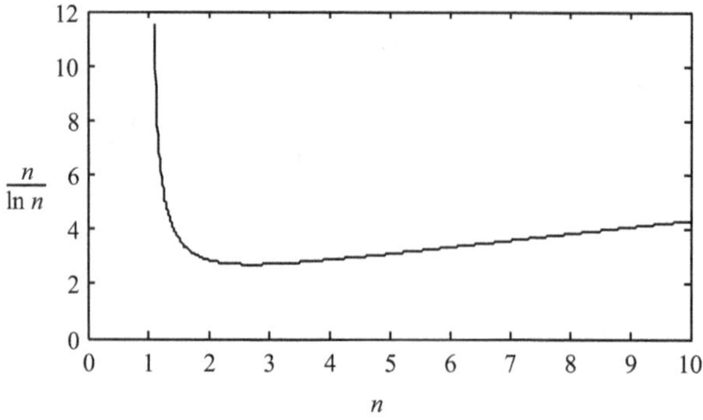

Figure 5.73 Variation of $n/\ln(n)$ versus n

Suppose that the peak voltage of the switches is represented by:

$$V_{switch} = \sum_{j=1}^{k} V_{switch,j} \tag{5.16}$$

In the above equation, $V_{switch,j}$ represents the peak voltage of the switches in unit j. Therefore, (5.16) can be considered as a criterion for comparison of different topology from the maximum voltage on the switches [16]. Lower criterion indicates that a smaller voltage is applied at the terminal of the switches of the topology, which is considered as advantage. Based on the above-mentioned points, the following equations can be obtained:

$$V_{switch,1} = P \times V_1 \tag{5.17}$$

$$V_{switch,2} = P \times V_2 \tag{5.18}$$

and finally:

$$V_{switch} = \sum_{j=1}^{k} V_{switch,j} \tag{5.19}$$

Therefore, (5.16) can be written as follows:

$$V_{switch} = P \times (V_1 + V_2 + \cdots + V_k) \tag{5.20}$$

In the above equations, P is calculated by:

$$\begin{gathered} p = 2\left[(n-1) + (n-2) + \cdots + \left(n - \frac{n}{2}\right)\right] \\ \textit{for } n = \textit{an even number} \\ p = 2\left[(n-1) + (n-2) + \cdots + \left(n - \frac{n+1}{2}\right)\right] - \frac{n-1}{2} \\ \textit{for } n = \textit{an old number} \end{gathered} \tag{5.21}$$

The above equations can be simplified as:

$$\begin{gathered} P = \frac{3n^2}{4} - \frac{n}{2} \textit{ for } n = \textit{an even number} \\ P = \frac{3n^2}{4} - \frac{n}{2} - \frac{1}{4} \textit{ for } n = \textit{an old number} \end{gathered} \tag{5.22}$$

According to (5.3)–(5.6) and (5.8) –(5.9), (5.20) can be simplified as follows:

$$V_{switch} = P \cdot V_{dc}\left(1 + n + n^2 + \cdots + n^{\log_n^{N_{level}} - 1}\right) \tag{5.23}$$

since

$$1 + n + n^2 + \cdots + n^{\log_n^{N_{step}} - 1} = \frac{N_{step} - 1}{n - 1} \tag{5.24}$$

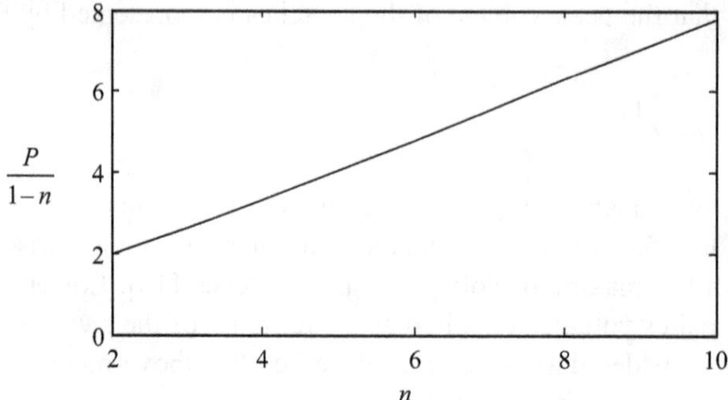

Figure 5.74 Variation of $P/(n-1)$ *versus* n

Thus, (5.23) can be written as:

$$V_{switch} = (N_{level} - 1) \cdot V_{dc} \cdot \left(\frac{P}{n-1}\right) \tag{5.25}$$

The variation of $P/(n-1)$ is shown in Figure 5.74. As illustrated in this figure, V_{switch} is minimized for $n = 2$. Thus, the optimal structure, from the minimum voltage of the switch point of view, consists of two switches.

5.7 Conclusion

In this chapter, different multilevel level inverters, which have been presented in the literature, were studied. These studies included the calculation of the key equations that can be considered as the characteristics of topology, such as the count of the switches, IGBTs, DC voltage sources, the variety in the magnitudes of the DC voltage sources, the total blocked voltage by the switches, and the maximum count of switches that are switched on to pass the output current. In the end, a comparative study was done using some of these topologies to determine the advantages and demerits of these topologies toward the conventional H-bridge inverter. The results of these comparisons were studied, and it has resulted that some of these topologies show better features in some parameters. However, they may show worse results in the other parameters. Moreover, the meaning of optimization in multilevel inverters was described and applied to a multilevel inverter topology.

5.8 Future directions

The studied multilevel inverter topologies have an open circuit control method and utilize the inductive-resistive local load; however, some can be specialized for use in different applications such as photovoltaic, ac motor drive, FACTS devices, and other practical ones. Also, it can be claimed that, nevertheless, the multilevel inverters with a reduced count of devices are popular, and many topologies have been presented in this era, they cannot be used in the mentioned applications. The cause of this claim might be the aim to minimize the number of required devices. Although reducing the devices' count may lead to lower costs for a cascaded multilevel inverter topology, it can lead to the loss of modularity and simplicity, which are vital factors for cascaded multilevel inverters. However, some topologies, such as the developed H-bridge cells, can be utilized in some applications under special circumstances.

References

[1] A. Nabae, I. Takahashi, and H. Akagi, "A new neutral-point clamped PWM inverter," *IEEE Trans. Ind. Appl.*, vol. IA-17, pp. 518–523, 1981.

[2] T. A. Meynard and H. Foch, "Multilevel choppers for high voltage applications," *Eur. Power Electron. Drives. J.*, vol. 2, no. 1, p. 41, 1992.

[3] F.Z. Peng, J.S. Lai, J. McKeever, and J. VanCoevering, "A multilevel voltage-source inverter with separate DC sources for static var generation," *IEEE Trans. Ind. Appl.*, vol. 32, no. 5, pp. 1130–1138, 1996.

[4] M. Vijeh, M. Rezanejad, E. Samadaei, and K. Bertilsson, "A general review of multilevel inverters based on main submodules: structural point of view," *IEEE Trans. Power Electron.*, vol. 34, no. 10, pp. 9479–9502, 2019.

[5] E. Babaei, S. H. Hosseini, G. B. Gharehpetian, M. T. Haque, and M. Sabahi, "Reduction of DC voltage sources and switches in asymmetrical multilevel converters using a novel topology," *Elsevier J. Electr. Power Syst. Res.*, vol. 77, no. 8, pp. 1073–1085, 2007.

[6] S. Laali, E. Babaei, and M. B. Bannae Sharifian, "Reduction the number of power electronic devices of a cascaded multilevel inverter based on new general topology," *J. Oper. Autom. Power Eng. (JOAPE)*, vol. 2, no. 2, pp. 81–90, 2014.

[7] G. Waltrich and I. Barbi, "Three-phase cascaded multilevel inverter using power cells with two inverter legs in series," *IEEE Trans. Ind. Appl.*, vol. 57, no. 8, pp. 2605–2612, 2010.

[8] E. Babaei, A. Dehqan, and M. Sabahi, "Improvement of the performance of the cascaded multilevel inverters using power cells with two series legs," *J. Power Electron.*, vol. 13, no. 2, pp. 223–231, 2013.

[9] A. Ajami, A. Mokhberdoran, M. R. Jannati Oskuee, and M. Toopchi Khosroshahi, "Cascade-multi-cell multilevel converter with reduced number of switches," *IET Power Electron.*, vol. 7, no. 3, pp. 552–558, 2014.

[10] A. Mokhberdoran and A. Ajami, "Symmetric and asymmetric design and implementation of new cascaded multilevel inverter topology," *IEEE Trans. Power Electron.*, vol. 29, no. 12, pp. 6712–6724, 2014.

[11] C. I. Odeh, E. S. Obe, and O. Ojo, "Topology for cascaded multilevel inverter," *IET Power Electron.*, vol. 9, no. 5, pp. 921–929, 2016.

[12] E. Samadaei, S. A. Gholamian, A. Sheikholeslami, and J. Adabi, "An envelope type (E-type) module: asymmetric multilevel inverters with reduced components," *IEEE Trans. Ind. Electron.*, vol. 63, no. 11, pp. 7148–7156, 2016.

[13] E. Samadaei, A. Sheikholeslami, S. A. Gholamian, and J. Adabi, "A square T-type (ST-type) module for asymmetrical multilevel inverters," *IEEE Trans. Power Electron.*, vol. 33, no. 2, pp. 987–996, 2018.

[14] M. Saeedian, J. Abadi, and M. Hosseini, "A cascaded multilevel inverter based on symmetric-asymmetric DC sources with reduced number of components," *IET Power Electron.*, vol. 10, no. 12, pp. 1468–1478, 2017.

[15] M. R. Banaei, M. R. Jannati Oskuee and H. Khounjahan, "Reconfiguration of semi-cascaded multilevel inverter to improve systems performance parameters," *IET Power Electron.*, vol. 7, no. 5, pp. 1106–1112, 2014.

[16] E. Babaei and M. Farhadi Kangarlu, "Cross-switched multilevel inverter: an innovative topology," *IET Power Electron.*, vol. 6, no. 4, pp. 642–652, 2013.

[17] E. Babaei, M. Farhadi Kangarlu, and M. Sabahi, "Cascaded cross-switched multilevel inverter in symmetric and asymmetric conditions," *IET Power Electron.*, vol. 6, no. 6, pp. 1041–1050, 2013.

[18] E. Babaei, S. Laali, and S. Alilu, "Cascaded multilevel inverter with series connection of novel H-bridge basic units," *IEEE Trans. Ind. Electron.*, vol. 61, no. 12, pp. 6664–6671, 2014.

[19] S. Alilu, E. Babaei, and S. B. Mozaffari, "A new general topology for multilevel inverters based on developed H-bridge," in *Proc. PEDSTC*, 2013, Tehran, Iran.

[20] E. Babaei and S. Laali, "Optimum structures of proposed new cascaded multilevel inverter with reduced number of components," *IEEE Trans. Ind. Electron.*, vol. 62, no. 11, pp. 6887–6895, 2015.

[21] R. Shalchi Alishah, S. H. Hosseini, E. Babaei, and M. Sabahi, "A new general multilevel converter topology based on cascaded connection of submultilevel units with reduced switching components, DC sources, and blocked voltage by switches," *IEEE Trans. Ind. Electron.*, vol. 63, no. 11, pp. 7157–7164, 2016.

[22] R. Shalchi Alishah, S. H. Hosseini, E. Babaei, and M. Sabahi, "Optimal design of new cascaded switch-ladder multilevel inverter structure," *IEEE Trans. Ind. Electron.*, vol. 64, no. 3, pp. 2072–2080, 2017.

[23] R. Shalchi Alishah, S. H. Hosseini, E. Babaei, and M. Sabahi, "Optimization assessment of a new extended multilevel converter topology," *IEEE Trans. Ind. Electron.*, vol. 64, no. 6, pp. 4530–4538, 2017.

[24] E. Babaei and S. H. Hosseini, "New cascaded multilevel inverter topology with minimum number of switches," *Elsevier J. Energy Convers. Manag.*, vol. 50, no. 11, pp. 2761–2767, 2009.

[25] J. Ebrahimi, E. Babaei, and G. B. Gharehpetian, "A new topology of cascaded multilevel converters with reduced number of components for high-voltage applications," *IEEE Trans. Power Electron.*, vol. 26, no. 11, pp. 3109–3118, 2011.

[26] J. Ebrahimi, E. Babaei, and G. B. Gharehpetian, "A new multilevel converter topology with reduced number of power electronic components," *IEEE Trans. Ind. Electron.*, vol. 59, no. 2, pp. 655–667, 2012.

[27] E. Babaei, "A cascade multilevel converter topology with reduced number of switches," *IEEE Trans. Power Electron.*, vol. 23, no. 6, pp. 2657–2664, 2008.

[28] E. Babaei, S. Laali, and S. Bahravar, "A new cascaded multilevel inverter topology with reduced number of components and charge balance control methods capabilities," *Electric Power Comp. Syst. J.*, vol. 43, no. 19, pp. 2116–2130, 2015.

[29] S. Laali, E. Babaei, and M. B. Bannae Sharifian, "A new basic unit for cascaded multilevel inverters with the capability of reducing the number of switches," *J. Power Electron.*, vol. 14, no. 4, pp. 671–677, 2014.

[30] M. Farhadi Kangarlu, E. Babaei, and S. Laali, "Symmetric multilevel inverter with reduced components based on non-insulated DC voltage sources," *IET Power Electron.*, vol. 5, no. 5, pp. 571–581, 2012.

[31] M. Farhadi Kangarlu and E. Babaei, "A generalized cascaded multilevel inverter using series connection of sub-multilevel inverters," *IEEE Trans. Power Electron.*, vol. 28, no. 2, pp. 625–636, 2013.

[32] E. Babaei, M. Farhadi Kangarlu, M. Sabahi, and M. R. Alizadeh Pahlavani, "Cascaded multilevel inverter using sub-multilevel cells," *Electr. Power Syst. Res.*, vol. 96, pp. 101–110, 2013.

[33] E. Babaei, M. Farhadi Kangarlu, and F. Najaty Mazgar, "Symmetric and asymmetric multilevel inverter topologies with reduced switching devices," *Electr. Power Syst. Res.*, vol. 86, pp. 122–130, 2012.

[34] R. Shalchi Alishah, M. Sabahi, D. Nazarpour, and S. H. Hosseini, "Novel multilevel inverter topologies for medium and high-voltage applications with lower values of blocked voltage by switches," *IET Power Electron.*, vol. 7, no. 12, pp. 3062–3071, 2014.

[35] R. Shalchi Alishah, D. Nazarpour, S. H. Hosseini, and M. Sabahi, "Reduction of power electronic elements in multilevel converters using a new cascade structure," *IEEE Trans. Ind. Electron.*, vol. 62, no. 1, pp. 256–269, 2015.

[36] E. Babaei, A. Dehqan, and M. Sabahi, "A new topology for multilevel inverter considering its optimal structures," *Electr. Power Syst. Res.*, vol. 103, no. 8, pp. 145–156, 2013.

[37] E. Babaei, S. Laali, and Z. Bayat, "A single-phase cascaded multilevel inverter based on a new basic unit with reduced number of power switches," *IEEE Trans. Ind. Electron.*, vol. 62, no. 2, pp. 922–929, 2015.

[38] M. R. Jannati Oskuee, E. Salary, and S. Najafi Ravadanegh, "Creative design of symmetric multilevel converter to enhance the circuit's performance," *IET Power Electron.*, vol. 8, no. 1, pp. 96–102, 2015.

[39] M. R. Jannati Oskuee, M. Karimi, S. Najafi Ravadanegh, and G. B. Gharehpetian, "An innovative scheme of symmetric multilevel voltage

source inverter with lower number of circuit devices," *IEEE Trans. Ind. Electron.*, vol. 62, no. 11, pp. 6065–6973, 2015.

[40] W. K. Choi and F. S. Kang, "H-bridge based multilevel inverter using PWM switching function," in *Proc. INTELEC*, 2009, pp. 1–5.

[41] M. Toopchi Khosroshahi, "Crisscross cascade multilevel inverter with reduction in number of components," *IET Power Electron.*, vol. 7, no. 12, pp. 2914–2924, 2014.

[42] Y. Hinago and H. Koizumi, "A single-phase multilevel inverter using switched series/parallel DC voltage sources," *IEEE Trans. Ind. Electron.*, vol. 57, no. 8, pp. 2643–2650, 2010.

[43] E. Babaei, S. Sheermohammadzadeh, and M. Sabahi, "Improvement of multilevel inverters topology using series and parallel connections of DC voltage sources," *Arab. J. Sci. Eng. (JSE)*, vol. 39, no. 2, pp. 1117–1127, 2013.

Chapter 6

GaN oscillator-based DC–AC converter for wireless power transfer applications

Anwar Jarndal[1]

This chapter reports a design for linear and nonlinear switching-mode oscillator based on Gallium nitride (GaN) high electron mobility transistors (HEMTs) for wireless power transfer (WPT) applications. The linear oscillator is based on a lower power depletion-type GaN-HEMT, while the nonlinear oscillator is implemented using a higher power enhancement-type GaN transistor. In-house advanced design system (ADS)-based model has been used for designing the lower power oscillator, while a commercial LTSpice model has been implemented for designing the higher power oscillator. The oscillators are designed and simulated using computer-aided software (CAD) and then realized and tested with transmitting and receiving coils. The realized linear-mode oscillators show good efficiency of dc-to-ac conversion (up to 45%) with lower total harmonic distortion (THD) in the order of 10%. This makes those oscillators proper for electromagnetic interference (EMI) sensitive applications. On the other side, the switching-mode oscillators provide higher efficiency up to 90%, which is optimal for designing high power WPT systems.

6.1 Introduction

Wireless power transfer (WPT) has gained significant interest of industries and researchers with the development of low power, battery-operated and compact devices like biomedical micro-implants [1] and wireless sensors [2]. This technology can also provide an optimal solution for wirelessly charging higher power-operated portable equipment and electric cars. WPT will also crucially need for the future autonomous driving electric vehicles [3].

One of the drawbacks of the WPT systems is the power efficiency that is contributed by dc-to-ac conversion rate of the ac source, coupling factor of the coils, and rectifier losses. The source conversion efficiency could be enhanced using improved device-technology and circuit design of lower power losses. Currently, there is a big interest on using GaN technology for power electronics

[1]Department of Electrical Engineering, University of Sharjah, United Arab Emirates

applications [4]. GaN has a higher power density about 9 W/mm [4]. The wide band-gap [5] and the high breakdown voltage [6] of GaN are also suitable for the high-voltage swings in wireless power systems. GaN also withstands more power than GaAs of the same size [5] allowing more efficient WPT systems.

At the circuit level, class-AB oscillator based on GaN HEMT represents a good alternative. This oscillator has a quietly higher efficiency (with respect to linear class-A oscillator) with simpler and lower harmonics (with respect to switching-mode oscillator) [7]. Another alternative, which is optimal for high power WPT application, is class-E oscillator-based GaN device. This oscillator provides very high efficiency [7,8], however, its high harmonic contents create strong electromagnetic interference (EMI) emitted from the transmission coil. This can cause problems for biomedical and communication systems. The EMI can be reduced by adding extra filters, but this will increase the size, cost, and complexity, while reduce the overall efficiency of the WPT system.

In this chapter, a class AB oscillator will be designed with a 4 W packaged GaN HEMT using its developed model in [9]. Another design for class-E oscillator based on 100 W GaN device will be presented. The next sections describe the operation of the oscillators and show their circuit analysis and design. In addition to the simulation results for the considered oscillators. Followed by the oscillator's implementation and characterization, and a conclusion will be drawn in the last section.

6.2 Class-AB oscillator

The first proposed Colpitts oscillator circuit is shown in Figure 6.1. The oscillator consists of a feedback system with an inverting (180°) amplifier and an inverting

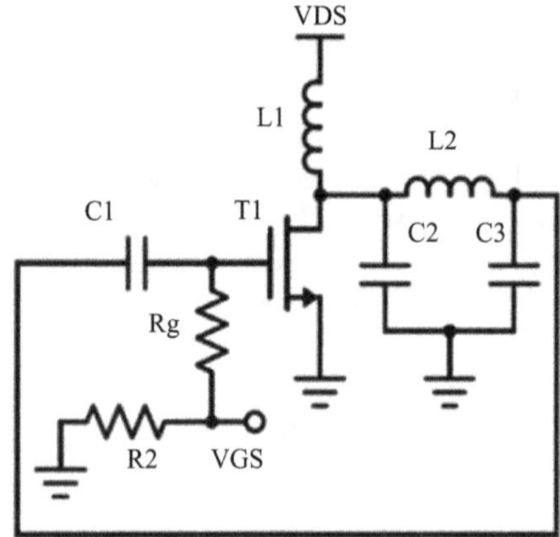

Figure 6.1 Class-AB oscillator circuit

LC feedback band-pass filter to provide a zero total phase shift. In Figure 6.1, the common source (CS) transistor amplifier with inductive load, consisting of R_g, T_1, and L_1, is the inverting amplifier and the filter, consisting of C_2, C_3, and L_2, provides the inverted feedback signal [10]. The passive elements are chosen, as it will be explained later, to provide an oscillation frequency of 4 MHz and their value are as follows: L_1 and L_2 are 10 μH, C_1 and C_2 are 1 nF and R_g is 47 Ω.

The amplifier is a simple CS inverting amplifier constituting the GaN HEMT T_1 with the inductive load L_1 (see Figure 6.1). The voltage gain is formulated, assuming no load for simplicity, at relatively low frequencies as follows:

$$G_v = \frac{V_d}{V_g} = -g_m s L_1. \tag{6.1}$$

As can be seen, the gain is negative hence it has a 180° of phase shift. The simulated gain and phase of the CS amplifier as shown in Figure 6.2, are obtained using 50 Ω S-parameter simulation in ADS with V_{DS} = 6 V and V_{GS} = −1.1 V. It is seen that the phase of the gain (S_{21}) is 180° and the gain is reasonably high (25 dB).

The feedback circuit of the oscillator consists of C_2, C_3, and L_2 in Figure 6.1. It acts as an inverting filter that feeds back the output to the input with a 180° phase shift and specifies the oscillation frequency. To obtain the phase shift of this feedback circuit, an equivalent circuit in Figure 6.3 is used and the gain from V_d to V_g is formulated (without L_1, T_1, and R_g) as follows:

$$G_{fb} = \frac{V_g}{V_d} = \frac{1}{1 - \omega^2 L_2 C_3}. \tag{6.2}$$

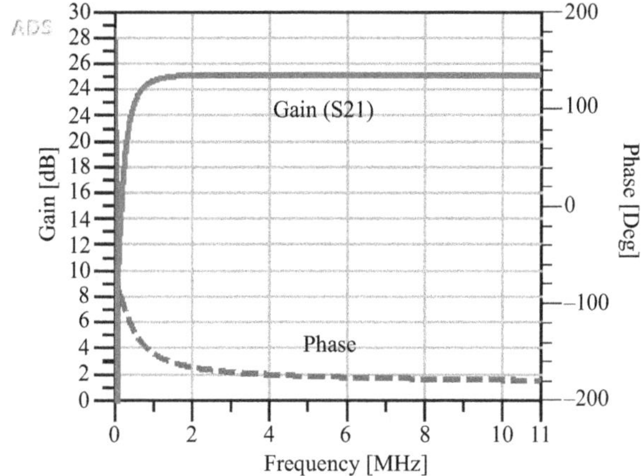

Figure 6.2 Gain and phase of the common-source amplifier

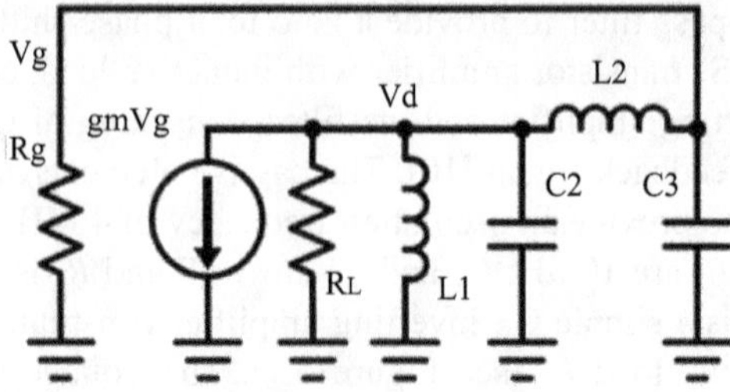

Figure 6.3 AC equivalent circuit for the proposed oscillator in Figure 6.1

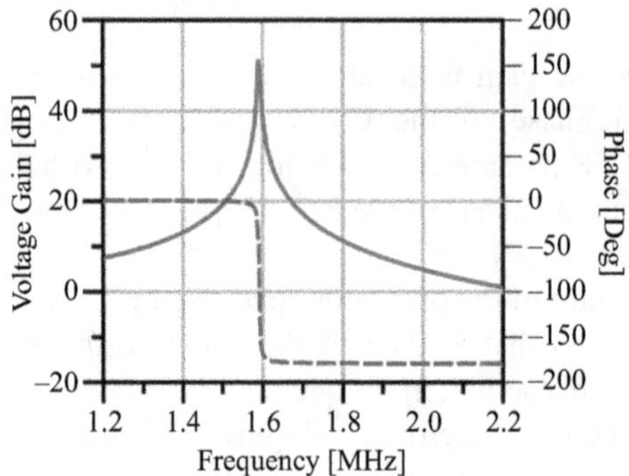

Figure 6.4 Gain and phase of the feedback filter

The resonance frequency (the minimum value of the denominator, theoretically zero) is

$$\omega_p = \frac{1}{\sqrt{L_2 C_3}}. \tag{6.3}$$

At ω_p, the phase shift of the feedback filter becomes 180° ($G_{fb} < 0$) therefore providing a total phase of 360° necessary for oscillation. The voltage gain and phase of the resonance circuit simulated in ADS is shown in Figure 6.4 where the peak gain is around 50 at which the phase is 180° afterward. The frequency of oscillation is not necessarily at this peak (the filter's resonance frequency) since the closed loop gain at this frequency might not be equal or greater than one.

The analysis is done based on the oscillator AC equivalent circuit as shown in Figure 6.3 where the load R_L is connected in parallel to L_1 through a 1:1 ideal transformer. The phasor characteristic equation of the oscillator is derived as follows:

$$\begin{aligned} &s^4(C_2C_3) + s^3\left(\frac{C_2}{R_g} + \frac{C_3}{R_L}\right) + s^2\left(\frac{C_3}{L_1} + \frac{C_3}{L_2} + \frac{C_2}{L_2} + \frac{1}{R_gR_L}\right) \\ &+ s\left(\frac{1}{L_1R_g} + \frac{g_m}{L_2} + \frac{1}{L_2R_g} + \frac{1}{L_2R_L}\right) + \frac{1}{L_2L_1} = 0. \end{aligned} \tag{6.4}$$

All terms in Eq. (6.4) are real except for s and s^3. For a system to oscillate, its imaginary characteristic equation should be zero. For the real part to be zero, all the real parts should be zero as follows:

$$\omega^4(C_2C_3) - \omega^2\left(\frac{C_3}{L_1} + \frac{C_3}{L_2} + \frac{C_2}{L_2} + \frac{1}{R_gR_L}\right) + \frac{1}{L_2L_1} = 0. \tag{6.5}$$

and the frequency of oscillation is then formulated by equating the real part to zero, which produces the following:

$$\omega_{osc} = \sqrt{\frac{\frac{C_2}{L_1} + \frac{C_2+C_3}{L_2} + \frac{1}{R_gR_L} \mp \sqrt{\left(\frac{C_2}{L_1} + \frac{C_2+C_3}{L_2} + \frac{1}{R_gR_L}\right)^2 - 4\frac{C_2C_3}{L_1L_2}}}{2C_2C_3}}. \tag{6.6}$$

Thus, the frequency of oscillation can be fixed by using proper values for the circuit passive elements. Performing dimensional analysis on Eq. (6.6), its unit turns out to be $\sqrt[-2]{HF}$ which is 1/sec implying that Eq. (6.6) is correct in its form. It can be noticed that Eq. (6.6) is the $\omega_{\rm osc}$ for the conventional Colpitts circuit but with the added effect of L_1 (reduces the frequency) and $R_{\rm L}$ (increases the frequency). Calculating $\omega_{\rm osc}$ using Eq. (6.6) provides two values. The one below $\omega_{\rm p}$ in Eq. (6.3) will be rejected since it will not produce the required 360° total phase shift (see Figure 6.4). The condition of oscillation is derived by setting the imaginary part of Eq. (6.4) to zero, which provides the following formula for condition of oscillation [11]:

$$\frac{g_m}{L_2} = \frac{-1}{R_g}\left(\frac{1}{L_1} + \frac{1}{L_2} + \omega^2 C_2\right) + \frac{-1}{R_L}\left(\frac{1}{L_2} + \omega^2 C_2\right) \tag{6.7}$$

In this equation, ω is the oscillation frequency from Eq. (6.6) which is fixed by the element values. The term $\omega^2 C_2$ has a unit of 1/H which makes the right side of Eq. (6.7) have the same unit as the left side. With $L_1 = L_2 = 10\ \mu$H, $C_1 = C_2 = 1$nF, $R_{\rm g} = 47\ \Omega$, and $R_{\rm L} = 50\ \Omega$ (the input resistance of the oscilloscope) the two frequencies from (6.6) are 4.17 MHz and 0.61 MHz. To know which is the oscillation frequency, the value lower than $\omega_{\rm p}$ (minimum frequency of oscillation) in Eq. (6.3) is rejected. The calculated value of $\omega_{\rm p}$ is 1.59 MHz, which is identical to the minimum frequency of 180° phase shift of the filter's response in Figure 6.4. Since 0.61 MHz $<$ 1.59 MHz, the frequency of oscillation should be 4.17 MHz. Using the condition of oscillation in Eq. (6.7), the transistor's transconductance $g_{\rm m}$ can be determined.

The HEMT T_1 is biased through the L_1 and $R_{\rm g}$, while C_1 acts as decoupling capacitors. It is crucial to follow the correct biasing sequence as described in [12]. The HEMT is biased at pinch-off, $V_{\rm DS}$ is set and then $V_{\rm GS}$ is increased until the desired $I_{\rm D}$ is obtained. $I_{\rm D}$ is defined by $g_{\rm m}$ from the start-up condition in (6.7). Since the HEMT will alternate close to the pinch-off and triode regions of operation during oscillation, the gate can sink and source current which requires a sink/source power supply. If that supply is not available, a normal supply is used with additional shunt resistor (R_2 in Figure 6.1) [12].

To bias the HMET for class-AB mode of operation, the DC operation point should be chosen so that the quiescent current is not zero as in the class B mode [11]. Since the mode of operation is defined by V_{GS} (or the drain current), V_{GS} is chosen to allow the HEMT to turn off when V_{GS} is low enough at the valleys of the input wave, this means that V_{GS} should be chosen just above the pinch-off region of the HEMT in class AB mode. In contrast, V_{GS} should not be made too high to prevent the operation in the class-A mode where the transistor never switches off. If V_{GS} was too high, the oscillation at the drain (feedback to the gate) will not have enough amplitude to switch-off the transistor even if it's driving the transistor very close to pinch-off at its valley-value [11]. Using in-house developed model [8], the DC performance was simulated and plotted in Figure 6.5. V_{GS} was chosen at −1.1 V as shown by the Q marker in Figure 6.5. This is close enough to the pinch-off to put the HEMT in class AB mode. As it has been mentioned, V_{DS} and V_{GS} are chosen to provide the transconductance gm necessary for oscillation.

The oscillator circuit is built in ADS and simulated using the harmonic balance (HB) simulation with oscillator analysis enabled. The same element values were used as in the analysis of the previous section and the transistor is biased at V_{GS} = −1.1 V, V_{DS} = 6 V, and I_D = 125 mA. The circuit oscillated at a frequency of 4.826 MHz and with an amplitude of 5.1 V as shown in Figure 6.6. In this simulation, an AC coupled (with a coupling capacitor or an ideal transformer) 50 Ω load resistor is used, the same as the oscilloscope's input resistance in the typically used co-axial cable. Although the waveform does not look like a pure sinusoid (because of the triode region nonlinearity), the spectrum shows that the second harmonic is about −5 dB and the fundamental is 15 dB. The total harmonic distortion (THD) is 10.6%, which is very low with respect to switching-mode oscillators. This output

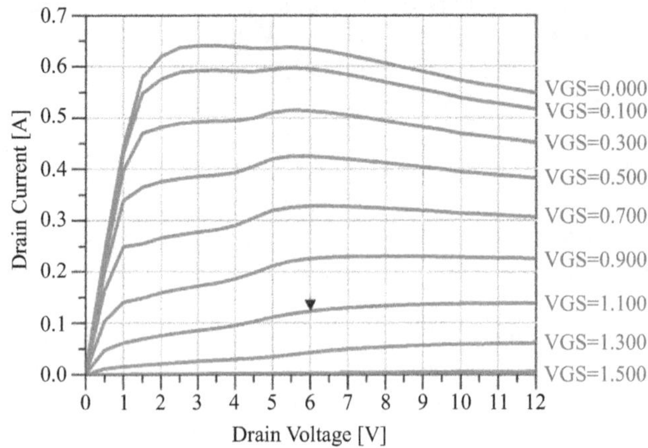

Figure 6.5 DC IV characteristics of the model transistor

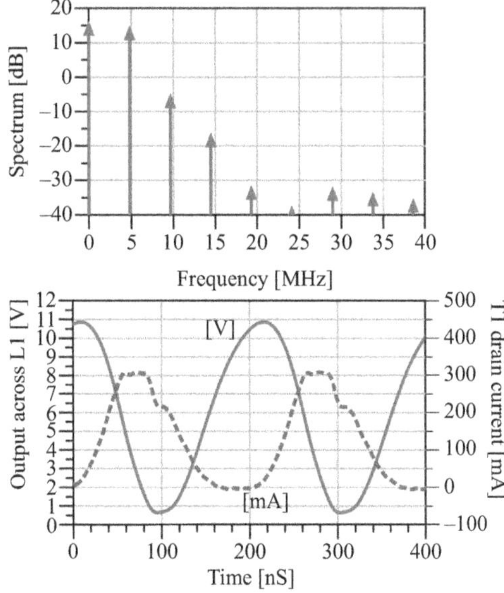

Figure 6.6 Simulated drain output signal and its spectrum

can easily be used for WPT since the coils can be designed to resonate with the frequency of oscillation. To calculate the efficiency, the AC output power of 271 mW divided by the DC input power of 750 mW results in a power efficiency of 36.2%. The circuit was implemented on a breadboard after soldering the SMD transistor to a small adapter. The implemented circuit oscillates at $V_{GS} = -0.7$ V, $V_{DS} = 6$ V, and $I_D = 120$ mA, which is close to the simulated ones. This bias condition provides the value of g_m that satisfies the condition of oscillation in Eq. (6.7).

After implementing the circuit and correctly biasing the HEMT, the oscilloscope's input resistance was set to 50 Ω and then it was connected to the drain of the HEMT through a coupling capacitor (2.2 μF). This simulates a 1:1 ideal transformer or an ideal transmitting and receiving coil setup. The measured output waveform is shown in Figure 6.7. Since the oscilloscope is AC coupled and the DC value of the output is 6V, then V_{min} at the valleys is about 0 V when the device is switched ON and over driven into the deep triode region. On the other side when the gate voltage approaching pinch-off to switch-off the device, the drain voltage raises to its peak value. Thus, the transistor in this case is partially acting as a switch, which confirms that the mode of operation is class AB. The FFT of the output is shown in Figure 6.8 where the THD is calculated to be 14.6%. The measured DC input power (V_D*I_D where $I_G \sim 0$) is 720 mW. The measured AC output power (V_{RMS}^2/R for each frequency component) is 310 mW. Therefore, the efficiency (P_{DC}/P_{AC}) is 43%. The measured frequency is 3.824 MHz. Table 6.1 summarizes all the circuit parameters and compares them to that of the simulated circuit.

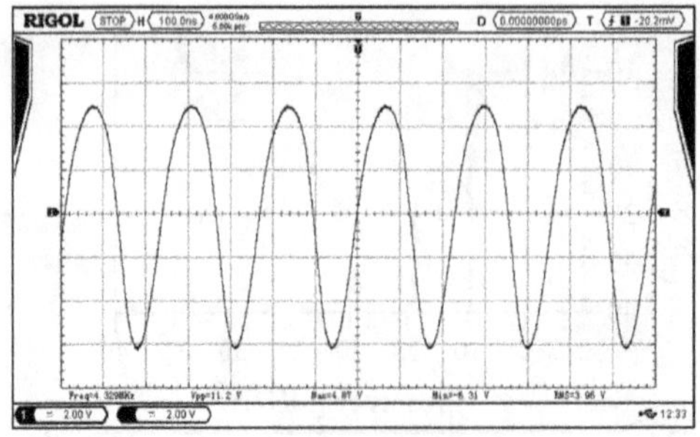

Figure 6.7 Measured AC drain voltage waveform with 50 Ω termination

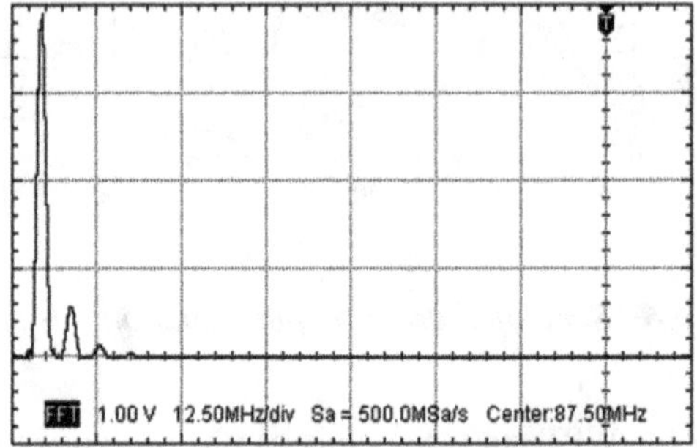

Figure 6.8 FFT of the measured AC drain voltage with 50 Ω termination

Table 6.1 Measured and simulated results

Parameter	Simulated	Measured
V_{GS}	−1.125	−0.7
V_{DS}	6	6
I_D	125 mA	120 mA
DC power input	750 mW	720 mW
Total AC power output	271 mW	310 mW
Efficiency	36.2%	43%
Output V_{PP}	10.22 V	11.1 V
Frequency	4.826 MHz	3.824 MHz
THD	10.631%	14.6%

As can be observed, good agreement is obtained between the simulated and measured parameters. The discrepancy in some parameters could be partially attributed to the small variation of device fabrication process, which affect the DC and AC transistor parameters. Thus, the developed model for a device may not give an accurate simulation for another device of the same process. This could be observed from the required value of V_{GS} to get the same drain current for the simulated and the implemented devices. It is clear also from the small difference between the simulated and the measured output peak voltage V_{PP}, which justifies the discrepancy between the measured and simulated output power and efficiency. Although the simulated and measured frequencies are different, they are still relatively close to the calculated value. This difference can be caused by the expected small shift of the transistor's pinch-off that occurs at the peaks of the wave as shown in Figure 6.6. It can be also attributed to the tolerances of the passive elements.

6.3 Class-E oscillator

The presented circuit in Figure 6.9 is formed by adopting the class-E oscillator based on the 100 W GaN transistor [13]. In this oscillator, the transistor is working as a switch and enforced by a positive feedback through an LC resonant circuit. As it is well known, zero-voltage and zero-derivative switching conditions must be satisfied for class-E operation. Under this condition, both the drain voltage and its derivative are zero when the transistor switch is turned ON [14] and this accordingly reduces the switching losses and maximizes the drain efficiency. Class-E oscillator consists of a DC voltage source V_{dd}, a DC-feed inductor L_2, a GaN HEMT as switching device, a capacitor C_4 shunting the device, a series resonant circuit L_3–C_1, two capacitors C_2 and C_3, and feedback inductor L_1. R_1 is a resistor for supplying the gate bias voltage, large enough to neglect current through it, and R_2 is load resistance [13].

The same approach presented in [14] has been followed to determine the values of the circuit components. The inductance of L_3 is calculated as:

$$L_3 = \frac{Q_L \cdot R_2}{2 \cdot \pi \cdot f} \tag{6.8}$$

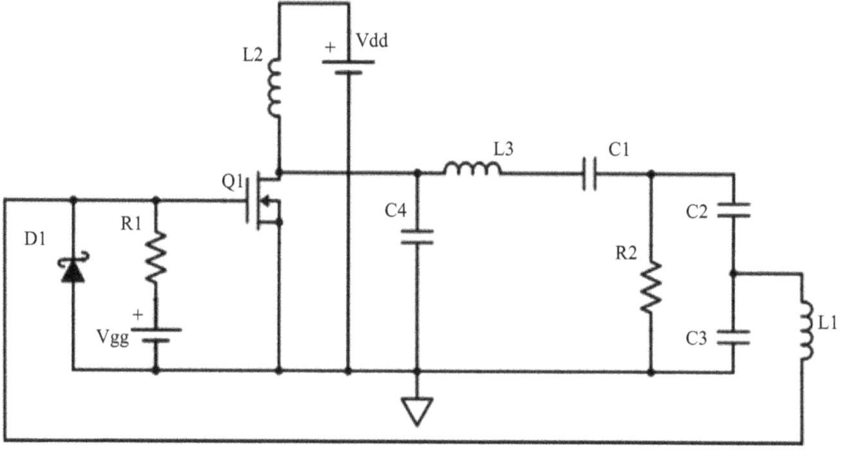

Figure 6.9 Class-E oscillator

Its value is selected to achieve a targeted quality factor Q_L. Using this value of Q_L, the capacitor C_4 can be computed as [14]:

$$C_4 = \frac{1}{34.2219 \cdot f \cdot R_2}\left(0.99866 + \frac{0.91424}{Q_L} - \frac{1.03175}{Q_L^2}\right) + \frac{0.6}{(2 \cdot \pi \cdot f)^2 \cdot L_1} \qquad (6.9)$$

From Eq. (6.9), at frequency 530 kHz and using L_1 of 99 μH, C_4 will have a value of 3.3 nF. C_1 can be calculated as [14]:

$$C_1 = \frac{1}{2 \cdot \pi \cdot f}\left(\frac{1}{Q_L - 0.104823}\right)\left(1.00121 + \frac{1.01468}{Q_L - 1.7879}\right) - \frac{0.2}{(2 \cdot \pi \cdot f)^2 \cdot L_1} \qquad (6.10)$$

From Eq. (6.10), the value for capacitor C_1 is calculated to be 549 pF. The circuit has been simulated using LTSpice and Figure 6.10 shows the output voltage of 53 V at a frequency of 530 KHz under bias voltages of V_{gg} = 2.1 V and V_{dd} = 16 V. Figure 6.11 shows the implemented class-E oscillator and Table 6.2 lists the values of the circuit components. The differences between the calculated and implemented values of the capacitors could be attributed to the circuit parasitic effects. For that reason, the circuit elements were tuned to compensate these effects and satisfy the oscillation condition. In Figure 6.11, the upper coil is used as a receiving coil; while the lower coil is used as a transmitter for the wireless power transmission circuit. As it is illustrated by the ruler, the distance between coils is shown. As it is well known, the amount of transferred power depends on the distance between the two coils. This circuit was tested on distance of 2 cm. Figure 6.12 shows the measured drain output voltage and current.

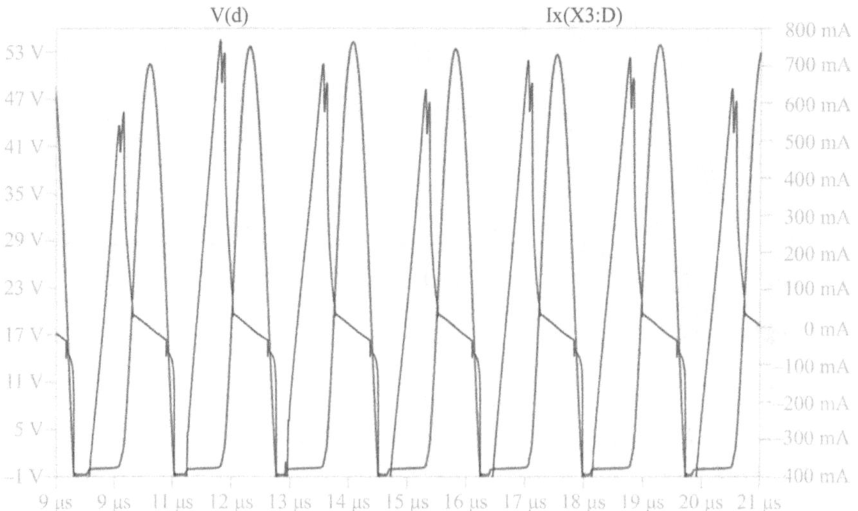

Figure 6.10 Class-E oscillator

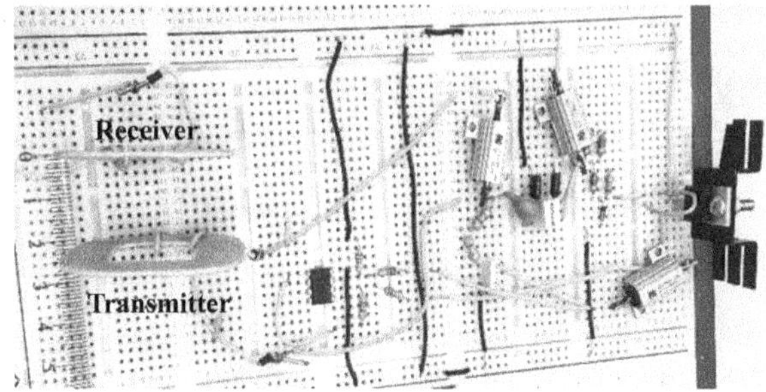

Figure 6.11 Implemented circuit of the class-E oscillator

Table 6.2 Components of the class-E oscillator

Component	**Value**
L_1	99 μH
L_2	7.2 μH
R_1	6.8 kΩ
C_1	453 pF
C_2	470 nF
C_3	10 nF
C_5	5.6 nF
L_3	100 μH

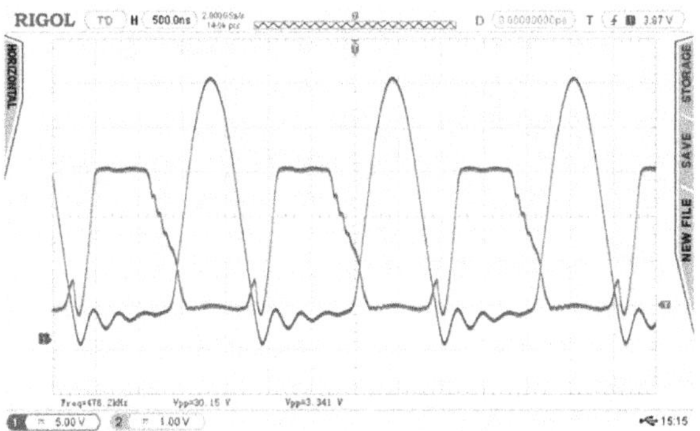

Figure 6.12 Measured drain voltage (blue) and current (red)

It shows the typical non overlapped signal of class-E oscillator, which results in higher DC-to-AC conversion efficiency. The estimated value of efficiency is a round 90%. The output voltage was also measured on the receiving coil. For testing the circuit, the bias gate voltage was 2.1 V and drain voltage is 16 V. Figure 6.13 shows output voltage at the receiving. As it can be seen, the obtained output voltage is 14.69 V at 2 cm distance between the coils.

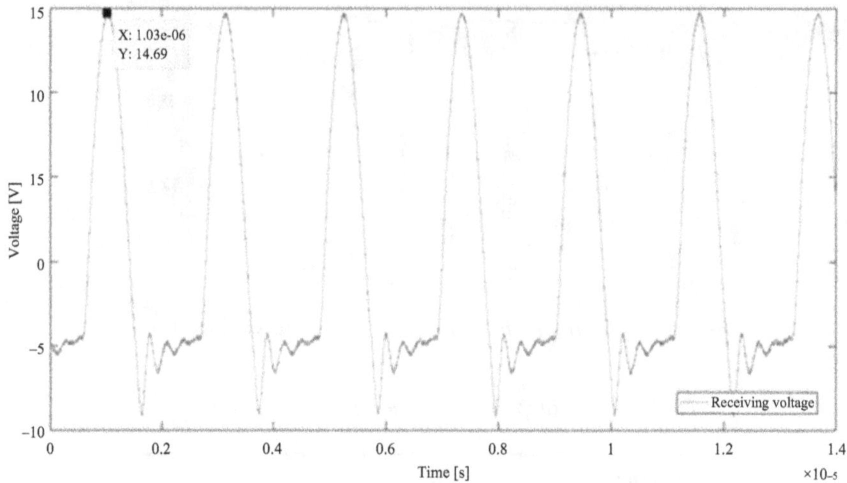

Figure 6.13 Measured voltage on the receiving coil for class-E WPT circuit

6.4 Conclusion

Different designs of class-AB and class-E oscillators based on enhancement- and depletion-type GaN transistors were successfully simulated and implemented. The oscillators were tested to achieve a measured overall efficiency of 43% with a THD of only 14.6% (without any filters) for 4 W class-AB oscillators. This low THD oscillators are suitable for EMI sensitive applications such biomedical equipment and implanted devices. The designed and implemented class-E oscillators showed high efficiency above 90%. These high efficiency oscillators are suitable for battery operated applications like cellphones and wireless sensors. The results of this paper showed the advantages of using GaN HEMTs for designing WPT system of lower THD or higher power efficiency.

Acknowledgment

The author gratefully acknowledges the support from the University of Sharjah, Sharjah, United Arab Emirates.

References

[1] S. R. Khan, S. K. Pavuluri, G. Cummins, and M. P. Y. Desmulliez, "Wireless power transfer techniques for implantable medical devices: a review," *Sensors (Switzerland)*, vol. 20, no. 12, pp. 1–58, 2020, doi: 10.3390/s20123487.

[2] L. Sun, D. Ma, and H. Tang, "A review of recent trends in wireless power transfer technology and its applications in electric vehicle wireless charging," *Renew. Sustain. Energy Rev.*, vol. 91, no. March, pp. 490–503, 2018, doi: 10.1016/j.rser.2018.04.016.

[3] V. D. Doan, H. Fujimoto, T. Koseki, T. Yasuda, H. Kishi, and T. Fujita, "Allocation of wireless power transfer system from viewpoint of optimal control problem for autonomous driving electric vehicles," *IEEE Trans. Intell. Transp. Syst.*, vol. 19, no. 10, pp. 3255–3270, 2018, doi: 10.1109/TITS.2017.2774013.

[4] M. D. Vecchia, S. Ravyts, G. Van den Broeck, and J. Driesen, "Gallium-nitride semiconductor technology and its practical design challenges in power electronics applications: an overview," *Energies*, vol. 12, no. 14, 2019, doi: 10.3390/en12142663.

[5] T. J. Flack, B. N. Pushpakaran, and S. B. Bayne, "GaN technology for power electronic applications: a review," *J. Electron. Mater.*, vol. 45, no. 6, pp. 2673–2682, 2016, doi: 10.1007/s11664-016-4435-3.

[6] H. Ishida, R. Kajitani, Y. Kinoshita, *et al.*, "GaN-based semiconductor devices for future power switching systems," in: *Tech. Dig. – Int. Electron Devices Meet. IEDM*, pp. 20.4.1–20.4.4, 2017, doi: 10.1109/IEDM.2016.7838460.

[7] L. Xue and J. Zhang, "Using high-voltage GaN power ICs for wireless charging applications," in: *IEEE Appl. Power Electron. Conf. Expo.*, no. 2014, pp. 3743–3750, 2017.

[8] J. Choi, D. Tsukiyama, Y. Tsuruda, and J. Rivas, "13.56 MHz 1.3 kW resonant converter with GaN FET for wireless power transfer," in: *2015 IEEE Wirel. Power Transf. Conf. WPTC 2015*, pp. 3–6, 2015, doi: 10.1109/WPT.2015.7140167.

[9] A. Jarndal and F. M. Ghannouchi, "Improved modeling of GaN HEMTs for predicting thermal and trapping-induced-kink effects," *Solid. State. Electron.*, vol. 123, pp. 19–25, 2016, doi: 10.1016/j.sse.2016.05.015.

[10] B. Razavi, *Design of Analog CMOS Integrated Circuits*, New York, NY: McGraw-Hill Education, 2016.

[11] E. Hegazi, J. Real, and A. Abidi, *The Designer's Guide to High-Purity Oscillators*, New York, NY: Springer, 2005.

[12] Nitronex, *AN-009: Bias Sequencing and Temperature Compensation for GaN HEMTs*, Technical N, 2008.

[13] D. V. Chernov, M. K. Kazimierczuk, and V. G. Krizhanovski, "Class-E MOSFET low-voltage power oscillator," *Proc. – IEEE Int. Symp. Circuits Syst.*, vol. 5, pp. V/509–V/512, 2002, doi: 10.1109/ISCAS.2002.1010752.

[14] M. K. Kazimierczuk, V. G. Krizhanovski, J. V. Rassokhina, and D. V. Chernov, "Class-E MOSFET tuned power oscillator design procedure," *IEEE Trans. Circuits Syst. I Regul. Pap.*, vol. 52, no. 6, pp. 1138–1147, 2005, doi: 10.1109/TCSI.2005.849127.

Chapter 7

Partial power processing and its emerging applications

Naser Hassanpour[1], *Andrii Chub*[1], *Andrei Blinov*[1] *and Samir Kouro*[2]

In this chapter, the idea of partial power processing is elaborated in detail. First, the concept is divided into two groups of differential power processing (DPP) and series partial power converters (S-PPCs). The focus of this chapter is mainly S-PPCs. Second, the S-PPCs are discussed considering the configuration of the converters. To analyze and demonstrate the behavior of S-PPCs, there are a variety of parameters like partiality, non-active power processing, and DC–DC topology requirements. The parameter's concept is explained and the methods of applying these parameters to S-PPCs are described. In the last section, numerous examples of S-PPCs application in different fields are presented and for each application at least one S-PPC topology is depicted and investigated.

Nomenclature

BES	battery energy storage
CSF	component stress factor
DAB	dual active bridge
DPC	differential power converter
DPP	differential power processing
ESS	energy storage system
FB	full bridge
FBPP	full bridge push pull
FPC	full power converter

[1]TalTech, Tallinn, Estonia
[2]Electronics Engineering Department, UTFSM, Chile

IGBT	insulated-gate bipolar transistor
IPOS	input parallel output series
ISOP	input series output parallel
LVDC	low-voltage DC
MOSFET	metal oxide semiconductor field effect transistor
MPPT	maximum power point tracking
PCB	printed circuit board
PFCC	power flow control converter
PPP	partial power processing
PV	photo voltaic
PWM	pulse width modulation
RMS	root mean square
SoC	state of charge
S-PPC	series partial power converter
TAB	triple active bridge
ZCS	zero current switching
ZVS	zero voltage switching

7.1 Overview of partial power processing (PPP)

Conventional DC–DC converters process all the power between input and output. They are called full power converters (FPC). Various technological barriers limit further improvements of conventional converters, and only relatively incremental efficiency improvements are possible. Theoretical feasibility of the direct power transfer, the fundamental principle of the partial power processing (PPP), was first demonstrated by T. Wilson in [1]. However, some of the first topologies were demonstrated much later in 1996 by J. Sebastian [2]. The PPP concept found its first practical application in the spacecraft industry, where the application of compact, high-efficiency, and high-power density power electronic converters is of significant importance [3]. The trend has spread to other areas. By utilizing the same approach, new PPP-based converters are developed for renewable energy systems, energy storage systems (ESSs), electric vehicle (EV) chargers, etc. The first practically applied PPP-based converters in renewable energy systems have been used in doubly fed induction generators-based wind energy systems. In this kind of wind generator, usually, the processed by the power converter is 30% of the total power generated by the turbine. As apparent from the PPP term, in this kind of converters, only a small share of the whole power between source/load or input/output is processed. The remaining power is delivered to output/load without processing by creating an almost lossless path with theoretically 100% efficiency. In Figure 7.1, both FPC and PPP converters are illustrated. As shown in Figure 7.1(a), the DC–DC converter efficiency is the determining factor of the system efficiency in FPCs. On the other hand, in PPP, the power loss in the DC–DC

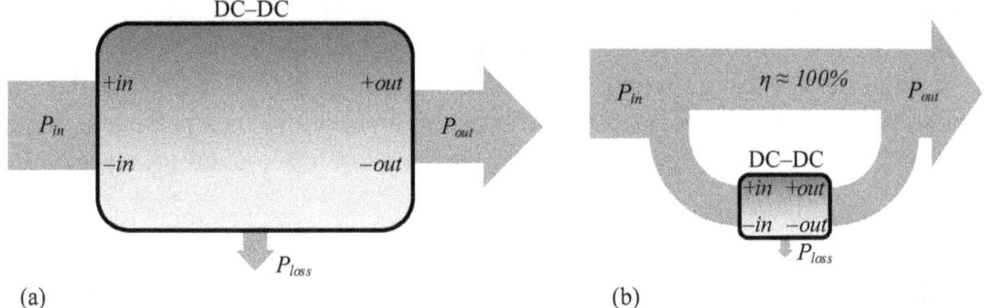

Figure 7.1 Power flows for (a) FPC and (b) PPP

converter is only a tiny percentage of the processed power, which can be seen from Figure 7.1(b). So, the system efficiency will not be equal to the DC–DC converter efficiency.

Besides efficiency improvement, the size and power rating of the converter elements will be reduced. The following equations can be obtained for the efficiency of both converters:

$$FPC : \eta_{Conv} = \frac{P_{out}}{P_{in}} \tag{7.1}$$

$$PPP : \eta_{sys} = f(\eta_{Conv}, P_P) \tag{7.2}$$

where η_{Conv} is the DC–DC converter efficiency and η_{sys} is PPP efficiency. P_P is processed power by the DC–DC converter. As seen from (7.1) and (7.2), the efficiency of PPP is related to both DC–DC stage efficiency and the amount of processed power by the DC–DC converter. In the literature, there are a variety of names and categorizations for PPP-based DC–DC converters. Considering all the connections, they can be categorized into two main groups. First, differential power converters (DPCs) are used for regulating current by creating a parallel current path to the main string. Second, series partial power converters (S-PPCs) are useful to regulate voltage difference between two DC ports (source/load or input/output). These two groups are further discussed in the following sections [4, 5].

7.1.1 DPCs

The primary aim of DPCs is to balance the current mismatch between two power sources connected in series. The elements can be photovoltaic (PV) modules or battery cells series connection to form a string feeding an inverter or a high voltage DC-bus. Hence, these converters are only responsible for processing a small share of the total PV or battery power to compensate for any imbalances. Since the introduction of the DPC concept, comparisons with FPC-based PV strings have shown a noticeable increase in the efficiency of the system as well as reduction of cost, passive components size, and required PCB area. In the following, the term "configuration" is used for the type of connection of the string and DPCs, and the term "topology" is used to describe the type of DC–DC converter utilized in a DPC.

A variety of topologies are introduced in the literature. From parallel-connected flyback modules to buck-boost current diverter, dual active bridge (DAB) converter, and buck-boost converter with shared core inductors. Considering the configuration of DPC-based systems, different classifications are presented in the literature. Most of them divide DPC-based systems into two main groups:

- element-to-bus connection where energy flows from each element to the bus or reverse;
- element-to-element connection (e.g., PV to PV) where the energy is transferred from one element to another one.

7.1.1.1 Element-to-bus configuration

This configuration is shown in Figure 7.2(a). Every DPC is connected between the corresponding string element and the DC-bus. The DPCs are in charge of balancing the current of the elements to keep them operating at their optimal point. For instance, in a PV string, DPCs are responsible for injecting or drawing current to maintain every PV element in its maximum power point (MPP). To fulfill this criterion, the topologies must be chosen from bidirectional DC–DC converters.

The string of n elements includes n DPCs, and the current equation can be written as

$$I_{ss} = I_{E,k} - I_{DPC,k} \tag{7.3}$$

for $k = 1, 2, 3, \ldots.., n$, where I_{ss} is the substring current, $I_{DPC,k}$ is the input current of kth DPC, and $I_{E,k}$ is the current kth PV element. Considering Figure 7.2(a), the return current of DPCs is

$$I_r = \sum_{k=1}^{n} \frac{V_{E,k} I_{DPC,k}}{V_{bus}} \tag{7.4}$$

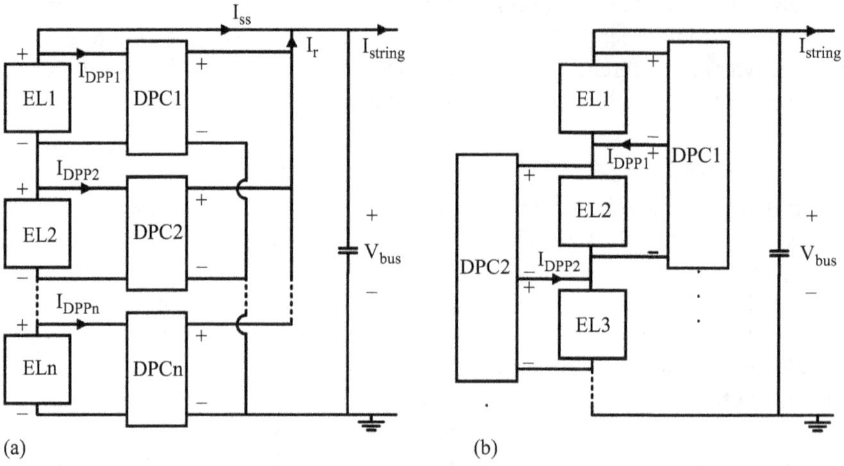

Figure 7.2 DPC configurations: (a) element-to-bus and (b) element-to-element

where $V_{E,k}$ is the voltage of kth PV element. Therefore, by adding the substring and return current, the string current can be obtained as

$$I_{string} = I_{ss} + I_r. \tag{7.5}$$

This configuration has some challenges that make it complicated to implement. The return current and string current are highly coupled, and a careful control scheme design is necessary. Furthermore, to avoid a non-isolated connection between source and load, the topology used in the DPC configuration must be an isolated DC–DC converter. For example, a bidirectional flyback topology is implemented to track the MPP of each PV module directly, and it can minimize the power processed by the DPC. The configuration seems modular but in fact, it is designed for a specific voltage gain and a certain number of elements. Then, the output voltage of the DPC is rated to a specified DC-bus voltage, and every change in the number of elements will result in redesigning DPCs. It must be noted that the voltage gain of DPC must be the same as the system voltage gain. Consequently, very high voltage gains cannot be obtained from this architecture due to DPC limitations.

7.1.1.2 Element to element configuration

This type of connection, shown in Figure 7.2(b), mainly focuses on correcting the current imbalance of the power sources by transferring power between adjacent sources. Similar to the element-to-bus configuration, the DPC must be bidirectional to add or remove current. For a string including n elements, $(n-1)$ converter is needed. If a buck-boost converter is selected, the current equation of the DPC can be written as

$$I_{L,k} = I_{E,k} - I_{E,k+1} + D_{k-1}.I_{L,k-1} + (1 - D_{k+1}).I_{L,k+1} \tag{7.6}$$

where $I_{L,k}$ is the inductor average current of kth converter, $I_{E,k}$ is the kth string element current, and D_k is the duty cycle of the high-voltage side switch of the kth DPC. As seen from (7.6), the current of each DPC is coupled to the current of the nearby converter. This imposes a limitation on the control strategy and must be considered carefully. The main benefit of this architecture is that every converter is rated based on the elements regardless of DC-bus voltage and current ratings. Consequently, this connection allows for flexible modular structure.

The main disadvantage of both DPC configurations is that the power rating of DPCs depends on how big the current mismatch should be compensated. For example, considering opaque shading of PV string elements would require DPCs rated for the full power.

7.1.2 S-PPCs

In comparison with DPC, the main goal of S-PPCs is to regulate the voltage difference between two dc ports. They can be suitable for various applications discussed in the following sections. Same as with DPCs, the S-PPC processes a small amount of the whole power, while the remaining part is transferred without processing and associated losses. There are different possibilities to connect input, output, and a DC–DC converter to compose an S-PPC configuration, including an almost lossless power path. Each configuration shows different characteristics in terms of voltage regulation and the amount of processed power. The basic configurations are depicted in Figure 7.3.

The first introduced connection is shown in Figure 7.3(a). It is called input parallel output series (IPOS). Then some authors utilized configurations in Figure 7.3(b) and (c) that are the same connection but the power flow inside the DC–DC converter is reversed. Following the same naming pattern, these two are called ISOP-I and ISOP-II, respectively. For a DC–DC converter with positive output voltage, one can see that all three configurations act as a step-up converter. To obtain a step-down converter, configurations of Figure 7.3(d–f) are introduced and named as ISOP, IPOS-I, and IPOS-II, respectively. As it is clear from all the connections, applying an isolated DC–DC converter is crucial to avoid short-circuits between input and output DC ports [6–8].

In all the configurations, a series path between input, output, and DC–DC converter terminal is created. Thus, in the desired application range, the voltage of the series connected port of the DC–DC converter is typically lower than that of the input and output dc ports. This is the central concept that defines the partial characteristics of S-PPC. For the IPOS configuration, the voltage and current equations comes as follows:

$$V_o = V_{in} + V_C, \tag{7.7}$$

$$I_{in} = I_o + I_{Conv}, \tag{7.8}$$

where V_o is the output voltage, V_{in} is the input voltage, and V_C is the DC–DC converter output terminal voltage. Also, I_{in} is the input current, I_o is the output current, and I_{Conv} is the DC–DC converter input terminal current. It can be seen from (7.7) that for positive V_C, this configuration is a step-up converter, and for negative V_C, it is a step-down converter. If the DC–DC converter voltage

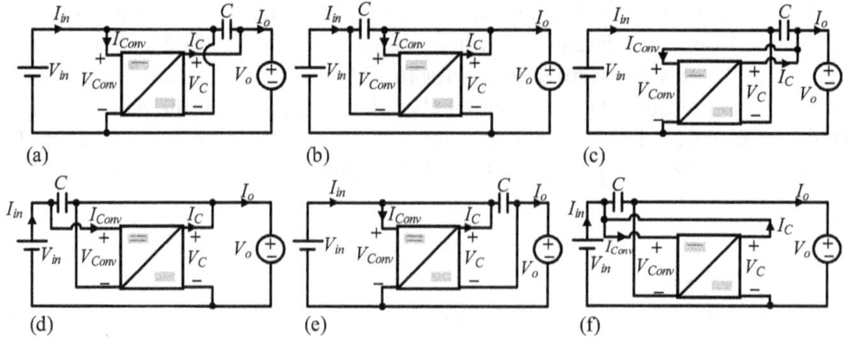

Figure 7.3 S-PPC configurations with isolated DC–DC converter topologies: (a) IPOS, (b) ISOP-I, (c) ISOP-II, (d) ISOP, (e) IPOS-I, and (f) IPOS-II

polarity can be changed, then step-up/down conversion can be achieved. The same logic is valid for the ISOP configuration, but it will act as a step-down converter with positive V_C and step-up with negative polarity. In case of no voltage imbalance between input and output, the DC–DC converter voltage will be zero, and it will not process any power. This is one of the essential features of S-PPC converters.

7.2 Requirements for S-PPCs

This section discusses in detail different aspects of the S-PPC operation and design. The first and foremost important factor is partiality. All the configurations must be compared to find the best candidate in which the DC–DC converter processes less active power. The IPOS and ISOP configurations are discussed in detail below. From the current and voltage equations of IPOS in Section 7.1.2, the voltage gain can be obtained [3,4]. It should be noted that in all equations G_C is the DC–DC converter gain and is defined as V_C/V_{Conv}. It is primarily assumed to be positive, but negative is possible for converters capable of V_C polarity reversing $-\infty \leq G_C \leq +\infty$:

$$G = \frac{V_o}{V_{in}} = \frac{V_{in} + V_C}{V_{in}} = 1 + G_C, \tag{7.8}$$

where G is the S-PPC gain. Regarding ISOP, in respect to Figure 7.3(d) of Section 7.1.2, equations are as follows:

$$V_{in} = V_o + V_C \tag{7.9}$$

$$I_o = I_{in} + I_{Conv} \tag{7.10}$$

From (7.9), the voltage gain for ISOP can be obtained as

$$G = \frac{1}{1 + G_C}. \tag{7.11}$$

In all cases, V_{Conv} is the voltage at the high-voltage (parallel) side of the DC–DC converter, and V_C is the voltage at the low-voltage (series) side. From voltage equations, it is clear that V_C regulates the voltage between the input and output DC ports, V_{in} and V_o, respectively.

7.2.1 Partiality and efficiency analysis and comparison

The coefficient of partiality is a parameter that is intended to show the amount of power processed by the DC–DC converter in relation to the total power of the S-PPC. It equals the ratio of the power handled by the DC–DC converter to the total input/output power. Concerning IPOS configuration, this is

$$P_C = V_C I_{out}, P_o = V_o I_o \tag{7.12}$$

where P_C is the processed power. The K_{Pr} is the coefficient of partiality, and it equals

$$K_{Pr} = \frac{P_C}{P_o} = \frac{G_C}{1 + G_C} \tag{7.13}$$

while for ISOP

$$P_C = V_C I_C,\ P_{in} = V_{in} I_{in},\ K_{Pr} = \frac{P_C}{P_{in}} = \frac{G_C}{1 + G_C} \tag{7.14}$$

The definitions of K_{Pr} are slightly different between (7.13) and (7.14) because it is more convenient to calculate them regarding the series port of the DC–DC converter regardless of the power flow direction in that port. From (7.11), (7.8), (7.13), and (7.14), it is apparent that if $G_C = 0$, the voltage gain of the S-PPC equals one and the DC–DC converter does not process any active power. It means that the only determining factor for the active power is the voltage regulation range between input and output. In respect to this fact, the voltage difference between input and output (Δv) will be equal to the voltage variations at the series/low-voltage side terminal of the DC–DC converter ($\Delta v = \Delta V_C \Rightarrow \Delta G_C = f(\Delta v)$). To better understand the voltage variations, the Δv could be normalized regarding either V_{in} or V_o ($\Delta\widehat{v}$). The partiality curve is shown in Figure 7.4. It describes both IPOS and ISOP configurations according to (7.13) and (7.14). For $\Delta\widehat{v} = 0$, the voltage gain of the DC–DC converter equals zero.

Furthermore, when the S-PPC gain equals one, the whole active power is delivered through the series path with almost 100% efficiency. Increasing $\Delta\widehat{v}$ results in increasing DC–DC converter gain and changing the S-PPC gain. This results in a higher K_{Pr} that increases with a steep slope. It must be noted that in Figure 7.4, the absolute value of K_{Pr} is plotted. Negative partiality means changing the active power direction inside the DC–DC converter. For IPOS configuration, the power flow of S-PPC and power flow direction inside the DC–DC converter has the same direction for $V_C \geq 0$. For $V_C \leq 0$, the power flow direction inside the DC–DC converter is reversed. Concerning voltage and current equations, the

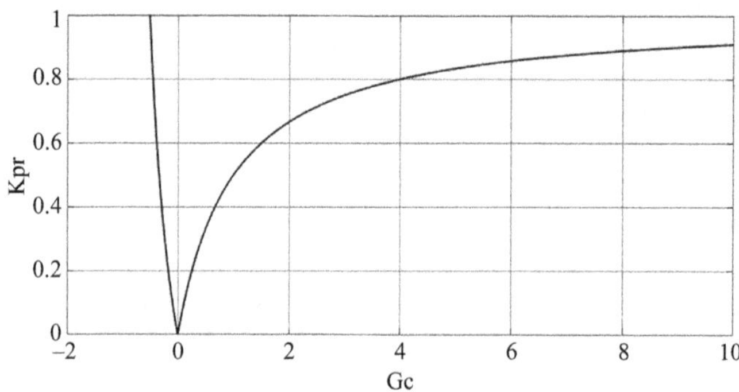

Figure 7.4 Partiality curve for a range of DC–DC converter gain

efficiency of the whole system can be obtained as

$$\eta_{sys} = \frac{P_o}{P_{in}} = 1 - K_{Pr}(1 - \eta_{Conv}) \tag{7.15}$$

in which η_{sys} is the S-PPC efficiency and η_{Conv} is the DC–DC converter efficiency. Understandably, the overall efficiency is higher than FPC efficiency. A lower partiality factor and a higher DC–DC converter efficiency combined result in much higher overall system efficiency, as illustrated in Figure 7.5.

To benefit from S-PPC advantages, it is crucial to reduce the processed power. Hence, it is appropriate to limit $\Delta\widehat{v}$ to around 30% regarding either the input or output side. In many DC applications such as battery, PV, and fuel cell, the output side is a DC-bus with constant voltage. So, the voltage variation will primarily be on the input side, but the DC-bus voltage also can change to implement a droop control or other purposes. The input side voltage variation must be 30% or less. Otherwise, wider voltage range will not allow the S-PPC to efficiently deliver the input power to the output, as exemplified in Figure 7.5. Thus, this design consideration must be taken into account, both for voltage variation range and S-PPC itself. Considering three possible options to implement the S-PPC, there will be different criteria for the input voltage range. For a step-up converter, the input voltage must be always lower than the output voltage. Then by selecting the maximum input voltage as $V_{in,\max} = V_o$ the minimum input voltage should be $0.7V_o$ or higher.

Furthermore, for the step-down converter, the minimum input voltage should equal the output voltage $V_{in,\min} = V_o$. Then its maximum value should not exceed $1.3V_o$. Finally, for a step-up/down S-PPC, the input voltage can be chosen so that the voltage gain of one is achieved in the middle of the voltage regulation range ($V_o = V_{in,r} = \frac{V_{in,\max}+V_{in,\min}}{2}$). The same logic is valid for voltages when the input voltage is fixed and the output voltage varies. Voltage limits required to achieve the same voltage range with three S-PPC configurations are shown in Table 7.1. It shows that step-up/down S-PPCs require a voltage regulation range two times less than that for two other configurations.

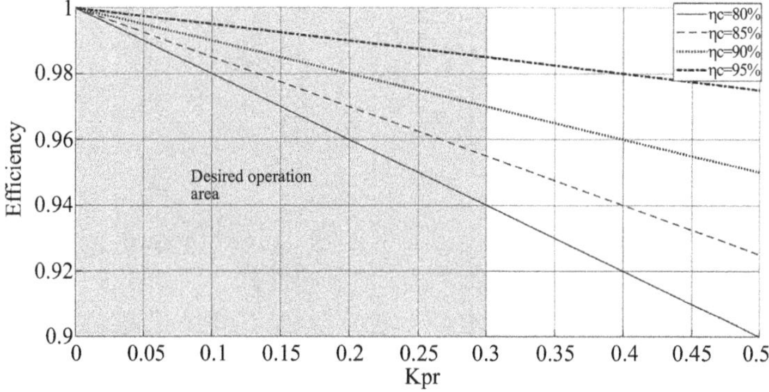

Figure 7.5 System efficiency regarding partiality and DC–DC converter efficiency

Table 7.1 Input and output voltage range for different voltage conversion types and constant voltages

Constant voltage	Voltage conversion type		
	Step-up	**Step-down**	**Step-up/down**
V_o	$0.7V_{in,\max} \leq V_{in} \leq V_{in,\max} = V_o$	$V_o = V_{in,\min} \leq V_{in} \leq 1.3V_{in,\min}$	$V_o = V_{in,r}, 0.85V_{in,r} \leq V_{in} \leq 1.15V_{in,r}$
V_{in}	$V_{in} = V_{o,\min} \leq V_o \leq 1.3V_{o,\min}$	$0.7V_{o,\max} \leq V_o \leq V_{o,\max} = V_{in}$	$V_{in} = V_{o,r}, 0.85V_{o,r} \leq V_o \leq 1.15V_{o,r}$

Regarding all the possible solutions, the processed power by the DC–DC converter is shown in Figure 7.6. For the IPOS configuration, the active power is as follows:

$$\frac{P_C}{P_o} = \frac{V_C I_o}{V_o I_o} = 1 - \frac{V_{in}}{V_o}. \tag{7.16}$$

The equation for ISOP configuration is similar to the previous one:

$$\frac{P_C}{P_{in}} = \frac{V_C I_{in}}{V_{in} I_{in}} = 1 - \frac{V_o}{V_{in}}. \tag{7.17}$$

The absolute values of (7.16) and (7.17) are evaluated in Figure 7.6. The shaded areas are for step-up/down S-PPC, while the dashed areas are for step-up and step-down S-PPCs in both Figure 7.6(a) and (b). Figure 7.6(a) demonstrates that the processed power is proportional to the voltage difference between the input and output ports of the S-PPC. Thus, the power processed by the DC–DC converter inside an IPOS configuration is the same for the same absolute value of the voltage V_C, regardless of its polarity. For ISOP configuration, the relationship is not symmetrical. If the input voltages higher is than the output voltage, the processed power will be lower than that in the IPOS configuration. On the other hand, for input voltages lower than the output voltage, the DC–DC converter in the ISOP configuration processes higher active power than that in the IPOS configuration under similar conditions. The maximum partiality in critical voltage points, i.e., 30% voltage variation, are collected in Table 7.2 for three S-PPC types and two basic configurations.

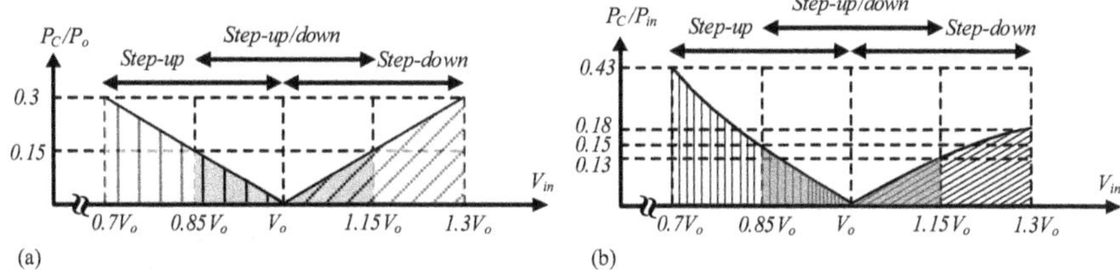

Figure 7.6 Processed power for 30% voltage variation with constant output voltage (a) IPOS and (b) ISOP

Table 7.2 Maximum partiality for different solutions

Configuration type	**Constant voltage**	**Voltage conversion type**		
		Step-up	**Step-down**	**Step-up/down**
IPOS	V_o	30%	30%	15%
ISOP		43%	23%	18%/13%
IPOS	V_{in}	23%	43%	13%/18%
ISOP		30%	30%	15%

Following the same concept, maximum partialities could be calculated for the case of constant input voltage and variable output voltage. Figure 7.6 and Table 7.2 reveal that for a specific voltage regulation range, the step-up/down S-PPC processes the least amount of active power among all the solutions. Besides voltage variation, it is crucial to consider the input side voltage-power distribution. For example, in PV applications, the MPP is approximately in the middle of the PV panel voltage range. This observation strengthens claims of the superior performance of the step-up/down S-PPC. Because in step-up/down converter, the partiality is zero at the middle of voltage regulation range where it is most likely to observe the maximum power generated by a PV panel. It would result in higher efficiency and lower current/voltage stress on converter components.

Analysis of the active power processing reveals its high dependency on the S-PPC configuration and voltage conversion type. The IPOS is the most promising candidate for all step-up and step-up/down applications where the output is constant. The ISOP, on the other hand, is suitable for all step-down applications and step-up/down with fixed input voltage. The conclusions are extracted from Table 7.2.

In order to classify a converter as a real partial processing converter, it must be compared with a non-isolated FPC, which has the same input-output features. It must be noted that partiality, i.e., active processing power relative to the full power, is one of the key parameters affecting the S-PPC efficiency. The other important factor is the DC–DC converter efficiency itself. Some converters fulfill the active power criteria to be considered as S-PPC, but they fail to decrease the nonactive power processing. This critical point significantly impacts the DC–DC converter and overall system efficiencies [9], as elaborated below.

7.2.2 Non-active power processing

In a DC–DC power electronics converter, the current and voltage waveform of all elements include a DC part and an alternating periodic part caused by the switching of semiconductors. The alternating current and voltage generate a kind of power that oscillates in the converter components and do not result in active power delivered from the input to output of a DC–DC converter. It can be stored in or released from the reactive components of the converter, i.e., capacitors and inductors. The IEEE Std.1459-2010 defines non-active power (N) and differentiates it with reactive power (Q), which is a swinging power in AC systems. The unit is similar for both non-active and reactive power and it is volt-amperes (var). Like reactive power, the non-active power leads to power loss in converter components and necessitates higher components voltage/current ratings. So, decreasing this swinging power is of great importance [10, 11].

In the steady-state converter operation, volt-second balance for inductors and charge balance for capacitors are valid. These elements do not store active power; however, they are the source of non-active swinging power that creates additional power loss in the converter. The total stored energy in a switching period is E_{LT} for an inductor and E_{CT} for a capacitor. Both of them are zero in a total switching period:

$$E_{LT} = \int_{t}^{t+T_S} v_L(t)i_L(t)dt = 0 \tag{7.18}$$

$$E_{CT} = \int_{t}^{t+T_S} v_C(t)i_C(t)dt = 0 \tag{7.19}$$

This means that the absorbed energy during DT_S ($D$is the duty cycle and T_S is switching period) is equal to released energy during $(1 - D)T_S$

$$\Delta E_L = \int_{t}^{t+DT_S} |v_L(t)i_L(t)dt| = \int_{t+DT_S}^{t+T_S} |v_L(t)i_L(t)dt| \tag{7.20}$$

$$\Delta E_C = \int_{t}^{t+DT_S} |v_C(t)i_C(t)dt| = \int_{t+DT_S}^{t+T_S} |v_C(t)i_C(t)dt| \tag{7.21}$$

According to (7.20) and (7.21), the non-active power of inductor (N_L) and capacitors (N_C) is calculated by the following equations:

$$N_L = \frac{2\Delta E_L}{T_S}, \tag{7.22}$$

$$N_C = \frac{2\Delta E_C}{T_S}, \tag{7.23}$$

By adding all inductors' and capacitors' non-active power of, the total non-active power inside the DC–DC converter can be calculated as

$$N_{DC-DC} = \sum_{i=1}^{n} N_{L,i} + \sum_{j=1}^{m} N_{C,j}, \tag{7.24}$$

where m and n are the total number of inductors and capacitors, respectively. Furthermore, the non-active power from both the input and output sides of the converter must be taken into account to obtain total non-active power. They can be obtained using the following equations:

$$N_{in} = \sqrt{S_{in}^2 - P_{in}^2}, \; N_{out} = \sqrt{S_{out}^2 - P_{out}^2} \tag{7.25}$$

where S is the apparent power for either input or output, N_{in} is the input side non-active power, and N_{out} is the output side non-active power. Then the total non-active power equals the sum of all the elements and can be calculated as

$$N_{tot} = N_{in} + N_{DC-DC} + N_{out} \tag{7.26}$$

In all steps of non-active power calculation, the small ripple assumption is taken into account for inductors and capacitors. Therefore, the non-active power is dependent on current and voltage waveforms that are related to the duty cycle and turns ratio of the transformer or coupled inductor. It means that the process is frequency-independent. The duty cycle range and the turns ratio needed to cover the target voltage range must be chosen carefully, since they define the maximum non-active power of the converter.

7.2.3 Basic requirements for DC–DC converter topology selection

In the previous sections, S-PPC configurations are analyzed, and features of each configuration are identified to select the most appropriate candidate for each application. In this section, a discussion about requirements for the DC–DC converter topology are given. As mentioned earlier, to implement an S-PPC, an isolated DC–DC converter must be used. Furthermore, there are additional requirements for the applied topologies.

7.2.3.1 Component stress factor (CSF)

It is critical to assess the voltage and current stress of the components for each topology. It is defined by dividing the apparent power of every component by the output power. The total CSF can be obtained by adding all the components CSF.

$$CSF = \frac{V_e I_e}{P_{out}} \tag{7.27}$$

where V_e is the maximum blocking voltage for switching devices like diodes and switches, average voltage for capacitors, and average AC voltage for inductors and transformers considering duty cycle and the voltage waveform on these components. Also, I_e is the RMS current for all the components, except for diodes and IGBTs that require the average current value to be used. The CSF is related to both topology and configuration type, but for the same configuration, it must be noted that the CSF is lower for step-up/down S-PPC. It results from the lower maximum voltage across the low side terminal of the DC–DC converter compared to the step-up and step-down S-PPCs.

7.2.3.2 Bidirectional active power flow in DC–DC converter

Low partiality of the step-up/down S-PPCs requires DC–DC topology to be capable of changing its voltage polarity at the low-voltage side. For a unidirectional power flow from input to output, the current direction at the low-voltage port will remain unchanged ($I_C = I_{in}$ or I_o), while the voltage polarity at this port can change. So, the active power flow inside the DC–DC converter will be bidirectional. Topology connection within S-PPC configurations shows that the high-voltage side polarity is not changing due to being connected directly to the input or output of the S-PPC ($V_{Conv} = V_{in}$ or V_o). Therefore, bidirectional current flow can be observed at the high-voltage terminal of the DC–DC converter. For both power flow directions, the unprocessed power and total power flow are unidirectional from input to output. In bidirectional power flow applications, the direction of $I_C = I_{in}$ or I_o must change as well as V_C polarity. Hence, the DC–DC converter has to operate in all four quadrants regarding V_C and I_C. Again, at the high-voltage port, the only changing parameter is the current direction because the voltage polarity depends on either the input or the output voltage of the S-PPC.

7.2.3.3 Topology selection criteria for step-up/down S-PPC

Regarding the step-up/down S-PPC, the most suitable candidate for a wide range of applications in terms of partiality, some requirements are essential for the DC–DC converter topology. The first one is the application of two- or four-quadrant switches on the low voltage side of the DC–DC to implement uni- or bidirectional step-up/down S-PPC, respectively. Second, the necessity of a transformer or coupled inductor providing galvanic isolation between high- and low-voltage sides. Third, the connection of an inductive impedance to the low-voltage side and zero inductive impedance to the high-voltage side.

A DC–DC topology corresponding to the requirements above can provide bipolar output voltage (V_C) and bidirectional input current (I_{Conv}) using modulation techniques that manipulate energy exchange between inductors and capacitors, discrete or embedded/parasitic. The goal is to vary voltage gain from negative to positive or vice versa. Magnetizing and leakage elements of the isolating magnetic component(s) must be considered when synthesizing commutation sequences of two- or four-quadrant switches to avoid overvoltage across them.

S-PPCs feature the common ground implementation and thus require DC–DC converter topology with galvanic isolation. However, even though the DC–DC converter topology is isolated, the S-PPCs lose this feature due to the series connection of the DC–DC converter ports. Therefore, S-PPCs are not suitable for applications where galvanic isolation is required. Due to the great impact of transformer or coupled inductor turns ratio on non-active power, the turns ratio selection is of great importance to guarantee the superior performance expected from S-PPCs in comparison with similar FPCs. It must be noted that the coupled inductor-based topologies generate more non-active power due to the sizeable magnetizing current of the coupled inductor.

Finally, the connection of an inductive impedance (circuit) to the low-voltage side also relates to the bipolar feature of the DC–DC topology. It is necessary to be able to change the polarity of V_C. On the high-voltage side, the current must be reversible so, any inductive reactance could hinder the current direction change. Following this approach, the bipolar and bidirectional operation of DC–DC topology can be achieved using special modulation techniques. Regarding all the mentioned criteria, most of the buck or buck-boost DC–DC topologies (SEPIC, Cuk, flyback, or DAB converter) cannot be used in a step-up/down S-PPC. An unfolding circuit implementation at the low-voltage side can solve this issue by changing the polarity of the voltage V_C [19].

7.3 Renewable energy applications

Due to their high-efficiency advantage, S-PPCs are among the most promising candidates in renewable energy applications. It must be taken into account that these converters are non-isolated DC–DC converters. If the isolation between input and output is an important issue, the S-PPCs will not be the best solution. In addition to the isolation, the discussion given in the previous section showed that S-PPCs reach their best performance in applications where the input and output voltage are close to each other. Thus, they are not suitable for high step-up or high step-down voltage conversion [12]. The S-PPC technology was elaborated the most in PV applications to outperform the mature FPC technology and provide better energy generation.

7.3.1 PV systems

The output power of PV modules and panels is heavily dependent on the environmental conditions at the site where they are installed. The statistical data of annual energy production is one of the determining factors in designing an appropriate converter. Besides the output power, the input voltage at which the maximum power is generated is an essential factor. In Figure 7.7, the statistical distribution of the PV module voltage is given as a histogram graph for a ten-year aggregation period. Converting the output power graph to a lognormal distribution reveal that this PV module has maximum power point at $\widehat{v} = 29.7$ V and standard deviation of$\sigma = 1.36$ V. Further analysis confirms that 99.9% of the total energy is generated within $\pm 3\sigma$ range. This means that almost all of the energy is generated between $V_{PV} = 25.6$ V and $V_{PV} = 33.8$ V. The chosen S-PPC must have a voltage regulation range wider than this range to deliver all the produced power to the DC-bus. Considering 29.7 V as the reference voltage, it becomes evident that an S-PPC with a voltage regulation range of $\Delta\widehat{v} = 30\%$, i.e., $\pm 15\%$ voltage regulation, is an ideal candidate for PV applications in the given climatic conditions. Applying PV modules with a higher operating voltage would result in a wider voltage regulation range of S-PPC in absolute numbers. Therefore, an S-PPC would process more power, and its feasibility declines [10, 11].

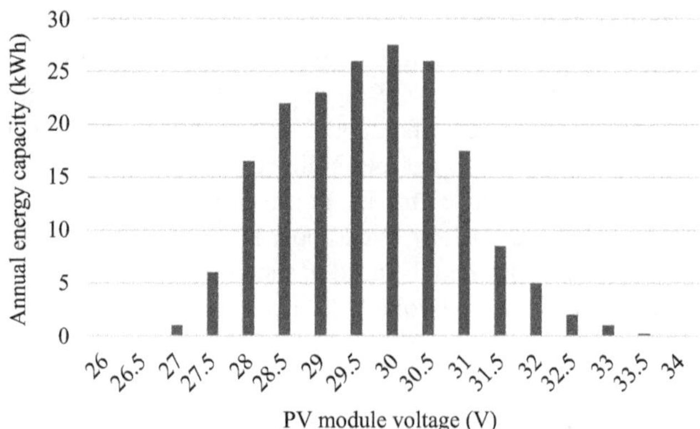

Figure 7.7 Histogram graph of the estimated average annual energy production for PV module SunEarth TPB 60-P for a site in Sao Martinho da Serra

To implement an S-PPC as an interface between PV modules/string and the DC-bus, there can be three voltage conversion types as discussed in the previous section. The number of PV modules must be selected appropriately for step-up, step-down, and step-up/down S-PPC. The data in Table 7.3 is derived according to the limits from Table 7.2 presented in the second section.

A full-bridge-based S-PPC (FB S-PPC) and a full-bridge push–pull S-PPC (FBPP S-PPC) were selected in [10] to carry out the evaluation. They are shown in Figure 7.8. In both cases, the S-PPC configuration is IPOS, and the DC-bus voltage equals 220 V. The voltage regulation range of 66 V was selected.

The full-bridge topologies can be used to synthesize high-performance S-PPCs due to high efficiency and power density. In the case of the FB S-PPC, the input side is controlled by the phase-shift modulation method, and the output side is the diode rectifier, passive or semi-active. In the step-up FB S-PPC from Figure 7.8(a), the V_C is always positive as the low-voltage side of the DC–DC converter. Non-active power calculations were carried out based on the formulations from the second section. It turns out that non-active power strongly depends on the turns ratio of the transformer. The full-bridge converter voltage gain is $V_C/V_{Conv} = n \cdot d$, and selection of the turn ratio is related to the desired duty cycle range of the converter. In other words, a lower turns ratio will result in a higher duty cycle variation range for a particular voltage regulation range. So, the minimum turns ratio can be calculated to avoid very high duty cycle values, which could be a

Table 7.3 The number of PV modules for three S-PPC voltage conversion types

Voltage conversion type	Number of modules
Step-up	$N_{PV} \le \frac{V_{DC-bus}}{{}^{1}V_{PV,\max}}$
Step-down	$N_{PV} \ge \frac{V_{DC-bus}}{{}^{2}V_{PV,\min}}$
Step-up/down	$N_{PV} \approx \frac{V_{DC-bus}}{{}^{3}V_{PV,nominal}}$

1. $V_{PV,\max} = \widehat{v} + \Delta v/2$
2. $V_{PV,\min} = \widehat{v} - \Delta v/2$
3. $V_{PV,nominal} = \widehat{v}$

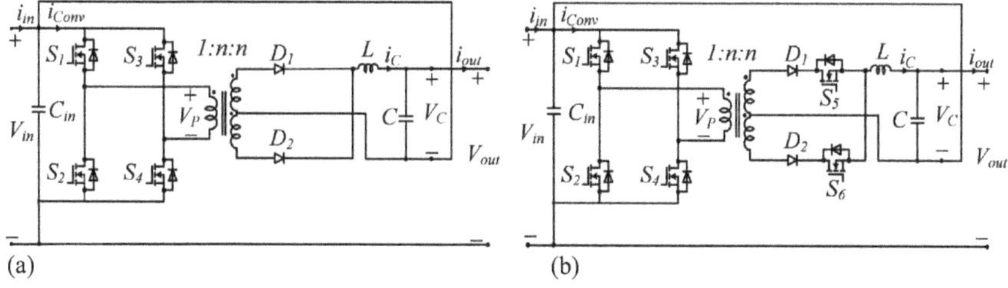

Figure 7.8 A FB S-PPC and a FBPP S-PPC

bottleneck for the control system, as

$$n_{\min} = \frac{V_{C,\max}}{V_{Conv,\min} d_{\max}} \tag{7.28}$$

For the maximum duty cycle of 85.7%, the minimum turns ratio *n* equals 0.5 [10]. Derived curves for non-active power and CSF are depicted in Figure 7.9. It can be seen that the amount of non-active power increases significantly by increasing the turns ratio. The results for CSF also show a similar increase for higher turn ratio values.

Utilization of the FBPP topology in S-PPC, shown in Figure 7.8(b), decreases the voltage regulation range of the DC–DC topology because it can change the voltage polarity at the low-voltage side (cf. Table 7.2). From (7.28), the minimum turn ratio equals 0.2. The results for non-active power and CSF confirm their increase by turn ratio that comes in Figure 7.10.

In both cases, the non-active power and CSF drop to zero when the S-PPC gain equals one. The reason is that the DC–DC converter is completely out of operation, and there is no switching. The series path transfers the full power. The results confirm that the step-up/down S-PPC significantly decreases the active and

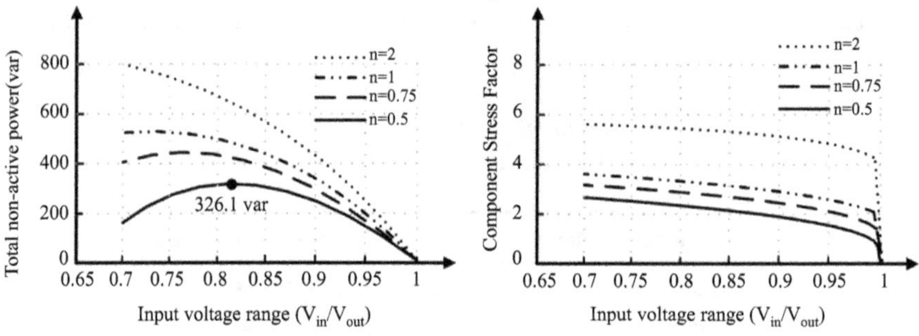

Figure 7.9 Total non-active power and CSF of FB S-PPC

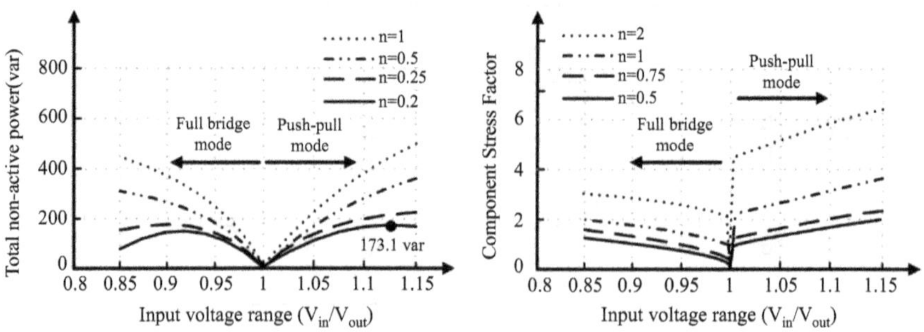

Figure 7.10 Total non-active power and CSF of FBPP S-PPC

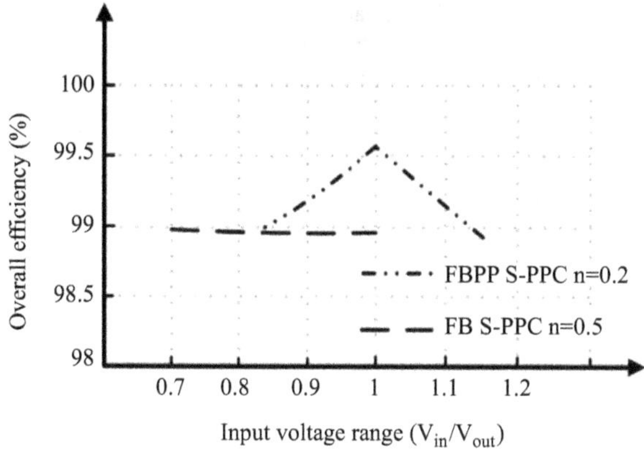

Figure 7.11 Efficiency curves for (a) FB S-PPC and (b) FBPP S-PPC

non-active power processing alongside stress on circuit elements. Moreover, in the step-up/down converter, the minimum partiality (processed power) occurs at the middle of the voltage range where the maximum power of the PV module is typically produced. This is the main reason allowing to elevate the efficiency of the whole system. The efficiency curve for output power of 750 W is shown in Figure 7.11. Predictably, the maximum efficiency occurs at $V_{in}/V_{out} = 1$.

7.3.2 Wind energy systems

To process the produced energy of wind turbines, a two-stage energy conversion system comprised of a rectifier and a DC–DC converter is advantageous due to decoupling between a rectifier and a DC–DC stage. The benefits are better MMPT performance and flexibility in comparison with single-stage conversion. To overcome the excessive loss of two-stage conversion, the application of S-PPC can be one of the profitable solutions. In [13], the connection of an S-PPC to a small wind turbine is analyzed. The nominal output of a wind turbine is 1.5 kW. It utilizes a permanent magnet synchronous generator. The mechanical power of the wind turbine is a function of wind speed and other parameters, as expressed below

$$P_{mech}(\lambda, v_\omega) = \frac{1}{2}\rho\pi r^2 v_\omega^3 C_P(\lambda), \; \omega_r(\lambda, v_\omega) = \frac{\lambda v_\omega}{r}, \tag{7.29}$$

where P_{mech}(W) is the mechanical power of the rotor shaft, λ(TSR) is the tip-speed ratio, C_P is the power coefficient, v_ω(m/s) is the wind speed, r(m) is the radius of rotor, and ρ(kg/m^3) is the air density.

There is a maximum power point (P_{mppt}) for each specific wind speed and variable rotor speed. In the given case, Figure 7.12 defines the output power curve of the rectifier regarding its output voltage. It specifies the S-PPC input voltage range and the amount of processed power. It can be observed that for a particular rotor speed, the output power of the three-phase rectifier has one MPP. Moreover,

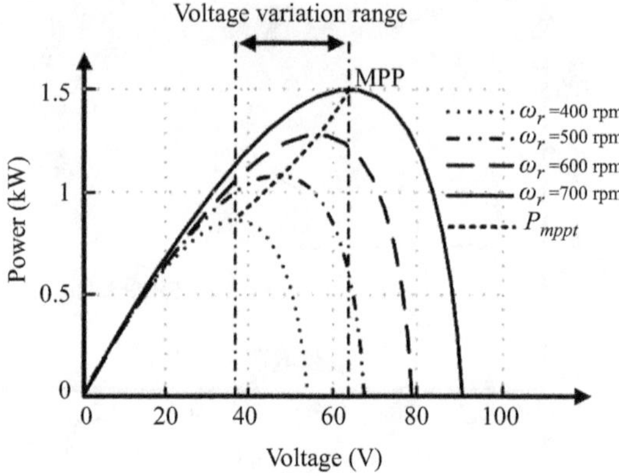

Figure 7.12 Rectifier output power curves for different rotor angular speeds

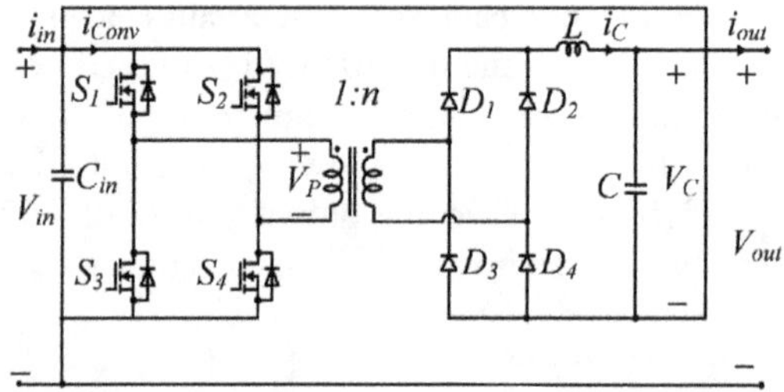

Figure 7.13 A step-up S-PPC applied to 1.5 kW small wind turbine comprising a permanent magnet synchronous generator and a diode rectifier

the voltage variation range can be selected to define the maximum processed power by the S-PPC. Hence, the input voltage variation range from 35 to 64 V corresponds to the maximum power points increasing gradually from 880 to 1,500 W. The full-bridge zero-voltage switching (ZVS) DC–DC converter topology was chosen to implement step-up S-PPC to interface the given wind turbine with a rectifier into a DC-bus of 220 V. This topology features high performance in terms of the power density of overall efficiency. The step-up requirement necessitates the IPOS connection of S-PPC, as illustrated in Figure 7.13.

The inverter side is controlled by constant switching frequency and phase shift modulation. The rectifier side is a passive diode rectifier with LC-filer. The topology benefits from ZVS in a wide operation range, further increasing overall efficiency. Considering constant DC-bus voltage and the input voltage variation of the S-PPC, it can be realized that the processed power is 1,050 W for nominal (out of 1.5 kW in $\omega_r = 700$ rpm). Therefore, power of 450 W is not processed, i.e., transferred directly, which results in S-PPC partiality of 70%. Although the partiality is high compared to the PV system discussed before, the direct transfer of 30%

of the total power can boost the overall system efficiency considerably. It is clear that for a wind turbine with a higher output voltage and lower voltage variation range, the partiality and processed power will be lower. The S-PPC mentioned above with peak power of 1.5 kW and the input voltage range of 35–64 V reaches the maximum efficiency of 96.5%, which validates the appropriateness of partial power converters in wind turbine applications.

7.3.3 Fuel cells

A step-up/down S-PPC implemented in [14] can interface a high-power fuel cell into a DC-bus with operating voltage in the middle of the fuel cell operating voltage range. It has three operation modes including step-up, step-down, and current limiting mode. A center-tapped transformer is used to regulate the voltage between the DC-bus and fuel cell and, consequently, to transfer power directly between those two ports. The converter circuit presented in Figure 7.14 contains a full-bridge inverter modulated by the PWM. By changing the PWM duty cycle from 0% to 100%, the V_S can be varied from zero to input voltage. The proposed converter works with 50% of the input voltage regulation range.

Based on the polarity of V_S, the step-up or step-down conversion can be designed and implemented. For both of the operation modes, voltage equations can be written as

$$boost : V_{out} = V_{in} + V_{in} * D * n \tag{7.30}$$

$$buck : V_{out} = V_{in} - V_{in} * D * n \tag{7.31}$$

where D is the PWM duty cycle of switches in the full-bridge inverter, and n is the transformer turns ratio. The selected fuel cell has a rated power of 10 kW, 340 V no-load voltage, and 200 V full load voltage at the current of 50 A. Fifty percent of the voltage regulation range and ten percent of the current regulation range must be considered in all points. The main design point is fuel cell voltage of 270 V and ±70 V regulation range of the S-PPC. Concerning these parameters, the maximum experimental efficiency reaches 98%. A closed-loop control system is designed to monitor the input and output voltage to determine the operation mode and duty cycle of the switches. Almost 65% of the overall power loss occurs in the full-bridge

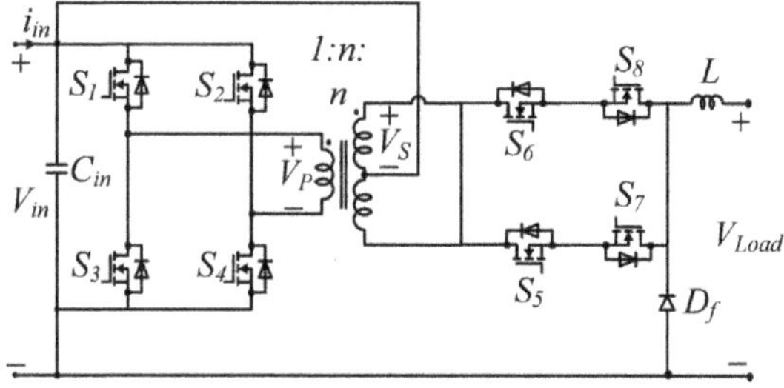

Figure 7.14 Step-up/down S-PPC applied as fuel cell interface

inverter (switching and conduction losses) and the isolation transformer. On the secondary side, mostly conduction losses in semiconductors occur.

7.4 DC microgrid and battery charging applications

7.4.1 EV chargers

The mobility sector is the source of nearly 23% of total CO_2 emissions worldwide. The tendency toward EVs rises dramatically to avoid carbon emission. To make EVs feasible and useful, a widespread installation of charging infrastructure is needed. Since the emergence of charging stations, the level 1 AC charging systems (<2 kW), level 2 AC systems (between 2 kW and 10 kW), and DC fast-charging stations (between 20 kW and 120 kW) have been designed and implemented. Recently, ultra-fast charging stations with capacities higher than 300 kW have gained a considerable share in added installations to reduce the charging time and driving range anxiety.

The typical attitude in the area is utilizing full power DC–DC converters after rectifier and PFC stage at the AC grid side. Processing this amount of active power requires high-voltage/current ratings of semiconductor devices that are costly. In [15], a novel ultra-fast charging infrastructure based on S-PPCs is introduced. The mentioned system exploits the local ESS connected to the DC-bus. Thus, the S-PPCs are considered unidirectional and the grid forming features like grid frequency regulation and peak shaving are handled by the central storage system. The desired features of the DC–DC charging stage are:

1. Power delivery from DC-bus to the EV with minimum losses and minimum partiality.
2. The charging units connected to the DC-bus must not create any circulating current.
3. The voltage and current ripple regulation based on IEC 61851-23 must be fulfilled.

Three configurations were discussed, and among them, the IPOS configuration is selected to implement a step-up S-PPC. The charging station contains six charging units, each rated for 350 kW. The considered battery pack features nominal voltage from the minimum voltage of 725 V at 20% of the state of charge (SoC) to the maximum voltage of 800 V at 80% of SoC. The high-voltage DC-bus facilitates the application of ultra-fast charging. This case study considered 650 V for step-up S-PPC. The partiality of S-PPC with mentioned parameters changes from 0.1 to 0.19.

A full-bridge phase-shifted DC–DC converter was selected and implemented in the given step-up S-PPC, as illustrated in Figure 7.15. The phase shift between the legs of the H-bridge is the main control variable defining the charging current. It is worth mentioning that in topologies like phase-shifted full-bridge, relatively large inductance inside DC–DC converter cell allows for simpler charging current control between the S-PPC ports.

As discussed earlier, the partiality of step-up S-PPC for IPOS configuration decreases with higher DC-bus voltage. However, there should be a limitation to

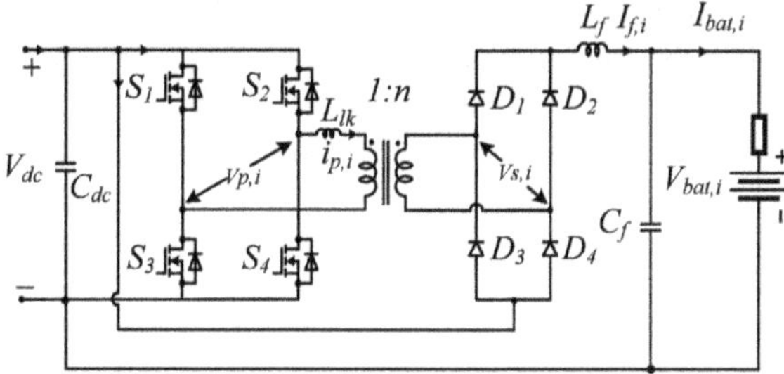

Figure 7.15 EV battery charger DC–DC stage using a step-up S-PPC

ensure that the DC-bus voltage is lower than the battery pack voltage:

$$V_{DC-bus,\max} \leq V_{bat-\min} \tag{7.32}$$

The selected DC–DC converter is a unidirectional/unipolar topology. So, fulfilling (7.32) is of significant importance. To avoid any violation of the criteria even in transient operation mode, the safety factor is considered to make sure the proper operation of the converter:

$$V_{DC-bus} = \frac{V_{bat,\min}}{1+x}, \tag{7.33}$$

where x is the safety factor that equals 0.2 in this application. The transformer turns ratio minimum value is determined by the maximum duty cycle of full-bridge switches (considering the phase-shift modulation):

$$n_{\min} = \frac{V_{bat,i-\max} - V_{DC-bus}}{D_{i,\max} V_{DC-bus}}, D_{i,\max} = \frac{\phi_{i,\max}}{\pi}, \tag{7.34}$$

in which $\phi_{i,\max}$ is the maximum phase shift between the bridge inverter legs in radians, which could be recommended to be set at 0.85π to leave 15% margin from the regulation range. The calculations result in $n_{\min} = 0.33$ for given parameters. The ZVS is also fulfilled for all switches in a wide range. The maximum efficiency of 99.2% was reported, which is relatively low and could result from overrated semiconductor components in the series port of the DC–DC converter. Unfortunately, the authors did not clarify this issue, but it is a common design approach for S-PPC prototypes without proper protection and transient control.

Another example of S-PPC application for EV charging can be found in [16]. Considering that most EV battery voltages are around 400 V, a step-down S-PPC was introduced to connect the battery charging system to 400 V_{ac} or 480 V_{ac} grid. In step-down S-PPCs, the ISOP configuration is advantageous as elaborated in the second section of this chapter. In general, the charging scheme of a battery can be divided into two modes. The first one is the constant current (CC) mode, which is used to deliver most of the power to the battery to reach an SoC of almost 80%. The second mode is the constant voltage (CV) mode when the battery is charged at the fixed voltage and gradually decreasing current. During CC mode, the charger injects a constant current into the battery regardless of the voltage drop on the DC–DC converter. Thus, the DC–DC converter output voltage is a fraction of the input voltage. Again, in this application, a full-bridge phase-shifted DC–DC converter is employed. It must be noted that providing galvanic isolation is critical due to IEC 851-23 and UL 2231 standards. The galvanic isolation between the grid and the whole system is provided by the grid frequency transformer. However, there is no galvanic isolation among charging points connected to the DC-bus as required by the IEC 61851-2. It can be one of the drawbacks of the S-PPC converter-based charging systems that need to be further researched. The employed S-PPC is depicted in Figure 7.16.

For this application, the DC-bus voltage, which is the output voltage of the rectifier that is fixed at the level of 450 V. A series connection of 98 battery cells with a rated voltage of 3.6 V and voltage range of 2.8–4.2 V was considered. The battery pack voltage will vary approximately from 274.4 to 411.6 V. With respect to these parameters, the theoretical maximum partiality factor for ISOP configuration will be 40% allowing direct (nearly lossless) transfer of 60% of the total power.

The behavior of the full-bridge DC–DC converter is similar to the previous converter in Figure 7.16. The phase shift between the inverter bridge legs and the transformer turns ratio determine the voltage difference between the S-PPC ports. Primary side transformer voltage V_P defines the difference between DC-bus voltage V_d and the battery pack voltage V_o. This is always valid if the transformer turns ratio is selected to ensure a sufficient voltage regulation range to control the DC–DC converter output current. The voltage conversion equation for the full-bridge DC–DC converter can be described as

$$V_r = \frac{\phi}{\pi} n V_P, \tag{7.35}$$

where ϕ is the phase shift between the inverter bridge legs. Considering (7.35) and the relationship between V_P, V_d, and V_o, the voltage gain and the partiality of S-PPC with the given DC–DC converter parameters can be calculated as

$$\frac{V_o}{V_d} = \frac{n\phi}{\pi + n\phi} = G_v, \tag{7.36}$$

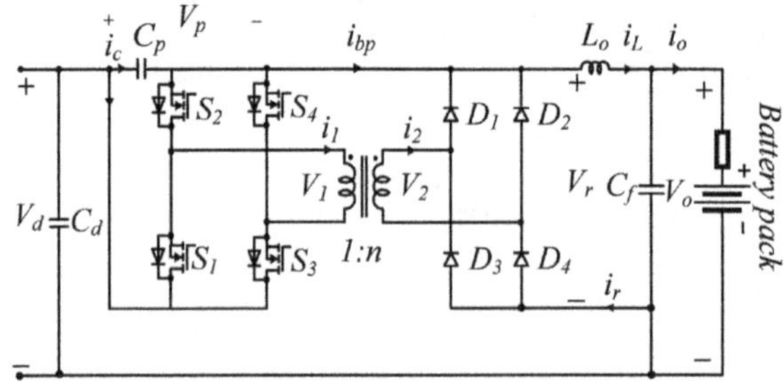

Figure 7.16 EV battery charger DC–DC stage using a step-up S-PPC

where G_v is the S-PPC voltage gain. Concerning the partiality formulations given in the second section, the partiality equation is

$$K_{Pr} = 1 - G_v = \frac{\pi}{\pi + n\phi} \tag{7.37}$$

Eq. (7.37) shows how the processed power of the DC–DC converter can be controlled. It must be noted that the phase shift range is directly related to the voltage regulation range at the low-voltage side of the DC–DC converter. An experimental prototype with 6.5 kW output power and input voltage range of 320–390 V was tested. The maximum efficiency of 99.1% was reported at the power of 5 kW. This prototype utilized high-voltage SiC MOSFETs at the low-voltage side, operating in hard-switching conditions. This means that this S-PPC requires further optimization towards using the low-voltage transistors and soft-switching to enhance the system efficiency further.

The third example of the S-PPCs application in battery charging is represented in [17]. A bidirectional S-PPC was proposed for interfacing battery energy storage into a DC-bus. This application example can be interpreted as the one capable not only of the grid-to-vehicle (G2V) but also the vehicle-to-grid (V2G) operation as an EV charger. Bidirectional operation is essential for any battery energy storage that needs to support the DC-bus and consume excessive energy. In the example provided in [17], battery energy storage is used for grid support to avoid voltage instability beyond an acceptable range. The S-PPC in the IPOS configuration acts as a step-up converter in all operation points as the DC-bus voltage is always lower than the battery voltage. Following the duality concept for the reverse power flow, this S-PPC can be regarded as a step-down converter for power flow from the battery to the DC-bus. A dual active bridge (DAB) converter with conventional phase shift modulation is employed as the DC–DC converter. The proposed converter is illustrated in Figure 7.17. Voltage polarity of the capacitors C_{in} and C_{out} does not change allowing for using switches with the unidirectional voltage blocking capability. At the same time, the bidirectional operation is achieved by changing the current polarity in those ports. Therefore, the given DC dual active bridge DC converter can operate in two quadrants.

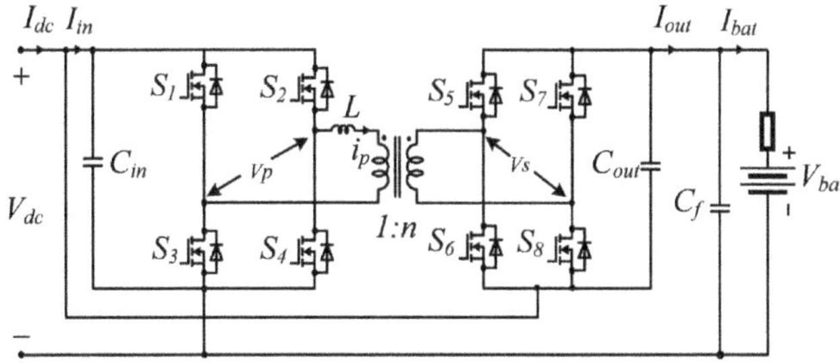

Figure 7.17 Step-up bidirectional S-PPC based on DAB converter

For the DC-bus voltage of 300 V and nominal battery voltage of 380 V (360–400 V), the performance metrics like active processed power, apparent power, non-active power, the RMS current, and the efficiency were compared for an FPC and the proposed S-PPC for both power flow directions. In the grid support mode, the reference current of the battery is defined by the DC-bus voltage mismatch caused by very low or very high load demand, resulting in forward and reverse power flow, respectively. In the dedicated battery charging mode, the reference current of the battery follows the CC charging mode at the rate defined by the battery parameters. Experimental results obtained using a 3.3 kW prototype reveal a significant current stress decrease: roughly 70% in peak current and 75% for RMS current of the components. The S-PPC handles only 25% of the total active power and 29% of the total apparent power. Moreover, an efficiency boost of 1.4–2.1% was reported through battery charge mode compared to an FPC. On the other hand, the efficiency was increased by 2.1–3% for battery discharge mode.

7.4.2 EV fast chargers

A multiport EV fast-charging station (FCS) is introduced in [18], which connects a low-voltage grid to the battery energy storage (BES) and EV charger. Without BES in the system, the low voltage grid must be designed for high peak power delivery to the FCS, which is not cost-optimal approach. A typical design approach toward integration of BES in a charging station requires a full-power DC–DC converter to connect the BES to the DC-link. The additional energy conversion stage is the main drawback of the typical approach. S-PPCs can be applied to the multiport DC–DC converters regarding some restrictions. As mentioned earlier, the galvanic isolation is essential between the grid and the charging station. Therefore, the series connection cannot be created for the grid-EV path.

On the other hand, the S-PPC concept can be implemented for the integration of BES into the FCS. The basic concept of 3P-PPC stems from the triple active bridge (TAB) converter with a medium voltage single core transformer that is depicted in Figure 7.18. Due to the reduction of energy conversion stages in comparison with FPC utilization in the DC-link, theoretical calculations reveal that the efficiency would be increased.

The output (EV charging) voltage of the converter is the sum of the third-port voltage of the TAB converter V_{out} and the BES voltage V_{S2}. As a consequence of this series connection, the voltage across the EV port of the TAB converter (V_{out}) decreased compared to the charging voltage V_L, which stems from the partiality concept. In addition to the lower processed power, the lower voltage across the converter ports results in the use of low-cost switches with a reduced voltage rating. The operation principle is based on phase-shift modulation between ports which enables active power regulation. The φ_{13} and φ_{23} are considered the positive phase shifts between grid-EV and BES-EV ports, respectively. The active power flow equation between ports i can be written as

$$P_{i,j} = \frac{v_i v_j}{2\pi f_s L_{i,j}} \varphi_{i,j}\left(1 - \frac{\varphi_{i,j}}{\pi}\right), \tag{7.38}$$

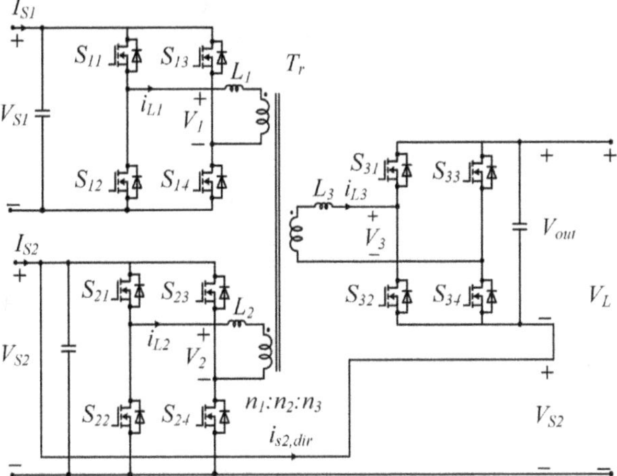

Figure 7.18 The 3P-PPC comprising a low-voltage grid port (V_{S1}), BES port (V_{S2}), and EV charging port (V_L)

where the v_i and v_j are voltages of the ith and jth ports, f_s is the switching frequency, $L_{i,j}$ is the inductance between ports, and $\varphi_{i,j}$ is the relative phase shift between ports in radians. There can be different power flow scenarios between the three ports. In the given application, it is essential to achieve simultaneous energy transfer from the BES and the low-voltage grid ports to the EV despite the complicated control required. Regarding P_L as the output power and P_{dir} as the direct power flow from BES to EV, P_{indir1} is the power flow from the low-voltage grid to EV, and P_{indir2} is the power processed from BES to EV, the partiality equations can be written as

$$k_{dir} = \frac{P_{dir}}{P_L} = \frac{V_{S2} I_L}{V_L I_L} = \frac{V_{S2}}{V_L}, \tag{7.39}$$

where k_{dir} is the parameter that defines the amount of total power transferred directly to the EV without processing. The closer the BES voltage V_{S2} is to the EV charging voltage V_L, the higher value k_{dir} is achieved. Considering the EV charging profile starting from the CC mode (20–80% of SoC), the parameter k_{dir} will decrease gradually following the increasing voltage of the EV battery. It will reach the minimum value at the end of the charging cycle. The BES voltage that determines the k_{dir} must be selected lower than the minimum EV battery voltage to avoid negative voltage across the output port of the converter. Assuming these constraints, the k_{dir} limits can be calculated:

$$k_{dir,\max} = \frac{V_{S2} - V_{safe}}{V_{L,\min}}, \; k_{dir,\min} = \frac{V_{S2}}{V_{L,\max}}, \tag{7.40}$$

where V_{safe} is the voltage safety margin allowing to prevent transient overvoltage for the safe operation of the 3P-PPC; $k_{dir,\min}$ is the determining parameter for the

3P-PPC design as it defines the maximum power processed by the converter components. Based on (7.38), the partiality factor for indirect powers (processed by the TAB converter) can be written as

$$k_{indir} = \frac{\frac{V_{S1}\frac{V_{out}}{n_1/n_3}}{2\pi f_s L_{1,3}}\varphi_{1,3}\left(1-\frac{\varphi_{1,3}}{\pi}\right)}{\frac{V_{S1}\frac{V_{out}}{n_1/n_3}}{2\pi f_s L_{1,3}}\varphi_{1,3}\left(1-\frac{\varphi_{1,3}}{\pi}\right)+\frac{V_{S2}\frac{V_{out}}{n_2/n_3}}{2\pi f_s L_{2,3}}\varphi_{2,3}\left(1-\frac{\varphi_{2,3}}{\pi}\right)} \tag{7.41}$$

As expected, the processed powers are directly related to the phase shifts between ports, which compose the system control variables. The BES should be dimensioned considering the peak power capacity of the low-voltage grid and maximum charging power at the EV port. The lower BES capacity means that most of the charging power/energy can be delivered from the low-voltage grid port, which, however, results in an increased fraction of the processed power.

A careful analysis must be performed to select the transformer turns ratio due to the wide EV voltage variation range. It is of great importance from several aspects. The first and most important one is non-active power processing which is directly related to the transformer turns ratio. Another issue is that unmatched input/output voltage of the transformer will increase the current/voltage stress of circuit components. Comparison of voltage matching at the middle of charge cycle (SoC of 50%, $n = n_{avg}$) and the end of charge cycle (SoC of 80%, $n = n_{end}$) reveal that the latter case provides optimal current waveforms and higher efficiency. From the BES perspective, the voltage range must be kept in a narrow range to fulfill the above-mentioned considerations. A control strategy can be designed to recharge the BES by a small share of grid power to maintain the BES voltage in the desired range.

A comparison between 3P-PPC and TAB converter is done for the charging current of $I_L = 50A$ in the CC mode. The results of this comparison are given in Table 7.4.

The results of the case study represent a significant reduction in active power processing of 3P-PPC in comparison with full power TAB. The k_{dir}decreases from 0.75 at the SoC of 20% to 0.58 at the SoC of 80%. The maximum power processed by the 3P-PPC equals 7.1 kW, which corresponds to 42% of the total power. In

Table 7.4 Comparison between 3P-PPC and full power TAB for EV charging applications with embedded BES

Parameter	**3P-PPC**	**TAB**
Input voltage V_{S1} (grid)	400 V	400 V
Input voltage V_{S2} (BES)	200 V	200 V
Output voltage V_{out} (EV)	60–140 V	260–340 V
Charging voltage V_L	260–340 V	260–340 V
Maximum charging power	17 kW	17 kW

comparison with the full-power TAB converter, the overall efficiency of the 3P-PPC is increased by 1.9% and reaches the peak value of 98.7%.

7.4.3 Power flow control in DC grids

Widespread electrification driven by the proliferation of renewable energy technologies requires a paradigm shift of power distribution from low-voltage AC to low-voltage DC grids to increase the efficiency of the energy sector. Consequently, DC microgrids were introduced to coordinate distributed energy resources, ESSs, and consumers in a distributed way to decrease the control burden of the main grid. In all DC systems, finding an optimal and efficient load-sharing method is of great importance. This issue is especially prominent in the meshed or ring architectures.

Various approaches were introduced for power flow control in the DC grid. One of the inefficient methods is changing the line resistance both in high- and low-voltage DC grids. The new approach is the application of DC–DC converters between DC buses to maximize the power transfer in the network. Conventionally, it was done using FPCs but the overall power loss and cost were the main drawbacks of such solutions.

Newly emerged partially rated power flow control converters (PFCCs) can be one of the most feasible solutions to overcome mentioned limitations. In [19], an S-PPC utilizing a DAB DC–DC converter coupled with an unfolding circuit is introduced. As shown in Figure 7.19, the IPOS configuration is used, where the DC–DC is connected in parallel to one of the ports and series to the other one. The DAB converter is advantageous in this application due to simple control and the low number of passive elements as well as zero voltage switching that can further increase system efficiency. The unfolder circuit enables the four-quadrant operation of the converter as it can change voltage polarity in the series port. The maximum processed power and operation range of the converter are directly related to the voltage of the DC-buses. Therefore, the functionality of the converter is defined by the grid parameters.

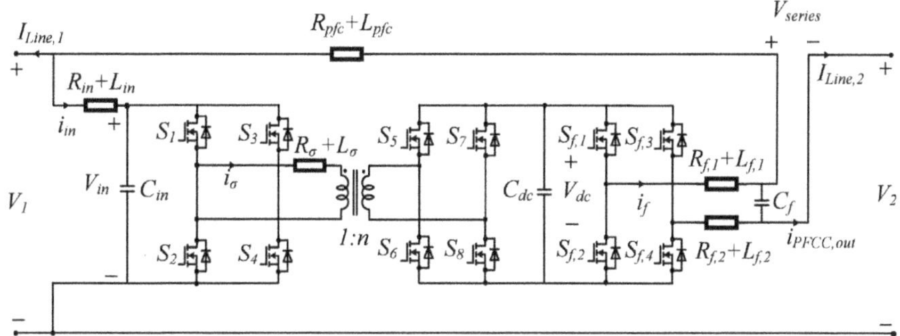

Figure 7.19 S-PPC as power flow controller between low-voltage DC-grids

In [19], the authors presented also averaged large and small-signal models of the PFCC, which is the topic poorly covered in the literature. The complete formulation for both models is done exploiting state-space models and finally validated by the experimental prototype for the meshed grid. The large-signal model of PFCC allows combination with the LVDC grid to simulate and algorithmize the whole system. The validation experiments were done in two steps to verify the power flow control capability of the proposed PFCC. Regarding the bus voltages 350 and 335 V, in the first step, the PFCC did not add any voltage and the transferred power between buses was equal to 3 kW. In the second step, the V_{Series} was 10 V, and the transferred power increases to 4.5 kW with $I_{Line,2} = 10A$. It means the DC–DC converter needed to process only 100 W to ensure delivery of the total power of 4.5 kW. It is a great reduction in processed power of the DC–DC converter as well as significant control flexibility. On the other hand, the authors failed to demonstrate any zero-crossing performance in the S-PPC series port, when the DAB converter could struggle to deliver any power close to zero voltage. This leaves some questions open for discussion.

7.5 Conclusions

The S-PPC technology gained most of the attention in the last decade. Since then, these converters have shown good performance in numerous applications. Some examples omitted in this chapter could be found in [9,20]. Numerous experimental studies demonstrate S-PPC efficiency of over 99% even without proper optimization of experimental prototypes. This only proves that the S-PPC technology has tremendous potential in applications where input and output voltages are of the same order of magnitude. However, it is not appropriate for high step-up and step-down applications. In general, this research field shall see an expansion stage for 5–10 years before reaching some maturity.

Even though the research on S-PPC has been accelerating for the past several years, numerous knowledge gaps still exist. Development of new topologies that better suit particular applications, multi-port topologies for multi-channel energy control, small- and large-signal modeling, protection, electromagnetic compatibility, and reliability could be considered as promising and underrepresented research direction in this field.

One of the main issues with S-PPC technology is low industry awareness about it, its advantages, and its most practical applications. To accelerate industry adoption of S-PPC, more application-oriented studies with strong experimental backgrounds are required, along with in-the-field show-casing by research groups.

Acknowledgment

This work was supported by the Estonian Research Council (grant PRG1086).

References

[1] Moore E, Wilson T. Basic considerations for DC to DC conversion networks. *IEEE Trans Magn*, 1966;2(3):620–624.

[2] Sebastian J, Villegas P, Nuno F, and Hernando MM. Very efficient two-input DC-to-DC switching post-regulators. In: PESC Record. 27th Annual IEEE Power Electronics Specialists Conference, 1996, pp. 874–880.

[3] Button RM. An advanced photovoltaic array regulator module. In: IECEC 96. Proceedings of the 31st Intersociety Energy Conversion Engineering Conference, 1996, pp. 519–524.

[4] Jeong H, Lee H, Liu Y-C, and Kim KA. Review of differential power processing converter techniques for photovoltaic applications. *IEEE Trans Energy Convers,* 2019;34(1):351–360.

[5] Shenoy PS, Kim KA, Johnson BB, and Krein PT. Differential power processing for increased energy production and reliability of photovoltaic systems. *IEEE Trans Power Electron,* 2013;28(6):2968–2979.

[6] Anzola J, Aizpuru I, Romero AA, *et al.* Review of architectures based on partial power processing for DC–DC applications. *IEEE Access,* 2020;8:103405–103418.

[7] Duan J, Zhang D, and Gu R. Partial-power post-regulated LLC resonant DC transformer. *IEEE Trans Indust Electron*, 2022;69(8):7909–7919, doi:10.1109/TIE.2021.3106000.

[8] Santos NGFd, Zientarski JRR and Martins MLdS. A two-switch forward partial power converter for step-up/down string PV systems. *IEEE Trans Power Electron*, 2022;37(6):6247–6252, doi:10.1109/TPEL.2021.3138299.

[9] Dos Santos NGF, Zientarski JRR, and Martins ML da S. A review of series-connected partial power converters for DC–DC applications. *IEEE J Emerg Sel Top Power Electron*, 2021;10(6):7825–7838.

[10] Zientarski JRR, Martins ML da S, Pinheiro JR, and Hey HL. Series-connected partial power converters applied to PV systems: a design approach based on stepup/down voltage regulation range. *IEEE Trans Power Electron,* 2018;33(9):7622–7633.

[11] Zientarski JRR, da Silva Martins ML, Pinheiro JR, and Hey HL. Evaluation of power processing in series-connected partial-power converters. *IEEE J Emerg Sel Top Power Electron,* 2019;7(1):343–352.

[12] Zapata JW, Kouro S, Carrasco G, Renaudineau H, and Meynard TA. Analysis of partial power DC–DC converters for two-stage photovoltaic systems. *IEEE J Emerg Sel Top Power Electron,* 2019;7(1):591–603.

[13] Balbino AJ, Tanca-Villanueva MC, and Lazzarin TB. A full-bridge partial-power processing converter applied to small wind turbines systems. In: 2020 IEEE 29th International Symposium on Industrial Electronics (ISIE), 2020, pp. 985–990.

[14] Birchenough AG. Series connected buck-boost regulator [Internet]. US7042199B1, 2006 [cited 2021 Nov 19]. Available from: https://patents.google.com/patent/US7042199B1/en.

[15] Iyer VM, Gulur S, Gohil G, and Bhattacharya S. An approach towards extreme fast charging station power delivery for electric vehicles with partial power processing. *IEEE Trans Ind Electron,* 2020;67(10):8076–8087.
[16] Rivera S, Rojas J, Kouro S, *et al.* Partial power converter topology of type II for efficient electric vehicle fast charging. *IEEE J Emerg Sel Top Power Electron*, 2021;10:1–1.
[17] Iyer VM, Gulur S, Bhattacharya S, and Ramabhadran R. A partial power converter interface for battery energy storage integration with a DC microgrid. In: 2019 IEEE Energy Conversion Congress and Exposition (ECCE), 2019, pp. 5783–5790.
[18] Hoffmann F, Person J, Andresen M, Liserre M, Freijedo FD, and Wijekoon T. A multiport partial power processing converter with energy storage integration for EV stationary charging. *IEEE J Emerg Sel Top Power Electron,* 2021;10:1–1.
[19] Purgat P, van der Blij NH, Qin Z, and Bauer P. Partially rated power flow control converter modeling for low-voltage DC grids. *IEEE J Emerg Sel Top Power Electron,* 2020;8(3):2430–2444.
[20] Jørgensen KL, Zhang Z, and Andersen MA. Next generation of power electronicconverter application for energy-conversion and storage units and systems. *Clean Energy,* 2019; 3(4):307–315.

Chapter 8

Matrix converters; topologies, control methods, and applications

Ebrahim Babaei[1] *and Mohammadamin Aalami*[1]

Electrical energy is the most common type of energy because it can be transferred easily and at a suitable cost. Although balanced three-phase electrical energy sources with a constant amplitude and frequency are the most economical kind for the production and transferring perspectives and are suitable for most customers, they are not appropriate for some applications. For instance, most chemical processes and electrical motors require controllable dc voltage sources and three-phase voltage with controllable amplitude and frequency. Besides, most appliances require a constant amplitude and constant frequency voltage source regardless of the unstable and fault conditions of the electrical grid. As a result, such applications require interfaces and converters in order to change the form of the electrical energy to the required sort. These converters are derived into different categories regarding the kind of conversation and their usability, whereas matrix converters gain more attraction as an emerging kind of power electronic converters.

The modulation techniques are some means to control switches of power electronic converters. The applicational advantages of the pulse width modulation (PWM) methods are widely known these days. These control methods' abilities and potentials had not been discovered since 1980 because the available analog and digital hardware had some practical limitations. Most of these applicational limitations are solved by the progress in manufacturing semiconductor parts, and thus, these control methods can be used. These techniques have lots of merits, such as reducing the low order harmonics compared to other methods. In this chapter, three PWM-based novel control methods, including the positive, negative, and combined control methods, are given as means to control matrix converters that are able to synthesize required output waveforms regardless of the type of input and output voltages (ac or dc) and the number of input and output phases. Also, these control methods can prevent short-circuiting of the input sources and open-circuiting of the output loads. The other feature of these control methods is that if the number of input phases is lesser or equal to output phases, the input currents will be continuous. Not only can these control methods be applied to matrix converters, but

[1]Faculty of Electrical and Computer Engineering, University of Tabriz, Iran

they can also be extended to most power electronic converters. Moreover, it will be shown that symmetric output waveforms can be achieved even by asymmetric input voltages.

8.1 Matrix converters

Two-stage ac/ac converters with a dc link are the most conventional converters that can be used to convert an ac voltage. Regarding Figure 8.1, this kind of topologies has two stages, and they are separated using a capacitor as a dc link. The input ac voltage is converted to a dc voltage using a rectifier in the first stage. After that, the dc voltage on this capacitor is converted to an ac voltage with variable amplitude and frequency. The power condition is from the input to the output side in the normal condition, but under the bidirectional power condition, the converters on the first and the second stage will work as an inverter and a rectifier, respectively.

Theoretically, a single-stage matrix converter can be used instead of a two-stage converter with a storage component. A matrix converter is a sort of power-electronic converter with variable amplitude and frequency that can convert *N* phases of input voltages to *m* phases of output voltages without any storage components. The general configuration of a matrix converter with *N* input and *m* output phases is demonstrated in Figure 8.2. As this figure states, this topology presents an approach for converting ac voltages, including only active components, excluding the usage of reactive power's storage components. This converter includes controllable bidirectional power switches where each switch connects an input phase to an output

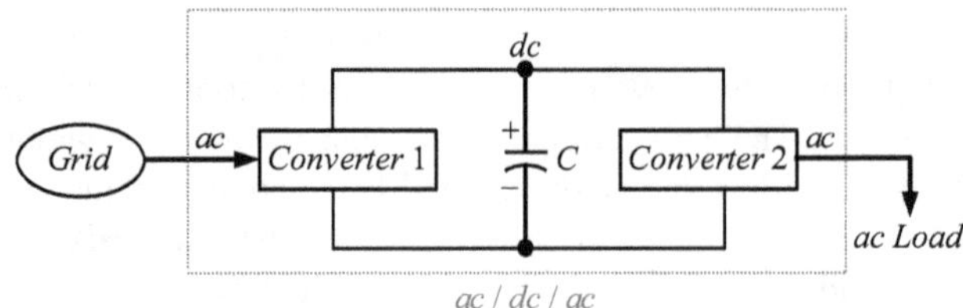

Figure 8.1 ac/ac converters with dc-link

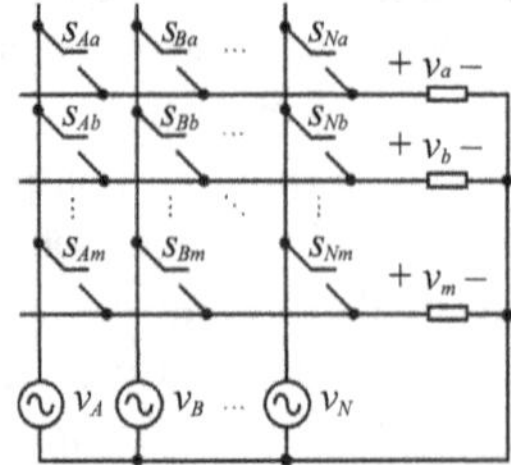

Figure 8.2 Power circuit of a matrix converter with N *input and* m *output phases*

phase at the intersection of these phases. These switches can block both positive and negative voltages and pass currents in both directions. Figure 8.3 shows different configurations of switches that can be used as bidirectional switches.

The outputs will be a current type if the inputs are voltage sourced. Besides, the converse condition is proper, which means the outputs will be voltage type if the inputs are current sources. For instance, if both the input and the output sides were voltage types, some operating states would lead to short-circuit. Generally, the operation of switches must guarantee that the switching states do not short circuit the voltage sources and open circuit the current sources. The input and output powers are equal when not using storage components and assuming ideal switches. However, the phase shift between input voltages and currents can be controlled, and there are no obligations to be the same as the output side. For example, the input and output reactive powers can be unequal. The matrix converters have fast response, small size, and the ability to design integrally regarding not using storage components. However, the number of required bidirectional switches is high. The storage components, including capacitors and inductors, are costly and occupy a large space. In some applications, especially electric vehicles and airplanes, where reducing weight and volume are vital, the matrix converters are more attractive than the two-stage ones.

The type of waveforms and their frequencies are independent of each other in the matrix converters. For example, although the input side can be three-phase ac, the output side can be dc ac. Moreover, the number of input and output phases can be different too. So, the matrix converters are general and can be used instead of the other conventional configurations. Figure 8.4 depicts some of the power electronic converters that are reconfigured by the matrix converters.

The main drawback of the matrix converters is that they require a high number of power switches; however, it is necessary to consider that high power switches with a low cost and easier contractility are available nowadays. As a result, the matrix converters are superior to the conventional two-stage ac/ac converters, which require sinusoidal input current and bidirectional power direction. Nevertheless, the matrix converters are not such developed ones regarding inherent commutation problems. Besides, these converters have some limitations associated with input and output amplitudes. However, it will result in matrix converters being such low cost and small if they are compared to conventional two-stage ac/ac converters. As mentioned before, the matrix converters suffer from an inherent commutation

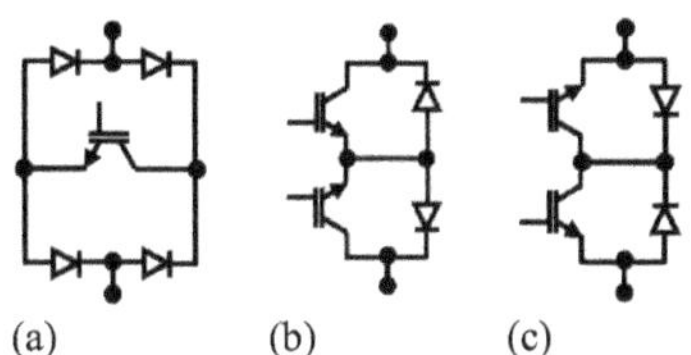

Figure 8.3 Different arrangements of bidirectional switches, (a) diode bridge, (b) common emitter, and (c) common collector

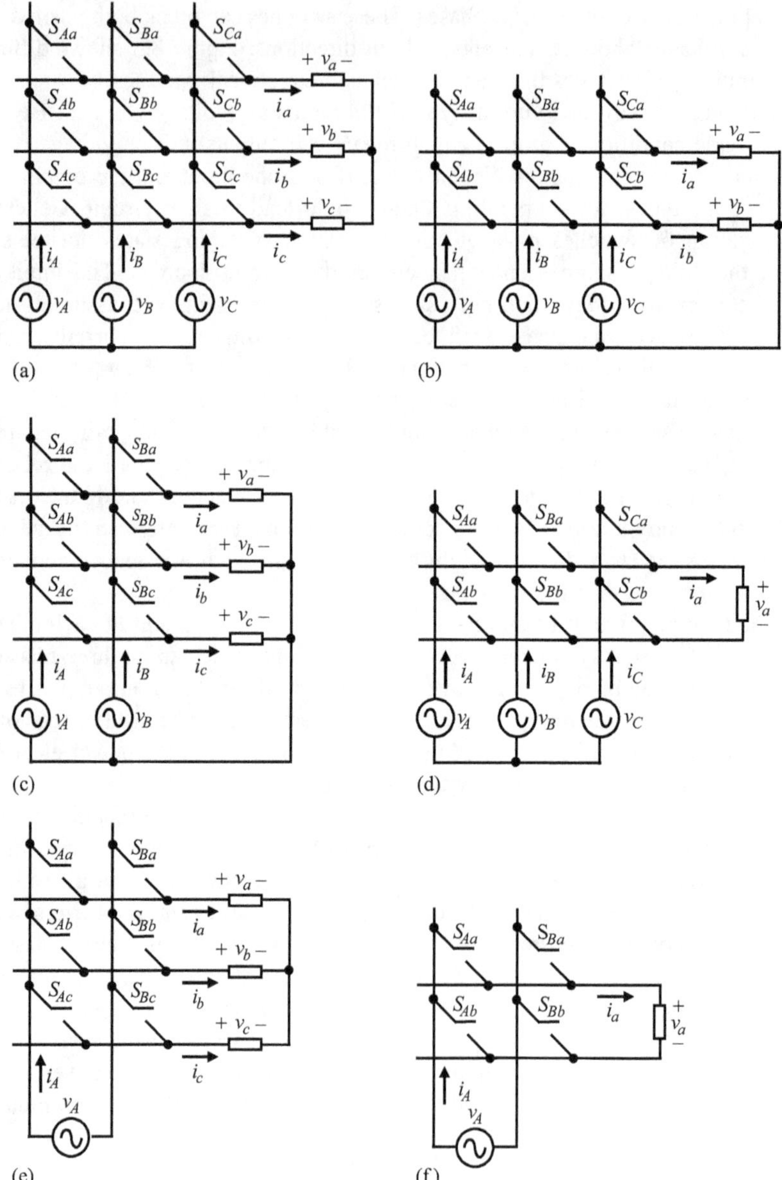

Figure 8.4 Rearranging conventional converters using the matrix converters, (a) three-phase to three-phase, (b) three-phase to two-phase [1], (c) two-phase to three-phase, (d) three-phase to single-phase or three-phase rectifier [2], (e) single-phase to three-phase, (f) single-phase to single-phase or single-phase rectifier [3], (g) three-phase inverter, and (h) single-phase inverter

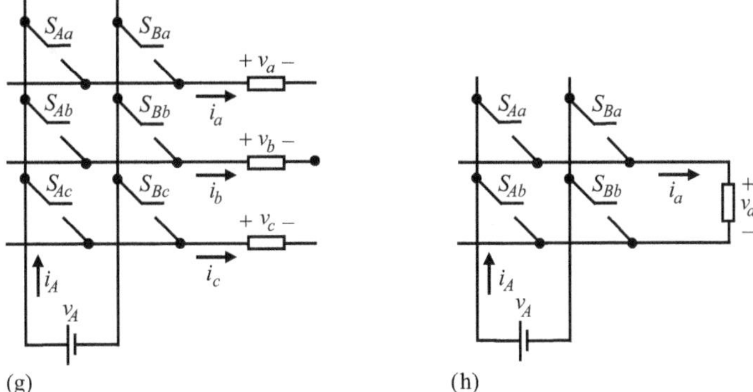

Figure 8.4 Continued

problem that adds such challenges to the control of these power electronic converters. Also, their control methods are generally complex and require so many mathematical calculations. Moreover, because the matrix converter is a direct type one, every slight disturbance on the input side, such as voltage unbalancing and sag/swells even for a short time, is directly transferred to the output side.

8.2 Applications of matrix converters

Regarding the reduced size and volume of the matrix converters, some of their applications, which the researchers in recent times appreciate, are their applications in electric vehicles, especially military vehicles and airplanes. Also, they can be used in submarine appliances. The conventional ac/ac converters that require large electrolyte capacitors only work under the atmosphere pressure if placed in some rooms full of air. In other words, the matrix converters, which do not include any capacitors, are suitable for submarine appliances that work under a pressure of more than 300 Bar and a depth of more than 300 m. Besides, the matrix converters have a fast response because they do not include any electrolyte capacitors. These converters are suitable for industrial applications that work at a high temperature because the conventional ac/ac converters that include large electrolyte capacitors lose their proper operation at high temperatures.

Moreover, the matrix converters can be used with both static and dynamic loads. Nowadays, so many publications utilize matrix converters in controlling ac motors. Hence, these converters have further potential to improve these converters' usage in this application.

8.3 Matrix converters modeling

A matrix converter is a single-stage one that includes $N \times m$ bidirectional power switches that directly connect N input phases to m output phases, as illustrated in Figure 8.2. In the following, capital words ($A, B, \ldots, N$) will be used for the input phases, and noncapital words (*a, b,* ..., *m*) will be used for the output phases. Besides, all switches are supposed to be ideal. The input voltages can be obtained as following under the balanced circumstances:

$$\begin{aligned} v_A &= V_{i,\max}\cos(\omega_i t) \\ v_B &= V_{i,\max}\cos\left(\omega_i t - \frac{2\pi}{N}\right) \\ &\vdots \\ v_N &= V_{i,\max}\cos\left[\omega_i t - (N-1)\frac{2\pi}{N}\right] \end{aligned} \tag{8.1}$$

where $V_{i,\max}$ and ω_i show the maximum amplitude and angular frequency of the input voltages.

The applied control method must be such that the primary harmonics of the output voltage will be in the following form:

$$\begin{aligned} v_a &= V_{o,\max}\cos(\omega_o t) \\ v_b &= V_{o,\max}\cos\left(\omega_o t - \frac{2\pi}{m}\right) \\ &\vdots \\ v_m &= V_{o,\max}\cos\left[\omega_o t - (m-1)\frac{2\pi}{m}\right] \end{aligned} \tag{8.2}$$

where $V_{o,\max}$ and ω_o show the maximum amplitude and the angular frequency of the output voltages.

The switching function can be addressed as follows:

$$S_{Kj} = \begin{cases} 1, & \text{for ON state} \\ 0, & \text{for OFF state} \end{cases} \quad \mathit{for} \quad \begin{matrix} K = A, B, \cdots, N \\ j = a, b, \cdots, m \end{matrix} \tag{8.3}$$

Observing that the input side of matrix converters is voltage type and most of the loads have an inductive characteristic, two following rules must be taken into account:

- Two lines of input voltages must not be connected to the same load simultaneously in order to prevent a short-circuit at the input side. In other words, a maximum of one switch can be turned on for each output line.
- The output lines must not be left open in order to prevent an open-circuit of output loads. In other words, a minimum of one switch must be turned on for each output line.

These limitations require that only one switch is turned on for each output line. For example, if we have the following condition for the first output line:

$$S_{Aa} = 1 \tag{8.4}$$

The other corresponding switches with the input lines must be turned off, which means:

$$S_{Ba} = 0 \ \& \ S_{Ca} = 0 \ \cdots \ S_{(N-1)a} = 0 \ \& \ S_{Na} = 0 \tag{8.5}$$

As another example, if:

$$S_{Ba} = 1 \tag{8.6}$$

Then, we must have:

$$S_{Aa} = 0 \ \& \ S_{Ca} = 0 \ \cdots \ S_{(N-1)a} = 0 \ \& \ S_{Na} = 0 \tag{8.7}$$

Ultimately, if:

$$S_{Na} = 1 \tag{8.8}$$

We must meet the following condition:

$$S_{Aa} = 0 \ \& \ S_{Ba} = 0 \ \& \ S_{Ca} = 0 \ \cdots \ S_{(N-1)a} = 0 \tag{8.9}$$

So, we can state the following for the first output line:

$$S_{Aa} + S_{Ba} + \cdots + S_{Na} = 1 \tag{8.10}$$

The following constraints can be stated for the other output lines in the same way:

$$\begin{array}{l} S_{Ab} + S_{Bb} + \cdots + S_{Nb} = 1 \\ S_{Ac} + S_{Bc} + \cdots + S_{Nc} = 1 \\ \quad\quad\vdots \\ S_{Am} + S_{Bm} + \cdots + S_{Nm} = 1 \end{array} \tag{8.11}$$

Finally, the limitations that are stated in (8.11) can be rewritten as follows:

$$S_{Aj} + S_{Bj} + \cdots + S_{Nj} = 1 \ j = a, \ b, \ \cdots, \ m \tag{8.12}$$

The first output line is considered in order to get familiar with the approach of extracting equations to relate input and output voltage to each other. If one corresponding switch to an input line and the first output line is turned on, the output voltage will be the same as the corresponding input voltage. Otherwise, this input voltage will not transfer to the first output line. Based on these descriptions, the relevant equations to the first output line can be obtained. If S_{Aa} is turned on, then:

$$v_a = v_A \ or \ v_a = S_{Aa}v_A \tag{8.13}$$

Similarly, if S_{Ba} is tuned on, then:

$$v_a = v_B \ or \ v_a = S_{Ba}v_B \tag{8.14}$$

Ultimately, if S_{Na} is turned on, then:

$$v_a = v_N \ or \ v_a = S_{Na}v_N \tag{8.15}$$

Regarding (8.13)–(8.15), a general equation associated with the first output line can be obtained as follows:

$$v_a = S_{Aa}v_A + S_{Ba}v_B + \cdots + S_{Na}v_N \tag{8.16}$$

In a similar way, it can be written for the other output lines:

$$\begin{aligned} v_b &= S_{Ab}v_A + S_{Bb}v_B + \cdots + S_{Nb}v_N \\ &\vdots \\ v_m &= S_{Am}v_A + S_{Bm}v_B + \cdots + S_{Nm}v_N \end{aligned} \tag{8.17}$$

The following result can be obtained by writing (8.16) and (8.17) in the matrix form:

$$\begin{bmatrix} v_a \\ v_b \\ \vdots \\ v_m \end{bmatrix} = \begin{bmatrix} S_{Aa} & S_{Ba} & \cdots & S_{Na} \\ S_{Ab} & S_{Bb} & \cdots & S_{Nb} \\ \vdots & \vdots & \ddots & \vdots \\ S_{Am} & S_{Bm} & \cdots & S_{Nm} \end{bmatrix} \begin{bmatrix} v_A \\ v_B \\ \vdots \\ v_N \end{bmatrix} \tag{8.18}$$

Regarding (8.18), the following matrixes can be defined:

$$[V_o]_{m\times 1} = \begin{bmatrix} v_a \\ v_b \\ \vdots \\ v_m \end{bmatrix} \tag{8.19}$$

$$[S]_{m\times N} = \begin{bmatrix} S_{Aa} & S_{Ba} & \cdots & S_{Na} \\ S_{Ab} & S_{Bb} & \cdots & S_{Nb} \\ \vdots & \vdots & \ddots & \vdots \\ S_{Am} & S_{Bm} & \cdots & S_{Nm} \end{bmatrix} \tag{8.20}$$

$$[V_i]_{N\times 1} = \begin{bmatrix} v_A \\ v_B \\ \vdots \\ v_N \end{bmatrix} \tag{8.21}$$

Eq. (8.18) can be rewritten as follows by substituting (8.19) to (8.21) in it:

$$[V_o]_{m\times 1} = [S]_{m\times N}[V_i]_{N\times 1} \tag{8.22}$$

where $[V_o]_{m\times 1}$, $[S]_{m\times N}$, and $[V_i]_{N\times 1}$ denote the output voltage, instantaneous transfer, and the input voltage matrixes, respectively.

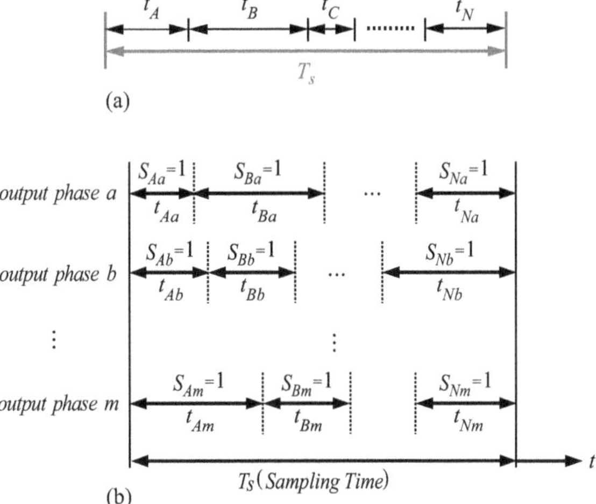

Figure 8.5 (a) Segmenting a switching period and (b) general form of switching pattern for matrix converters

It is required to consider used switching methods to obtain switches' turning on and off rules. The general switching pattern for matrix converters is depicted in Figure 8.5.

Every switching period is divided into N time segments of t_A, t_B, …, and t_N regarding Figure 8.5(a). Regarding Figure 8.5(b), S_{Aa}, S_{Ba}, and S_{Na} turn on in the period of t_{Aa}, t_{Ba}, and t_{Na}, respectively. The following results can be obtained using Figure 8.5(b):

$$\begin{aligned} & t_{Aa} + t_{Ba} + \cdots + t_{Na} = T_s \\ & t_{Ab} + t_{Bb} + \cdots + t_{Nb} = T_s \\ & \qquad \vdots \\ & t_{Am} + t_{Bm} + \cdots + t_{Nm} = T_s \end{aligned} \tag{8.23}$$

The three-phase to three-phase matrix converter, which was depicted in Figure 8.4(a) can be considered to get more familiar with the given switching pattern. Regarding Figure 8.6, each switching period is divided into three periods as it is depicted in the proposed switching pattern.

A low-frequency output voltage with variable amplitude and frequency can be created by modulation of the duty cycle of switches and corresponding switching functions considering a high-frequency operation of switches. The duty cycles of switches can be defined as follows:

$$m_{Aa} = \frac{t_{Aa}}{T_s};\ m_{Ba} = \frac{t_{Ba}}{T_s};\ \cdots\ m_{Na} = \frac{t_{Na}}{T_s};\ \cdots\ ;\ m_{Ab} = \frac{t_{Ab}}{T_s};\ \cdots\ m_{Nm} = \frac{t_{Nm}}{T_s} \tag{8.24}$$

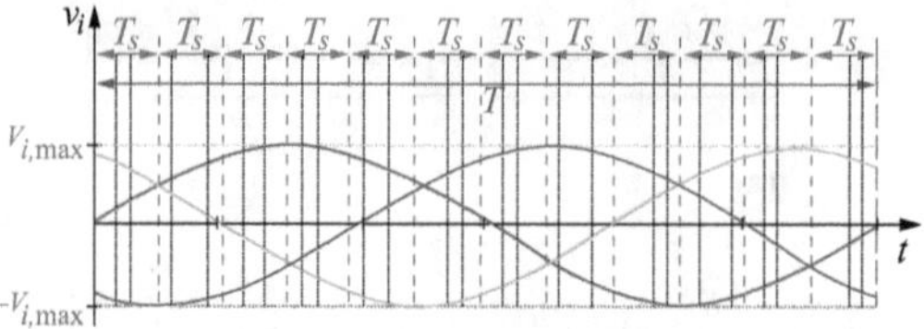

Figure 8.6 Waveforms of three-phase input voltage in one switching period

If the turned on time of the switch of S_{Kj} in each switching period of T_s is denoted by t_{Kj}, the duty cycle of the switch of S_{Kj} will be defined as follows:

$$m_{Kj} = \frac{t_{Kj}}{T_s} \quad \text{for} \quad \begin{matrix} K = A, B, \cdots, N \\ j = a, b, \cdots, m \end{matrix} \tag{8.25}$$

Regarding Figure 8.5, it can be obtained that each t_{Kj} must apply to the following equation and inequalities:

$$0 \leq t_{Kj} \leq T_s \quad \text{for} \quad \begin{matrix} K = A, B, \cdots, N \\ j = a, b, \cdots, m \end{matrix} \tag{8.26}$$

$$t_{Aj} + t_{Bj} + \ldots + t_{Nj} = T_s \quad \text{for} \quad j = a, b, \cdots, m \tag{8.27}$$

Some limitations for the duty cycles of switches can be obtained by dividing (8.26) and (8.27) to T_s:

$$0 \leq m_{Kj} \leq 1 \quad \text{for} \quad \begin{matrix} K = A, B, \cdots, N \\ j = a, b, \cdots, m \end{matrix} \tag{8.28}$$

$$m_{Aj} + m_{Bj} + \ldots + m_{Nj} = 1 \quad \text{for} \quad j = a, b, \cdots, m \tag{8.29}$$

If the switching frequency is assumed much more significant than the input and output frequencies, the input and output voltage can be considered constant in a switching period. Regarding this explanation, the voltage of the first output line can be written as follows using input voltages and the times at which they are transferred to the output stage:

$$v_a = \frac{1}{T_s}(t_{Aa}v_A + t_{Ba}v_B + \cdots + t_{Na}v_N) \tag{8.30}$$

In the same way, the following equations can be obtained for the other output voltages:

$$\begin{aligned} v_b &= \frac{1}{T_s}(t_{Ab}v_A + t_{Bb}v_B + \cdots + t_{Nb}v_N) \\ &\vdots \\ v_m &= \frac{1}{T_s}(t_{Am}v_A + t_{Bm}v_B + \cdots + t_{Nm}v_N) \end{aligned} \tag{8.31}$$

Regarding (8.25), (8.30) and (8.31) can be rewritten as follows:

$$\begin{aligned} v_a &= m_{Aa}v_A + m_{Ba}v_B + \cdots + m_{Na}v_N \\ v_b &= m_{Ab}v_A + m_{Bb}v_B + \cdots + m_{Nb}v_N \\ &\vdots \\ v_m &= m_{Am}v_A + m_{Bm}v_B + \cdots + m_{Nm}v_N \end{aligned} \tag{8.32}$$

Eq. (8.32) can be written in matrix form as follows:

$$\begin{bmatrix} v_a \\ v_b \\ \vdots \\ v_m \end{bmatrix} = \begin{bmatrix} m_{Aa} & m_{Ba} & \cdots & m_{Na} \\ m_{Ab} & m_{Bb} & \cdots & m_{Nb} \\ \vdots & \vdots & \ddots & \vdots \\ m_{Am} & m_{Bm} & \cdots & m_{Nm} \end{bmatrix} \begin{bmatrix} v_A \\ v_B \\ \vdots \\ v_N \end{bmatrix} \tag{8.33}$$

The transition matrix can be defined as follows:

$$[M]_{m\times N} = \begin{bmatrix} m_{Aa} & m_{Ba} & \cdots & m_{Na} \\ m_{Ab} & m_{Bb} & \cdots & m_{Nb} \\ \vdots & \vdots & \ddots & \vdots \\ m_{Am} & m_{Bm} & \cdots & m_{Nm} \end{bmatrix} \tag{8.34}$$

Regarding (8.34), (8.33) can be rewritten as follows:

$$[V_o]_{m\times 1} = [M]_{m\times N}\,[V_i]_{N\times 1} \tag{8.35}$$

The prominent harmonics of the input and output currents can be obtained as follows:

$$[I_i]_{N\times 1} = \begin{bmatrix} i_A \\ i_B \\ \vdots \\ i_N \end{bmatrix} \tag{8.36}$$

$$[I_o]_{m\times 1} = \begin{bmatrix} i_a \\ i_b \\ \vdots \\ i_m \end{bmatrix} \tag{8.37}$$

where $[I_i]_{N\times 1}$ and $[I_o]_{m\times 1}$ depict the arrays of the input and output currents, respectively. The input and output powers are identical in an ideal matrix converter:

$$P_i = P_o \tag{8.38}$$

By using (8.38), it can be obtained that:

$$[I_i]^T_{1\times N}[V_i]_{N\times 1} = [I_o]^T_{1\times m}[V_o]_{m\times 1} \tag{8.39}$$

By substituting (8.35) into (8.39), it can be obtained that:

$$[I_i]^T_{1\times N}[V_i]_{N\times 1} = [I_o]^T_{1\times m}[M]_{m\times N}[V_i]_{N\times 1} \tag{8.40}$$

Regarding (8.40), the main harmonic of the input current can be obtained as follows:

$$[I_i]_{N\times 1} = [M]^T_{N\times m}[I_o]_{m\times 1} \tag{8.41}$$

8.4 Positive, negative, and combined control methods

Controlling output voltages by switching between the valid states in the PWM technique is done using an approach in which the average value of input voltages is the same as the desired waveforms. The output voltage is synthesized with the sampled segments of the input voltages in this approach. Three control methods of positive, negative, and combined ones for the matrix converters are described in the following. A matrix converter with N inputs and m outputs is considered as obtained in (8.1) and (8.2). Every switching period is divided into ℓ periods of t_1, t_2, …, and t_ℓ in this approach, as shown in Figure 8.7. If N is greater than m, ℓ will be equal to N. Otherwise, ℓ will be equal to m.

The relation between T_s and t_1, t_2, …, and t_ℓ is as follows for each switching period:

$$T_s = t_1 + t_2 + \cdots + t_\ell \tag{8.42}$$

Regarding (8.42), the following constraints can be obtained:

$$0 \le t_k \le T_s \quad for \quad k = 1, 2, \cdots, \ell \tag{8.43}$$

First, the switches' matrix can be formed as follows:

$$[S]_{m\times N} = \begin{bmatrix} S_{Aa} & S_{Ba} & \cdots & S_{Na} \\ S_{Ab} & S_{Bb} & \cdots & S_{Nb} \\ \vdots & \vdots & \ddots & \vdots \\ S_{Am} & S_{Bm} & \cdots & S_{Nm} \end{bmatrix} \tag{8.44}$$

If $[S]_{m\times N}$ is a nonsquare matrix that its columns are greater than its rows, the output currents will be discontinuous. In this condition, the switches' matrix can be turned into a square matrix by repeating the rows in need (from the first row to the

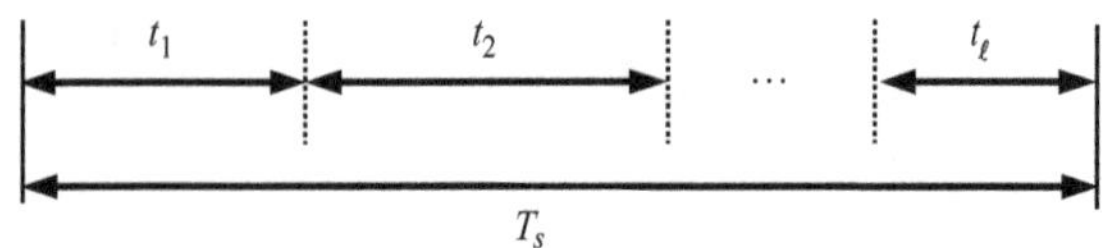

Figure 8.7 Dividing a switching period

last one), in order to prevent discontinuous currents on the output stage. It is required to note that repeating the rows must not lead to the short-circuit of the input voltage sources and the open-circuit of the output loads. For instance, the switches' matrix of a three-phase inverter is nonsquare, so it should be turned into a square one by repeating the other rows or columns appropriately.

An appropriate combination of the output voltages can be created by using a combination of suitable input voltages with the periods of t_1, t_2, …, and t_ℓ. The following three approaches are proposed in order to synthesize these output voltages:

- Positive control method
- Negative control method
- Combined control method

8.4.1 Positive control method

Each switching period is divided into ℓ periods of $t_{1,p}$, $t_{2,p}$, …, and $t_{\ell,p}$ in the positive control method, as Figure 8.8 illustrates.

As Figure 8.9 illustrates, in the positive control method, the first $N-1$ columns of the switches' matrix are repeated after the Nth column, the same as the approach to calculating the determinant of a matrix. Regarding this figure, in the period of $t_{1,P}$ the switches on the main diagonal are turned on, and the others left off. For the period of $t_{2,P}$, the switches on the diagonal right after the main one are turned on. The same approach repeats for the other periods. Figure 8.9 depicts the

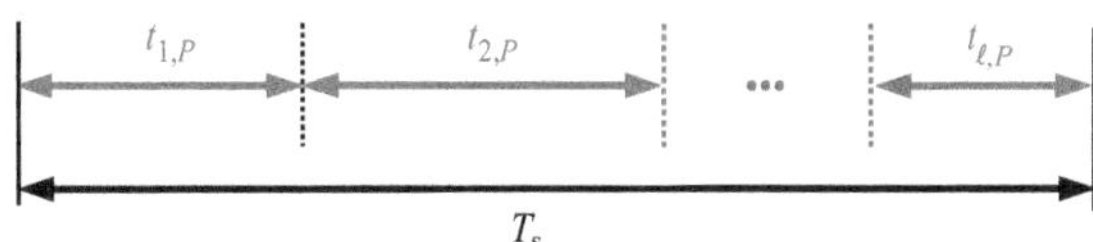

Figure 8.8 Dividing a switching period in the positive control method

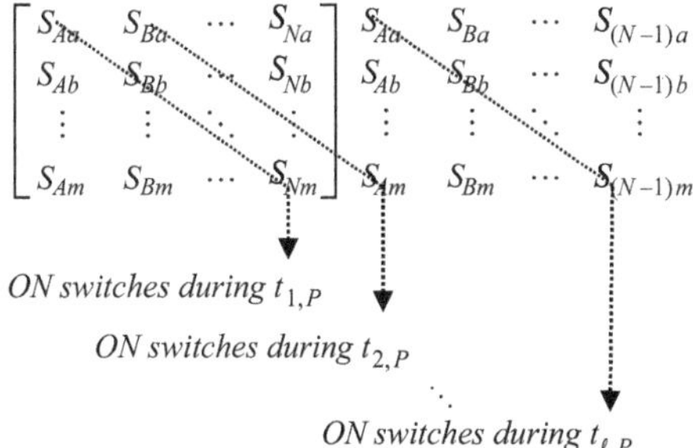

Figure 8.9 Switching pattern using the positive control method

positive switching pattern where $N = m = \ell$. Regarding this figure, this switching method is as follows:

- Switches of S_{Aa}, S_{Bb}, S_{Cc}, $\ldots$, and S_{Nm} are turned on for the period of $t_{1,P}$ and the other switches are left off
- Switches of S_{Ba}, S_{Cb}, S_{Dc}, $\ldots$, and S_{Am} are turned on for the period of $t_{2,P}$ and the other switches are left off
- $\vdots$
- Switches of S_{Na}, S_{Ab}, S_{Bc}, $\ldots$, and $S_{(N-1)m}$ are turned on for the period of $t_{\ell,P}$ and the other switches are left off.

Regarding Figure 8.9, it can result in that only one switch is turned on in every period, and it guarantees that there are no open-circuits at the output and short-circuits at the input side. From Figures 8.2 and 8.9, the following results can be obtained:

$$\begin{aligned} &During\ t_{1,P} : v_a = v_A \\ &During\ t_{2,P} : v_a = v_B \\ &\qquad\qquad \vdots \\ &During\ t_{\ell,P} : v_a = v_N \end{aligned} \tag{8.45}$$

When the switching frequency (f_s) is supposed to be far greater than the input and output frequencies ($f_s >> f_i$ and $f_s >> f_o$), the input voltages can be considered constant values during a switching period. The output voltages can be obtained as follows under these circumstances:

$$\begin{aligned} v_a &= \frac{1}{T_s}\left[v_A t_{1,P} + v_B t_{2,P} + \cdots + v_N t_{\ell,P}\right] \\ v_b &= \frac{1}{T_s}\left[v_B t_{1,P} + v_C t_{2,P} + \cdots + v_A t_{\ell,P}\right] \\ &\ \ \vdots \\ v_m &= \frac{1}{T_s}\left[v_N t_{1,P} + v_A t_{2,P} + \cdots + v_{N-1} t_{\ell,P}\right] \end{aligned} \tag{8.46}$$

It is required to remind that these equations are valid for a condition of $N = m = \ell$. Regarding (8.46), it can be obtained that the desired output voltages can be synthesized by combining the input voltages and the different periods of $t_{1,p}$, $t_{2,p}$, $\ldots$, and $t_{\ell,p}$. For instance, the input voltage of v_A, v_B, $\ldots$, and v_N is only combined with the periods of $t_{1,p}$, $t_{2,p}$, $\ldots$, and $t_{\ell,p}$ in order to synthesize the output voltage of v_a, respectively. Moreover, (8.46) can be rewritten in a matrix form as follows:

$$\begin{bmatrix} v_a \\ v_b \\ \vdots \\ v_m \end{bmatrix} = \frac{1}{T_s}\begin{bmatrix} v_A & v_B & \cdots & v_N \\ v_B & v_C & \cdots & v_A \\ \vdots & \vdots & \ddots & \vdots \\ v_N & v_A & \cdots & v_{N-1} \end{bmatrix}\begin{bmatrix} t_{1,P} \\ t_{2,P} \\ \vdots \\ t_{\ell,P} \end{bmatrix} \tag{8.47}$$

Regarding periods of $t_{1,p}$, $t_{2,p}$, $\ldots$, and $t_{\ell,p}$, it can be obtained that they must meet the following conditions:

$$T_s = t_{1,P} + t_{2,P} + \cdots + t_{\ell,P} \tag{8.48}$$

$$0 \le t_{k,P} \le T_s \quad k = 1,\ 2,\ \cdots,\ \ell \tag{8.49}$$

Some new parameters can be defined as follows:

$$P_1 = \frac{t_{1,P}}{T_s}, \quad P_2 = \frac{t_{2,P}}{T_s}, \quad \cdots \quad P_\ell = \frac{t_{\ell,P}}{T_s} \tag{8.50}$$

Eqs. (8.48) and (8.49) can be rewritten as follows by substituting (8.50) into them:

$$P_1 + P_2 + \cdots + P_\ell = 1 \tag{8.51}$$

$$0 \le P_k \le 1\ k = 1,\ 2,\ \cdots,\ \ell \tag{8.52}$$

Moreover, the following results can be obtained by substituting (8.50) into (8.47):

$$\begin{aligned} v_a &= P_1 v_A + P_2 v_B + \cdots + P_\ell v_N \\ v_b &= P_1 v_B + P_2 v_C + \cdots + P_\ell v_A \\ &\vdots \\ v_m &= P_1 v_N + P_2 v_A + \cdots + P_\ell v_{N-1} \end{aligned} \tag{8.53}$$

Eq. (8.53) can be rewritten in the matrix form as follows:

$$\begin{bmatrix} v_a \\ v_b \\ \vdots \\ v_m \end{bmatrix} = \begin{bmatrix} P_1 & P_2 & \cdots & P_\ell \\ P_\ell & P_1 & \cdots & P_{\ell-1} \\ \vdots & \vdots & \ddots & \vdots \\ P_2 & P_3 & \cdots & P_1 \end{bmatrix} \begin{bmatrix} v_A \\ v_B \\ \vdots \\ v_N \end{bmatrix} \tag{8.54}$$

Regarding (8.54), the following matrixes can be defined:

$$[V_o]_{m\times 1} = [P]_{\ell\times\ell}[V_i]_{N\times 1} \tag{8.55}$$

The following equations can be obtained by considering (8.53):

$$\begin{bmatrix} v_a \\ v_b \\ \vdots \\ v_m \end{bmatrix} = \begin{bmatrix} v_A & v_B & \cdots & v_N \\ v_B & v_C & \cdots & v_A \\ \vdots & \vdots & \ddots & \vdots \\ v_N & v_A & \cdots & v_{N-1} \end{bmatrix} \begin{bmatrix} P_1 \\ P_2 \\ \vdots \\ P_\ell \end{bmatrix} \tag{8.56}$$

In other words,

$$[V_o]_{m\times 1} = \left[V_{i,P}\right]_{N\times N} \begin{bmatrix} P_1 \\ P_2 \\ \vdots \\ P_\ell \end{bmatrix} \tag{8.57}$$

A unique answer for the matrix equation of (8.57) will be available only if the following constraint is valid:

$$\det\left[V_{i,P}\right]_{N\times N} \neq 0 \tag{8.58}$$

Otherwise, one of the duty cycles such as P_1 must be replaced by:

$$P_1 = 1 - P_2 - \cdots - P_\ell \tag{8.59}$$

Only one valid answer for each of the duty cycles of P_1, P_2 …, and, P_ℓ is obtained by substituting (8.59) into (8.52). The following equation associated with the input and output currents is obtained by using (8.41) and (8.55):

$$[I_i]_{N\times 1} = [P]^T_{\ell\times\ell}[I_o]_{m\times 1} \tag{8.60}$$

8.4.2 Negative control method

Each switching period is divided into ℓ periods of $t_{1,N}$, $t_{2,N}$, …, and $t_{\ell,N}$ in the negative control method, as Figure 8.10 illustrates.

As Figure 8.11 illustrates, in the negative control method, the first $N-1$ columns of the switches' matrix are repeated after the Nth column, the same as the approach to calculating the determinant of a matrix. It is required to note that this figure depicts the negative switching pattern in where $N = m = \ell$. Regarding this figure, this switching method is as follows:

- Switches of S_{Na}, $S_{(N-1)b}$, $S_{(N-2)c}$, …, and S_{Am} are turned on for the period of $t_{1,N}$ and the other switches are left off
- Switches of S_{Aa}, S_{Nb}, $S_{(N-1)c}$, …, and S_{Bm} are turned on for the period of $t_{2,N}$ and the other switches are left off
- ⋮
- Switches of $S_{(N-1)a}$, $S_{(N-2)b}$, $S_{(N-3)c}$, …, and S_{Nm} are turned on for the period of $t_{\ell,N}$ and the other switches are left off.

From Figures 8.2 and 8.10, the following results can be obtained:

$$\begin{aligned} &During\ t_{1,N} : v_a = v_N \\ &During\ t_{2,N} : v_a = v_A \\ &\qquad\vdots \\ &During\ t_{\ell,N} : v_a = v_{N-1} \end{aligned} \tag{8.61}$$

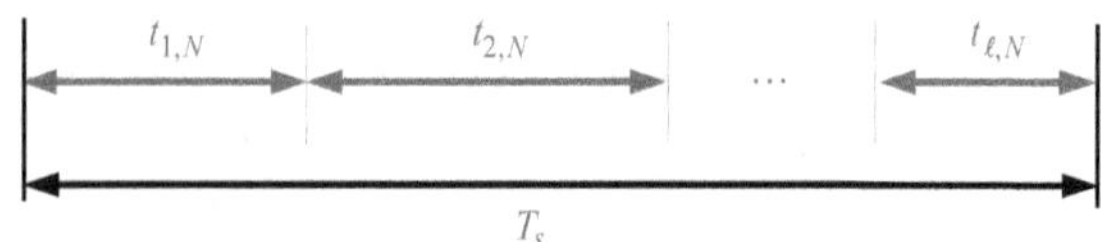

Figure 8.10 Dividing a switching period in the negative control method

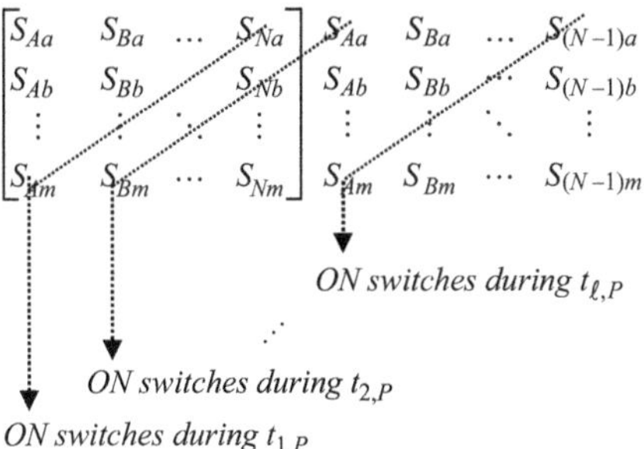

Figure 8.11 Switching pattern using the negative control method

When the switching frequency (f_s) is supposed to be so greater than the input and output frequencies ($f_s >> f_i$ and $f_s >> f_o$), the input voltages can be considered constant values during a switching period. The output voltages can be obtained as follows under these circumstances:

$$\begin{aligned} v_a &= \frac{1}{T_s}\left[v_N t_{1,N} + v_A t_{2,N} + \cdots + v_{N-1} t_{\ell,N}\right] \\ v_b &= \frac{1}{T_s}\left[v_{N-1} t_{1,N} + v_N t_{2,N} + \cdots + v_{N-2} t_{\ell,N}\right] \\ &\vdots \\ v_m &= \frac{1}{T_s}\left[v_A t_{1,N} + v_B t_{2,N} + \cdots + v_N t_{\ell,N}\right] \end{aligned} \tag{8.62}$$

Eq. (8.62) can be rewritten in a matrix form as follows:

$$\begin{bmatrix} v_a \\ v_b \\ \vdots \\ v_m \end{bmatrix} = \frac{1}{T_s}\begin{bmatrix} v_N & v_A & \cdots & v_{N-1} \\ v_{N-1} & v_N & \cdots & v_{N-2} \\ \vdots & \vdots & \ddots & \vdots \\ v_A & v_B & \cdots & v_N \end{bmatrix}\begin{bmatrix} t_{1,N} \\ t_{2,N} \\ \vdots \\ t_{\ell,N} \end{bmatrix} \tag{8.63}$$

Similar to the positive control method, it can be obtained that the desired output voltages can be synthesized by combining the input voltages and the different periods of $t_{1,N}$, $t_{2,N}$, ..., and $t_{\ell,N}$. For instance, the input voltage of v_N, v_A, ..., and v_{N-1} are only combined with the periods of $t_{1,N}$, $t_{2,N}$, ..., and $t_{\ell,N}$ in order to synthesize the output voltage of v_a, respectively. Also, regarding periods of $t_{1,N}$, $t_{2,N}$, ..., and $t_{\ell,N}$, it can be obtained that they must meet the following conditions:

$$T_s = t_{1,N} + t_{2,N} + \cdots + t_{\ell,N} \tag{8.64}$$

$$0 \le t_{k,N} \le T_s \quad k = 1, 2, \cdots, \ell \tag{8.65}$$

Some new parameters can be defined as follows:

$$N_1 = \frac{t_{1,N}}{T_s}, \quad N_2 = \frac{t_{2,N}}{T_s}, \quad \cdots \quad N_\ell = \frac{t_{\ell,N}}{T_s} \tag{8.66}$$

Eqs. (8.64) and (8.65) can be rewritten as follows by substituting (8.66) into them:

$$N_1 + N_2 + \cdots + N_\ell = 1 \tag{8.67}$$

$$0 \leq N_k \leq 1 \quad k = 1,\ 2,\ \cdots,\ \ell \tag{8.68}$$

Moreover, the following results can be obtained by substituting (8.66) into (8.63):

$$\begin{aligned} v_a &= N_1 v_N + N_2 v_A + \cdots + N_\ell v_{N-1} \\ v_b &= N_1 v_{N-1} + N_2 v_N + \cdots + N_\ell v_{N-2} \\ &\vdots \\ v_m &= N_1 v_A + N_2 v_B + \cdots + N_\ell v_N \end{aligned} \tag{8.69}$$

Eq. (8.69) can be rewritten in the matrix form as follows:

$$\begin{bmatrix} v_a \\ v_b \\ \vdots \\ v_m \end{bmatrix} = \begin{bmatrix} N_2 & N_3 & \cdots & N_1 \\ N_3 & N_4 & \cdots & N_2 \\ \vdots & \vdots & \ddots & \vdots \\ N_1 & N_2 & \cdots & N_\ell \end{bmatrix} \begin{bmatrix} v_A \\ v_B \\ \vdots \\ v_N \end{bmatrix} \tag{8.70}$$

Regarding (8.70), the following matrixes can be defined:

$$[V_o]_{m\times 1} = [N]_{\ell\times\ell}[V_i]_{N\times 1} \tag{8.71}$$

The following equations can be obtained by considering (8.70):

$$\begin{bmatrix} v_a \\ v_b \\ \vdots \\ v_m \end{bmatrix} = \begin{bmatrix} v_N & v_A & \cdots & v_{N-1} \\ v_{N-1} & v_N & \cdots & v_{N-2} \\ \vdots & \vdots & \ddots & \vdots \\ v_A & v_B & \cdots & v_N \end{bmatrix} \begin{bmatrix} N_1 \\ N_2 \\ \vdots \\ N_\ell \end{bmatrix} \tag{8.72}$$

In other words,

$$[V_o]_{m\times 1} = \left[V_{i,N}\right]_{N\times N} \begin{bmatrix} N_1 \\ N_2 \\ \vdots \\ N_\ell \end{bmatrix} \tag{8.73}$$

A unique answer for the matrix equation of (8.73) will be available only if the following constraint is valid:

$$\det\left[V_{i,N}\right]_{N\times N} \neq 0 \tag{8.74}$$

Otherwise, one of the duty cycles such as N_1 must be replaced by:

$$N_1 = 1 - N_2 - \quad - N_l \tag{8.75}$$

Only one valid answer for each of the duty cycles of N_1, N_2, …, and, N_ℓ is obtained by substituting (8.75) into (8.69). The following equation associated with the input and output currents is obtained by using (8.41) and (8.71):

$$[I_i]_{N\times 1} = [N]^T_{\ell\times\ell}[I_o]_{m\times 1} \tag{8.76}$$

8.4.3 Combined control method

The other control method for the matrix converters proposed in this section is the combined control method. The switching pattern for this control method is depicted in Figure 8.12. As this figure shows, the output voltages are synthesized using the positive and negative control methods during α and β times of a switching period.

Regarding that, the same output voltages can be obtained using either positive or negative control methods, and so these output voltages can be obtained by combining these control methods. Regarding (8.55) and (8.71), the following equation can be written:

$$[V_o]_{m\times 1} = \alpha[P]_{\ell\times\ell}[V_i]_{N\times 1} + \beta[N]_{\ell\times\ell}[V_i]_{N\times 1} \tag{8.77}$$

where:

$$\alpha + \beta = 1 \tag{8.78}$$

Besides, an equation associated with the input and output voltages can be written as follows using (8.41):

$$[I_i]_{N\times 1} = \alpha[P]^T_{\ell\times\ell}[I_o]_{m\times 1} + \beta[N]^T_{\ell\times\ell}[I_o]_{m\times 1} \tag{8.79}$$

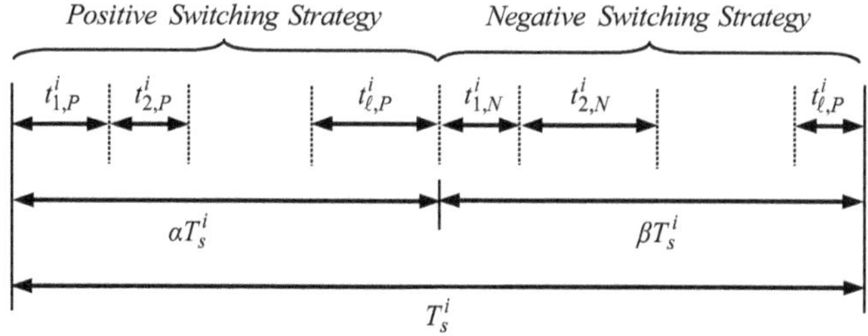

Figure 8.12 Combined control method

It is required to remind that the done calculations are valid when $N = m = \ell$. The subsequent analysis will show that the coefficients of α and β can highly affect different characteristics of matrix converters, such as controlling the input power factor regardless of the load's type and mitigating low-order harmonics for the input currents and so, reducing the input currents' total harmonic distortion (THD).

Example 8.1 Obtain switches' matrix for a three-phase inverter.

Solution: The power circuit of the three-phase inverter is depicted in Figure 8.13.

Regarding Figure 8.13, the switches' matrix can be formed as follows:

$$S = \begin{bmatrix} S_1 & S_3 & S_5 \\ S_2 & S_4 & S_6 \end{bmatrix}_{2\times 3}$$

This matrix is a nonsquare one, so it is required to turn it into a square one in order to prevent short-circuits at the input side and open-circuits at the output side. After that, the control of this converter can be done using any of the proposed control methods, such as the negative control method.

The turned-on switches in every period can be obtained as follows by using Figure 8.14:

$$\begin{aligned} t_{1,N} &: S_5 \quad S_4 \quad S_1 \\ t_{2,N} &: S_1 \quad S_6 \quad S_3 \\ t_{3,N} &: S_3 \quad S_2 \quad S_5 \end{aligned}$$

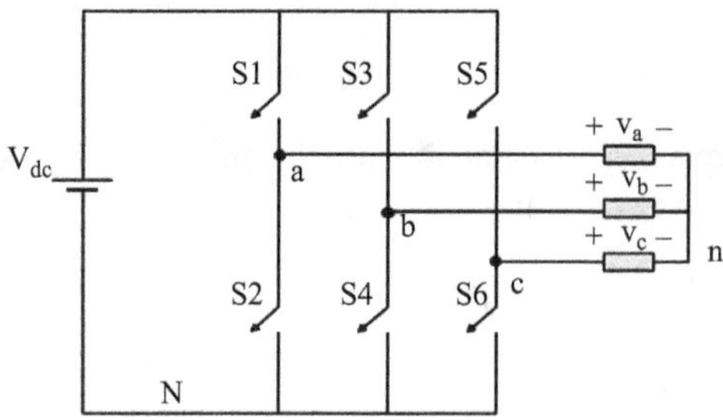

Figure 8.13 Power circuit of the three-phase inverter

$$\begin{bmatrix} S_1 & S_3 & S_5 \\ S_2 & S_4 & S_6 \\ S_1 & S_3 & S_5 \end{bmatrix}_{3\times 3} \begin{matrix} S_1 & S_3 \\ S_2 & S_4 \\ S_1 & S_3 \end{matrix}$$

$$t_{1,N} \quad t_{2,N} \quad t_{3,N}$$

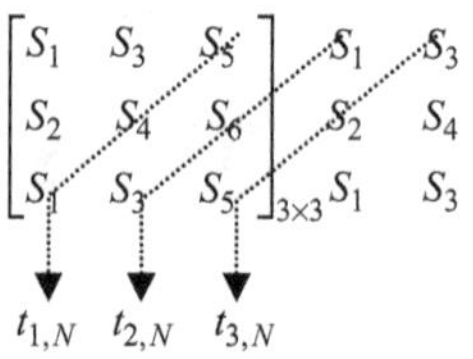

Figure 8.14 Applying the negative control method on the three-phase inverter

First, the switches of S_1, S_4, and S_5 are turned on during the period of $t_{1,N}$. Second, the switches of S_1, S_3, and S_6 are turned on during the period of $t_{2,N}$. Finally, the switches of S_2, S_3, and S_5 are turned on during the period of $t_{3,N}$. It can be seen that this switching pattern does not lead to a short-circuit at the input side. However, if the switches' matrix did not turn into a square one, some open-circuit states would occur on the output stage that are invalid states.

Example 8.2 Obtain switches' matrix for a three-phase to two-phase matrix converter.

Solution: The power circuit of the three-phase to two-phase matrix converter is depicted in Figure 8.15.

The switches' matrix can be formed for this converter as follows:

$$S = \begin{bmatrix} S_{Aa} & S_{Ba} & S_{Ca} \\ S_{Ab} & S_{Bb} & S_{Cb} \end{bmatrix}_{2\times 3}$$

It can be seen that the obtained matrix cannot be converted into a square one because it leads to the short-circuit states on the input stage using either the positive or negative control methods. Thus, controlling this converter is done without turning the switches' matrix to a square one, as shown in Figure 8.16.

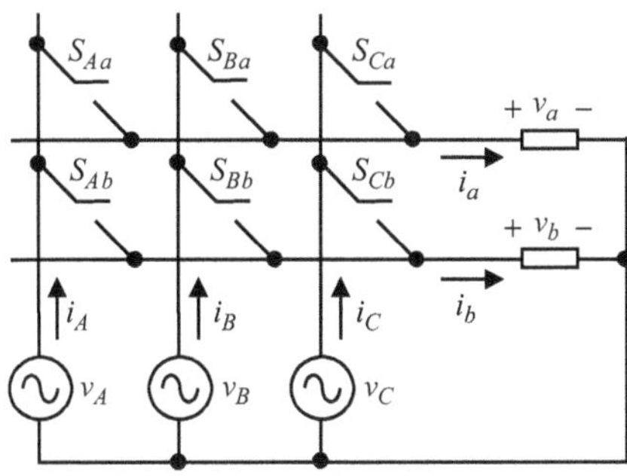

Figure 8.15 The three-phase to two-phase converter

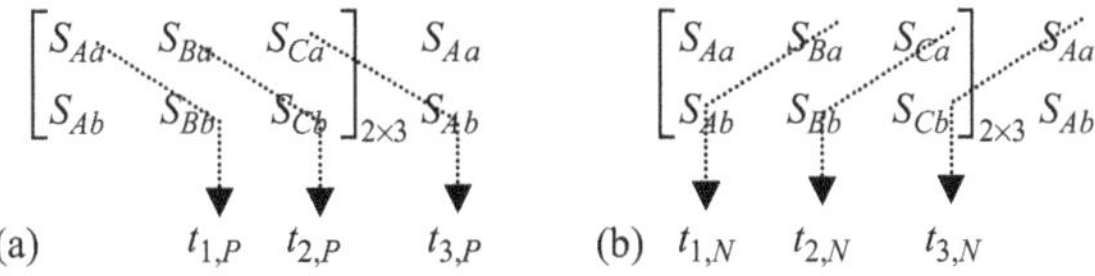

Figure 8.16 Applying the positive and negative control methods on the three-phase to two-phase matrix converter, (a) the positive control method and (b) the negative control method

8.5 Implementation of positive, negative, and combined control methods on single-phase matrix converter

In this section, the results of implementing the three different control methods on the single-phase matrix converter depicted in Figure 8.4(f) will be described. The input voltage and the main harmonic of the desired output voltage are assumed as follows:

$$v_i = V_{i,\max} \sin(\omega_i t) \tag{8.80}$$

$$v_o = V_{o,\max} \sin(\omega_o t) \tag{8.81}$$

where $V_{i,\max}$, ω_i, $V_{o,\max}$, and ω_o are the peak amplitude and the angular frequency of the input and output voltages, respectively. The switches' matrix of a converter is required in order to control it. Regarding the single-phase matrix converter, this matrix can be obtained as follows:

$$S = \begin{bmatrix} S_{11} & S_{21} \\ S_{12} & S_{22} \end{bmatrix} \tag{8.82}$$

As (8.82) shows, the switches' matrix of this converter is a square one with a size of 2 × 2, so a switching period of T_s is divided into two times of t_1 and t_2. The results of implementing each control method on this converter will be expressed in the next subsections.

8.5.1 Positive control method

Using this control method, a switching period will be divided, as Figure 8.17 shows. Besides, the switching pattern and the control pulses using this control method are shown in Figure 8.18, where 1 and 0 mean the on-state and off-state of a

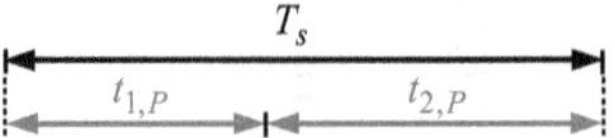

Figure 8.17 Dividing a switching period using the positive control method for the single-phase matrix converter

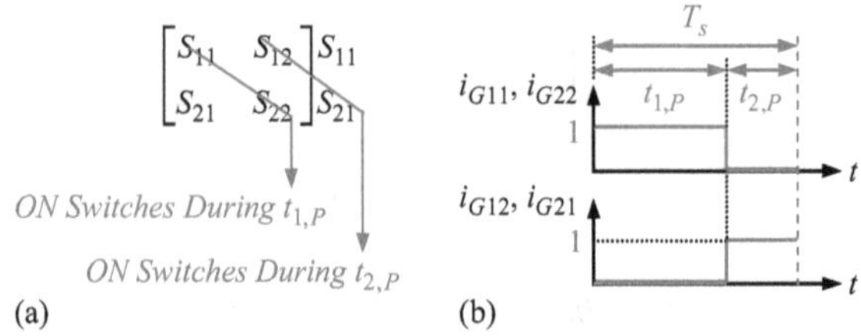

Figure 8.18 Implementing the positive control method on the single-phase matrix converter, (a) switching pattern and (b) gate pulses

switch. Regarding Figure 8.18(b), this control method guarantees no short-circuits at the input side and no open-circuits at the output side will happen.

Regarding Figure 8.18, the following constraints can be obtained:

$$\begin{aligned} &0 \le t_{1,P} \le T_s \\ &0 \le t_{2,P} \le T_s \end{aligned} \tag{8.83}$$

$$t_{1,P} + t_{2,P} = T_s \tag{8.84}$$

Regarding the configuration of the single-phase matrix converter, the equivalent circuits can be obtained for each of the operational stages as follows:

By having Figure 8.19, the output voltage can be written as follows:

$$v_o = \begin{cases} v_i & \textit{During } t_{1,P} \\ -v_i & \textit{During } t_{2,P} \end{cases} \tag{8.85}$$

Regarding (8.85), the output voltage alters between two values of $+ v_i$ and $-v_i$ in a switching period. Regarding (8.85), the average value of the output voltage for a switching period can be obtained from:

$$v_o = \frac{1}{T_s}(t_{1,P} - t_{2,P})v_i \tag{8.86}$$

As (8.86) shows, if $t_{1,P}$ and $t_{2,P}$ are almost equal, the average value of output voltage will tend to be zero. Moreover, the following results can be obtained if the switching period is assumed to be so higher than the input and output frequencies ($f_s >> f_i$ and $f_s >> f_o$):

- The input voltage is almost constant in each switching period.
- The average value of the output voltage traces (8.86).

Now, the duty cycles using the positive control method can be expressed as:

$$\begin{aligned} P_1 &= \frac{t_{1,P}}{T_s} \\ P_2 &= \frac{t_{2,P}}{T_s} \end{aligned} \tag{8.87}$$

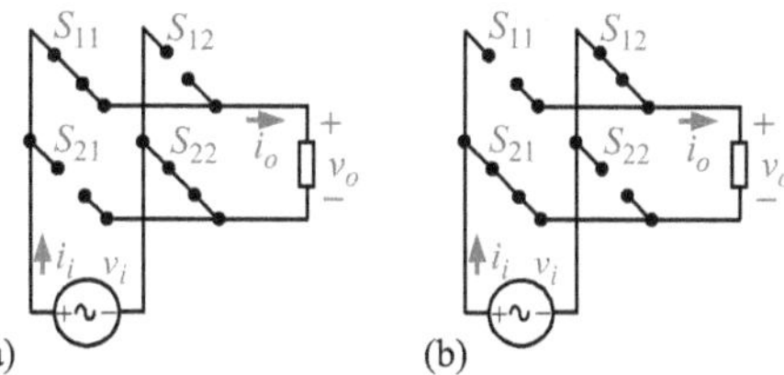

Figure 8.19 Equivalent circuits using the positive control method, (a) in $t_{1,P}$ and (b) in $t_{2,P}$

Regarding (8.83), (8.84), and (8.87), the following constraints are always true:

$$\begin{aligned} &0 \leq P_1 \leq 1 \\ &0 \leq P_2 \leq 1 \end{aligned} \tag{8.88}$$

$$P_1 + P_2 = 1 \tag{8.89}$$

The following result can be obtained by substituting (8.8) into (8.7):

$$v_o = (P_1 - P_2)v_i \tag{8.90}$$

The unique answers for P_1 and P_2 will obtain as follows if (8.10) and (8.11) are solved simultaneously:

$$\begin{aligned} P_1 &= \frac{1}{2}\left(1 + \frac{v_o}{v_i}\right) \\ P_2 &= \frac{1}{2}\left(1 - \frac{v_o}{v_i}\right) \end{aligned} \tag{8.91}$$

The following results can be obtained by substituting (8.80) and (8.81) into (8.91):

$$\begin{aligned} P_1 &= \frac{1}{2}\left[1 + \frac{V_{o,\max}\sin(\omega_o t)}{V_{i,\max}\sin(\omega_i t)}\right] \\ P_2 &= \frac{1}{2}\left[1 - \frac{V_{o,\max}\sin(\omega_o t)}{V_{i,\max}\sin(\omega_i t)}\right] \end{aligned} \tag{8.92}$$

The following inequality can result from the constraints that are described in (8.92):

$$\left|\frac{V_{o,\max}\sin(\omega_o t)}{V_{i,\max}\sin(\omega_i t)}\right| \leq 1 \tag{8.93}$$

Regarding that, (8.93) must be valid each time, when $\sin(\omega_i t)$ tends to zero, $\sin(\omega_o t)$ must also to zero, as well. Thus, input and output waveforms must have an equal phase shift which means they must cross zero simultaneously. As a result, although the output waveform cannot have a lower frequency than the input waveform, the output waveform frequency can be many folds of the frequency of the input waveform. This description is depicted in Figure 8.20. Regarding this figure, the single-phase matrix converter can be considered a frequency booster converter that can multiply the input voltage's frequency using as follows:

$$\omega_o = k\omega_i \quad \textit{for} \quad k = 1, 2, 3, \cdots \tag{8.94}$$

The following result can be obtained by substituting $\omega = 2\pi f$:

$$f_o = kf_i \quad \textit{for} \quad k = 1, 2, 3, \cdots \tag{8.95}$$

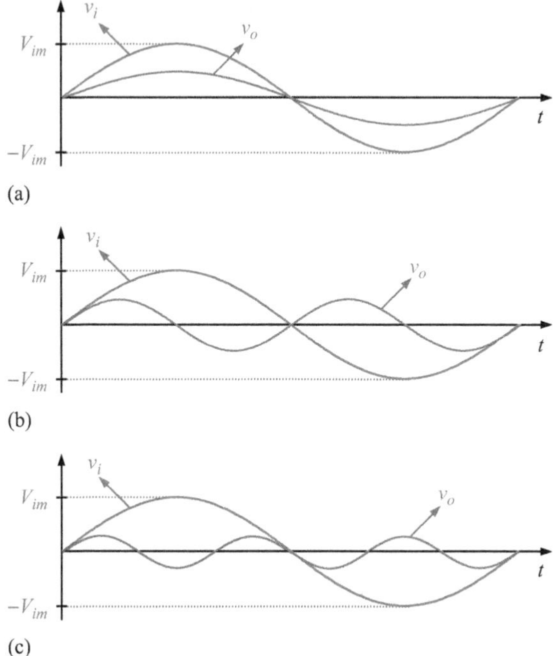

Figure 8.20 Illustrating the limitations of the output voltages' frequency and amplitude to the input voltage, (a) $f_o = f_i$, (b) $f_o = 2f_i$, and (c) $f_o = 3f_i$

When $\sin(\omega_i t)$ equals zero, the values of P_1 and P_2 can be calculated as follows using L hospital's rule:

$$\begin{aligned} P_1 &= \frac{1}{2}\left(1 + \frac{V_{o,\max}\omega_o}{V_{i,\max}\omega_i}\right) \\ P_2 &= \frac{1}{2}\left(1 - \frac{V_{o,\max}\omega_o}{V_{i,\max}\omega_i}\right) \end{aligned} \tag{8.96}$$

The amplitude of the output voltage must meet the following inequality regarding (8.96):

$$V_{o,\max} \leq \frac{\omega_i}{\omega_o} V_{i,\max} \tag{8.97}$$

The following equation can result by using (8.94) and (8.97):

$$V_{o,\max} \leq \frac{1}{k} V_{i,\max} \quad for \quad k = 1, 2, 3, \cdots \tag{8.98}$$

Regarding (8.98), from the amplitude perspective, the single-phase matrix converter is a buck converter. Besides, as expressed before, this converter is a frequency booster. Besides, regarding (8.90), the relation between the input and output currents can be expressed as:

$$i_o = (P_1 - P_2)i_i \tag{8.99}$$

It is required to note that the input current contains some high-order harmonics with the input voltage's frequency, and they can be eliminated using a low-pass filter on the input stage. Finally, $t_{1,P}$ and $t_{2,P}$ can be obtained by substituting (8.91) into (8.87):

$$\begin{aligned} t_{1,P} &= \frac{T_s}{2}\left[1 + \frac{V_{o,\max}\sin(\omega_o t)}{V_{i,\max}\sin(\omega_i t)}\right] \\ t_{2,P} &= \frac{T_s}{2}\left[1 - \frac{V_{o,\max}\sin(\omega_o t)}{V_{i,\max}\sin(\omega_i t)}\right] \end{aligned} \tag{8.100}$$

Example 8.3 Consider a single-phase matrix converter with an input voltage of $v_i = 100\sin(100\pi t)$. It is controlled using the positive control method to synthesize an output voltage of $v_o = V_{o,\max}\sin(\omega_o t)$.

(a) What frequencies can be obtained on the output stage?

(b) Calculate the relevant output voltage for each of these output frequencies.

Solution:

(a) Regarding the input voltage, its peak magnitude and frequency are equal $V_{i,\max} = 100$ V and $f_i = 50$ Hz, respectively. Because the single-phase matrix converter is a frequency multiplier, it can produce different frequencies in the output stage, including 50 Hz, 100 Hz, 150 Hz, 200 Hz, and so on.

(b) Regarding (8.95) and (8.98), the following results can be obtained for different k:

$$k = 1: f_o = 50\text{ Hz}, V_{o,\max} \le 100\text{ V}$$

$$k = 2: f_o = 100\text{ Hz}, V_{o,\max} \le 50\text{ V}$$

$$k = 3: f_o = 150\text{ Hz}, V_{o,\max} \le 33.33\text{ V}$$

$$k = 4: f_o = 200\text{ Hz}, V_{o,\max} \le 25\text{ V}$$

The output voltage's magnitude decreases by increasing its frequency regarding these results.

Example 8.4 Consider a single-phase matrix converter with an input voltage of $v_i = 100\sin(100\pi t)$. This converter uses the positive control method in order to synthesize an output voltage of $v_o = 50\sin(200\pi t)$. Assume the switching frequency as 5 kHz.

(a) Sketch the power circuit of this converter.
(b) Obtain the numerical values for $t_{1,P}$ and $t_{2,P}$ for each time.
(c) Analyze the effect of the switching frequency on the converter's operation.

Solution:

(a) The power circuit of this converter is depicted in Figure 8.4(f).
(b) Regarding the given data, the following results can be written:

$$T_i = \frac{1}{f_i} = \frac{1}{50} = 20 \text{ ms}$$

$$T_o = \frac{1}{f_o} = \frac{1}{100} = 10 \text{ ms}$$

$$T_s = \frac{1}{f_s} = \frac{1}{5000} = 0.2 \text{ ms}$$

Regarding the obtained results:

$$T_i = 100T_s$$
$$T_o = 50T_s$$

So, the input and output waveforms can be depicted in Figure 8.21.

Regarding Figure 8.21, the results can be obtained for one of the sampling periods. For instance, the results can be written as follows for 4–4.2 ms:

Start of sampling:

$$v_i = 100 \sin[100\pi(0.004)] = 95.11 \text{ V}$$

$$v_o = 50 \sin[200\pi(0.004)] = 29.39 \text{ V}$$

End of sampling:

$$v_i = 100 \sin[100\pi(0.0042)] = 96.86 \text{ V}$$

$$v_o = 50 \sin[200\pi(0.0042)] = 24.09 \text{ V}$$

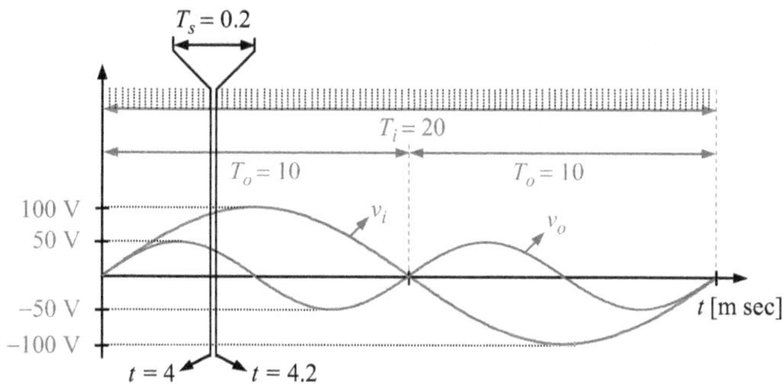

Figure 8.21 Input and output waveforms, Example 9.4

Regarding these results, it can be seen that both input and output voltage can be considered constant during a sampling period. It is required to note that a high switching frequency brings this result. So, it is enough to substitute the time when a sampling starts to calculate $t_{1,P}$ and $t_{2,P}$ using (8.84) and (8.100)

$$t_{1,P} = \frac{0.2}{2}\left[1 + \frac{50\sin(200\pi t)}{100\sin(100\pi t)}\right]$$

$$t_{2,P} = T_s - t_{1,P}$$

The first sampling period starts at $t = 0$. So:

$$t_{1,P} = 0.2\ \text{ms},\ t_{2,P} = 0$$

The second sampling period starts at $t = 0.2$ ms. So:

$$t_{1,P} = 0.1998\ \text{ms},\ t_{2,P} = 0.0002\ \text{ms}$$

The third sampling period starts at $t = 0.4$ ms. So:

$$t_{1,P} = 0.1992\ \text{ms},\ t_{2,P} = 0.0008\ \text{ms}$$

Ultimately, the hundredth sampling period starts at $t = 19.8$ ms. So:

$$t_{1,P} = 0.1998\ \text{ms},\ t_{2,P} = 0.0002\ \text{ms}$$

(c) Regarding the obtained results, the switching frequency must be chosen large enough in order to obtain valid values for $t_{1,P}$ and $t_{2,P}$. Not only does this assumption increase the accuracy of calculations but also the magnitude of the low-order harmonics decreases. Besides, the size of the required filter minimizes, and the quality of the output voltage increases. In contrast, increasing the switching frequency negatively affects the switching losses and the cost of required switches.

Example 8.5 Consider a single-phase matrix converter with an input voltage of $v_i = 200\sin(100\pi t)$. This converter uses the positive control method in order to synthesize an output voltage of $v_o = 80\sin(200\pi t)$. Assume the switching frequency as 500 Hz.

(a) Obtain the numerical values for $t_{1,P}$ and $t_{2,P}$ for each time.
(b) Sketch gate pulses and the main harmonic of the output voltage waveforms.
Solution:
(a) Regarding the given data, the following results can be written:

$$T_i = \frac{1}{f_i} = \frac{1}{50} = 20\ \text{ms}$$

$$T_o = \frac{1}{f_o} = \frac{1}{100} = 10\ \text{ms}$$

$$T_s = \frac{1}{f_s} = \frac{1}{500} = 2\ \text{ms}$$

Regarding the obtained results:

$$T_i = 10T_s$$
$$T_o = 5T_s$$

Now, we can calculate $t_{1,P}$ and $t_{2,P}$ using (8.84) and (8.100)

$$t_{1,P} = \frac{2}{2}\left[1 + \frac{50\sin(200\pi t)}{100\sin(100\pi t)}\right]$$

$$t_{2,P} = T_s - t_{1,P}$$

The first sampling period starts at $t = 0$. So:

$$t_{1,P} = 1.8 \text{ ms}, t_{2,P} = 0.2 \text{ ms}$$

The second sampling period starts at $t = 2$ ms. So:

$$t_{1,P} = 1.647 \text{ ms}, t_{2,P} = 0.353 \text{ ms}$$

The third sampling period starts at $t = 4$ ms. So:

$$t_{1,P} = 1.247 \text{ ms}, t_{2,P} = 0.753 \text{ ms}$$

The fourth sampling period starts at $t = 6$ ms. So:

$$t_{1,P} = 0.753 \text{ ms}, t_{2,P} = 1.247 \text{ ms}$$

The fifth sampling period starts at $t = 8$ ms. So:

$$t_{1,P} = 0.353 \text{ ms}, t_{2,P} = 1.647 \text{ ms}$$

The sixth sampling period starts at $t = 10$ ms. So:

$$t_{1,P} = 0.2 \text{ ms}, t_{2,P} = 1.8 \text{ ms}$$

The seventh sampling period starts at $t = 12$ ms. So:

$$t_{1,P} = 0.353 \text{ ms}, t_{2,P} = 1.647 \text{ ms}$$

The eighth sampling period starts at $t = 14$ ms. So:

$$t_{1,P} = 0.753 \text{ ms}, t_{2,P} = 1.247 \text{ ms}$$

The ninth sampling period starts at $t = 16$ ms. So:

$$t_{1,P} = 1.247 \text{ ms}, t_{2,P} = 0.753 \text{ ms}$$

Ultimately, the tenth sampling period starts at $t = 18$ ms. So:

$$t_{1,P} = 1.647 \text{ ms}, t_{2,P} = 0.353 \text{ ms}$$

(b) Regarding the obtained results, waveforms of the input voltage, switches' gate pulses, and the main harmonic of the output voltage can be sketched in Figure 8.22.

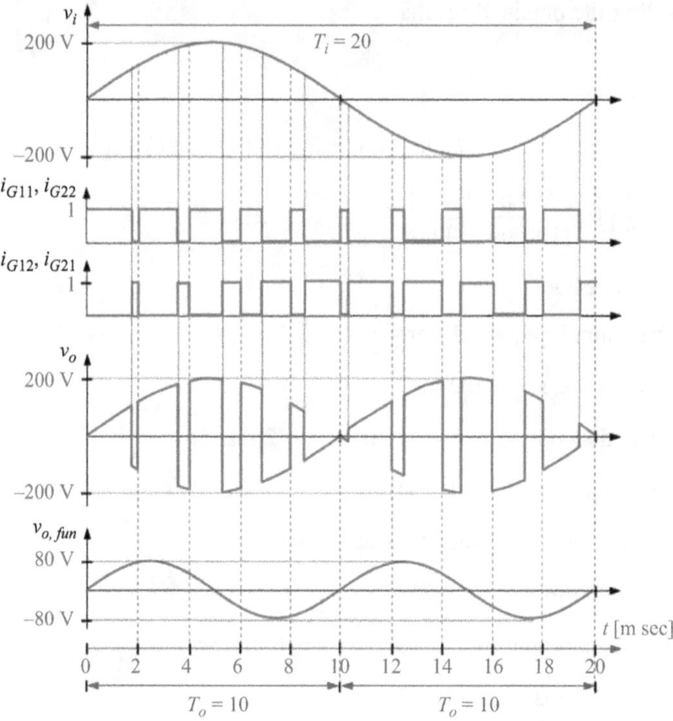

Figure 8.22 Waveforms of the input voltage, the switches' gate pulses, and the output voltage in Example 9.5

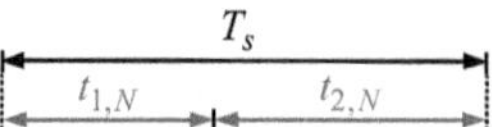

Figure 8.23 Dividing a switching period using the negative control method for the single-phase matrix converter

8.5.2 Negative control method

Using this control method, a switching period will be divided, as Figure 8.23 shows. Besides, the switching pattern and the control pulses using this control method are shown in Figure 8.24.

Regarding Figure 8.24, the following constraints can be obtained:

$$\begin{aligned} &0 \leq t_{1,N} \leq T_s \\ &0 \leq t_{2,N} \leq T_s \end{aligned} \tag{8.101}$$

$$t_{1,N} + t_{2,N} = T_s \tag{8.102}$$

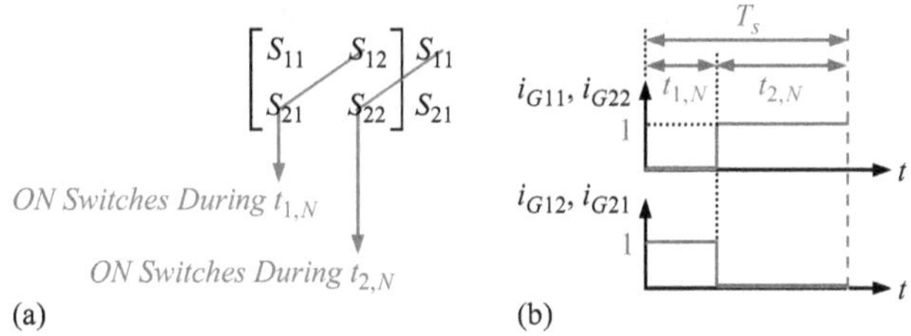

Figure 8.24 Implementing the positive control method on the single-phase matrix converter, (a) switching pattern and (b) gate pulses

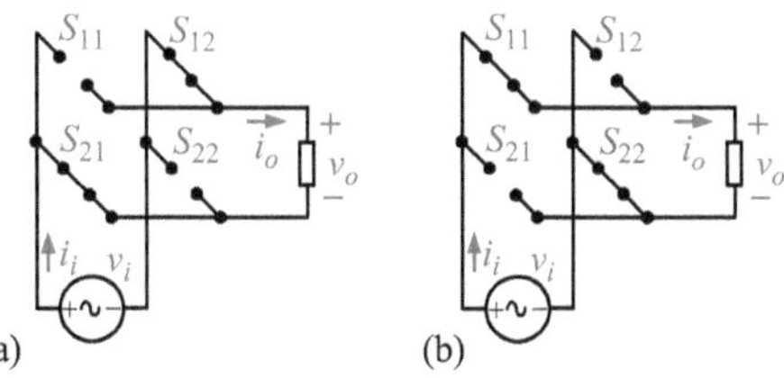

Figure 8.25 Equivalent circuits using the negative control method: (a) in $t_{1,N}$ and (b) in $t_{2,N}$

Regarding the configuration of the single-phase matrix converter, the equivalent circuits can be obtained for each of the operational stages as follows:

By having Figure 8.25, the output voltage can be written as follows:

$$v_o = \begin{cases} -v_i & \text{during } t_{1,N} \\ v_i & \text{during } t_{2,N} \end{cases} \tag{8.103}$$

Regarding (8.103), the average value of the output voltage for a switching period can be obtained from:

$$v_o = \frac{1}{T_s}(-t_{1,N} + t_{2,N})v_i \tag{8.104}$$

As (8.104) shows, if $t_{1,N}$ and $t_{2,N}$ are almost equal, the average value of output voltage will tend to be zero. Now, the duty cycles using the positive control method can be expressed as:

$$\begin{aligned} N_1 &= \frac{t_{1,N}}{T_s} \\ N_2 &= \frac{t_{2,N}}{T_s} \end{aligned} \tag{8.105}$$

Regarding (8.101), (8.102), and (8.105), the following constraints are always true:

$$\begin{aligned} 0 \leq N_1 \leq 1 \\ 0 \leq N_2 \leq 1 \end{aligned} \tag{8.106}$$

$$N_1 + N_2 = 1 \tag{8.107}$$

The following result can be obtained by substituting (8.105) into (8.104):

$$v_o = (-N_1 + N_2)\, v_i \tag{8.108}$$

The unique answers for N_1 and N_2 will obtain as follows if (8.107) and (8.108) are solved simultaneously:

$$\begin{aligned} N_1 &= \frac{1}{2}\left(1 - \frac{v_o}{v_i}\right) \\ N_2 &= \frac{1}{2}\left(1 + \frac{v_o}{v_i}\right) \end{aligned} \tag{8.109}$$

The following results can be obtained by substituting (8.80) and (8.81) into (8.109):

$$\begin{aligned} N_1 &= \frac{1}{2}\left[1 - \frac{V_{o,\max}\sin(\omega_o t)}{V_{i,\max}\sin(\omega_i t)}\right] \\ N_2 &= \frac{1}{2}\left[1 + \frac{V_{o,\max}\sin(\omega_o t)}{V_{i,\max}\sin(\omega_i t)}\right] \end{aligned} \tag{8.110}$$

Regarding the described constraints in (106), the magnitude and frequency of the output voltage must meet (8.105) and (8.107), respectively. In other words, the matrix converter is a buck converter from the voltage perspective; however, it can boost the output frequency.

Besides, regarding (8.108), the relation between the input and output currents can be expressed as:

$$i_o = (-N_1 + N_2) i_i \tag{8.111}$$

Finally, $t_{1,N}$ and $t_{2,N}$ can be obtained by substituting (8.109) into (8.105):

$$\begin{aligned} t_{1,N} &= \frac{T_s}{2}\left[1 - \frac{V_{o,\max}\sin(\omega_o t)}{V_{i,\max}\sin(\omega_i t)}\right] \\ t_{2,N} &= \frac{T_s}{2}\left[1 + \frac{V_{o,\max}\sin(\omega_o t)}{V_{i,\max}\sin(\omega_i t)}\right] \end{aligned} \tag{8.112}$$

8.5.3 Comparison of the positive and negative control methods

Regarding Figure 8.26, (8.100), and (8.112), it can be obtained that for the single-phase matrix converter $t_{1,P} = t_{2,N}$ and $t_{1,P} = t_{2,N}$ are always true. Thus, it can be described that both of these control methods lead to the same result. In other words, the positive control method has no advantages or disadvantages compared to the negative one when they are applied to the single-phase matrix converter. Moreover, as Figure 8.27 shows, the results of applying the negative control method to the single-phase matrix converter can be obtained from the results of applying the positive one to the same converter.

8.5.4 Combined control method

The combined control method is a combination of the positive and negative ones. In this control method, each sampling period is divided into two periods of αT_s and βT_s as Figure 8.27 shows. The positive and negative control methods are implemented in the periods of αT_s and βT_s, respectively.

The gate signals obtained from this control method are depicted in Figure 8.28. Considering that both positive and negative control methods give the same result

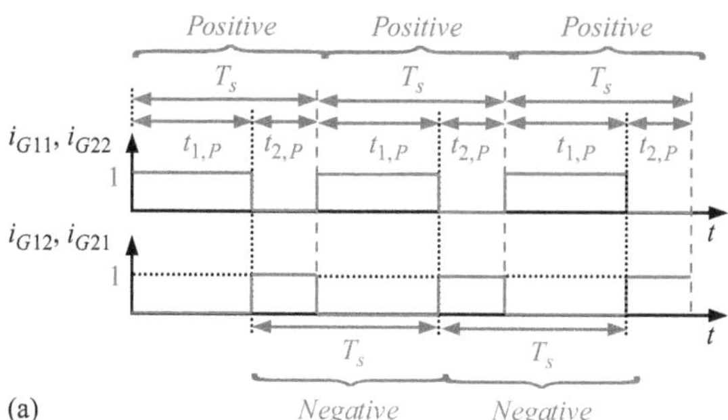

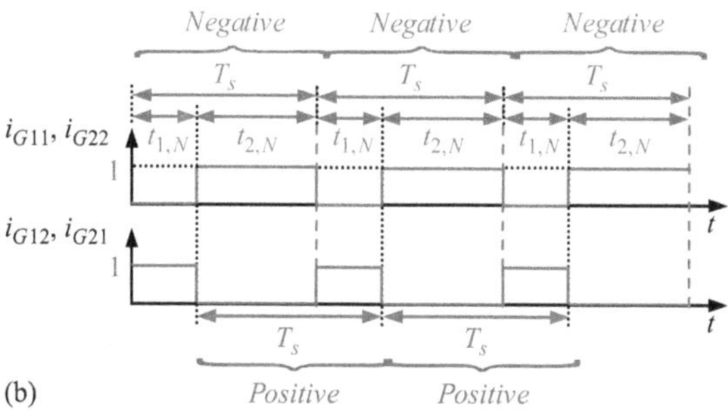

Figure 8.26 Comparison of the positive and negative control methods: (a) obtaining the negative control method from the positive one and (b) obtaining the positive control method from the negative one

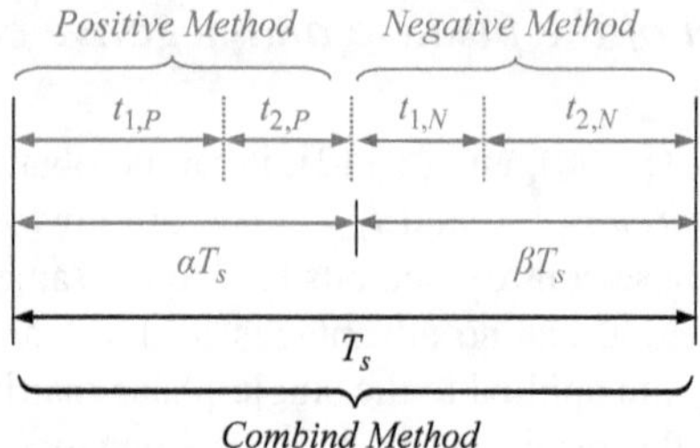

Figure 8.27 Using the combined control method on the single-phase matrix converter and dividing a sampling time into two periods

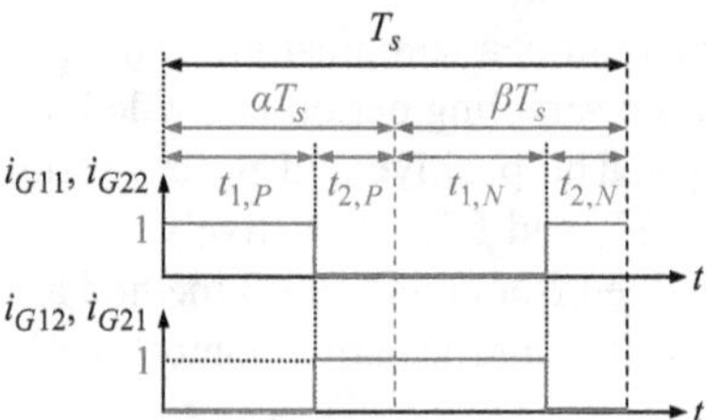

Figure 8.28 Gate signals using the combined control method

for the single-phase matrix converter and the combined control method is a combination of these methods, it is expected that the combined one gives the same result, as well. Figure 8.28 shows the result of this exception.

From Figure 8.27, we can have:

$$\alpha T_s + \beta T_s = T_s \tag{8.113}$$

So:

$$\alpha + \beta = 1 \tag{8.114}$$

In this control method, the output voltage can be obtained as:

$$v_o = \alpha(P_1 - P_2)v_i + \beta(-N_1 + N_2)v_i \tag{8.115}$$

Similarly, we can write:

$$i_i = \alpha(P_1 - P_2)i_o + \beta(-N_1 + N_2)i_o \tag{8.116}$$

In order to obtain the voltage and current formulas, we can suppose the output load as an ohmic-inductive kind, as Figure 8.30 demonstrates. By having this assumption, we have:

$$v_o = v_R + v_L \tag{8.117}$$

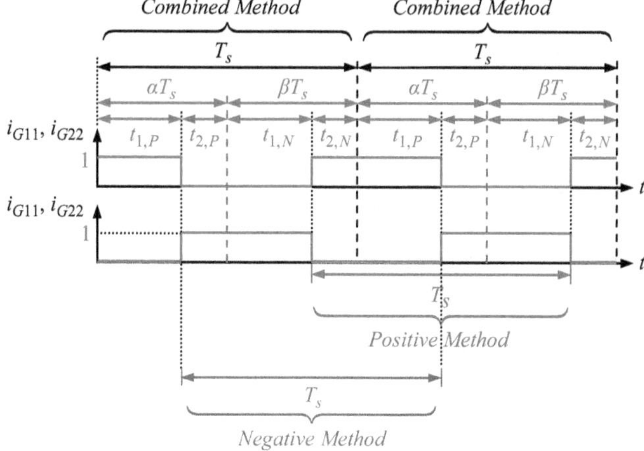

Figure 8.29 Comparing the combined control methods with the positive and negative ones

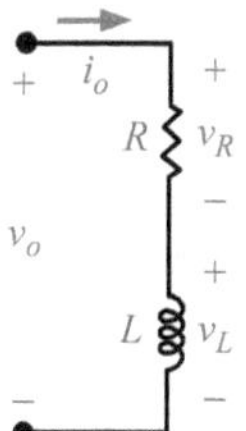

Figure 8.30 The ohmic-inductive load

So:

$$v_o = R\, i_o + L\frac{di_o}{dt} \tag{8.118}$$

The main harmonic of the output current can be calculated as follows by substituting (8.81) in (8.118):

$$i_o = \frac{V_{o,\max}}{\sqrt{R^2 + (L\omega_o)^2}} \sin\left[\omega_o t - \tan^{-1}\left(\frac{L\omega_o}{R}\right)\right] \tag{8.119}$$

Eq. (8.119) can be rewritten as:

$$i_o = I_{o,\max} \sin\left(\omega_o t + \phi_o\right) \tag{8.120}$$

where:

$$I_{o,\max} = \frac{V_{o,\max}}{\sqrt{R^2 + (L\omega_o)^2}} \tag{8.121}$$

$$\phi_o = -\tan^{-1}\left(\frac{L\omega_o}{R}\right) \tag{8.122}$$

Regarding (8.99), (8.111), (8.120), the instantaneous input current by using positive, negative, and combined control methods can be obtained as follows:

$$i_i = (P_1 - P_2)I_{o,\max}\sin(\omega_o t + \phi_o) \tag{8.123}$$

$$i_i = (-N_1 + N_2)I_{o,\max}\sin(\omega_o t + \phi_o) \tag{8.124}$$

$$i_i = \alpha(P_1 - P_2)I_{o,\max}\sin(\omega_o t + \phi_o) + \beta(-N_1 + N_2)I_{o,\max}\sin(\omega_o t + \phi_o) \tag{8.125}$$

Example 9.6 Consider a single-phase matrix converter with an input voltage of $v_i = 220\sqrt{2}\sin(100\pi t)$. It is controlled using the positive control method to synthesize an output voltage of $v_o = 50\sqrt{2}\sin(200\pi t)$. Assume the switching frequency as 5 kHz and the output load as an ohmic-inductive one with $R = 40\Omega$ and $L = 55$ mH. Obtain the main harmonic of the output current.

Solution: Regarding (8.117), we can write:

$$i_o = \frac{50\sqrt{2}}{\sqrt{40^2 + (0.055 \times 200\pi)^2}}\sin\left[200\pi t - \tan^{-1}\left(\frac{0.055 \times 200\pi}{40}\right)\right]$$

If this equation is simplified, it will result in that:

$$i_o = 1.34\sin(200\pi t - 40.82^\circ)$$

8.6 Implementation of positive, negative, and combined control methods on three-phase matrix converter

The configuration of the three-phase matrix converter is depicted in Figure 8.4(a). Regarding this figure, the switches' matrix can be obtained as:

$$S = \begin{bmatrix} S_{Aa} & S_{Ba} & S_{Ca} \\ S_{Ab} & S_{Bb} & S_{Cb} \\ S_{Ac} & S_{Bc} & S_{Cc} \end{bmatrix} \tag{8.126}$$

Since the switches' matrix is square, the input currents will always be continuous. Moreover, as Figure 8.31 shows, the ith sampling period is divided into three periods of t_1^i, t_2^i, and t_3^i.

The following subsections will be given the results of implementing each of the positive, negative, and combined control methods on this topology.

8.6.1 Positive control method

The switching pattern and the gate pulse signals can be shown in Figures 8.32 and 8.33 using the positive control on the three-phase matrix converter.

The average values of the output voltages can be written as follows:

$$\begin{bmatrix} v_a(t) \\ v_b(t) \\ v_c(t) \end{bmatrix} = \begin{bmatrix} v_A(t) & v_B(t) & v_C(t) \\ v_B(t) & v_C(t) & v_A(t) \\ v_C(t) & v_A(t) & v_B(t) \end{bmatrix} \begin{bmatrix} P_1 \\ P_2 \\ P_3 \end{bmatrix} \tag{8.127}$$

In other words,

$$[V_o(t)]_{3\times1} = \left[V_{i,P}(t)\right]_{3\times3} \begin{bmatrix} P_1 \\ P_2 \\ P_3 \end{bmatrix} \tag{8.128}$$

8.6.2 Negative control method

The switching pattern and the gate pulse signals can be shown in Figures 8.34 and 8.35 using the negative control on the three-phase matrix converter.

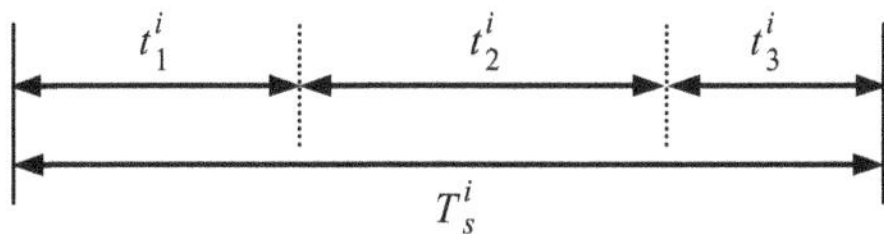

Figure 8.31 Dividing a sampling period into different parts for the three-phase matrix converter

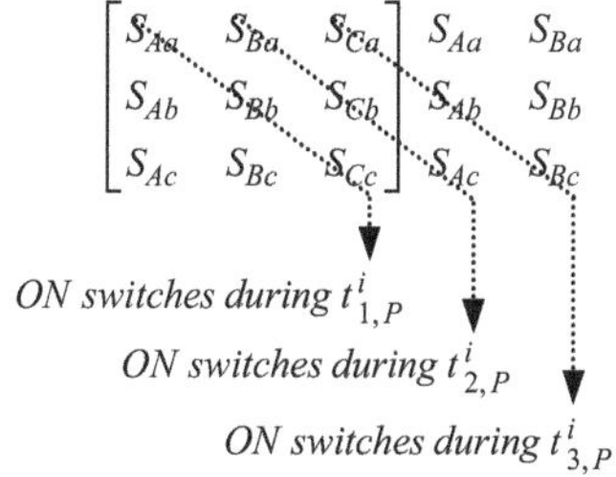

Figure 8.32 Implementing the positive control method on the three-phase matrix converter

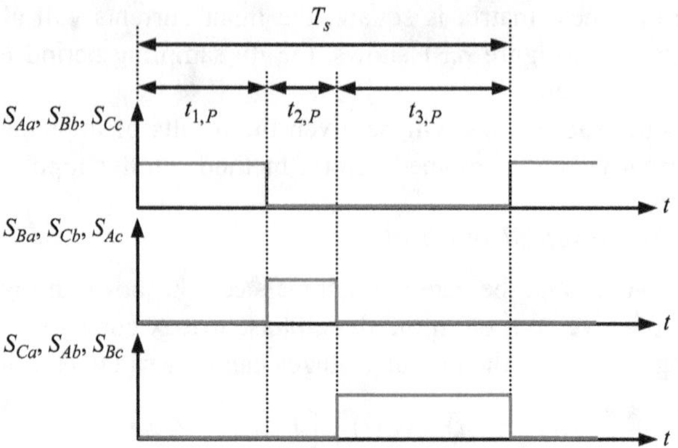

Figure 8.33 Gate pulse signals obtained from implementing the positive control method on the three-phase matrix converter

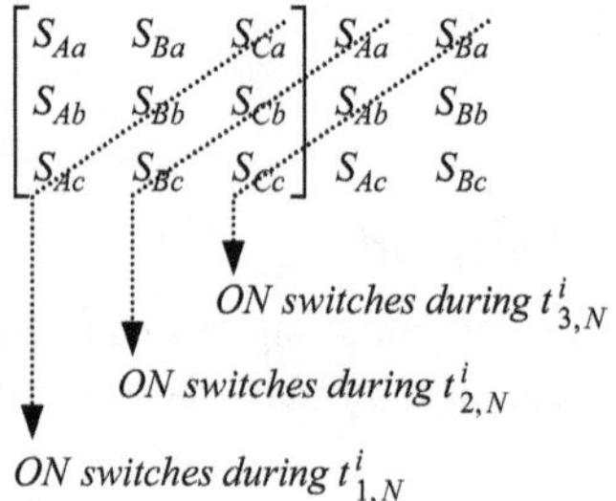

Figure 8.34 Implementing the negative control method on the three-phase matrix converter

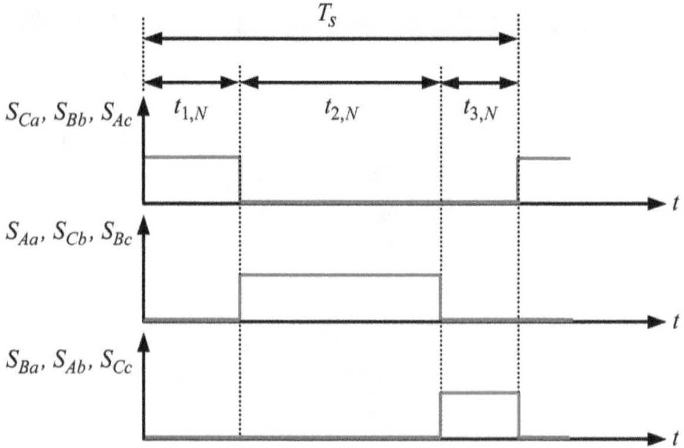

Figure 8.35 Gate pulse signals obtained from implementing the negative control method on the three-phase matrix converter

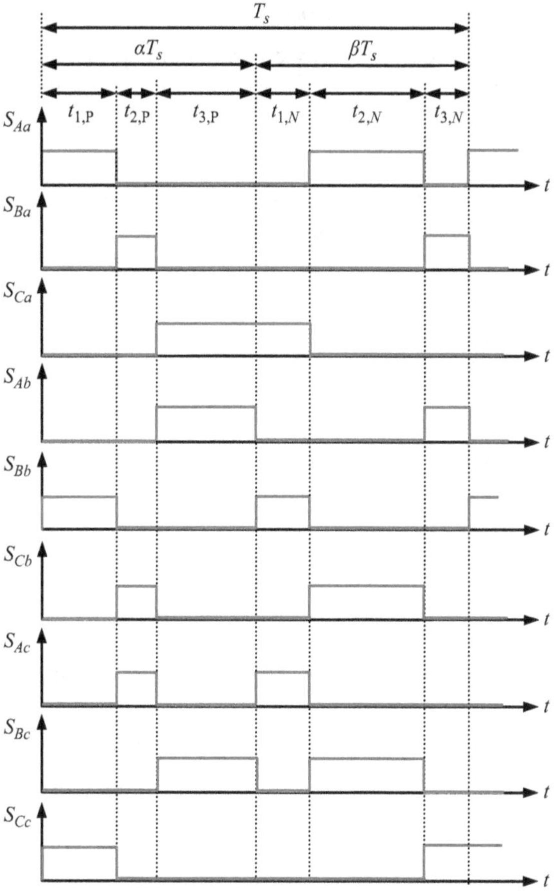

Figure 8.36 Obtained gate signals by implementing the combined control method on the three-phase matrix converter

The average values of the output voltages can be written as follows:

$$\begin{bmatrix} v_a(t) \\ v_b(t) \\ v_c(t) \end{bmatrix} = \begin{bmatrix} v_C(t) & v_A(t) & v_B(t) \\ v_B(t) & v_C(t) & v_A(t) \\ v_A(t) & v_B(t) & v_C(t) \end{bmatrix} \begin{bmatrix} N_1 \\ N_2 \\ N_3 \end{bmatrix} \tag{8.129}$$

In other words,

$$[V_o(t)]_{3\times 1} = \left[V_{i,N}(t)\right]_{3\times 3} \begin{bmatrix} N_1 \\ N_2 \\ N_3 \end{bmatrix} \tag{8.130}$$

8.6.3 Combined control method

The output voltages can be obtained by combining the previous results obtained in (8.127)–(8.130). Figure 8.36 illustrates the gate signals using the combined control method.

8.6.4 Implementing the positive control method with the balanced inputs and outputs

The input voltages can be assumed as follows in the balanced condition:

$$[V_i(t)]_{3\times1} = \begin{bmatrix} v_A(t) \\ v_B(t) \\ v_C(t) \end{bmatrix} = V_{i,\max}\begin{bmatrix} \cos(\omega_i t) \\ \cos(\omega_i t - 120^\circ) \\ \cos(\omega_i t + 120^\circ) \end{bmatrix} \tag{8.131}$$

The main harmonic of the output voltage will be as follows if the output voltages are assumed to be balanced:

$$[V_o(t)]_{3\times1} = \begin{bmatrix} v_a(t) \\ v_b(t) \\ v_c(t) \end{bmatrix} = V_{o,\max}\begin{bmatrix} \cos(\omega_o t) \\ \cos(\omega_o t - 120^\circ) \\ \cos(\omega_o t + 120^\circ) \end{bmatrix} \tag{8.132}$$

In this condition, the input and output currents can be defined as:

$$[I_i(t)]_{3\times1} = \begin{bmatrix} i_A(t) \\ i_B(t) \\ i_C(t) \end{bmatrix} = I_{i,\max}\begin{bmatrix} \cos(\omega_i t + \phi_i) \\ \cos(\omega_i t + \phi_i - 120^\circ) \\ \cos(\omega_i t + \phi_i + 120^\circ) \end{bmatrix} \tag{8.133}$$

$$[I_o(t)]_{3\times1} = \begin{bmatrix} i_a(t) \\ i_b(t) \\ i_c(t) \end{bmatrix} = I_{o,\max}\begin{bmatrix} \cos(\omega_o t + \phi_o) \\ \cos(\omega_o t + \phi_o - 120^\circ) \\ \cos(\omega_o t + \phi_o + 120^\circ) \end{bmatrix} \tag{8.134}$$

where ϕ_i and ϕ_o stand for the phase shift between the input voltage with current and output voltage with current, respectively. The duty cycles of P_1, P_2, and P_3 can be obtained as follows by substituting (8.131) and (8.132) into (8.127):

$$P_n = \frac{1}{3} + \frac{2}{3}\frac{V_{o,\max}}{V_{i,\max}}\cos\left[(\omega_i - \omega_o)t - (n-1)\frac{2\pi}{3}\right] \quad n = 1,\ 2,\ 3 \tag{8.135}$$

The following limitation can be obtained regarding the constraints of P_1, P_2, and P_3:

$$\frac{V_{o,\max}}{V_{i,\max}} \leq \frac{1}{2} \tag{8.136}$$

Ultimately, the output voltages can be obtained as:

$$\begin{bmatrix} v_a \\ v_b \\ v_c \end{bmatrix} = \begin{bmatrix} P_1 & P_2 & P_3 \\ P_3 & P_1 & P_2 \\ P_2 & P_3 & P_1 \end{bmatrix} \begin{bmatrix} v_A \\ v_B \\ v_C \end{bmatrix} \tag{8.137}$$

Regarding the relation between the input voltage and the current, the input currents can be obtained as follows:

$$\begin{bmatrix} i_A \\ i_B \\ i_C \end{bmatrix} = \begin{bmatrix} P_1 & P_3 & P_2 \\ P_2 & P_1 & P_3 \\ P_3 & P_2 & P_1 \end{bmatrix} \begin{bmatrix} i_a \\ i_b \\ i_c \end{bmatrix} \tag{8.138}$$

The following result can be obtained by substituting P_1, P_2, and P_3 in (8.137) and (8.138) and equal input and output powers:

$$\phi_i = \phi_o \tag{8.139}$$

Eq. (8.139) shows that the phase shift between input and output voltages and currents is positive using the positive control method.

8.6.5 Implementing the negative control method with the balanced inputs and outputs

The duty cycles of N_1, N_2, and N_3 can be obtained as follows by substituting (8.131) and (8.132) into (8.129):

$$N_n = \frac{1}{3} + \frac{2}{3}\frac{V_{o,\max}}{V_{i,\max}} \cos\left[(\omega_i + \omega_o)t - (n-2)\frac{2\pi}{3}\right] \qquad n = 1,\ 2,\ 3 \tag{8.140}$$

The same limitation as (8.136) can be obtained regarding the constraints of N_1, N_2, and N_3. Ultimately, the output voltages can be obtained as:

$$\begin{bmatrix} v_a \\ v_b \\ v_c \end{bmatrix} = \begin{bmatrix} N_2 & N_3 & N_1 \\ N_3 & N_1 & N_2 \\ N_1 & N_2 & N_3 \end{bmatrix} \begin{bmatrix} v_A \\ v_B \\ v_C \end{bmatrix} \tag{8.141}$$

Regarding the relation between the input voltage and the current, the input currents can be obtained as follows:

$$\begin{bmatrix} i_A \\ i_B \\ i_C \end{bmatrix} = \begin{bmatrix} N_2 & N_3 & N_1 \\ N_3 & N_1 & N_2 \\ N_1 & N_2 & N_3 \end{bmatrix} \begin{bmatrix} i_a \\ i_b \\ i_c \end{bmatrix} \tag{8.142}$$

The following result can be obtained by substituting N_1, N_2, and N_3 in (8.131) and (8.132) and equal input and output powers:

$$\phi_i = -\phi_o \tag{8.143}$$

Eq. (8.143) shows that the phase shifts between input and output voltages and currents are opposite using the negative control method.

8.6.6 Implementing the combined control method with the balanced inputs and outputs

Either positive or negative control methods can synthesize the output voltages. The only difference between these control methods is that for the positive and negative control methods, we have $\phi_i = \phi_o$ and $\phi_i = -\phi_o$, respectively. The output voltage using the combined control method can be obtained as:

$$\begin{bmatrix} v_a \\ v_b \\ v_c \end{bmatrix} = \alpha \begin{bmatrix} P_1 & P_2 & P_3 \\ P_3 & P_1 & P_2 \\ P_2 & P_3 & P_1 \end{bmatrix} \begin{bmatrix} v_A \\ v_B \\ v_C \end{bmatrix} + \beta \begin{bmatrix} N_2 & N_3 & N_1 \\ N_3 & N_1 & N_2 \\ N_1 & N_2 & N_3 \end{bmatrix} \begin{bmatrix} v_A \\ v_B \\ v_C \end{bmatrix} \tag{8.144}$$

Besides, for the input current:

$$\begin{bmatrix} i_A \\ i_B \\ i_C \end{bmatrix} = \alpha \begin{bmatrix} P_1 & P_3 & P_2 \\ P_2 & P_1 & P_3 \\ P_3 & P_2 & P_1 \end{bmatrix} \begin{bmatrix} i_a \\ i_b \\ i_c \end{bmatrix} + \beta \begin{bmatrix} N_2 & N_3 & N_1 \\ N_3 & N_1 & N_2 \\ N_1 & N_2 & N_3 \end{bmatrix} \begin{bmatrix} i_a \\ i_b \\ i_c \end{bmatrix} \tag{8.145}$$

where P_1, P_2, and P_3 can be obtained from (8.135) and N_1, N_2, and N_3 can be calculated using (8.140). The following equation can be obtained by supposing the equal input and output voltages and substituting P_1, P_2, P_3, N_1, N_2, and N_3 in (8.144) and (8.145):

$$\tan \phi_i = (2\alpha - 1)\tan \phi_o \tag{8.146}$$

The input phase shift can be obtained as follows using (8.146) and regarding different α:

$$\phi_i = \begin{cases} \phi_o & \text{for } \alpha = 1 \\ 0 & \text{for } \alpha = \dfrac{1}{2} \\ -\phi_o & \text{for } \alpha = 0 \end{cases} \tag{8.147}$$

In order to utilize the three-phase matrix inverter with a power factor of one, the following limitation must be true:

$$\alpha = \beta = 0.5 \tag{8.148}$$

Regarding (8.148), it can be seen that a power factor of one can be obtained by using both positive and negative control methods. The positive and negative control methods are implemented in the first and second segments for the ith sampling period. Also, it is obtained that, unlike the conventional control methods, there is no need to control and measure the input currents in order to have an operation with a power factor of one.

8.6.7 Implementing the positive control method with the unbalanced outputs

It is supposed that the output voltage is as follows:

$$[V_o(t)]_{3\times 1} = \begin{bmatrix} v_a(t) \\ v_b(t) \\ v_c(t) \end{bmatrix} = \begin{bmatrix} V_{a,\max}\cos(\omega_o t) \\ V_{b,\max}\cos(\omega_o t - 120°) \\ V_{c,\max}\cos(\omega_o t + 120°) \end{bmatrix}$$

$$= V_{o,\max}\begin{bmatrix} \cos(\omega_o t) \\ (1+\lambda_b)\cos(\omega_o t - 120°) \\ (1+\lambda_c)\cos(\omega_o t + 120°) \end{bmatrix} \tag{8.149}$$

where λ_b and λ_c stand for the degrees of unbalancing in two phases of the output voltage. The output current can be expressed as follows for this kind of operation:

$$\begin{aligned} i_a &= I_{a,\max}\cos(\omega_o t + \phi_a) \\ i_b &= I_{b,\max}\cos(\omega_o t + \phi_b - 120°) \\ i_c &= I_{c,\max}\cos(\omega_o t + \phi_c + 120°) \end{aligned} \tag{8.150}$$

The duty cycles of P_1, P_2, and P_3 can be obtained as follows by supposing (8.131) and (8.149) into (8.127):

$$\begin{aligned} P_1 &= \frac{1}{3} + \frac{2\sqrt{3}}{9}\frac{V_{b,max}}{V_{i,\max}}\{\cos[(\omega_i - \omega_o)t - 30°] - \sin(\omega_i + \omega_o)t\} \\ &\quad + \frac{2\sqrt{3}}{9}\frac{V_{c,\max}}{V_{i,\max}}\{\sin(\omega_i + \omega_o)t + \cos[(\omega_i - \omega_o)t + 30°]\} \\ P_2 &= \frac{1}{3} + \frac{2\sqrt{3}}{9}\frac{V_{b,\max}}{V_{i,\max}}\{\cos[(\omega_i + \omega_o)t - 30°] - \cos[(\omega_i - \omega_o)t + 30°]\} \\ &\quad + \frac{2\sqrt{3}}{9}\frac{V_{c,\max}}{V_{i,\max}}\{\sin(\omega_i - \omega_o)t - \cos[(\omega_i + \omega_o)t - 30°]\} \\ P_3 &= \frac{1}{3} + \frac{2\sqrt{3}}{9}\frac{V_{b,\max}}{V_{i,\max}}\{-\cos[(\omega_i + \omega_o)t + 30°] - \sin(\omega_i - \omega_o)t\} \\ &\quad + \frac{2\sqrt{3}}{9}\frac{V_{c,\max}}{V_{i,\max}}\{\cos[(\omega_i + \omega_o)t + 30°] - \cos[(\omega_i - \omega_o)t - 30°]\} \end{aligned} \tag{8.151}$$

If the output voltages are supposed to be balanced, in other words, λ_b and λ_c equal to zero, (8.151) will be the same as (8.135). The following constraint can be obtained regarding the limitation of P_1, P_2, and P_3:

$$\sqrt{q_b{}^2 + q_c{}^2 + q_b q_c} + |q_b - q_c| \le \frac{\sqrt{3}}{2} \tag{8.152}$$

where

$$\begin{aligned} q_b &= \frac{V_{b,\max}}{V_{i,\max}} \\ q_c &= \frac{V_{c,\max}}{V_{i,\max}} \end{aligned} \tag{8.153}$$

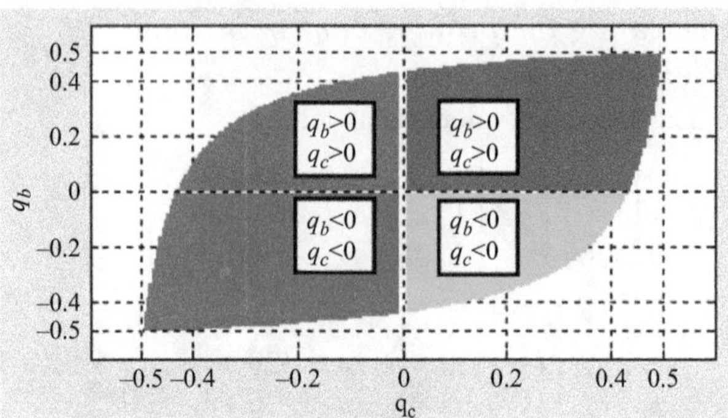

Figure 8.37 Variation of the output voltage gain in the three-phase matrix converter

Figure 8.37 illustrates the variation of q_b and q_c. It is required to note that some unbalancing in the phases of the output voltages can be applied by choosing negative q_b and q_c. Besides, if λ_b and λ_c equal to zero, (8.152) will be the same as (8.136). The output voltages and input currents can be obtained using (8.137) and (8.138), respectively. The main harmonic of the input current of the phase A can be obtained as follows using (8.138), (8.150), and (8.151):

$$
\begin{aligned}
i_A(t) =& \frac{1}{3}\{I_{a,\max}\cos(\omega_o t+\phi_a)+I_{b,\max}\cos(\omega_o t+\phi_b-120^\circ)+I_{c,\max}\cos(\omega_o t+\phi_c+120^\circ)\}\\
&+\frac{\sqrt{3}}{9}\frac{V_{b,\max}}{V_{i,\max}}I_{a,\max}\{\cos(\omega_i t+\phi_a-30^\circ)+\cos[(\omega_i-2\omega_o)t-\phi_a-30^\circ]\\
&\qquad -\sin[(\omega_i+2\omega_o)t+\phi_a]-\sin(\omega_i t-\phi_a)\}\\
&+\frac{\sqrt{3}}{9}\frac{V_{c,\max}}{V_{i,\max}}I_{a,\max}\{\sin[(\omega_i+2\omega_o)t+\phi_a]+\sin(\omega_i t-\phi_a)+\cos(\omega_i t+\phi_a+30^\circ)\\
&\qquad +\cos[(\omega_i-2\omega_o)t-\phi_a+30^\circ]\}\\
&+\frac{\sqrt{3}}{9}\frac{V_{b,\max}}{V_{i,\max}}I_{b,\max}\{-\sin[(\omega_i+2\omega_o)t+\phi_b]+\cos(\omega_i t-\phi_b-30^\circ)\\
&\qquad +\cos(\omega_i t+\phi_b-30^\circ)-\cos[(\omega_i-2\omega_o)t-\phi_b+30^\circ]\}\\
&+\frac{\sqrt{3}}{9}\frac{V_{c,\max}}{V_{i,\max}}I_{b,\max}\{\sin[(\omega_i+2\omega_o)t+\phi_b]-\cos(\omega_i t-\phi_b-30^\circ)\\
&\qquad +\cos(\omega_i t+\phi_b+30^\circ)+\sin[(\omega_i-2\omega_o)t-\phi_b]\}\\
&+\frac{\sqrt{3}}{9}\frac{V_{b,\max}}{V_{i,\max}}I_{c,\max}\{-\sin[(\omega_i+2\omega_o)t+\phi_c]-\cos(\omega_i t-\phi_c+30^\circ)\\
&\qquad +\cos(\omega_i t+\phi_c-30^\circ)-\sin[(\omega_i-2\omega_o)t-\phi_c]\}\\
&+\frac{\sqrt{3}}{9}\frac{V_{c,\max}}{V_{i,\max}}I_{c,\max}\{\cos(\omega_i t+\phi_c+30^\circ)-\cos[(\omega_i-2\omega_o)t-\phi_c-30^\circ]\\
&\qquad +\sin[(\omega_i+2\omega_o)t+\phi_c]+\cos(\omega_i t-\phi_c+30^\circ)\}
\end{aligned}
\tag{8.154}
$$

As seen in (8.154), the input currents include harmonics of ω_i, ω_o, $\omega_i-2\omega_o$, and $\omega_i+2\omega_o$.

8.6.8 Implementing the negative control method with the unbalanced outputs

The duty cycles of N_1, N_2, and N_3 can be obtained as follows by supposing (8.131) and (8.149) into (8.129):

$$\begin{aligned}
N_1 &= \frac{1}{3} + \frac{2\sqrt{3}}{9}\frac{V_{b,\max}}{V_{i,\max}}\{\cos[(\omega_i-\omega_o)t+30^\circ] - \cos[(\omega_i+\omega_o)t-30^\circ]\} \\
&\quad + \frac{2\sqrt{3}}{9}\frac{V_{c,\max}}{V_{i,\max}}\{-\sin(\omega_i+\omega_o)t - \cos[(\omega_i-\omega_o)t+30^\circ]\} \\
N_2 &= \frac{1}{3} + \frac{2\sqrt{3}}{9}\frac{V_{b,\max}}{V_{i,\max}}\{\cos[(\omega_i+\omega_o)t+30^\circ] + \sin(\omega_i-\omega_o)t\} \\
&\quad + \frac{2\sqrt{3}}{9}\frac{V_{c,\max}}{V_{i,\max}}\{\cos[(\omega_i+\omega_o)t-30^\circ] - \sin(\omega_i-\omega_o)t\} \\
N_3 &= \frac{1}{3} + \frac{2\sqrt{3}}{9}\frac{V_{b,\max}}{V_{i,\max}}\{\sin(\omega_i+\omega_o)t - \cos[(\omega_i-\omega_o)t-30^\circ]\} \\
&\quad + \frac{2\sqrt{3}}{9}\frac{V_{c,\max}}{V_{i,\max}}\{\cos[(\omega_i-\omega_o)t-30^\circ] - \cos[(\omega_i+\omega_o)t+30^\circ]\}
\end{aligned} \tag{8.155}$$

The same constraint as (8.152) can be obtained regarding the limitation of N_1, N_2, and N_3. The output voltages and input currents can be obtained using (8.141) and (8.142), respectively. The main harmonic of the input current of the phase A can be obtained as follows using (8.142), (8.150), and (8.155):

$$\begin{aligned}
i_A(t) &= \frac{1}{3}\{I_{a,\max}\cos(\omega_o t+\phi_a) + I_{b,\max}\cos(\omega_o t+\phi_b-120^\circ) + I_{c,\max}\cos(\omega_o t+\phi_c+120^\circ)\} \\
&\quad + \frac{\sqrt{3}}{9}\frac{V_{b,\max}}{V_{i,\max}}I_{a,\max}\{\cos[(\omega_i+2\omega_o)t+\phi_a+30^\circ] + \cos(\omega_i t-\phi_a+30^\circ) \\
&\qquad + \sin(\omega_i t+\phi_a) + \sin[(\omega_i-2\omega_o)t-\phi_a]\} \\
&\quad + \frac{\sqrt{3}}{9}\frac{V_{c,\max}}{V_{i,\max}}I_{a,\max}\{\cos[(\omega_i+2\omega_o)t+\phi_a-30^\circ] + \cos(\omega_i t-\phi_a-30^\circ) - \sin(\omega_i t+\phi_a) \\
&\qquad - \sin[(\omega_i-2\omega_o)t-\phi_a]\} \\
&\quad + \frac{\sqrt{3}}{9}\frac{V_{b,\max}}{V_{i,\max}}I_{b,\max}\{-\cos[(\omega_i+2\omega_o)t+\phi_b-30^\circ] + \cos(\omega_i t-\phi_b+30^\circ) \\
&\qquad + \cos(\omega_i t+\phi_b+30^\circ) + \sin[(\omega_i-2\omega_o)t-\phi_b]\} \\
&\quad + \frac{\sqrt{3}}{9}\frac{V_{c,\max}}{V_{i,\max}}I_{b,\max}\{-\cos(\omega_i t+\phi_b+30^\circ) - \sin[(\omega_i-2\omega_o)t-\phi_b] \\
&\qquad - \sin[(\omega_i+2\omega_o)t+\phi_b] + \cos(\omega_i t-\phi_b-30^\circ)\} \\
&\quad + \frac{\sqrt{3}}{9}\frac{V_{b,\max}}{V_{i,\max}}I_{c,\max}\{-\cos(\omega_i t+\phi_c-30^\circ) + \sin[(\omega_i-2\omega_o)t-\phi_c] \\
&\qquad + \sin[(\omega_i+2\omega_o)t+\phi_c] + \cos(\omega_i t-\phi_c+30^\circ)\} \\
&\quad + \frac{2\sqrt{3}}{9}\frac{V_{c,\max}}{V_{i,\max}}I_{c,\max}\{-\cos[(\omega_i+2\omega_o)t+\phi_c+30^\circ] + \cos(\omega_i t-\phi_c-30^\circ) \\
&\qquad + \cos(\omega_i t+\phi_c-30^\circ) - \sin[(\omega_i-2\omega_o)t-\phi_c]\}
\end{aligned} \tag{8.156}$$

As seen in (8.154), the input currents include harmonics of ω_i, ω_o, $\omega_i-2\omega_o$, and $\omega_i+2\omega_o$.

8.6.9 Implementing the combined control method with the unbalanced outputs

The output voltages and input currents can be obtained using the combined control method from (8.144) and (8.145), where duty cycles of P_1, P_2, P_3, N_1, N_2, and N_3 can be obtained from (8.151) and (8.155).

8.7 Conclusion

In this chapter, first, some applications for the N-phase to m-phase matrix converter were expressed. Then, the procedure of modeling matrix converters was described. After, three different control methods of the positive, negative, and combined, based on the PWM techniques, were proposed. These control methods can synthesize the desired even if there are unbalances at the input or output sides. If the switching frequency is chosen so higher than the input and output frequencies, the generated harmonics will be almost around the switching frequency. So they can be eliminated by using a low-pass filter.

8.8 Future directions

Due to the interesting features of the matrix converters, including bidirectional power flow, not using any passive devices as energy storage elements, sinusoidal current waveforms on the input and output sides, and the possibility of controlling the power factor on the input side, many publications using these converters in different applications have been reported in the literature. Some applications, such as driving multi-phase motors, military, and aircraft systems, are some of the applications in which this sort of power electronic converters can be used.

References

[1] E. Babaei, S.H. Hosseini, G.B. Gharehpetian, and M. Sabahi, "Development of modulation strategies for 3-phase to 2-phase matrix converters," *Int. J. Power Electron.*, vol. 2, no. 1, pp. 82–106, 2010.

[2] E. Babaei, "Control of direct three-phase to single-phase converters under balanced and unbalanced operations," *Energy Convers. Manag.*, vol. 52, no. 1, pp. 66–74, 2011.

[3] E. Babaei, S.H. Hosseini, and G.B. Gharehpetian, "Reduction of THD and low order harmonics with symmetrical output current for single-phase ac/ac matrix converters," *Int. J. Electric. Power Energy Syst.*, vol. 32, no. 3, pp. 225–235, 2010.

Chapter 9

Modelling, simulation and validation of average current and constant voltage operations in non-ideal buck and boost converters

Sumukh Surya[1], *S. Mohan Krishna*[2] *and Sheldon Williamson*[3]

DC–DC converters play a major role in a various applications in automobile engineering, portable electronics and LED drivers. In this work, basic converters like Buck and Boost converters operating in continuous conduction mode (CCM) considering the non-ideal parameters are modelled using volt-sec and amp-sec balance equations. The equations were simulated using MATLAB®/Simulink® software using appropriate step time, solver and the transients in inductor current and capacitor voltage were observed. Later, using the state space averaging (SSA) technique the transfer function of inductor current to duty ratio (G_{id}) and output voltage to duty ratio (G_{vd}) were derived. The parameters like low-frequency gain, gain margin (GM), phase margin (PM), crossover frequency, and stability were analysed using MATLAB software. It was found the non-ideal boost converter showed instability under constant voltage operation due to the presence of right half plane (RHP) zero. In order to validate the obtained transfer function using SSA, a new control technique called circuit averaging technique was used. The validation was performed using LTspice software tool. The frequency response of G_{id} and G_{vd} obtained using MATLAB and LTspice software tools showed a perfect match.

9.1 Introduction

The power electronic modulators used for low- and high-frequency switching applications are highly non-linear and complex. Modelling the steady-state and

[1]Bosch Global Software Technologies Private Limited (BGSW), India
[2]Department of Electrical and Electronics, Alliance University, India
[3]Department of Electrical, Computer and Software Engineering, Ontario Tech University, Canada

dynamic behaviour of these systems present a challenging task. State-space modelling simplifies and aids the formulation of estimation and control algorithms for these switching converters. There are several control strategies considered for converters used for low-power applications, charging electric vehicles (EVs) and power factor correction (PFC). Switching converters like buck, boost and buck–boost topologies are popularly used for the above applications. Researchers have focused on reducing the number of switches, optimising the control strategy as well as modelling these topologies [1,2] for real-time control. Due to the non-linear and time variant nature of these switching converters (mainly due to high switching, power devices, and passive elements like inductors, and capacitors), the traditional linear control techniques would not be sufficient for analysis. The dynamic model [3] of the switching converter needs to be taken into consideration for designing a suitable non-linear control strategy to achieve desired performance [4]. There is also a need to ensure the operating power factor of these converters is maintained near to unity for the purpose of power factor correction [5]. Figure 9.1 illustrates the different control methods to ensure the same.

DC–DC converter current control is categorised into Average Current Control (ACC) and Peak Current Mode (PCM) controls. In the former, the variation of the inductor current perturbation with respect to the duty ratio perturbation is taken into account. ACC is more popular owing to its improved immunity to noise and wide range of control. Constant frequency control techniques have also been reported in many works, their main objective being reduced variation in the switching frequency and less recovery time for a load transient [6]. The SSA methods have been commonly used for small signal modelling of DC–DC converters for the purpose of analysing stability [7]. From the stability perspective, many techniques exist in the literature for estimating the open-loop transfer function namely, small signal modelling, circuit averaging and SSA [8]. In [9], a generalised circuit averaging technique is proposed for Cuk and SEPIC converters which are primarily used for charging batteries. It is used to determine the frequency response of open-loop

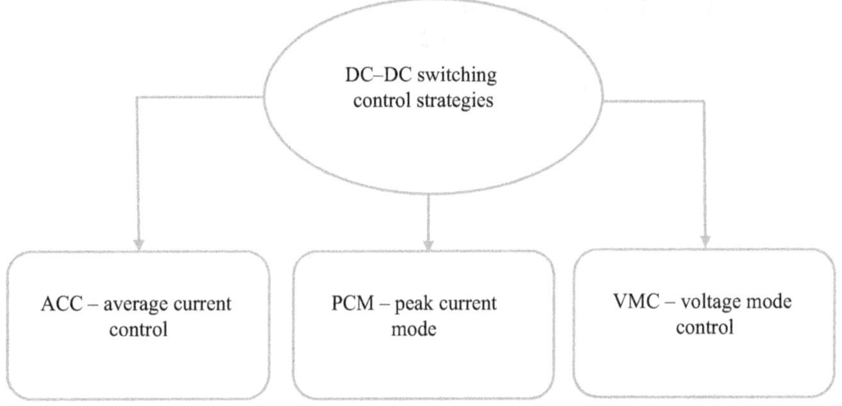

Figure 9.1 Switching control strategies for DC–DC converters

transfer function. The averaged converter model enables lesser design and simulation time. A new switch network is proposed in [10] utilising the averaged switch model which comprises of the power devices like MOSFET and diode thereby giving a generalised modelling approach for DC–DC converters. In this work, the following non-ideal features of power devices and circuit components are considered which include the turn-on resistance of power devices, DC resistance (DCR) of inductor and equivalent series resistance (ESR) of capacitor. For faster transient response and good tracking performance, the switched power converter compensation must be adjusted. The goal is to ensure sufficient gain margin (10 dB) and phase margin (45°), slope of −20 dB/decade (while passing through 0 dB) and bandwidth of 1/5 to 1/10 of switching frequency, to prevent oscillations and maintain optimal stability [11].

The switching frequency also significantly affects the performance of the DC–DC switching converters in terms of efficiency, thermals, ripple and transient response. In [12], it has been inferred that the greater the switching frequency, there is reduction in the size of the inductor and capacitor, with improved dynamic performance but at the same time, it reduces the efficiency. The performance analysis of non-isolated DC–DC converters [13] and a three-phase PWM controlled rectifier under ideal conditions was presented in [14]. The mathematical foundation was established by means of the volt-sec and amp-sec balance equations. Bidirectional converters continue to occupy considerable research space owing to their reduced size and complexity in control [15,16]. There is also an increased emphasis on designing an optimal control strategy for better dynamic response, low or zero steady-state error, and reduced distortion [17].

In this work

(a) The modelling of the non-ideal two switched PWM converters namely buck and boost are modelled using the basic principles of volt-sec and amp-sec balance equations.
(b) 'Commonly Used Blocks' were used for the simulation and hence, the dependence on the Sim Power System tool box is reduced.
(c) The transfer function of the G_{vd} and G_{id} are obtained using the SSA technique and validated using a new control technique called circuit averaging.
(d) The frequency responses obtained from two different software tools showed a perfect match in terms of stability, low-frequency gain, GM and PM

The chapter is organised as follows.

9.2 Mathematical model of non-ideal boost converter

The dynamic equation for the basic converters is shown in [18].

In this work, the mathematical models for the ideal and non-ideal buck converters operating in CCM are derived and modelled using MATLAB/Simulink. Figure 9.2 depicts a schematic of a non-ideal Boost converter operating in CCM.

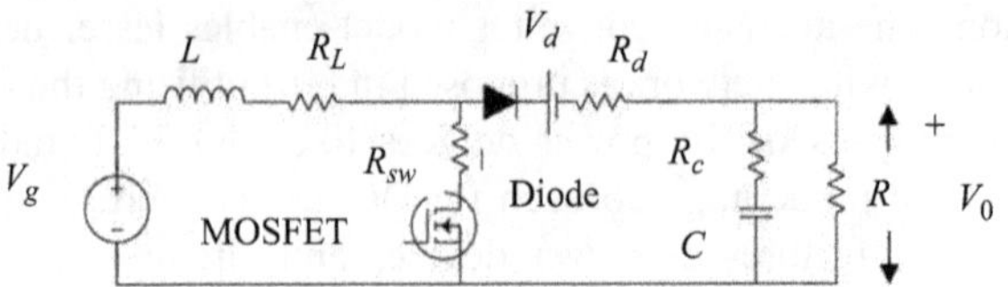

Figure 9.2 Non-ideal boost converter

The operation of the converter can modelled in two stages of the switch when (a) MOSFET OFF and (b) MOSFET ON.

When the MOSFET is closed,

$$V_L = V_g - i_L(R_L + R_{sw}) \tag{9.1}$$

$$i_c = -V_c/(R + R_c) \tag{9.2}$$

where V_L is the voltage drop across the inductor (V), V_g is the supply voltage (V), i_L is the current in the inductor (A), R_L is the ESR of the inductor (Ω) and R_{sw} is the switch resistance (Ω), i_c is the capacitor current (A), V_c is the voltage across the capacitor (V), R is the load resistance (Ω), R_c is the capacitor ESR (Ω) and V_0 is the output voltage (V).

When the MOSFET is opened

$$V_L = V_g - i_L(R_L + R_d + RR_c/(R + R_c)) - V_d - RV_c/(R + R_c) \tag{9.3}$$

$$i_c = i_L R/(R + R_c) - V_c/(R + R_c) \tag{9.4}$$

where V_d is the diode drop (V) and R_d is dynamic resistance of the diode (Ω). The mathematical model for the converter can be obtained by combining (9.1) and (9.3) with (9.2) and (9.4)

$$\begin{aligned} V_L = L\frac{di_L}{dt} &= (s * (V_g - i_L(R_L + R_{sw})) + (1 - s) \\ &* (V_g - i_L(R_L + R_d + RR_c/(R + R_c)) - V_d - RV_c/(R + R_c)) \end{aligned} \tag{9.5}$$

where s is the instantaneous duty cycle

$$\begin{aligned} i_c &= C\frac{dV_c}{dt} \\ &= (s * (-V_c/(R + R_c))) + (1 - s) * (i_L R/(R + R_c) - V_c/(R + R_c)) \end{aligned} \tag{9.6}$$

To capture the dynamics of i_L and V_c, Eqs (9.5) and (9.6) were modelled using MATLAB/Simulink.

9.3 Mathematical model of non-ideal buck converter

The mathematical model for an ideal buck converter is shown in [18]. For a non-ideal buck converter, the volt-sec and amp-sec balance equations are shown below:

When the switch is closed,

$$V_L = \frac{Ldi_L}{dt} = V_g - i_L(R_{sw} + R_L) - V_0 \tag{9.7}$$

$$i_c = \frac{Cdv_c}{dt} = i_L - \frac{V_0}{R} \tag{9.8}$$

$$V_0 = V_c + i_c R_c \tag{9.9}$$

(9.9) in (9.8),

$$V_0 = \frac{R(Vc + i_L R_c)}{R + R_c} \tag{9.10}$$

(9.10) in (9.7),

$$V_L = \frac{Ldi_L}{dt} = V_g - i_L\left(R_{sw} + R_L + \frac{RR_c}{R + R_c}\right) - \frac{RV_c}{R + R_c} \tag{9.11}$$

(9.10) in (9.8),

$$i_c = \frac{Cdv_c}{dt} = \frac{i_L R}{R + R_c} - \frac{V_c}{R + R_c} \tag{9.12}$$

Similarly when the switch is opened,

$$V_L = \frac{Ldi_L}{dt} = -i_L\left(R_L + R_d + \frac{RR_c}{R + R_c}\right) - V_d - \frac{RV_c}{R + R_c}$$

$$i_c = \frac{Cdv_c}{dt} = \frac{i_L R}{R + R_c} - \frac{V_c}{R + R_c} \tag{9.13}$$

The mathematical model can be represented as shown in (9.15) and (9.16):

$$\begin{aligned} V_L = \frac{Ldi_L}{dt} &= s * \left(V_g - i_L\left(R_{sw} + R_L + \frac{RR_c}{R + R_c}\right) - \frac{RV_c}{R + R_c}\right) \\ &+ (1 - s) * \left(-i_L\left(R_L + R_d + \frac{RR_c}{R + R_c}\right) - V_d - \frac{RV_c}{R + R_c}\right) \end{aligned} \tag{9.14}$$

$$i_c = \frac{Cdv_c}{dt} = s * \left(\frac{i_L R}{R + R_c} - \frac{V_c}{R + R_c}\right) + (1 - s) * \left(\frac{i_L R}{R + R_c} - \frac{V_c}{R + R_c}\right) \tag{9.15}$$

To capture the dynamics of i_L and V_c, Eqs (9.14) and (9.15) were modelled using MATLAB/Simulink.

9.4 SSA

Control techniques like small signal model, circuit averaging and SSA are used to obtain the various transfer functions of the DC–DC converters like output voltage to duty ratio $\left(G_{vd} = \frac{\widehat{V_0}}{\widehat{d}}\right)$, inductor current to duty ratio $\left(G_{id} = \frac{\widehat{i_L}}{\widehat{d}}\right)$ etc. In this work, SSA technique is used for estimating the (G_{vd}) and (G_{id}) transfer functions for non-ideal buck and boost converters.

(a) Non-ideal buck converter

$$A_1 = \begin{bmatrix} \frac{-1}{L}\left(R_{sw} + R_L + \frac{RR_c}{R+R_c}\right) & \frac{-R}{L(R+R_c)} \\ \frac{R}{C(R+R_c)} & \frac{-1}{L(R+R_c)} \end{bmatrix} \tag{9.16}$$

$$B_1 = \begin{bmatrix} \frac{1}{L} & 0 \\ 0 & 0 \end{bmatrix} \tag{9.17}$$

$$A_2 = \begin{bmatrix} \frac{-1}{L}\left(R_L + R_d + \frac{RR_c}{R+R_c}\right) & \frac{-R}{L(R+R_c)} \\ \frac{R}{C(R+R_c)} & \frac{-1}{C(R+R_c)} \end{bmatrix} \tag{9.18}$$

$$B_2 = \begin{bmatrix} 0 & \frac{-1}{L} \\ 0 & 0 \end{bmatrix} \tag{9.19}$$

$$Y(s) = C[(sI - A)^{-1} * [(A_1 - A_2)X + (B_1 - B_2)U]]\widehat{d}(s) \tag{9.20}$$

$$\text{Where } A = A_1 D + A_2(1-D) \tag{9.21}$$

$$A = \begin{bmatrix} a_{11} & a_{12} \\ a_{21} & a_{22} \end{bmatrix} \tag{9.22}$$

$$a_{11} = \frac{-DR_{sw}}{L} - \frac{DR_L}{L} - \frac{DRR_c}{R+R_c} - \frac{(1-D)}{L}\left(R_L + R_d + \frac{RR_c}{R+R_c}\right) \tag{9.23}$$

$$a_{12} = \frac{-RD}{L(R+R_c)} + (1-D)\left(\frac{-R}{L(R+R_c)}\right) \tag{9.24}$$

$$a_{21} = \frac{RD}{C(R+R_c)} + (1-D)\left(\frac{R}{C(R+R_c)}\right) \tag{9.25}$$

$$a_{22} = \frac{-D}{C(R+R_c)} - (1-D)\left(\frac{1}{C(R+R_c)}\right) \tag{9.26}$$

$$(A_1 - A_2)X + (B_1 - B_2)U = \begin{bmatrix} \frac{-I_L}{L}(R_{sw} - R_d) + \frac{V_g + V_d}{L} \\ 0 \end{bmatrix} \tag{9.27}$$

$$\text{Where } I_L = \frac{V_0}{R} \tag{9.28}$$

$$(sI - A)^{-1} * \Delta = \begin{bmatrix} b_{11} & b_{22} \\ b_{21} & b_{22} \end{bmatrix} \tag{9.29}$$

$$b_{11} = s + \frac{1}{C(R + R_c)} \tag{9.30}$$

$$b_{12} = -\frac{R}{L(R + R_c)} \tag{9.31}$$

$$b_{21} = -\frac{R}{L(R + R_c)} \tag{9.32}$$

$$b_{22} = s + \frac{1}{L}(DR_{sw} + R_L + R_d(1 - D)) \tag{9.33}$$

$$\text{Where } \Delta = \left(s + \frac{1}{C(R + R_c)}\right)\left(s + \frac{DR_{sw} + R_L + R_d * (1 - D)}{L}\right) + \frac{R^2}{LC(R + R_c)^2} \tag{9.34}$$

$$\frac{\widehat{V_0}}{\widehat{d}} = \frac{1}{\Delta}\left(\frac{R}{C(R + R_c)}\left(\frac{-I_L}{L}(R_{sw} - R_d) + \frac{V_d + V_g}{L}\right)\right) \tag{9.35}$$

$$\text{Where } C = [0 \quad 1] \tag{9.36}$$

$$\text{Considering } C = [1 \quad 0] \tag{9.37}$$

$$\frac{\widehat{i_L}}{\widehat{d}} = \frac{1}{\Delta}\left(s + \frac{1}{C(R + R_c)}\right)\left(\frac{-I_L}{L} * (R_{sw} - R_d) + \frac{V_g + V_d}{L}\right) \tag{9.38}$$

(a) Non-ideal boost converter

$$A_1 = \begin{bmatrix} \frac{-1}{L}(R_{sw} + R_L) & 0 \\ 0 & \frac{-1}{C(R + R_c)} \end{bmatrix} \tag{9.39}$$

$$B_1 = \begin{bmatrix} \frac{1}{L} & 0 \\ 0 & 0 \end{bmatrix} \tag{9.40}$$

$$A_2 = \begin{bmatrix} \frac{-1}{L}\left(R_d + R_L + \frac{RRc}{R+R_c}\right) & \frac{-1}{L}\left(\frac{R}{R+R_c}\right) \\ \frac{R}{C(R+R_c)} & \frac{-1}{C(R+R_c)} \end{bmatrix} \tag{9.41}$$

$$B_2 = \begin{bmatrix} \frac{1}{L} & -\frac{1}{L} \\ 0 & 0 \end{bmatrix} \tag{9.42}$$

$$(A_1 - A_2)X + (B_1 - B_2)U = \begin{bmatrix} \left(\frac{R_d - R_{sw}}{L} + \frac{RR_c}{L(R+R_c)}\right)I_L + \frac{V_d}{L} + \frac{V_c R}{L(R+R_c)} \\ \frac{-RI_L}{C(R+R_c)} \end{bmatrix} \tag{9.43}$$

$$(sI - A)^{-1} * \Delta = \begin{bmatrix} f_{11} & f_{22} \\ f_{21} & f_{22} \end{bmatrix} \tag{9.44}$$

$$f_{11} = s + \frac{1}{C(R+R_c)} \tag{9.45}$$

$$f_{12} = -\frac{(1-D)R}{L(R+R_c)} \tag{9.46}$$

$$f_{21} = \frac{R(1-D)}{C(R+R_c)} \tag{9.47}$$

$$f_{22} = s + \frac{1}{L}\left(-DR_{sw} - R_L - R_d * (1-D) - \frac{RR_c(1-D)}{R+R_c}\right) \tag{9.48}$$

$$\frac{\widehat{i_L}}{\widehat{d}} = \frac{1}{\Delta}\left(\left(s + \frac{1}{C(R+R_c)}\right) * \left(\left(\frac{R_d - R_{sw}}{L} + \frac{RRc}{L(R+Rc)}\right)I_L + \frac{V_d + V_0 R}{L(R+R_c)}\right)\right) + \frac{R^2(1-D)I_L}{LC(R+R_c)^2} \tag{9.49}$$

Where

$$\Delta = \left(s + \frac{1}{L}\left(-DR_{sw} - R_L - R_d(1-D) - \frac{RR_c(1-D)}{R+R_c}\right)\right)\left(s + \frac{1}{C(R+Rc)}\right) + \frac{R^2(1-D)^2}{LC(R+Rc)^2} \tag{9.50}$$

$$\frac{\widehat{V_0}}{\widehat{d}} = \frac{1}{\Delta}\left(\left(\frac{R(1-D)}{C(R+R_c)}\right) * \left(\left(\frac{R_d - R_{sw}}{L} + \frac{RRc}{L(R+Rc)}\right)I_L + \frac{V_d + V_0R}{L(R+R_c)}\right)\right) + X \tag{9.51}$$

Where

$$X = \left(s + \frac{1}{L}(-DR_{sw} - R_L - R_d(1-D) - \frac{RR_c(1-D)}{R+R_c}\right) * \left(\frac{-RI_L}{C(R+R_c)}\right) \tag{9.52}$$

9.5 Converter specifications

Buck converter

S. no.	Parameter	Value
1	Input voltage, V_g	16 V
2	Output voltage, V_0	12 V
3	Output current, I_0	1 A
4	Inductor, L	1.1 mH
5	Inductor, ESR	0.18 Ω
6	Switch resistance, R_{sw}	0.044 Ω
7	Diode drop, V_d	0.65 V
8	Capacitor, C	84 μF
9	Capacitor, ESR	0.3 Ω
10	Duty ratio, D	0.78
11	Switching frequency, f_s	25 kHz

Boost converter

S. no.	Parameter	Value
1	Input voltage, V_g	5 V
2	Output voltage, V_0	12 V
3	Output current, I_0	1 A
4	Inductor, L	4.7 μH
5	Inductor, ESR	0.071 Ω
6	Switch resistance, R_{sw}	0.024 Ω
7	Diode drop, V_d	0.555 V
8	Capacitor, C	9.66 μF
9	Capacitor, ESR	0.16 Ω
10	Duty ratio, D	0.6285
11	Switching frequency, f_s	500 kHz

9.6 Results and conclusion

'Commonly Used Blocks' in Simulink were used to model Eqs (9.5), (9.6) and (9.14), (9.15). An auto solver with an appropriate step time was used. Figure 9.2 shows the open loop responses of i_L and V_0 for a non-ideal buck converter.

Figure 9.3 shows the open loop responses of i_L and V_0 for a non-ideal boost converter. As seen from Figure 9.3, the maximum values of V_0 and i_L were around 16 V and 12.5 A, respectively. However, the steady-state values were 12 V and 2.6A.

The manually obtained transfer function using SSA technique was modelled using MATLAB software and the parameters like stability, low-frequency gain, and gain and phase margins were analysed. Figure 9.4 shows the frequency response of G_{id} for a non-ideal buck converter. It is observed from Figure 9.4 that the converter is highly stable for average current operation in CCM. The gain and phase margins were infinity and 91.9°. The low-frequency gain was about 2.73 dB.

Eq. (53) shows the transfer function of inductor current to the duty cycle.

$$G_{id\,buck} = \frac{29295 * (s + 967.9)}{s^2 + 1877s + 2.082 * 10^7} \tag{9.53}$$

Figure 9.5 shows the bode plot of G_{vd} for non-ideal buck converter. As observed from Figure 9.5, it is found that the converter was stable for constant

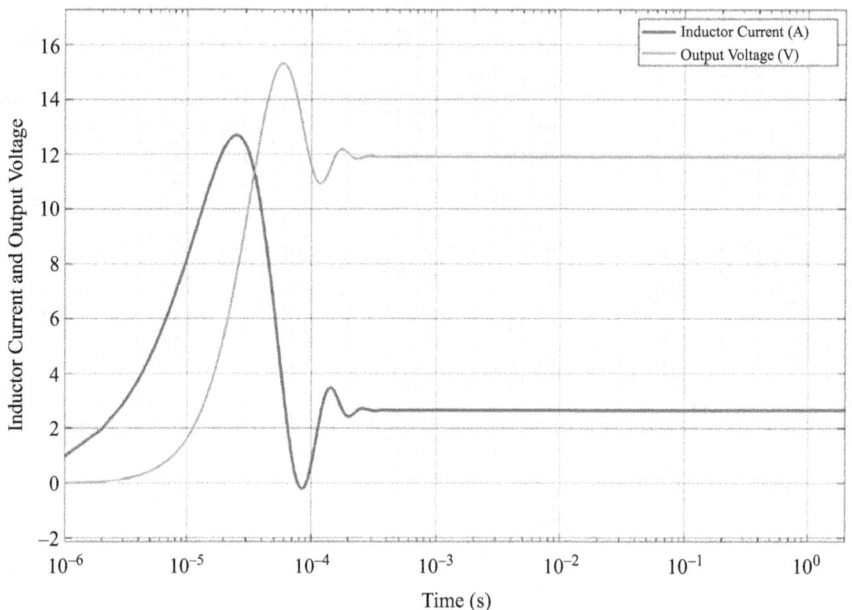

Figure 9.3 i_L *and* V_0 *vs. time*

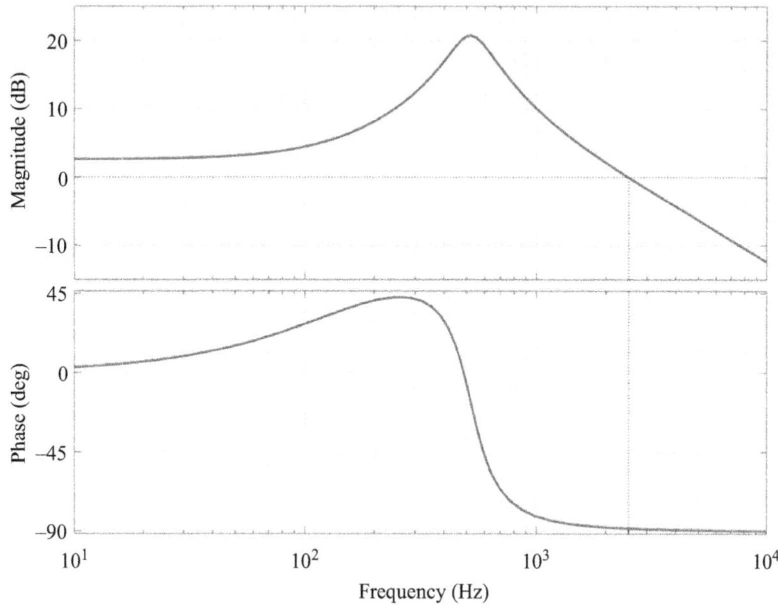

Figure 9.4 Frequency response of G_{id} *for a non-ideal buck converter*

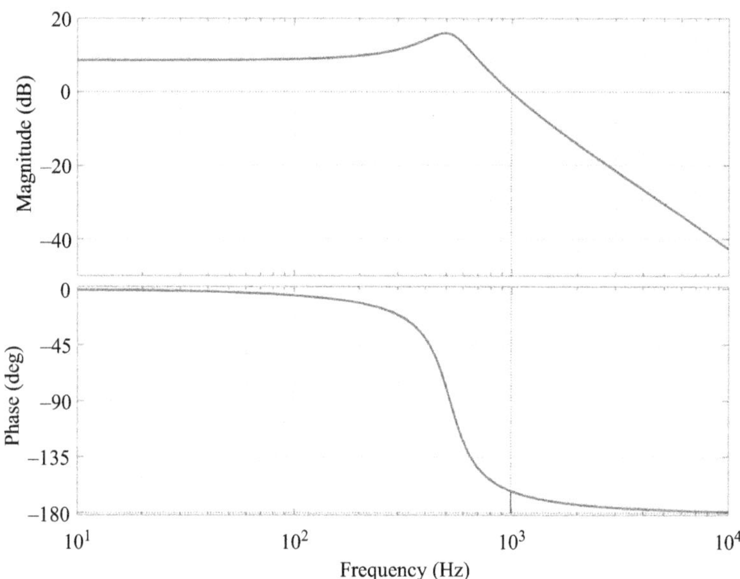

Figure 9.5 Frequency response of G_{vd} *for a non-ideal buck converter*

voltage operation in CCM. The gain and phase margins were infinity and 6°. The low-frequency gain was about 24 dB. Eq. (54) shows the transfer function the transfer function of output voltage to the duty cycle.

$$G_{vd\,buck} = \frac{3.4025 * 10^8}{s^2 + 1877s + 2.082 * 10^7} \tag{9.54}$$

Figure 9.6 shows the bode plot of combined responses of G_{id} and G_{vd} of the non-ideal buck converter. It is observed that the low-frequency gain for G_{vd} is greater than that of G_{id}. However, the crossover frequency of G_{id} is greater than that of G_{vd}.

Parameters like stability, low-frequency gain, gain and phase margins were analysed for the non-ideal boost converter operating in CCM. Figure 9.7 shows the frequency response of G_{id} for a non-ideal boost converter. It is observed from Figure 9.7 that the converter is highly stable for average current operation in CCM. The gain and phase margins were infinity and 90.5°. The low-frequency gain was about 23 dB.

Eq. (55) shows the transfer function of inductor current to the duty cycle:

$$G_{id\,boost} = \frac{2.7175 * 10^6 * (s + 1.641 * 10^4)}{s^2 + 4.123 * 10^4 s + 3.239 * 10^9} \tag{9.55}$$

Figure 9.8 shows the frequency response of G_{vd} for a non-ideal boost converter. It is observed from Figure 9.8 that the converter is unstable for the constant voltage operation in CCM. The gain and phase margins were −16.3 dB and −39.3°. The low-frequency gain was about 30 dB.

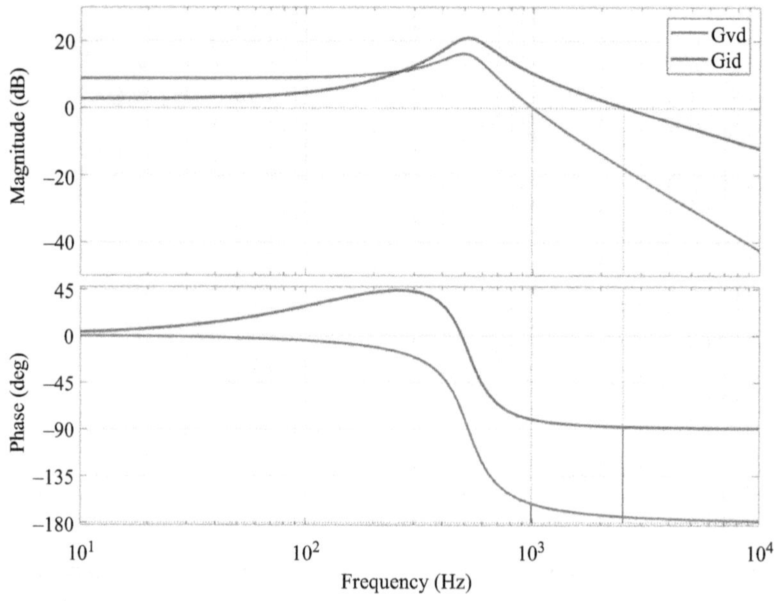

Figure 9.6 G_{id} *and* G_{vd} *for non-ideal buck converter*

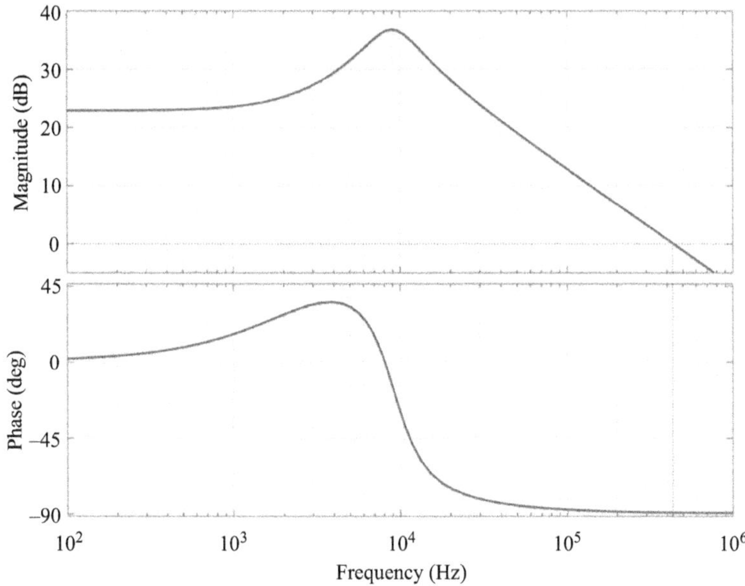

Figure 9.7 G_{id} *for non-ideal Boost converter*

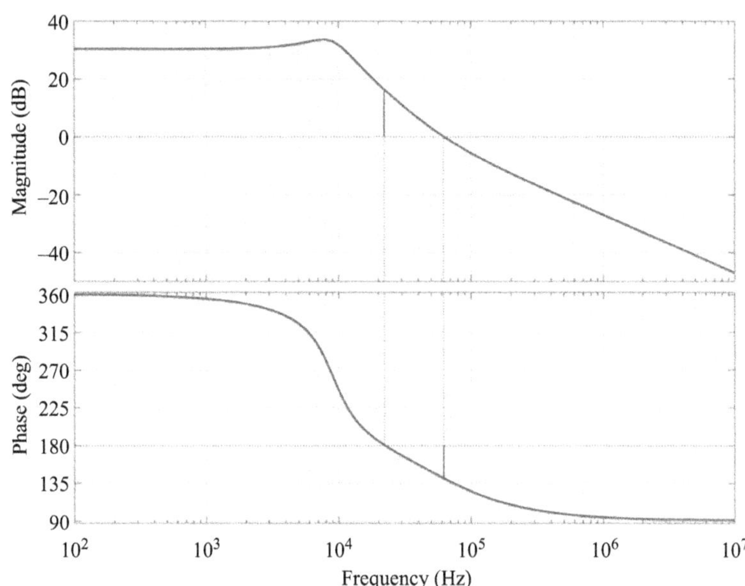

Figure 9.8 *Frequency response of* G_{vd} *for a non-ideal boost converter*

Figure 9.9 shows the root locus of open loop poles and zeros of G_{vd} for a practical boost converter.

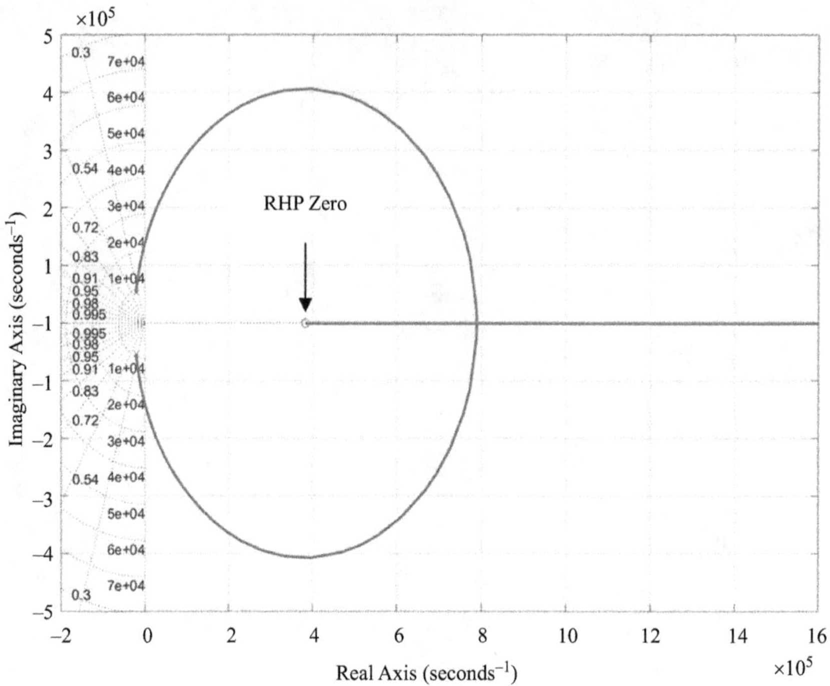

Figure 9.9 Root locus of G_{vd} *for boost converter*

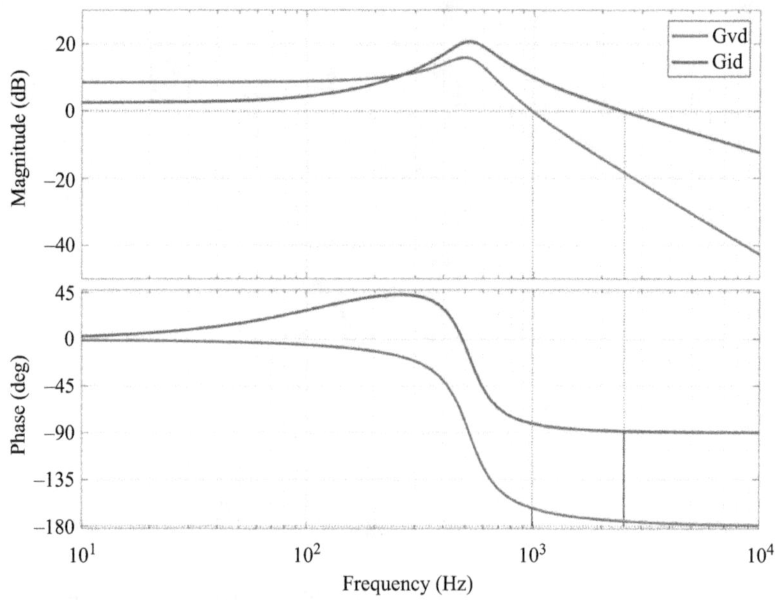

Figure 9.10 Frequency response of G_{vd} *for non-ideal boost converter*

Eq. (56) shows the transfer function of inductor current to the duty cycle. As seen from Eq. (56), the instability was due to the right half plane (RHP) zero. Figure 9.9 shows the root locus of G_{vd}

$$G_{id\,boost} = \frac{-2.7499 * 10^5 * (s - 3.822 * 10^5)}{s^2 + 4.214 * 10^4 s + 3.247 * 10^9} \tag{9.56}$$

Figure 9.10 shows the bode plot of combined responses of G_{id} and G_{vd} of the non-ideal boost converter. Observations similar to that of non-ideal buck converter are made.

9.7 Validation using LTspice software

The derived transfer function using SSA technique was modelled using MATLAB software and the stability analysis was carried out. In order to validate the obtained response, LTspice software tool was used for the system level simulation. It was found that the frequency obtained using the two software tools perfectly matched in terms of low-frequency gain, GM and PM.

The equivalent switch network was derived using circuit averaging technique and was generalised for two switch PWM converters. CCM2 block from average library was chosen for the validation.

The equivalent switch network of any two switch PWM converter is shown in Figure 9.11 [8].

Where $\frac{N_2}{N_1} = \frac{1}{D}$, for ideal buck converter

$\frac{N_2}{N_1} = \frac{1}{1 - D}$, for ideal boost converter

Figure 9.12 shows the developed model using LTspice software tool for validating the obtained G_{id} and G_{vd} for non-ideal buck converter. The specifications are chosen as shown in Section 9.5. Figure 9.13 shows the bode plot of G_{id} and G_{vd} for the selected converter.

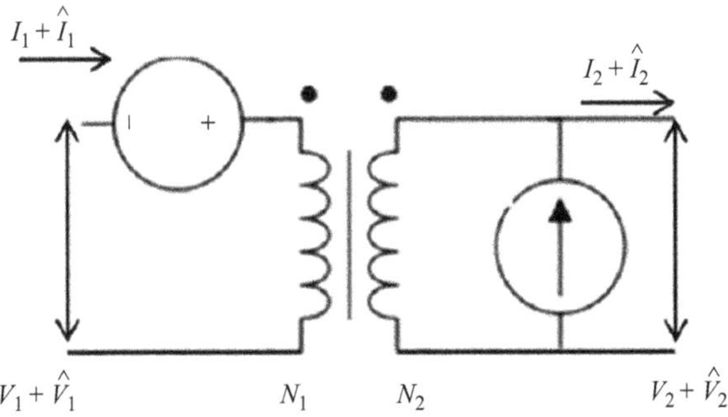

Figure 9.11 Equivalent switch network

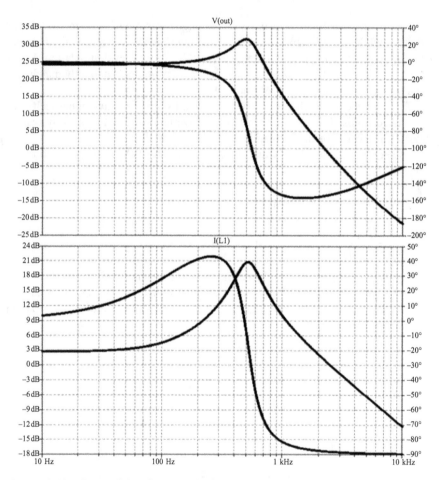

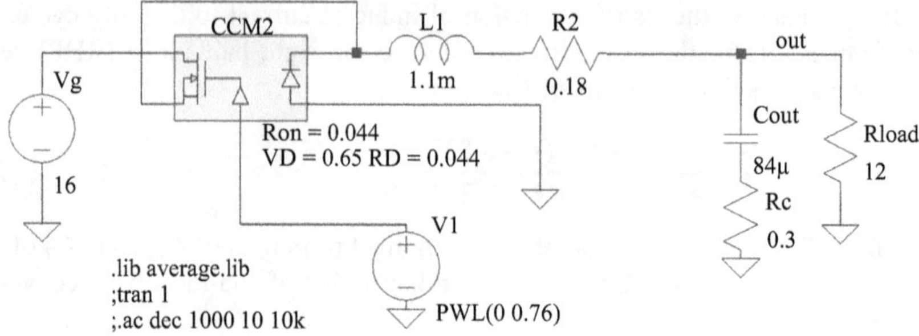

Figure 9.12 LTspice model of the non-ideal buck converter

Figure 9.13 G_{vd} *and* G_{id} *for non-ideal buck converter using circuit averaging technique*

Figure 9.13 shows the frequency response of G_{id} and G_{vd} for the non-ideal buck converter. Comparing Figure 9.13 with Figures 9.4 and 9.5, it is inferred that bode plots show a perfect match comparing the low-frequency gain, cut-off frequencies, GM and PM. Hence, the transfer function obtained using the SSA technique is validated against circuit averaging technique using LTspice software.

Figure 9.14 shows the frequency response of G_{id} and G_{vd} for the non-ideal boost converter. Comparing Figure 9.14 with Figures 9.7 and 9.8, it is inferred that

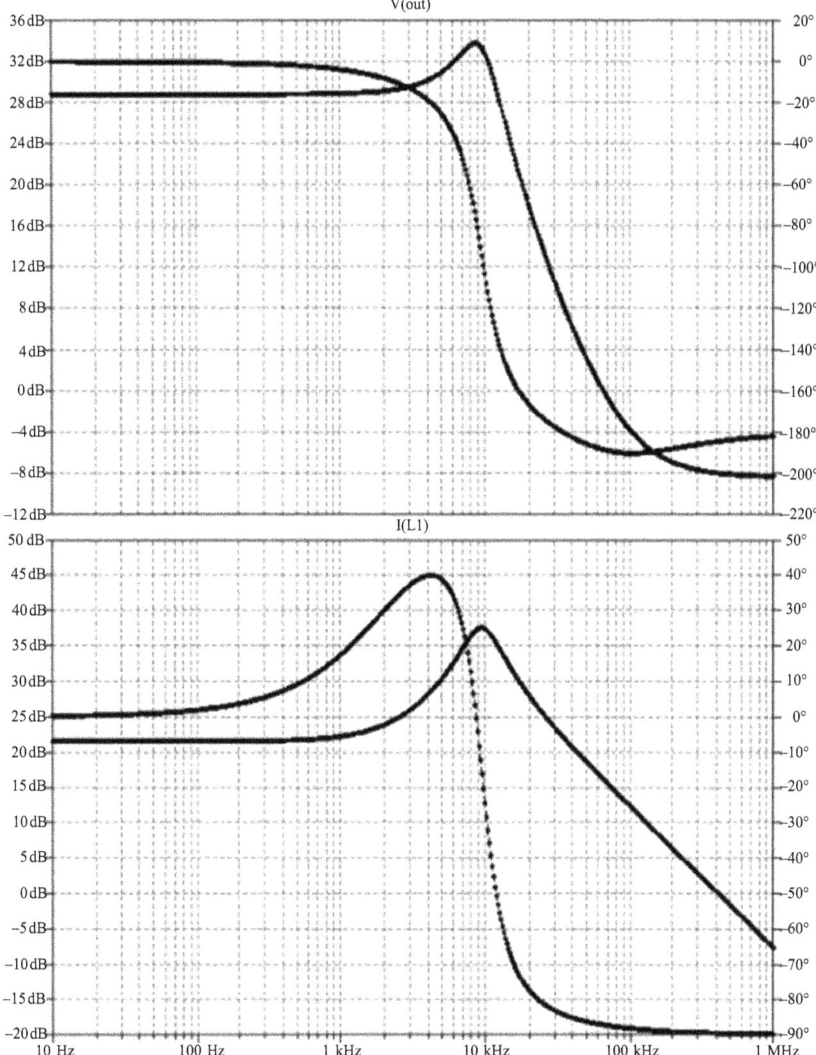

Figure 9.14 G_{vd} *and* G_{id} *for non-ideal boost converter using circuit averaging technique*

bode plots show a perfect match comparing the low-frequency gain, cut-off frequencies, GM and PM. Hence, the transfer function obtained using the SSA technique is validated against circuit averaging technique using LTspice software.

Figure 9.15 shows the frequency response of G_{id} and G_{vd} for the change in R_L for non-ideal boost converter. R_L was changed from 0.023 Ω to 0.071 Ω in steps of 0.023 Ω and bode plots were analysed for G_{id} and G_{vd}. It was found that the highest resonant frequency was seen for the highest switch resistance. Similar observations were seen for the non-ideal buck converter.

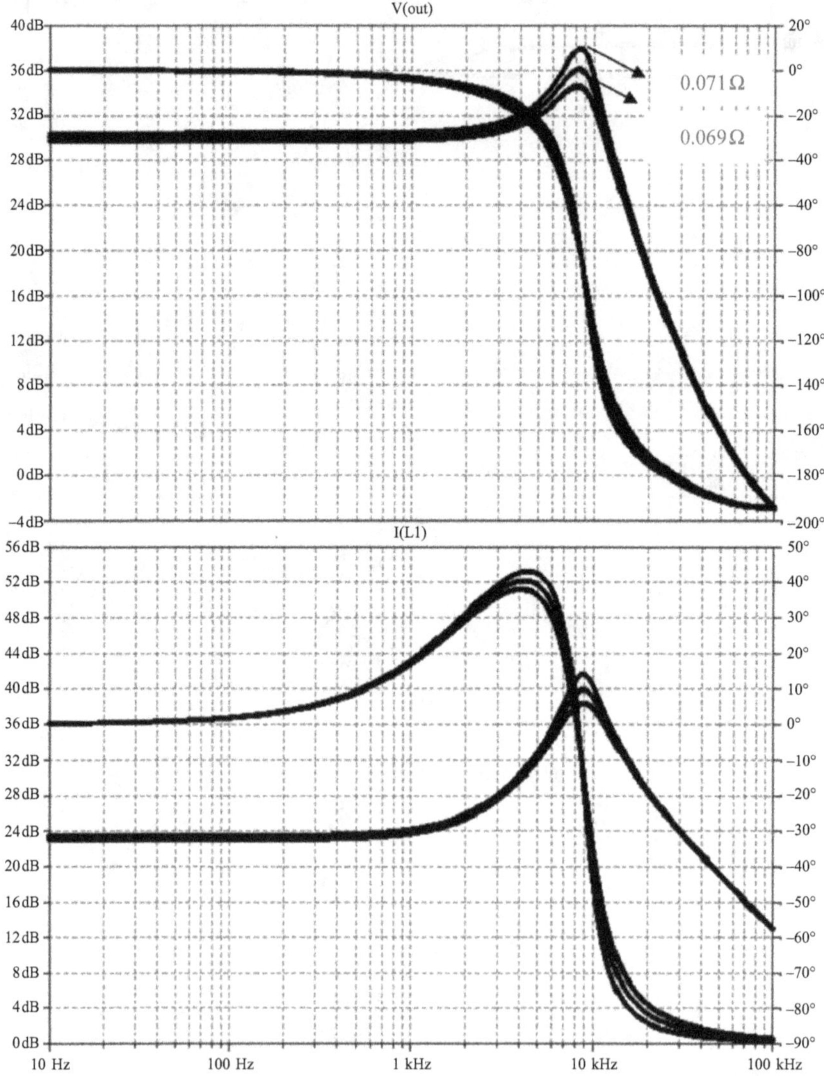

Figure 9.15 Effect of R_L *on* G_{vd} *and* G_{id} *for non-ideal boost converter*

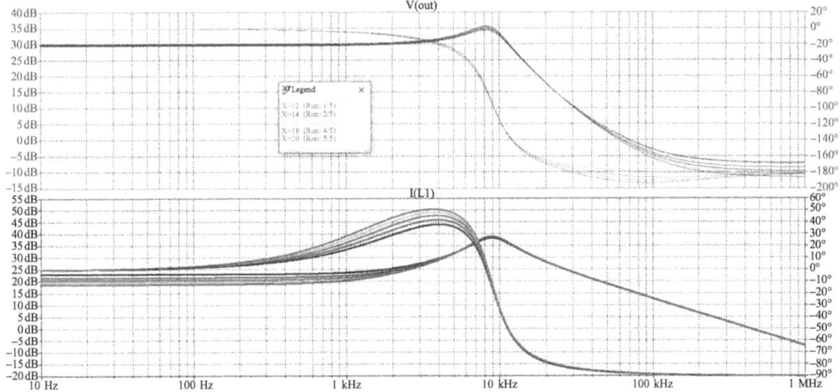

Figure 9.16 Effect of R *on* G_{id} *and* G_{vd}

Similar observations are made as seen in Figure 9.15 for the changes in R_c for non-ideal boost converter and non-ideal buck converter.

In Figure 9.16, the load resistance R is varied from 12 to 20 Ω and frequency response of G_{id} and G_{vd} for the non-ideal boost converter is analysed. It is observed that as the value of R increases, the magnitude of resonant frequency increases. Similar observations were seen for the non-ideal buck converter.

MATLAB code

A sample code for a non-ideal buck converter in CCM

```
clc;clear;
vg=16;
v0=12;
r=12;
d=0.76;
l=1.1e-3;
rl=0.18;
c=84e-6;
rc=0.3;
rd=0.044;
rsw=rd;
vd=0.65;
fs=25*10^3;
```

```
il=v0/r;
s=tf('s')
delta=(s+(1/l))*((d*rsw)+rd*(1-d)+rl+((r*rc)/
    (r+rc)))*(s+1/(c*(r+rc)))+(r^2/(l*c*(r+rc)^2))
%%ilcap/dcap%%
sys1=((s+1/(c*(r+rc)))*((il/l)*(rd-rsw)+((vg+vd)/
    l)))/delta
%%vocap/dcap%%
sys3=(r/(c*(r+rc)))*(((il/l)*(rd-rsw))+((vg+vd)/l))
    +(s+((d*rsw)/l)+(rd*(1-d)/l)+(rl/l)+((r*rc)/
    (r+rc)))
sys4=sys3/delta
```

Code for non-ideal boost converter

```
clc;clear;
vg=5;
v0=12;
ln=4.7e-6;
rl=0.071;
c=9.66e-6;
rc=0.08+0.08;
dn=0.6285;
ddashn=1-dn;
r=12;
rsw=31e-3
vd=0.555
rd=0.024
il=v0/(r*ddashn)
s=tf('s')
delta=(s+(1/(c*(r+rc))))*(s+(1/ln)*((dn*rsw)+rl
    +(rd*ddashn)+((r*rc*ddashn)/(r+rc))))
    +((r^2*ddashn^2)/(ln*c*(r+rc)^2))
%%ilcap/dcap%%
sys2=((s+(1/(c*(r+rc))))*((il/ln)*(rd-rsw+((r*rc)/
    (r+rc)))+((r*v0)/(ln*(r+rc)))+vd/ln)
    +(r^2*ddashn*il)/(ln*c*(r+rc)^2))/delta
%%vocap/dcap%%
sys3=((((rd-rsw)/ln+((r*rc)/(r+rc)))*(il)+((vd
    +(v0*r))/(ln*(r+rc))))*((r*ddashn)/(c*(r+rc)))
    +((-r*il)/(c*(r+rc)))*(s-((dn*rsw)/ln)-
    ((ddashn*rd)/ln)-(rl/ln)-(r*rc*ddashn)/
    (ln*(r+rc))))/delta
```

9.8 Conclusion

In this work, mathematical modelling of two switched non-ideal PWM DC–DC converters like buck and boost converters operating in CCM is performed using the volt-sec and amp-sec balance equations. Simulation of these converters was performed using MATLAB/Simulink software tool using appropriate step time and auto solver. The transients in inductor current and output voltage were observed. In order to capture the dynamics of G_{id} and G_{vd} operations, the transfer functions of the converters were derived using SSA technique. The obtained transfer functions were validated using a new control strategy called 'Circuit Averaging' using the LTspice software tool. The bode plots of G_{id} and G_{vd} showed a perfect match in terms of low-frequency gain, GM, PM and stability. It was shown that as the ESRs of inductor and capacitor increase, the resonant frequency increased. LTspice provided the frequency responses of G_{id} and G_{vd} using lesser computation time and memory requirement. Similar analyses based on DCM operation for isolated and non-isolated converters can be performed and validated using LTspice.

Disclaimer

The authors declare no conflict of interest and no data / information has been taken from Bosch Global Software Technologies Pvt. Limited (BGSW), Bangalore, India.

References

[1] Rahimi, A. M., Parto, P., and Asadi, P. (2010). Compensator design procedure for buck converter with voltage-mode error-amplifier. In *AN-1162, International Rectifier*.

[2] Ma, H., Qi, F., and Zhang, N. (2007). Modeling and control for DC-DC converters based on hybrid system. *Proc. – Chinese Soc. Electric. Eng.*, *27*(36), 92.

[3] Corti, F., Laudani, A., Lozito, G. M., and Reatti, A. (2020). Computationally efficient modeling of DC–DC converters for PV applications. *Energies*, *13* (19), 5100.

[4] Chander, S., Agarwal, P., and Gupta, I. (2010, October). Design, modeling and simulation of DC–DC converter. In *2010 Conference Proceedings IPEC* (pp. 456–461). IEEE.

[5] Surya, S., Srinivasan, M. K., and Williamson, S. (2021). Modeling of average current in non-ideal buck and synchronous buck converters for low power application. *Electronics*, *10*(21), 2672.

[6] Chou, H. H. and Chen, H. L. (2021). A novel buck converter with constant frequency controlled technique. *Energies*, *14*(18), 5911.

[7] Hassan, M. S. and Elbaset, A. A. (2015). Small-signal MATLAB/Simulink model of dc-dc buck converter using state-space averaging method. *MPECON, Egypt, December*.

[8] Surya, S. and Williamson, S. (2021). Generalized circuit averaging technique for two-switch PWM DC–DC converters in CCM. *Electronics*, *10*(4), 392.

[9] Surya, S., Channegowda, J., and Naraharisetti, K. (2020). Generalized circuit averaging technique for two switch DC–DC converters. In *2020 8th International Conference on Power Electronics Systems and Applications (PESA)* (pp. 1–6). IEEE.

[10] Lee, C. M. and Lai, Y. S. (2007). Averaged switch modeling of dc/dc converters using new switch network. In *2007 7th International Conference on Power Electronics and Drive Systems* (pp. 1427–1430). IEEE.

[11] Sheehan, R. and Diana, L. (2016). *Switch-Mode Power Converter Compensation Made Easy*. Dallas, TX: Texas Instruments.

[12] Xie, H. and Guo, E. (2019). *How the Switching Frequency Affects the Performance of a Buck Converter*. Dallas, TX: *Texas Instruments*.

[13] Surya, S. and Arjun, M. N. (2021). Mathematical modeling of power electronic converters. *SN Comput. Sci.*, *2*(4), 1–9.

[14] Tan, R. H. G. and Hoo, L. Y. H. (2015). DC–DC converter modeling and simulation using state space approach. In *2015 IEEE Conference on Energy Conversion (CENCON)*.

[15] Sreelakshmi, S., Krishna, M., and Deepa, K. (2019). Bidirectional converter using fuzzy for battery charging of electric vehicle. In *2019 IEEE Transportation Electrification Conference (ITEC-India)*.

[16] Daya John Lionel, F., Dias, J., Krishna Srinivasan, M., Parandhaman, B., and Prabhakaran, P. (2021). A novel non-isolated dual-input DC–DC boost converter for hybrid electric vehicle application. *Int. J. Emerging Electric Power Syst.*, 22(2), 191–204. https://doi.org/10.1515/ijeeps-2020-0229

[17] Prabhakaran, P., Krishna, S. M., Febin, D. J. L., and Perumal, T. (2021). A novel PR controller with improved performance for single-phase UPS inverter. In *2021 4th Biennial International Conference on Nascent Technologies in Engineering (ICNTE)*.

[18] Hinov, N. L. (2018). Mathematical modeling of transformerless DC–DC converters. In *2018 IEEE XXVII International Scientific Conference Electronics – ET*. IEEE.

Chapter 10

Artificial intelligent-based modified direct torque control strategy: enhancing the dynamic torque response of permanent magnet electric traction

Dattatraya kalel[1], *Harshit Mohan*[2] *and R. Raja Sing*[3]

Electric traction drive prefers direct torque control due to its simplicity and easy implementation on permanent magnet machines (PMMs). Variable switching frequency, as well as more torque and flux ripples, are the key challenges of conventional direct torque control. A modified switching table-based direct torque control has been widely adapted for controlling the PMM drives. Artificial intelligent-based switching table substitutes the switching table and hysteresis comparator provides a significant reduction in current harmonic distortion, torque, and flux ripple, which shows a greater advantage in speed control for smart electric vehicles. In this chapter, artificial intelligence-based multisector direct torque control is analyzed for a suitable voltage vector selection to minimize torque and stator flux ripple. To demonstrate, a comparison of the intended switching tables shows the virtues of each switching table on the performance of the multisector direct torque control strategy. This premises on the theory of keeping the divergence between the commanded torque and the calculated torque as small as possible and does not provide information on the conduction time mode of three-phase switching. It adapts changes in the three phase-current waveform to keep electromagnetic torque consistent, eliminating the commutation torque ripple that would have occurred with conventional direct torque control (CDTC). Simulation results are taken in MATLAB®/Simulink®, and it is observed that the PMM ripples are reduced, particularly at high rotational speeds.

[1]School of Electrical Engineering, Vellore Institute of Technology (VIT), India
[2]Department of Electrical Engineering, Indian Institute of Technology (IIT), India
[3]Department of Energy and Power Electronics, Vellore Institute of Technology (VIT), India

Nomenclature

F_{load}	forces exerted just on vehicle (N)
d_{wheel}	the tyre diameter (m)
i_g	gear ratio
F_{grade}	gravitational force (N)
$F_{rolling}$	resistance force to rolling (N)
M_g	vehicle's weight (kg)
ß	gradient angle
ρ	air density 1.225 kg/m (kg/m^3)
C_w	drag coefficient
A_d	frontal area (m^2)
V	wind velocity (m/s)
E_b	back EMF voltage (V)
T_e	electromagnetic torque (Nm)
T_{load}	load torque (Nm)
J	moment of inertia (kg m^2)
F	coefficient of friction
ω_r	motor shaft speed (rad/sec)
V_{ds}, V_{qs}	stator voltage on the *d*- and the *q*-axis (V)
I_{ds}, I_{qs}	stator current on the *d*- and the *q*-axis (A)
$V_{\alpha s}$, $V_{\beta s}$	stator voltage on the *α*- and the *β*-axis (V)
$I_{\alpha s}$, $I_{\beta s}$	stator current on the *α*- and *β*-axis (A)
L_{ds}, L_{qs}	inductances in the *d*- and the *q*-axis (H)
R_s	stator resistance (Ω)
p_n	number of poles
Ψ_{ds}, Ψ_{qs}	*d* and *q*-axis stator flux (Wb)
Ψ_f	permanent magnet flux (Wb)
R_i	core loss resistance (Ω)
∂	angle between the stator and the rotor reference frames
ω_e	angular speed (rad/sec)
$\Psi_{s\|t=0}$	preliminary value of flux at t=0 (Wb)
Ψ_s	stator flux (Wb)
$\ominus$	sector angle

10.1 Introduction

Transportation consumes almost one-fifth of the world's energy, and around 10% of energy-related CO_2 emissions come from passenger transport vehicles. Various green and fuel transport technologies are being developed and used, but traction drives (TD) are one of the most promising technologies. TD has a lot of potential for enhancing energy security, minimizing carbon dioxide emissions, and preventing pollution. The evolution from the steam drive, internal combustion drive to dc drive, and ac drive, every improvement is smooth and efficient work of traction drives. The key components of the propulsion system in TD are the electric motor, power converter, and digital electronic controller. Permanent magnet synchronous motors (PMSMs) are commonly employed in traction systems due to their excellent torque response and efficiency, as well as their good reliability [1,2]. According to the literature, PMSM drives can be classified as Volts/Hz (V/f) control, vector control or Field-Oriented Control (FOC), and direct torque control (DTC). The V/F control system is proposed for wide frequency operation to maintain the stator flux linkage in PMSM, this control system has limitations such as low dynamic response and applications like fan and pump. For stable operation of V/F control, methods are proposed in [3]. FOC proposed in 1970 is controlled by coordinate rotational transformation and field orientation in the same way as a dc motor is [1,2]. FOC can control the torque and the flux component via quadrature axis and direct axis stator currents, respectively [4] and converge the actual components' actual values to their reference values, proportional-integral (PI) regulators have always been used. Thus, rotor speed and electromagnetic torque can execute efficiently; however, the drawbacks FOC is the electromagnetic torque response which is limited by the armature winding's time constant and the complexity of decoupling transformation.

Induction motor drives were the first application of DTC. Because of its applicability to any AC drive, DTC is now being used for PMSM drives. DTC is a general term that incorporates a variety of techniques, including DTC-SVM (space vector modulation) and switching table-based DTC (ST-DTC). ST-DTC is the most well-known technique [5] because the torque and flux component change patterns are regulated via six active voltage vectors. However, nonlinear switching patterns of frequency, large torque, and flux ripples are the disadvantages of the ST-DTC. These drawbacks are overcome in the SVM-DTC scheme, which would be made up of a small number of voltage vectors with specified amplitudes and fixed switching frequency. Constant switching frequency and amplitude result the lower torque and flux ripple. However, SVM-DTC required knowledge of motor parameters, rotary coordinate transformation, three proportional-integral (PI) controllers, and a lot of computing power, which negates the simplicity and reliability of traditional DTC [6]. To overcome complexity and torque, flux ripple in ST-DTC has been employed with modified DTC such as sliding mode controller, model predictive control, and duty cycle control [7, 8]. In the last decades, in the artificial-intelligence-based controllers, the use of artificial neural network (ANN), fuzzy logic control (FLC), and adaptive neural-fuzzy inference system (ANFIS) in power electronics and electric drives has grown. Particularly in DTC, FLC replaced the PI controller, and switching table. DTC for ac motor is the nonlinear control system, and other nonlinear control systems, such as sliding mode controllers and duty cycle controllers, are dependent on the machine's parameters. Whereas AI-based algorithm does not depend on machine parameters. For complicated and imprecise processes, FLC is suited to nonlinear control commands and very effective procedures because it behaves as approximate human expert's [10–12].

In ST-DTC for induction motor, zero voltage vector employed for stator flux (Ψ_s) stays at its original position or does not change, whereas in PMSM, the magnets revolve with the rotor, even when zero voltage vectors are applied, Ψ_s will vary. Therefore, there is no need for zero voltage vectors for controlling PMSM. So, there is no requirement for three-level torque hysteresis comparators [13]. In this chapter, the proposed modified FLC-based multisector direct torque control is analyzed for the optimal choice of voltage vectors to reduce the torque and stator flux ripple. This chapter is classified as follows: Section 10.2 represents mathematical modelling of traction and PMSM. Section 10.3 discusses the conventional DTC strategy. Section 10.4 discusses fuzzy logic controlled – DTC, followed by the simulation of ST-DTC and FLC-DTC in Section 10.5 and conclusion in Section 10.6.

10.2 Mathematical modeling

In this section, the mathematical modeling of mechanical traction and PMSM is provided.

10.2.1 Traction dynamic modeling

Had used principles of vehicle mechanics and aerodynamics, someone could investigate both the driving power and the energy required to keep the vehicle running in the case of an EV [14]. The mechanical load and electromagnetic torque equation are as follows:

$$T_e - T_{load} = J\frac{d\omega_m}{dt} + F\omega_m \tag{10.1}$$

The torque of a mechanical load is derived in Eq. (10.2) as

$$T_{load} = F_{load}\frac{d_{wheel}}{2i_g} \tag{10.2}$$

where F_{load} is the force applied on the traction. The wheels have a diameter of d_{wheel}. Likewise, Eqs (10.3) and (10.4) are used to determine the force created:

$$F_{load} = F_{grade} + F_{rolling} + F_{ad} \tag{10.3}$$

$$F_{grade} = \pm mg\sin ß \tag{10.4}$$

where mg denotes the vehicle's mass, and ß is the angle of trend line (gradient), $F_{rolling}$ rolling resistance force causes the tyres to flatten at the tarmac mating surfaces:

$$F_{rolling} = \pm\mu mg\cos ß \tag{10.5}$$

where, μ is a nonlinear function of vehicle speed considering the properties of the tarmac mating surface, shape and strength of tyres. It increases in direct correlation toward vehicle's speed while it is running. Owing to the possession, the rolling resistance force can be reduced by keeping the axles as fully enlarged as conceivable, i.e.

$$F_{sf} = K_a V \tag{10.6}$$

The drag generated by the pressure difference of the air toward the vehicle is known as aerodynamic force F_{ad} i.e.

$$F_{ad} = \frac{1}{2}\rho C_w A_d (V + V_0) \tag{10.7}$$

10.2.2 Dynamic modeling of PMSM

The rotor synchronous coordinate model is the most popular and preferred for PMSM since all of the parameters become constant. In a synchronous frame, the following are the PMSM's machine equations: (the components indicated by dq). The most practical technique of evaluating sinusoidal PMSMs is to use flux linkage phasors, instantaneous current, and voltage in a synchronously rotating reference frame locked to the rotor. As shown in Figure 10.1, the *d*-axis component is associated with the PM flux linkage Ψ_f, and the resultant back-emf E_b phasor is associated with the orthogonal *q*-axis component. The magnitude of the back-emf phasor can be written as

$$E_b = \omega_r \psi_f \tag{10.8}$$

Park's transformation is used to decouple stator and rotor fluxes by rotating in an arbitrary frame of reference into a rotating rotor flux reference frame. Using the park transformation, the stator voltage equation is represented as dq reference

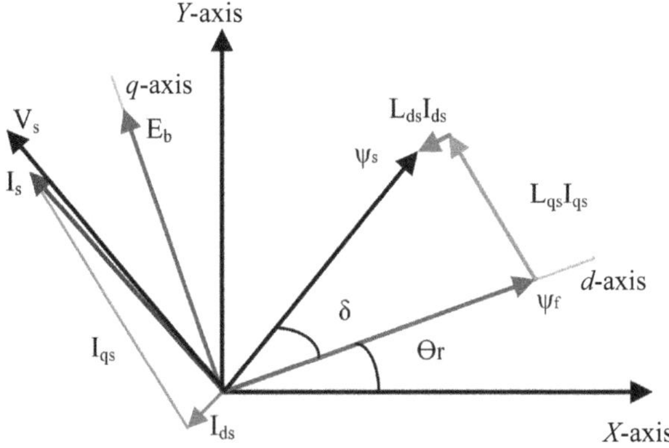

Figure 10.1 Phasor diagram of PMSM

frame in equation (10.9) and (10.10),

$$V_{ds} = R_s I_{ds} + \frac{d\Psi_{ds}}{dt} - \omega_e \Psi_{qs} \tag{10.9}$$

$$V_{qs} = R_s I_{qs} + \frac{d\Psi_{qs}}{dt} + \omega_e \Psi_{ds} \tag{10.10}$$

$$\Psi_{qs} = L_{qs} I_{qs} \tag{10.11}$$

$$\Psi_{ds} = L_{ds} I_{ds} + \Psi_f \tag{10.12}$$

The PMSM electromagnetic torque (T_e) is given as

$$T_e = \frac{3}{2}\frac{p_n}{2}\left(\Psi_{ds} I_{qs} - \Psi_{qs} I_{ds}\right) \tag{10.13}$$

10.3 DTC control strategy

In the DTC approach, the angle between the stator and rotor reference frames directly controls torque [15]. The voltage vector is chosen for controlling the drives speed and torque is based on the switching pattern. Figure 10.2 depicts the ST-DTC for PMSM. The difference between the reference and the measured speed is fed into the speed (PI) controller, and the reference torque component is the output of the speed controller block. Obtained reference torque is then evaluated by comparing with the calculated torque via flux and torque estimator block. This information is fed into the three-level hysteresis controller called as torque controller. Similarly, the computed flux (Ψ_s) is compared with the reference flux (Ψ_{ref}) for providing the error to the two-level hysteresis controller. The flux is computed using equation (10.16).

The DTC's primary premise is to choose the voltage vectors that will be delivered into the machine in the next switching period based on differences between the reference and calculated torque and stator flux linkage values. In flux – torque estimator (FTE) block, the three-phase voltage and current (V_{abc} and I_{abc}) is transformed into two phase stationary frame ($V_{\alpha\beta}$ and $I_{\alpha\beta}$) using Clarke transformation. In FTE, to derive the magnitude of stator flux space vector Ψs and its specific angle $\ominus$(s), to integrate the machine back emf in the synchronous reference frame, two integrators are used. Response of stator flux (Ψ_s) is significantly quicker than the permanent magnet flux response (Ψ_f) [8]. At all rotational speeds, the stator flux and its angle (∂) to the rotor flux can be controlled. In PMSM, Ψ_f is constant, and the desired torque can be obtained by varying the voltage vectors of stator. To achieve the appropriate torque, the torque or load angle is modulated, which is directly controlled by the DTC system's angle formed between the fluxes of the stator and the rotor [9]. The synchronous reference frame is specified by Eq. (10.13). The torque can be calculated using the amplitude of the stator and rotor flux, and angle as [10]:

$$T_e = \frac{3}{4}\frac{P_n|\Psi_s|}{L_d L_q}\left[2\Psi_f L_q \sin(\partial) - |\Psi_s|\left(L_q - L_q\right)\sin(2\partial)\right] \tag{10.14}$$

Here ∂ is the load or torque angle. In PMSM, for a salient pole machine, $L_d = L_q = Ls$, the equation can be simplified as:

$$T_e = \frac{3}{4}\frac{P_n}{L_s}|\psi_s|\psi_f \sin(\partial) \tag{10.15}$$

In Eq. (10.15), the torque is controlled directly by changing the torque angle ∂. Keeping constant amplitude of the $\Psi_{s,}$ furthermore, as ∂ increases, so does the torque. For controlling stator flux magnitude, the flux linkages of stator in the synchronous stationary reference frame, torque equation of PMSM, the stator flux linkage, and sector angle θs are expressed as

$$\left.\begin{aligned} \Psi_s &= \int (V_s - R_s I_s)dt \\ T_e &= \frac{3}{2}\frac{p_n}{2}\left(\Psi_{\alpha s} I_{\beta s} - \Psi_{\beta s} I_{\alpha s}\right) \\ \Psi_s = \sqrt{\Psi_{\alpha s}^2 + \Psi_{\beta s}^2} &\quad \text{and} \quad \ominus_s = \arctan\left(\frac{\Psi_{\beta s}}{\Psi_{\alpha s}}\right) \end{aligned}\right\} \tag{10.16}$$

where $V_s = \begin{bmatrix} V_{\alpha s} & V_{\beta s} \end{bmatrix}^T$ and $I_s = \begin{bmatrix} I_{\alpha s} & I_{\beta s} \end{bmatrix}^T$, $\Psi_s = \begin{bmatrix} \Psi_{\alpha s}^2 & \Psi_{\beta s}^2 \end{bmatrix}$ $R_s I_s$ product voltage value is fraction so it is neglected. $\Psi_s = V_s t + \Psi_{s|t=0}$, where $\Psi_{s|t=0}$ is the preliminary value of flux at $t = 0$. An FTE can provide feedback signals for estimated torque T_e and stator flux linkages Ψ_s, this also gives information of the six regions, Along with the torque (T_e), and the stator flux (Ψ_s), the FTE provide the angle $\ominus_s$ to the switching table for determining the sectors between $\ominus_1$ and $\ominus_6$, shown in Figure 10.3. According to information of sector and the status of the hysteresis comparator, the most suitable of the eight voltage vectors selected from a two-level VSI, so in order to obtain the torque and flux-linkage transitions that are desired, is shown in Figure 10.2. In conventional ST-DTC, flux and torque controllers use two-level and three hysteresis comparator, respectively, and Table 10.1 shows the ST and the available voltage vectors for a two-level inverter. In ST-DTC for induction motor, zero voltage vector employed for stator flux (Ψ_s) stays at its original position or does not change, whereas, in PMSM, magnets rotate with the rotor so Ψ_s even though zero voltage vectors can be used. As a result, in PMSM, control does not employ zero voltage vectors. So there is no requirement for a three-level torque hysteresis comparator [12]. Table 10.2 shows the modified switching table voltage vector without applying a zero-voltage vector.

In the conventional ST-DTC, every sector has two cases. Within a 60-degree span, this causes torque and flux to be unpredictable. For a more stable switching by using all six active vectors within a sector, the flux trajectory is divided into sectors instead of six sectors. This division is depicted in Figure 10.3. Similarly, Figure 10.3 shows the sector diagram, between red colour dotted lines are shown the sectors I to VI. Each sector has an angle of 60 degree between them, −30 to 30 degree for sector I, 30 degree to 90 degree for sector II, and so on for the conventional sector. Whereas, black dotted line shows the ⊖(1) to ⊖(12) for 12 sectors, each sector has a 30 degree angle between them like −15 degree to 15 degree for sector I, 15–45 degrees for sector II and so on. Table 10.3 shows switching table for 12 sectors.

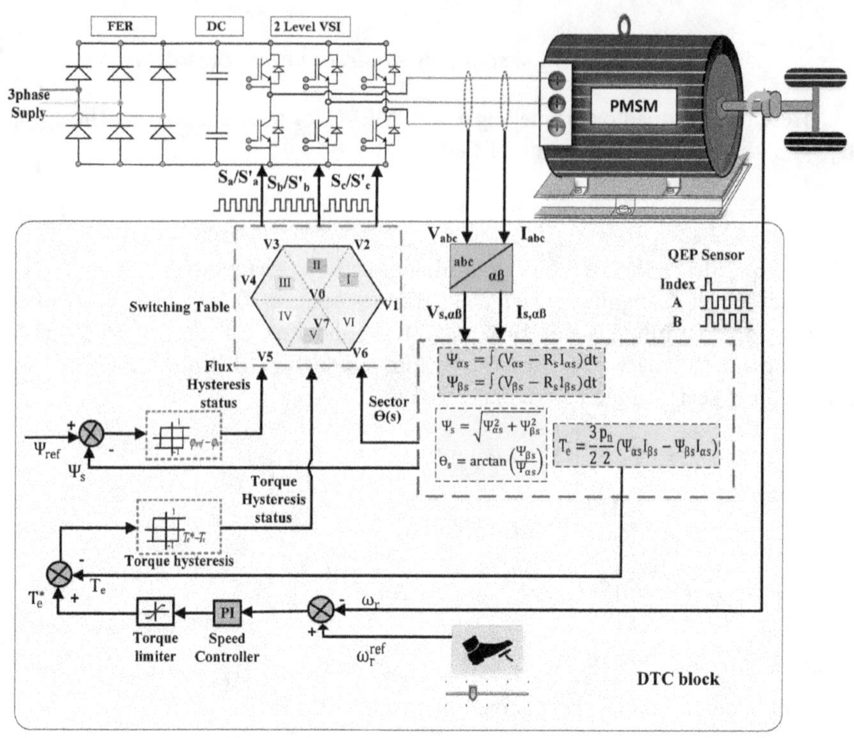

Figure 10.2 Conventional direct control of PMSM

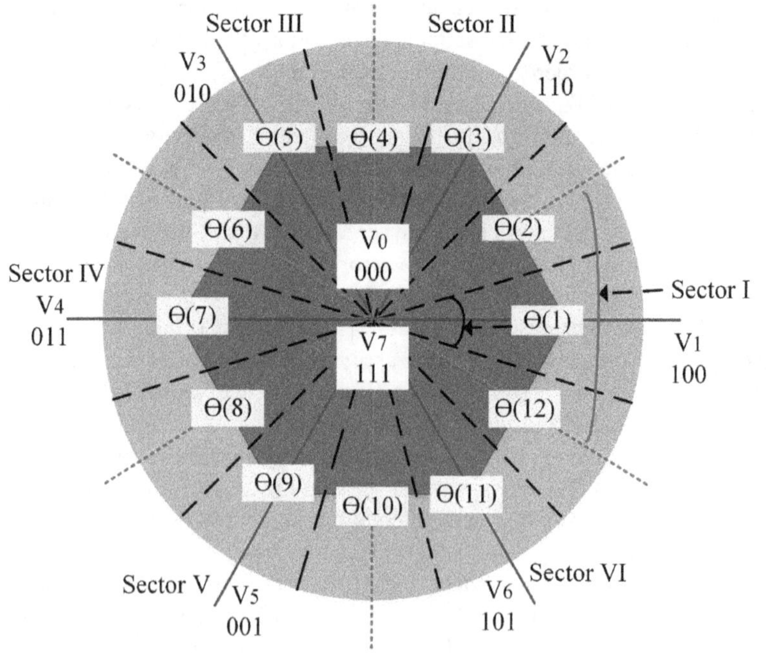

Figure 10.3 Vectors of stator voltages in the case of table with 6 sectors and 12 sectors

Table 10.1 Conventional switching table for voltage vector selection

Flux	Torque	⊖ (1)	⊖ (2)	⊖ (3)	⊖ (4)	⊖ (5)	⊖ (6)
X	X	V2 (XXO)	V3 (OXO)	V4 (OXX)	V5 (OOX)	V6 (XOX)	V1 (XOO)
	O	V0 (OOO)	V7 (XXX)	V0 (OOO)	V7 (XXX)	V0 (OOO)	V7 (XXX)
	−X	V6 (XOX)	V1 (XOO)	V2 (XXO)	V3 (OXO)	V4 (OXX)	V5 (OOX)
−X	X	V3 (OXO)	V4 (OXX)	V5 (OOX)	V6 (XOX)	V1 (XOO)	V2 (XXO)
	O	V7 (XXX)	V0 (OOO)	V7 (XXX)	VO (OOO)	V7 (XXX)	V0 (OOO)
	−X	V5 (OOX)	V6 (XOX)	V1 (XOO)	V2 (XXO)	V3 (OXO)	V4 (OXX)

"X" indicates switch ON and "O" indicates switch OFF.

Table 10.2 Conventional switching table without zero voltage vector

Flux	Torque	⊖ (1)	⊖ (2)	⊖ (3)	⊖ (4)	⊖ (5)	⊖ (6)
X	X	V2 (XX0)	V3 (0X0)	V4 (0XX)	V5 (00X)	V6 (X0X)	V1 (X00)
	−X	V6 (X0X)	V1 (X00)	V2 (XX0)	V3 (0X0)	V4 (0XX)	V5 (00X)
	X	V3 (0X0)	V4 (0XX)	V5 (00X)	V6 (X0X)	V1 (X00)	V2 (XX0)
−X	−X	V5 (00X)	V6 (X0X)	V1 (X00)	V2 (XX0)	V3 (0X0)	V4 (0XX)

"X" indicates switch ON and "O" indicates switch OFF.

Table 10.3 Twelve sector switching table without zero voltage vector for voltage vector selection

Ψ_S	1		−1	
Te	1	−1	1	−1
⊖ (1)	V2 (XX0)	V6 (X0X)	V3 (0X0)	V5 (00X)
⊖ (2)	V3 (0X0)	V1 (X00)	V4 (0XX)	V6 (X0X)
⊖ (3)	V3 (0X0)	V1 (X00)	V4 (0XX)	V6 (X0X)
⊖ (4)	V4 (0XX)	V2 (XX0)	V5 (00X)	V1 (X00)
⊖ (5)	V4 (0XX)	V2 (XX0)	V5 (00X)	V1 (X00)
⊖ (6)	V5 (00X)	V3 (0X0)	V6 (X0X)	V2 (XX0)
⊖ (7)	V5 (00X)	V3 (0X0)	V6 (X0X)	V2 (XX0)
⊖ (8)	V6 (X0X)	V4 (0XX)	V1 (X00)	V3 (0X0)
⊖ (9)	V6 (X0X)	V4 (0XX)	V1 (X00)	V3 (0X0)
⊖ (10)	V1 (X00)	V5 (00X)	V2 (XX0)	V4 (0XX)
⊖ (11)	V1 (X00)	V5 (00X)	V2 (XX0)	V4 (0XX)
⊖ (12)	V2 (XX0)	V6 (X0X)	V3 (0X0)	V5 (00X)

"X" indicates switch ON and "O" indicates switch OFF.

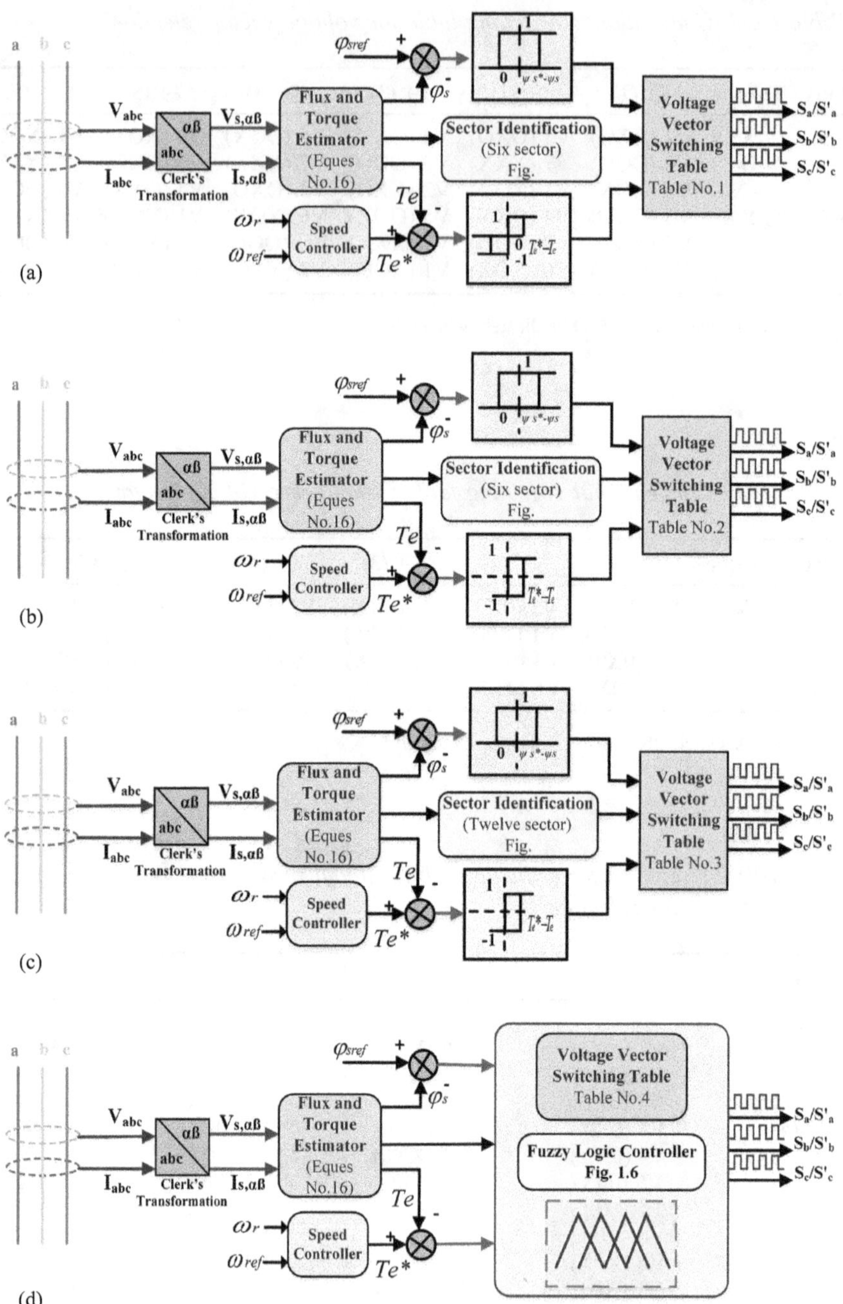

Figure 10.4 Different DTC structure: (a) conventional DTC, (b) modified switching table with six sectors, (c) modified switching table with six sectors, and (d) FL-based DTC with 12 sectors

Figure 10.4 shows the comparison DTC structure, (a) shows that the CDTC have zero voltage vector in switching table, (b) shows that as per the MDTC for PMSM, there is role of zero voltage vector, so ST have without zero voltage vector, (c) shows that the 12 sector MDTC, in conventional, have six sectors, whereas in Figure 10.3, six sectors without zero voltage vector are shown and (d) shows the fuzzy logic-based DTC, in these three inputs, torque error, flux error, 12 sectors and output are the voltage vector for switching controller.

10.4 Fuzzy logic controller for DTC

Various investigations are carried out using fuzzy logic to see if it can be used as a decision rule in the design process. Unlike conventional controllers, the FL does not require mathematical models of systems, for this reason, it has attracted a tremendous amount of focus on every aspect of electromechanical device control [3]. In [4–6] cases, in conventional ST-DTC, an FLC is employed to choose voltage vectors. In [7], the FLC is used to estimate the stator resistance under steady state. Also, the resistance changes are determined with temperature variations under various operating conditions. For the duty ratio control method, a FLC is used to calculate the time of the output voltage vector at each sample period [8]. These FLCs are capable of delivering excellent dynamic performance as well as robustness. By constructing membership functions for each input variable, the fuzzification process transforms deterministic input variables into linguistic variables. In the DTC induction motor drive, torque and flux ripples are a significant challenge even though there are no any inverter switching vectors that can provide the exact voltage necessary to achieve the desired torque and flux changes. This section proposes a fuzzy strategy for reducing torque ripples. The FLC achieves this goal by appropriately selecting the vector state of the inverter that is desirable. Figure 10.3 depicts the fuzzy technique of the DTC with the PMSM that has been proposed. The hysteresis controller and switching table have been replaced by an FLC to improve control performance. The FLC design is based on instinct and data are gathered through system simulations. Three inputs and a single output are proposed in this scheme. Torque error, stator flux error, and sector information-based angle are the inputs of FLC. Four steps are required for FLC execution and they are the following: (1) the first step is a fuzzifier, which is a process that transforms crisp data into fuzzy data. Those same variables are generated to use membership functions (MF); (2) a fuzzy rule base that includes a list of fuzzy rules that describe the workings of the fuzzy system; (3) a fuzzy reasoning algorithm that links input variables with fuzzy rules to do approximation reasoning; and (4) a defuzzifier is a process that transforms the FLC's fuzzy output to a crisp value when it is applied to the target. The FLC's performance is influenced by the MF's shape, fuzzy reasoning rules, and defuzzification approach. Figure 10.5(a) depicts four triangular MF that are designated and the fuzzy sets are used to fuzzify the torque error, NL indicates negative large, NS signify negative small, PS signify positive small, and PL for positive large. Fuzzification of flux error in the stator to

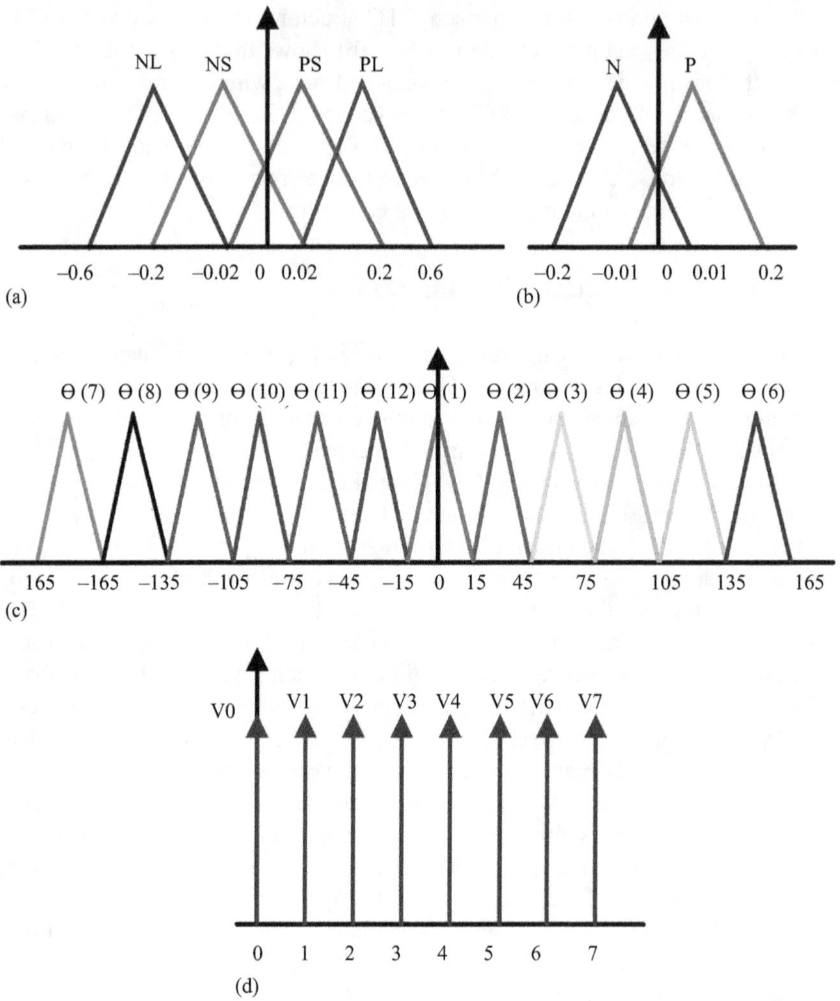

Figure 10.5 Membership function inputs of FLC: (a) torque error, (b) flux error, (c) sector, and (d) MF of output of FLC switching voltage vector

fuzzify the stator flux error and two trapezoidal membership functions are employed, as a result, two fuzzy sets emerge: N for negative and P for positive. Figure 10.5(b) depicts these two functions.

The flux angle has a discourse space of 360° (0, 2). For better certainty of the stator flux angles $\theta(1)$ to $\theta(12)$, the discourse universe of this fuzzy variable can be splitted into 12 fuzzy sets, labelled 1–12, as can be seen in Figure 10.5(c). The MFs that correlate to the flux and torque errors are defined in the same way. The MFs are made up of 12 equidistant equilateral triangular MFs, V_0 and V_7, two zero voltage vectors, and V_1 to V_6, six non-zero voltage vectors that are assigned to the fuzzy controller's output variable. The MFs of the output voltage vectors are described in Figure 10.5(d). The fuzzy controller's behavior is governed by the fuzzy control rules. It keeps track of technical competence on how to keep the plant under control by practicing 12 sectors. The voltage vector diagram in Figure 10.3 can be used to derive fuzzy control rules. For example, if the control desirable slowly decreases the torque while quickly increasing the flux, then the desirable decision is V1 if the positioning angle of the stator's flux is placed in domain ⊖2. The regulation aim is to keep the stator flux constant while the torque responds quickly. Using these identical rules, 180 fuzzy membership functions are obtained as described in Table 10.4. The stator flux angle is determined, and the switching state is determined based on the torque and flux errors. The motor is formed using the fuzzy AND operation, which selects the minimum of the membership values of E_t, Ψ_e, and $\theta(s)$. Torque and flux have reached their target values, $(3 \times 5 \times 12)$ 180 rules based on membership functions of three inputs. e.g.: "Rule: If E_T is PL and E_ψ is P and θ is $\theta 1$ then output is V1." Each rule's weight is established using Fuzzy "Min," which symbolizes logical "AND" operation, and the fuzzified output is created using the "Min–Max" fuzzy inference method. Rules are based on switching as shown in Table 10.4. FLSC is implemented using a "Mamdani type fuzzy inference system (FIS)." Figure 10.6 depicts the FLSC implementation flow diagram [16]. Although crisp variables cannot be applied straight to Fuzzy rule viewer, the fuzzification procedure converts three inputs into fuzzy variables utilizing fuzzy membership functions (MF's). Triangular and trapezoidal MFs are chosen from the available membership functions because they are basic and easy to implement.

Table 10.4 Fuzzy logic rule-based switching table for voltage vector selection

Flux	**P**				**N**			
Torque	**PL**	**PS**	**NS**	**NL**	**PL**	**PS**	**NS**	**NL**
⊖ (1)	V2 (XX0)	V2 (XX0)	(X0X)	V1 (X00)	V3 (0X0)	V4 (0XX)	V5 (00X)	V5 (00X)
⊖ (2)	V3 (0X0)	V2 (XX0)	V1 (X00)	V1 (X00)	V4 (0XX)	V4 (0XX)	V5 (00X)	V6 (X0X)
⊖ (3)	V3 (0X0)	V3 (0X0)	V1 (X00)	V2 (XX0)	V4 (0XX)	V5 (00X)	V6 (X0X)	V6 (X0X)
⊖ (4)	V4 (0XX)	V3 (0X0)	V2 (XX0)	V2 (XX0)	V5 (00X)	V5 (00X)	V6 (X0X)	V1 (X00)
⊖ (5)	V4 (0XX)	V4 (0XX)	V2 (XX0)	V3 (0X0)	V5 (00X)	V6 (X0X)	V1 (X00)	V1 (X00)
⊖ (6)	V5 (00X)	V4 (0XX)	V3 (0X0)	V3 (0X0)	V6 (X0X)	V6 (X0X)	V1 (X00)	V2 (XX0)
⊖ (7)	V5 (00X)	V5 (00X)	V3 (0X0)	V4 (0XX)	V6 (X0X)	V1 (X00)	V2 (XX0)	V2 (XX0)
⊖ (8)	V6 (X0X)	V5 (00X)	V4 (0XX)	V4 (0XX)	V1 (X00)	V1 (X00)	V2 (XX0)	V3 (0X0)
⊖ (9)	V6 (X0X)	V6 (X0X)	V4 (0XX)	V5 (00X)	V1 (X00)	V2 (XX0)	V3 (0X0)	V3 (0X0)
⊖ (10)	V1 (X00)	V6 (X0X)	V5 (00X)	V5 (00X)	V2 (XX0)	V2 (XX0)	V3 (0X0)	V4 (0XX)
⊖ (11)	V1 (X00)	V1 (X00)	V5 (00X)	V6 (X0X)	V2 (XX0)	V3 (0X0)	V4 (0XX)	V4 (0XX)
⊖ (12)	V2 (XX0)	V1 (X00)	V6 (X0X)	V6 (X0X)	V3 (0X0)	V3 (0X0)	V4 (0XX)	V5 (00X)

"X" indicates switch ON and "O" indicates switch OFF.

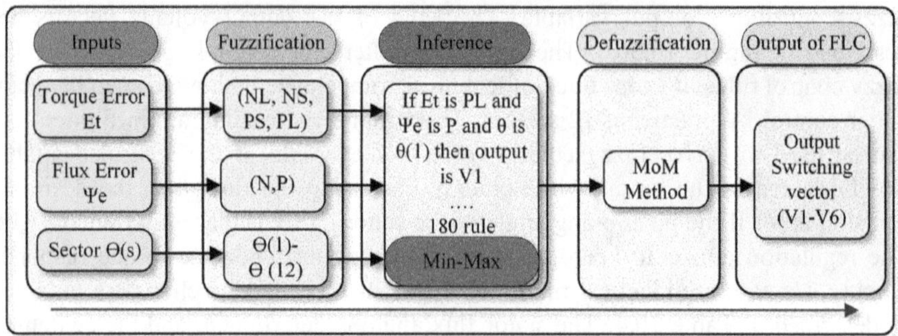

Figure 10.6 FLSC implementation flow diagram

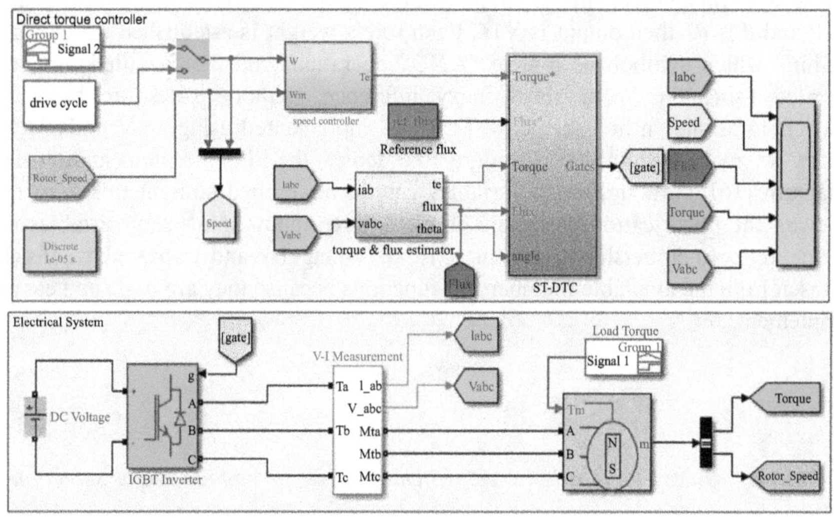

Figure 10.7 Overall simulation of DTC-based induction motor

10.5 MATLAB/Simulink simulation

The simulations in this study were carried out using MATLAB/Simulink, which is able to handle extremely complex simulations. Figure 10.7 depicts simulation model of DTC-based induction motor. This is divided into two parts: one is electrical system and another is DTC block. Electrical system consists of DC voltage source, two level IGBT-based voltage source inverter, three-phase voltage current measurement, and signal builder consists of load torque profile and three-phase PMSM. PMSM parameters are given in Table 10.1. Another is direct torque controller block that consists of outer loop as speed controller, torque and flux estimator block, switching table for direct torque control (ST-DTC), and data monitoring and display.

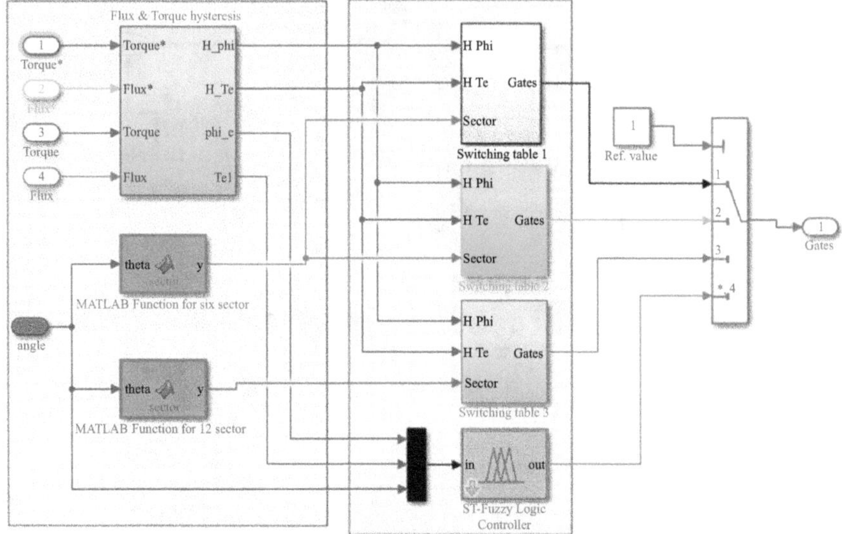

Figure 10.8 Different types of switching table

Figure 10.8 shows the details of different types of switching table for DTC. It consists of torque and flux hysteresis, MATLAB functions, lookup table (ST-DTC), and multiport switch. The input of torque and the flux hysteresis are reference torque (Torque*), reference flux (Flux*), estimated torque, and flux by using Eq. (10.16). Angle is given to the MATLAB function to find the sector, and two MATLAB functions are used: (1) 6 and (2) 12 sectors. Figure 10.8 shows the different types of switching interval systems such as switching Tables 1, 2, and 3 and ST-Fuzzy logic controller. Switching Table 1 for conventional switching, switching Table 2 for six sectors switching without zero voltage vector, switching Table 3 for 12-sector switching without zero voltage vector. Finally, the Fuzzy logic rule-based switching table for 12- sectors. The Multi-port Toggle Simulink box specifies whether any of the block's numerous input signals is routed toward the output. This decision is based on the first input's value. The control input is the first input, and the rest are inputs as dat. Control input's signals determine which data input goes to the output. For this, four data inputs are used for different switching tables and output of the multiport switch is gate pulses for voltage source inverter.

10.6 Results and discussions

The simulation model was developed in MATLAB/Simulink software. PMSM drives models simulated with a different controller such as CDTC, MDTC for 6 sectors, MDTC with 12 sectors, and FL-DTC with 12 sectors.

Figure 10.9 shows the steady-state response of CDTC and MDTC for 6 sectors. Figure 10.9(a) and (b) shows the speed response of CDTC and MDTC,

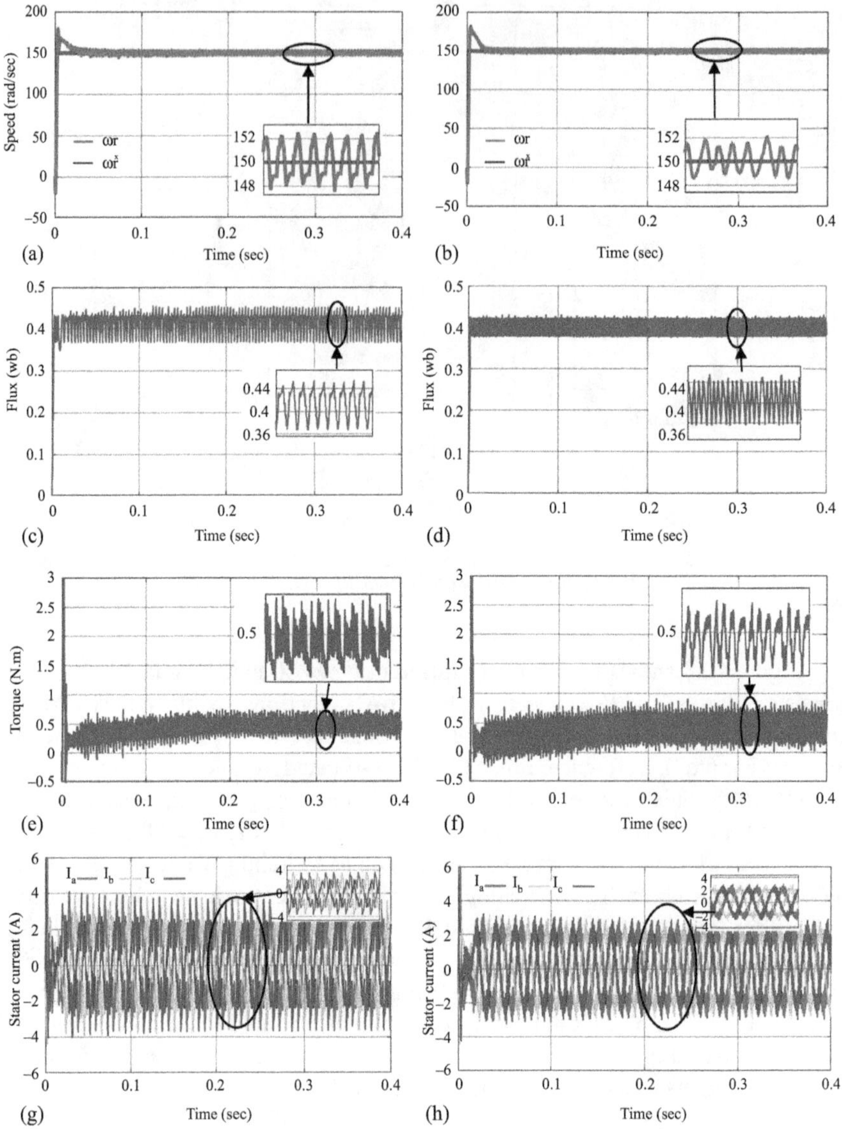

Figure 10.9 Comparison of CDTC and MDTC for six sectors: (a) and (b) speed response of CDTC and MDTC, respectively, (c) and (d) flux response of CDTC and MDTC, respectively, (e) and (f) torque response of CDTC and MDTC, respectively, (g) and (h) stator current response of CDTC and MDTC, respectively

respectively, at 150 rad/sec. It can be observed that PMSM have a better response as compared to IM. In CDTC, the variation of actual speed between 148 rad/sec and 152 rad/sec and, in MDTC, the variation of actual speed between 148 rad/sec and 152 rad/sec are noted.

Figure 10.9(c) and (d) shows the flux response of CDTC and MDTC, respectively. In CDTC, the flux variation between 0.36 Wb and 0.44 Wb and MDTC flux

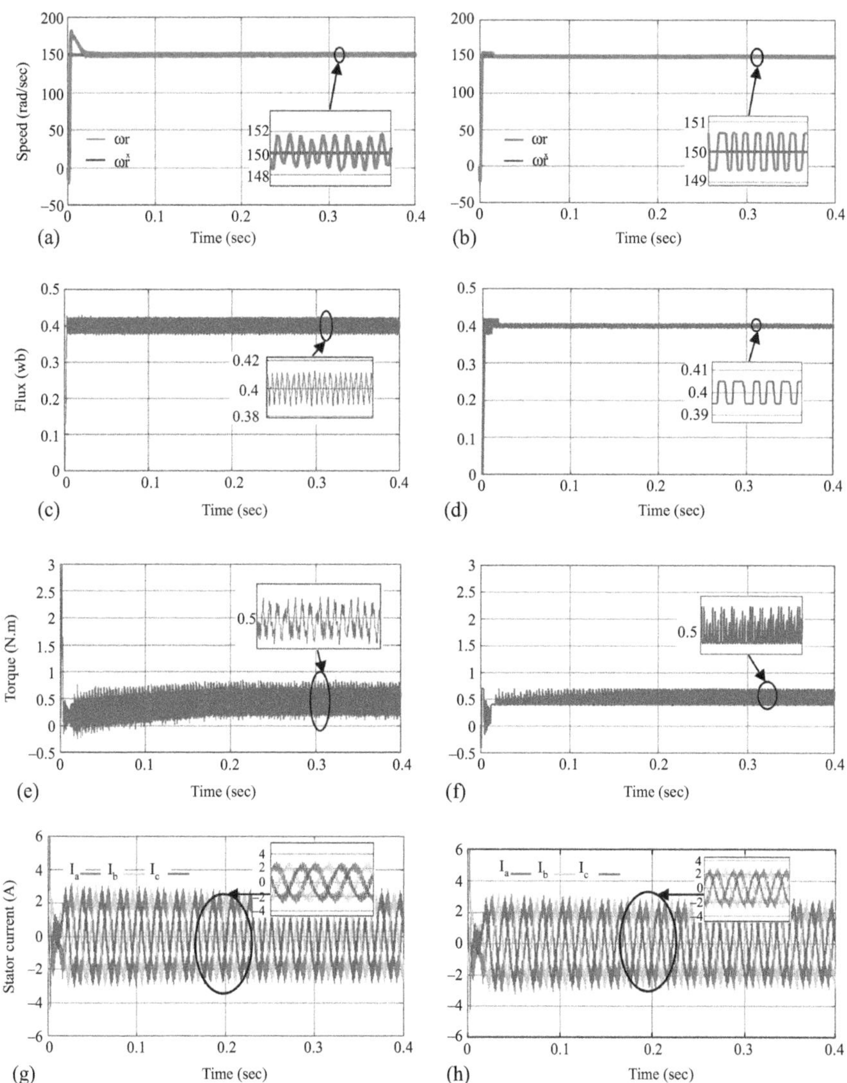

Figure 10.10 Comparison of MDTC and FL-DTC for 12 sectors: (a) and (b) speed response of MDTC and FL-DTC, respectively, (c) and (d) flux response of MDTC and FL-DTC, respectively, (e) and (f) torque response of MDTC and FL-DTC, respectively, (g) and (h) stator current response of MDTC and FL-DTC, respectively

variation between 0.38 Wb and 0.42 Wb are noted. It can be observed that CDTC have more ripple as compared to MDTC.

Figure 10.9(e) and (f) shows the torque response of CDTC and MDTC, respectively. The more torque ripples are observed in CDTC than in MDTC, in CDTC, the torque variation between 0.3 N.m. and 0.8 N.m., whereas in MDTC 0.1 N.m. and 0.8 N.m.

Figure 10.9(g) and (h) illustrates the three-phase stator current drawn by PMSM using CDTC and MDTC schemes, respectively. The CDTC current is not pure sinusoidal and it includes more harmonic content. However, the current MDTC contains fewer harmonics as compared to CDTC.

Figure 10.10 shows the steady-state response of MDTC and FL-DTC for 12 sectors. Figure 10.10(a) and (b) shows speed response of MDTC and FL-DTC, respectively, at 150 rad/sec. In MDTC, the variation of actual speed is between 149 rad/sec and 152 rad/sec and, in FL-DTC, the variation of actual speed is between 149 rad/sec and 151 rad/sec.

Figure 10.10(c) and (d) shows flux response of MDTC and FL-DTC respectively. In MDTC, the flux variation between 0.39 Wb and 0.41 Wb and FL-DTC flux variation is approximately 0.4 Wb. While compared to other systems, results reveal that the proposed FL-DTC method reduces flux ripples significantly under steady-state.

Figure 10.10(e) and (f) shows the torque response of MDTC and FL-DTC, respectively. The more torque ripples are observed in CDTC, MDTC with 6 sectors, MDTC with 12 sectors as compared to FL-DTC. In MDTC, the torque variation is noted between 0.3 N.m. and 0.7 N.m., whereas in FL-DTC, 0.4 N.m. and 0.7 N.m. This demonstrates that the proposed FL-DTC technique reduces torque ripples significantly in the steady-state when compared to previous schemes.

Figure 10.10(g) and (h) depicts the three-phase stator current drawn by PMSM using MDTC and FL-DTC schemes, respectively. It also demonstrates that the stator current has a lower harmonic content in the FL-DTC scheme as compared to CDTC.

Table 10.5 depicts comparison table different switching table for DTC-based induction motor. In this table, the variation of speed response, stator flux, and torque ripple is significantly reduced in DTC with FLC. In MDTC (6 sectors), torque variation increases as compared to others because of the absence of zero voltage vector but flux and speed variation reduce. In FLC (12 sectors), it clearly indicates less variation speed response, flux, and torque ripple as compared to others.

Table 10.5 Comparison table of different ST with parameter variation

Sr. no.	Parameters methods	Speed (rad/sec)	Flux (Wb)	Torque (Nm)
1	CDTC	148–152	0.36–0.44	0.3–0.8
2	MDTC (six sector)	149–152	0.38–0.42	0.1–0.8
3	MDTC (six sector)	149–152	0.39–0.41	0.3–0.7
4	Fuzzy logic controller	149–151	0.4	0.4–0.7

10.7 Conclusion

The DTC strategy for PMSMs is described in this chapter. The dynamic response of a DTC-based PMSM was compared with performances under conventional, multi-sector modified switching table and Fuzzy logic-based multisector. The performance of the PMSM's DTC algorithm has been improved by using the FLC. Switching table selection blocks and two-hysteresis controllers are replaced by the FLC. Simulation results show that the proposed technique outperforms the traditional DTC. In reality, it enables torque and flux ripples to be significantly reduced in the stator, as well as excellent commencing performance. The choosing of the voltage vector becomes more convenient when using intelligent controller strategies. The simulation results support the validity of the proposed controller strategy. The FLC's architecture makes use of a smaller number of rules, which helps to lower the FLC's computation time. Finally, the FLC-based DTC technique provides for a more dynamic response than the conventional DTC strategy. In this proposed scheme, flux response is approximately 0.4 Wb whereas as torque variation in between 0.4 and 0.7 Nm, it clearly indicates that the variation of flux and torque is reduced.

References

[1] X. Lin, W. Huang, W. Jiang, Y. Zhao, and S. Zhu, "A Stator Flux Observer with Phase Self-Tuning for Direct Torque Control of Permanent Magnet Synchronous Motor," *IEEE Trans. Power Electron.*, vol. 35, no. 6, pp. 6140–6152, 2020, doi: 10.1109/TPEL.2019.2952668.

[2] Y. Wang, L. Geng, W. Hao, and W. Xiao, "Control Method for Optimal Dynamic Performance of DTC-Based PMSM Drives," *IEEE Trans. Energy Convers.*, vol. 33, no. 3, pp. 1285–1296, 2018, doi: 10.1109/TEC.2018.2794527.

[3] P. D. Chandana Perera, F. Blaabjerg, J. K. Pedersen, and P. Thøgersen, "A Sensorless, Stable V/F Control Method for Permanent-Magnet Synchronous Motor Drives," *IEEE Trans. Ind. Appl.*, vol. 39, no. 3, pp. 783–791, 2003, doi: 10.1109/TIA.2003.810624.

[4] S. Wang, C. Li, C. Che, and D. Xu, "Direct Torque Control for 2L-VSI PMSM Using Switching Instant Table," *IEEE Trans. Ind. Electron.*, vol. 65, no. 12, pp. 9410–9420, 2018, doi: 10.1109/TIE.2018.2815995.

[5] C. Xia, S. Wang, X. Gu, Y. Yan, and T. Shi, "Direct Torque Control for VSI-PMSM Using Vector Evaluation Factor Table," *IEEE Trans. Ind. Electron.*, vol. 63, no. 7, pp. 4571–4583, 2016, doi: 10.1109/TIE.2016.2535958.

[6] Z. Zhang and X. Liu, "A Duty Ratio Control Strategy to Reduce Both Torque and Flux Ripples of DTC for Permanent Magnet Synchronous Machines," *IEEE Access*, vol. 7, no. 10, pp. 11820–11828, 2019, doi: 10.1109/ACCESS.2019.2892121.

[7] S. A. Saleh and A. Rubaai, "Extending the Frame-Angle-Based Direct Torque Control of PMSM Drives to Low-Speed Operation," *IEEE Trans.*

Ind. Appl., vol. 55, no. 3, pp. 3138–3150, 2019, doi: 10.1109/TIA.2018.2890060.

[8] F. Niu, B. Wang, A. S. Babel, K. Li, and E. G. Strangas, "Comparative Evaluation of Direct Torque Control Strategies for Permanent Magnet Synchronous Machines," *IEEE Trans. Power Electron.*, vol. 31, no. 2, pp. 1408–1424, 2016, doi: 10.1109/TPEL.2015.2421321.

[9] D. Kalel, H. Mohan, M. Baranidharan and R. R. Singh, "ANN Deployed DTC for Enhanced Torque Performance on IM based EV Drive," in: *2021 XVIII International Scientific Technical Conference Alternating Current Electric Drives* (*ACED*), 2021, pp. 1–6, doi:10.1109/ACED50605.2021.9462263.

[10] J. J. Justo, F. Mwasilu, E. K. Kim, J. Kim, H. H. Choi, and J. W. Jung, "Fuzzy Model Predictive Direct Torque Control of IPMSMs for Electric Vehicle Applications," *IEEE/ASME Trans. Mechatronics*, vol. 22, no. 4, pp. 1542–1553, 2017, doi: 10.1109/TMECH.2017.2665670.

[11] S. Gdaim, A. Mtibaa, and M. F. Mimouni, "Design and Experimental Implementation of DTC of an Induction Machine Based on Fuzzy Logic Control on FPGA," *IEEE Trans. Fuzzy Syst.*, vol. 23, no. 3, pp. 644–655, 2015, doi: 10.1109/TFUZZ.2014.2321612.

[12] L. Romeral, A. Arias, E. Aldabas, and M. G. Jayne, "Novel Direct Torque Control (DTC) Scheme with Fuzzy Adaptive Torque-Ripple Reduction," *IEEE Trans. Ind. Electron.*, vol. 50, no. 3, pp. 487–492, 2003, doi: 10.1109/TIE.2003.812352.

[13] M. Zigliotto, "Permanent Magnet Synchronous Motor Drives," *Power Electron. Convert. Syst. Front. Appl.*, vol. 12, no. 3, pp. 313–332, 2016, doi: 10.1049/PBPO074E_ch10.

[14] Habibzadeh, M. and S. M. Mirimani. "A Novel Implementation of SVM-DTC: Integrated Control of IPM Motor and Hybrid Energy Storage System for Electric Vehicle Application," *Iran. J. Electric. Electron. Eng.*, vol. 17, no. 4, pp. 1931–1931, 2021.

[15] R. Sitharthan, S. Krishnamoorthy, P. Sanjeevikumar, J. Holm-Nielsen, R. Raja Singh, and M. Rajesh, "Torque Ripple Minimization of PMSM Using an Adaptive Elman Neural Network-Controlled Feedback Linearization-Based Direct Torque Control Strategy," *Int. Trans. Electric. Energy Syst.*, vol. 31, e12685, 2021, doi: 10.1002/2050-7038.126850.

[16] C. Wang and Z. Q. Zhu, "Fuzzy Logic Speed Control of Permanent Magnet Synchronous Machine and Feedback Voltage Ripple Reduction in Flux-Weakening Operation Region," *IEEE Trans. Ind. Appl.*, vol. 56, no. 2, pp. 1505–1517, 2020, doi:10.1109/TIA.2020.2967673.

Chapter 11

Non-parametric auto-tuning of PID controllers for DC–DC converters

Ahmed Shehada[1], *Abdul R. Beig*[1] *and Igor Boiko*[1]

Nomenclature

a or a_0	amplitude of MRFT oscillations as measured in the plant output
d	duty-cycle input to the dc–dc converter
e	error signal (difference between reference and plant output)
f_s	switching (PWM) frequency of the dc–dc converter, Hz
u	controller output (or input to the plant)
v_o	output voltage of the converter, V
y	plant output
c_1, c_2, c_3	coefficients of the PID tuning rules of the MRFT tuning method
$e_{\max}$	last recorded maximum of e during the test stage of the MRFT
$e_{\min}$	last recorded minimum of e during the test stage of the MRFT
h	magnitude of the MRFT relay/algorithm – a perturbation added to D
C or C_o	output filter capacitance, F
D	nominal or steady-state duty-cycle
G	Laplace transfer function of the dc–dc converter
K_c	proportional gain of the PID controller
K_u	ultimate gain
L	filter inductor of the dc–dc converter, H
$N(a)$	describing function approximation of the MRFT relay/algorithm
R or R_o	output load resistance of the dc–dc converter, Ohm
T_i	integral time constant of the PID controller, s
T_d	derivative time constant of the PID controller, s
T_s	sampling time period of the digital controller of the dc–dc converter, s
T_u	ultimate time period, s

[1]Department of Electrical Engineering and Computer Science, Khalifa University, United Arab Emirates

V_s	nominal input voltage of the dc–dc converter, V
W_c	frequency response of the controller
W_p	frequency response of the plant
β	a parameter of the test stage of the MRFT tuning method
γ_m	gain margin
τ_m	dead-time or time-delay in system frequency response, s
ϕ_m	phase margin, rad
ω	angular frequency, rad/s
ω_π	phase cross-over $(-180°)$ frequency, rad/s
Ω_0	frequency of the MRFT oscillations, rad/s

11.1 Introduction

The Ziegler and Nichols (Z–N) open-loop and closed-loop tuning methods [1] were a major success in terms of providing a standard for the tuning of PID controllers. While neither of them guaranteed stability (except for the closed-loop method in the case of a P-only controller), in most cases, they provided a reliable initial tuning that brought the system to stability where it could be further fine-tuned manually. In the closed-loop method, which will be further considered, Ziegler and Nichols used the continuous cycling test in order to identify the system's ultimate gain (K_u) and ultimate time period (T_u) that plug into the Z–N tuning rules. Other tests for obtaining the same were subsequently proposed, but for decades the continuous cycling test was perhaps the most widely adopted. However, a drawback of the continuous cycling test was that it relied on a trial-and-error approach to drive the system to the state of limit-cycle oscillation, and so was not suitable for use in automatic online tuning (or auto-tuning) applications. The introduction of the relay feedback test (RFT) [2] by Åström and Hägglund in 1984 provided a more convenient method for obtaining K_u and T_u that did not involve trial-and-error, and thus was a good candidate for auto-tuning. The RFT largely replaced the continuous cycling test as a means of obtaining K_u and T_u, but for the tuning part (i.e. calculating the PID parameters), the Z–N tuning rules and its derivatives continued to be the method of choice in industrial applications. Variations of the original RFT also later emerged that enabled exciting oscillations at frequencies other than the phase cross-over frequency, which can be advantageous in some situations. So while several methods of obtaining K_u and T_u became available, what remained lacking were tuning rules that can guarantee stability, and perhaps further provide a sufficiently good level of performance. Yet of more significance would be tuning rules that can be specifically tailored to a certain class of plant dynamics. The aim of the work in this chapter is to provide such an auto-tuning for the class of digital voltage-mode PID-controlled dc–dc buck converters. An existing auto-tuning technique is considered for this purpose; the scope of this chapter is to provide the necessary background related to this auto-tuning technique, and explain how it is adapted to the considered class of converters. But first a review of related literature is provided in the following subsection.

11.1.1 A review of auto-tuning applied to digitally-controlled dc–dc power electronic converters

Many control methods have been proposed and successfully applied to dc–dc converters. Taking the example of the simple dc–dc buck converter, besides the conventional voltage-mode control, there exists average-current-mode and peak-current-mode control [3], constant on-time control, V^2 control [4], and others. Yet, the standard single-loop PID voltage-mode control of a dc–dc buck converter continues to be a solution of choice in many applications. Both analog and digital means have been used to implement PID control in dc–dc converters. The analog means are dominant, especially in high-volume very-low-cost applications, due to their simplicity and lower cost. However, the continuous improvement in the speed and performance of digital controllers, along with the steady drop in their cost, has made them increasingly popular in switching power converter applications [5]. Some of the advantages offered by digital controllers are their immunity to the tolerance and ageing that affect the components of an analog controller, and that they allow for the implementation of advanced control techniques. The rise of digital control has also provided more opportunities for the application of auto-tuning. Auto-tuning, unlike the traditional controller design methods that are usually based on simplified analytical models, has the advantage of being performed on an actual system, which accounts for several phenomena not captured by the typically used analytical models. These include:

1. Accounting for the actual value of the converter's LC components as opposed to using estimated values and substituting them in a simplified analytical model; estimated values tend to differ from the actual ones due to the tolerance and ageing phenomena of the LC components, and it is not feasible to, say, measure the actual value of the LC components for every converter on a manufacturing line in order to get an accurate model.
2. Capturing parasitic effects associated with the components, as well as smaller delays and non-linearities, which introduce additional dynamics that are unaccounted for in typical simplified models.
3. Accounting for the actual value of the connected load and the influence of the input capacitance of a next-stage converter.

Auto-tuning methods may be broadly categorized into parametric and non-parametric methods. Parametric methods are those in which a system identification is first performed, and the identified system is used to tune the controller. Non-parametric methods on the other hand do not involve a system identification step. But both approaches start with a test stage in which the dynamics of the plant are excited at one or more frequencies.

Several methods of controller auto-tuning for dc–dc converters have been reported in literature. In [6–8], the test stage consists of injecting a pseudo-random binary sequence (PRBS), which is a digital approximation of white noise. Post-processing of the output using cross-correlation and discrete Fourier Transform (DFT) techniques provides the frequency response (FR) of the converter. Such technique is used in [6] to auto-tune a forward dc–dc converter, where a full parametric identification is performed based on the obtained FR data; the controller is then tuned using a direct digital design technique. A remark regarding this work is that a cost function is minimized online during the system identification process, which is likely to take appreciable time. In [7], a buck converter is auto-tuned also using PRBS injection, and the FR data obtained after post-processing is directly used to tune the controller (without identifying the converter's physical parameters). However, the tuning of the controller consists of several stages, with some requiring several iterations. This may be of concern in practical applications due to the level of complexity involved. It is also noted that in [7] the system is operated in open-loop mode during the test stage. In [9], a different test stage is employed; small signal sinusoidal injection around the nominal duty-cycle is used, and a DFT post-processing step is used to obtain the system's FR data. However, the duration of the test stage using such frequency-sweep-based methods is considerably long.

Other reported dc–dc converter auto-tuning methods include RFT-based auto-tuning [10–13]. The RFT begins with a test stage where ON/OFF control using a relay is applied to excite limit-cycle oscillations in a system. The frequency and the amplitude of these oscillations are then used to tune the controller. In [10], the relay is applied as a perturbation (h) around the steady-state duty-cycle, D, of a buck converter; the duty-cycle is $D + h$ when the relay is ON and $D–h$ when relay is OFF. The ultimate gain and ultimate time period are measured from the oscillations that appear in the output voltage, and the Z–N tuning rules are used to tune a PID controller using these measurements. However, as earlier mentioned, the Z–N tuning rules do not guarantee stability, and even if the system is stable it is not possible to specify a certain gain or phase margin. In [11,12], modified forms of the RFT are used for the auto-tuning of a PID-controlled buck converter. First, the LC resonant frequency is identified by adding an integrator in series with the relay in order to excite oscillations at the $-90°$ frequency instead of the $-180°$ frequency. Several tuning stages are then used to place the zeroes of the PID such that certain performance criteria (like the phase margin and bandwidth) are met. However, the use of several stages with iterations may not be preferred in practical situations. Another RFT-based parametric auto-tuning of a PID-controlled buck converter is reported in [13], where limit-cycle oscillations are excited in a dc-dc buck converter by reducing the resolution of the digital PWM, which imitates the relay action. Integral-only control is maintained during the test stage. System identification is then performed, and a stored lookup table is used to select the controller coefficients. Though the tuning is simpler than methods in [7, 11, 12], a tradeoff exists between the size of the stored lookup table and the number of possible discrete control laws.

Yet another approach that has been reported is that of continuous adaptive tuning such as in [14], where small perturbations are added to the duty-cycle of a buck converter during the closed-loop operation, and PID gains are continuously adjusted in order to achieve certain performance criteria. While such approach has the advantage of providing continued regulation despite the continuous tuning in the background, some prior knowledge of the converter's parameters may be required in order to ensure that the controller gain update rates do not lead to instability of the adaptive loop. Also, with such methods, there may be situations where the desired performance criteria cannot be met, and so a way to handle such situations needs to be included.

It is seen from the review above that the reported methods vary widely in terms technique, duration, etc. However, it is also evident that there are certain disadvantages that need to be overcome. For example, while methods [6,7] and [11–13] claim good performance of the tuning, the tuning either takes significant time and/or has some level of complexity. A practical auto-tuning should have short and simple test and tuning stages. The RFT-based method of [10] indeed has short and simple test and tuning stages, but as noted earlier the tuning is quite simplistic, with no guarantee on stability. A method that combines simplicity (goal 1) and guaranteed stability (goal 2), and that can yet provide a minimum level of dynamic performance (goal 3), is the aim of this work.

11.1.2 The MRFT and coordinated test and tuning

A method that addresses PID tuning, allows for the specification of a stability margin, and that is also suitable for auto-tuning, is the modified relay feedback test (MRFT) tuning method [15–17]. It uses a single test stage and has simple PID tuning rules. The method gains its strength from its *coordinated test and tuning concept*; the *test* stage uses a parameter (β) that is *coordinated* with the *tuning* stage in that β and the coefficients of the PID tuning rules (c_1, c_2, c_3) are related through constraints that guarantee a specified gain margin or a phase margin – though the two cannot be specified simultaneously in this method. First the test (the MRFT) is run, then the PID parameters are computed through special tuning rules (similar in form to the Z–N rules) that embed the stability margin specification. This already achieves goal 1 (simplicity) and goal 2 (stability). The third goal of ensuring good dynamic performance is achieved with the help of a prior offline optimization of the set (β, c_1, c_2, c_3). This optimization is done only once for a certain class of plant dynamics, e.g. the class of PID-digitally controlled voltage-mode dc–dc buck converters. Since this design (selection) of the tuning set (β, c_1, c_2, c_3) is done in advance, the online tuning effort is substantially reduced, which results in a fast and simple tuning. Other advantages of this method are that it does not require any information on the parameters of the system and that the oscillations excited in this method are quite small and so hardly affect the system's normal operation. This chapter discusses the details of applying such method to the class of voltage-mode digitally controlled dc–dc buck converters, which constitutes two major tasks. The first task is the adaptation of the test stage of the MRFT tuning method to suit switching power converters. The second task is the design (selection) of a tuning set (β, c_1, c_2, c_3) that results in a near-optimal dynamic performance for the PID controller obtained after tuning with the MRFT method. The explanation provided in this chapter can also be used as a guide for applying the MRFT tuning method to any class of PID-controlled systems.

To conclude this introduction, a summary is given below of the desired characteristics in an auto-tuning and how the MRFT tuning method helps achieve them:

1. The auto-tuning should guarantee stability.
   The MRFT tuning method allows for the specification of a gain or phase margin and so stability is inherently guaranteed in this method.
2. The auto-tuning should work for a wide range of plant dynamics; for example, in the case of dc–dc buck converters with a load of resistive behavior, the auto-tuning should accommodate a wide range of designs, i.e., wide ranges of the filter inductor (L), output capacitor (C), load (R), and switching frequency (f_s).
   In the MRFT tuning method, as will be seen later, a wide range of plant dynamics is considered as part of the process of designing the tuning rules.
3. There should be provision for ensuring a minimum/satisfactory dynamic performance.
   The tuning rules in this method may be optimized for a given class of plant dynamics with the objective of meeting a certain performance criterion or a combination of criteria and constraint(s).
4. The auto-tuning should be practical in the sense that it should not result in a large disturbance that would affect the converter's operation, or require that the converter be run below rated conditions for the duration of the (re)tuning.
   Oscillations excited in the MRFT tuning method are typically small and should not cause any appreciable disturbance to the converter.
5. The auto-tuning should not depend on or require knowledge of any parameter of the plant.
   This is true for the MRFT tuning method, which makes it simple to implement, and also makes it quite insensitive to parameter variation.

This chapter is organized as follows: Section 11.2 provides an overview of the MRFT tuning method and explains how the test can be adapted for switching power converters that work on the principle of pulse-width modulation (PWM). Section 11.3 gives the procedure for designing tuning rules for the class of digitally controlled voltage-mode dc–dc buck converters. Section 11.4 provides experimental results for the MRFT tuning method which validate the adapted test-stage described in Section 11.2 and the tuning rules derived in Section 11.3. Finally, a conclusion is given in Section 11.5.

11.2 Overview of the MRFT tuning method

While it is usually not possible to design a single controller that works for a wide range of converter designs, since different L, C, R, and f_s require significantly different PID parameters, it is indeed possible to design tuning rules that fit a wide range of designs. The MRFT tuning method is a non-parametric tuning technique for PID-type controllers that are based on such a tuning rules approach [15]. The first step in this method is a test stage in which oscillations are excited at a specific, pre-determined phase lag of the plant frequency response; amplitude and time period measurements of the oscillations are taken. The second step is to simply calculate the new PID parameters using these measurements with pre-derived tuning rules. The PID controller in this method must be of the form below, where $K_c > 0$ is the proportional gain, $T_i \geq 0$ is the integral time constant and $T_d \geq 0$ is the derivative time constant:

$$W_c(s) = K_c\left(1 + \frac{1}{T_i s} + T_d s\right). \tag{11.1}$$

Figure 11.1(a) shows the block diagram of the test stage of the MRFT tuning method, where $W_p(s)$ is the plant to be controlled. The test is performed by basically replacing the PID controller with the MRFT algorithm (only for the duration of the test).

In the original formulation of the MRFT tuning method [15], the MRFT algorithm is given by the discontinuous control

$$u(t) = \begin{cases} h, \ if & \begin{matrix} e(t) \geq -\beta e_{\min} \\ or \\ \{e(t) \geq -\beta e_{\max} \ \& \ u(t-) = h\} \end{matrix} \\ -h, \ if & \begin{matrix} e(t) \leq -\beta e_{\max} \\ or \\ \{e(t) \leq -\beta e_{\min} \ \& \ u(t-) = -h\} \end{matrix} \end{cases} \tag{11.2}$$

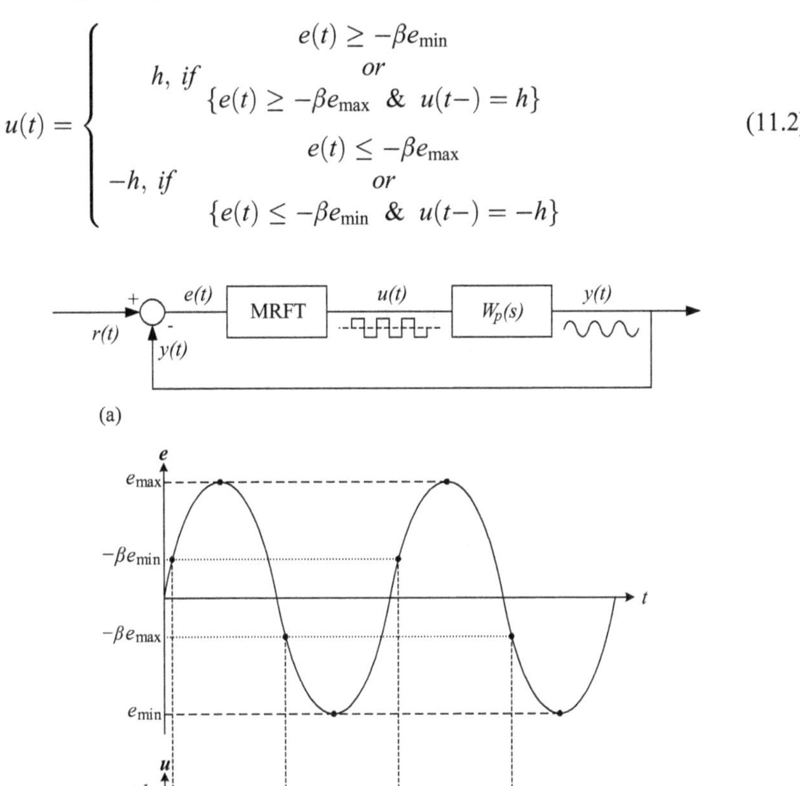

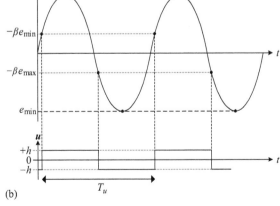

Figure 11.1 Block diagram and illustrative waveforms of the MRFT test stage

where β is a test-stage parameter that is restricted to $-1 < \beta < 1$, h is the value of the control input, $u(t-)$ is the control output immediately prior to time t, and $e_{\max}$ and $e_{\min}$ represent the last maximum and the last minimum of the error signal $e(t)$, respectively. Figure 11.1(b) provides a diagrammatic illustration of the waveforms of $e(t)$ and $u(t)$, for the duration of the test. The test is started with $e_{\max}$ and $e_{\min}$ set to zero. Since $u(t)$ is always non-zero, oscillations start to develop in $y(t)$ and $e(t)$, at a frequency decided by the value of the test-stage parameter β. Every time a minimum ($e_{\min}$) or maximum ($e_{\max}$) is recorded in the oscillations, the corresponding switching condition ($-\beta e_{\max}$ or $-\beta e_{\min}$) for the upcoming half-cycle is updated. The oscillations typically stabilize after a few transient cycles such that the amplitude $a_0 = |e_{\max}| = |e_{\min}|$; the ultimate gain ($K_u$) is then computed using

$$K_u = \frac{4h}{\pi a_0}. \tag{11.3}$$

The time period of the oscillations (T_u) and K_u are then used to calculate the new PID parameters using the tuning rules

$$K_c = c_1 K_u, T_i = c_2 T_u, T_d = c_3 T_u, \tag{11.4}$$

where the coefficients c_1, c_2, and c_3 are positive constants [15]. The design (selection) of the set (β, c_1, c_2, c_3) is dealt with in the next section. In the remainder of this section, the adaptation of the test stage of this method to suit PWM converters is explained, followed by a mathematical analysis of the conditions at which the oscillations occur.

11.2.1 Adapting the test stage of the MRFT tuning method for PWM converters

The general schematic of a synchronous dc-dc buck converter with digital voltage-mode control is given in Figure 11.2, where DPWM stands for digital PWM. Switches Q_1 and Q_2 operate in a complementary fashion, so that when one is ON the other is OFF. The *actuator* in this converter is the combination of the modulator (PWM) and the switches. The DPWM accepts a positive duty-cycle, and translates it into width-modulated pulses that control the switches. In order to perform the test

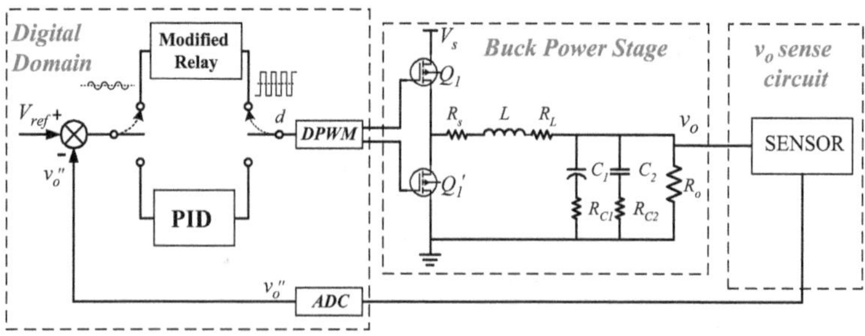

Figure 11.2 Schematic diagram of the digitally controlled dc–dc buck converter

stage of the MRFT tuning method (as illustrated in Figure 11.1), it may seem intuitive to eliminate the PWM part for the duration of the test, so that when the output of the MRFT relay is positive, the main switch (Q_1) is turned ON, and when the MRFT relay output is negative, the main switch (Q_2) is turned OFF. In other words, a straightforward implementation of the test would be to allow the MRFT to directly control the converter switches. While such approach would indeed excite oscillations at the intended point in the frequency response (as decided by the value of β), the converter will switch at the MRFT frequency, which is typically much lower than the designed PWM frequency of the converter. This will result in an increased ripple in the inductor current and output voltage. A more appropriate and elegant approach is that of double-modulation, where the PWM is still maintained during the test stage, and the MRFT modulation is only added to the duty-cycle that is input to the PWM [18]. This is done by setting the relay output $u(t)$ to either $D + h$ (when the relay is ON) or $D–h$ (when the relay is OFF), where D is the nominal duty-cycle output by the controller at steady-state just prior to application of the MRFT, and where h is typically a small percentage (e.g. 3%) of D. With this approach, the MRFT oscillations appear superposed on top of the output voltage. In terms of gate pulses, the MRFT manifests as a series of slightly wider PWM pulses corresponding to the $D + h$ duty-cycle and lasting for $T_u/2$, followed by a series of slightly narrower PWM pulses corresponding to the $D–h$ duty-cycle and also lasting for $T_u/2$. Besides the obvious advantage of keeping all variables within their linear zone of operation, this approach also ensures that even during the test stage the switching frequency for which the converter's components were designed for is maintained. This preserves the normal ripple levels, thus minimizing the impact of the MRFT auto-tuning on the converter's voltages and currents, making it possible to execute during normal operation of the converter (in most applications). The modified MRFT algorithm in this approach is expressed as

$$u(t) = \begin{cases} D + h, \text{if} & \begin{matrix} \{e(t) \geq -\beta e_{\min} \;\&\; u(t-) = D - h\} \\ or \\ \{e(t) \geq -\beta e_{\max} \;\&\; u(t-) = D + h\} \end{matrix} \\ & \\ D - h, \text{if} & \begin{matrix} \{e(t) \leq -\beta e_{\max} \;\&\; u(t-) = D + h\} \\ or \\ \{e(t) \leq -\beta e_{\min} \;\&\; u(t-) = D - h\} \end{matrix} \end{cases} \tag{11.5}$$

where it is also noted that the MRFT algorithm of (11.5) has been modified from its original form in [15] to accommodate negative values of β as well.

11.2.2 Specification of the gain margin or phase margin

Limit-cycle oscillations for tuning purposes are normally excited at the phase crossover ($-180°$) frequency, ω_π, of the plant. Along with its corresponding plant gain, ω_π can be used directly, for example, with the Z–N tuning rules to calculate PID controller parameters. However, the Z–N rules guarantee stability (with a gain margin of 2) only if a proportional-only controller is used – which is a consequence of the plant and the open-loop system (consisting of the controller and plant) having the same ω_π. But if a PI, PD, or PID is used, ω_π of the resulting open-loop system is different from that of the plant alone. Then, not only the gain margin of 2 is not obtained, but stability itself cannot be guaranteed. It would be quite significant if the test can be modified such that the oscillations occur at what would be ω_π of the open-loop system. This is achieved through the holistic tuning approach of the MRFT tuning method, where the test is coordinated with the tuning rules in such a manner that allows for the specification of the either the gain or the phase margin. Some relevant mathematical analysis is first presented before the tuning rule derivation is taken up in the following section.

Let $\Omega_0 = 2\pi/T_u$ be the frequency of the MRFT oscillations (in rad/s). The harmonic balance, described by the equation below, should be satisfied at the frequency Ω_0 for the plant W_p and the MRFT relay, where the latter is described by its describing function (DF) approximation, $N(a)$:

$$W_p(j\Omega_0)N(a_0) = -1 \rightarrow W_p(j\Omega_0) = -\frac{1}{N(a_0)} \tag{11.6}$$

Note that $N(a)$ is a function of the amplitude of the oscillations, a. Let b be the hysteresis of the modified relay, where $b = |\beta e_{\max}| = |-\beta e_{\min}| = \beta a$, for any amplitude of oscillations, a. The general expressions for $N(a)$ and $-1/N(a)$ are given by:

$$N(a) = \frac{4h}{\pi a}\sqrt{1-\left(\frac{b}{a}\right)^2} - j\frac{4h}{\pi a}\left(\frac{b}{a}\right) = \frac{4h}{\pi a}\left(\sqrt{1-\beta^2} - j\beta\right)$$

$$\rightarrow -\frac{1}{N(a)} = -\frac{\pi a}{4h}\left(\sqrt{1-\beta^2} + j\beta\right)$$

The magnitude and phase delay expressions of $-1/N(a)$ are

$$\left|\frac{1}{N(a)}\right| = \left|-\frac{\pi a}{4h}\right|\sqrt{\left(\sqrt{1-\beta^2}\right)^2 + (\beta)^2} = \frac{\pi a}{4h}, \tag{11.7}$$

$$\angle\left(-\frac{1}{N(a)}\right) = \angle\left(-\frac{\pi}{4}\frac{a}{h}\right) + \angle\left(\left(\sqrt{1-\beta^2} + j\beta\right)\right)$$

$$\angle\left(-\frac{1}{N(a)}\right) = -\pi + \tan^{-1}\frac{\beta}{\sqrt{1-\beta^2}} = -\pi + \sin^{-1}\beta \tag{11.8}$$

Figure 11.3 shows the Nyquist plots of the plant $W_p(j\omega)$ and $-1/N(a)$. The plot of $-1/N(a)$ is a ray starting from the origin and extending in proportion with a, forming an angle of $\sin^{-1}\beta$ with the negative real axis. The state of sustained

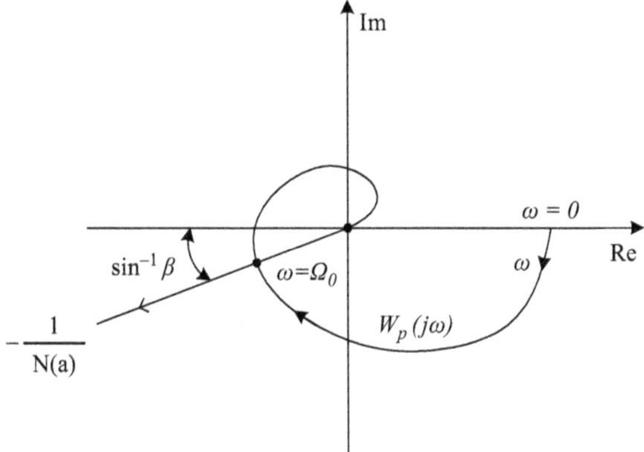

Figure 11.3 Nyquist of the plant and the negative reciprocal of the relay's DF

oscillations is represented by the intersection of the two plots, occurring at frequency $\omega = \Omega_0$, and expressed mathematically by

$$\left|W_p(j\Omega_0)\right| = \left|-\frac{1}{N(a_0)}\right| = \frac{\pi a_0}{4h} = \frac{1}{K_u}, \tag{11.9}$$

$$\angle W_p(\mathrm{j}\Omega_0) = \angle\left(-\frac{1}{N(a_0)}\right) = -\pi + \sin^{-1}\beta \tag{11.10}$$

where (11.9) is obtained using (11.3) and (11.7). It is therefore obvious from this analysis and from Figure 11.3 that by varying β one can control the phase lag (and thus indirectly the frequency) at which oscillations are excited.

With the theory laid above, gain or phase margin specifications may be now explained. The theory of the MRFT tuning method specifies that selecting the tuning set (β, c_1, c_2, c_3) as per the following constraints would guarantee the specified gain margin, γ_m, for any arbitrary system [15]:

$$\begin{aligned} \gamma_m &= \frac{1}{c_1\sqrt{1+\xi^2}} \\ \beta &= -\frac{\xi}{\sqrt{1-\xi^2}} \end{aligned}, \text{ where } \xi = 2\pi c_3 - \frac{1}{2\pi c_2}. \tag{11.11}$$

This is a consequence of the MRFT accounting for the phase lag introduced by the PID controller so that the specified gain margin already takes that into account; further detailed reasoning may be found in [16]. The tuning rules of the MRFT tuning method may alternatively be designed (selected) in order to guarantee a

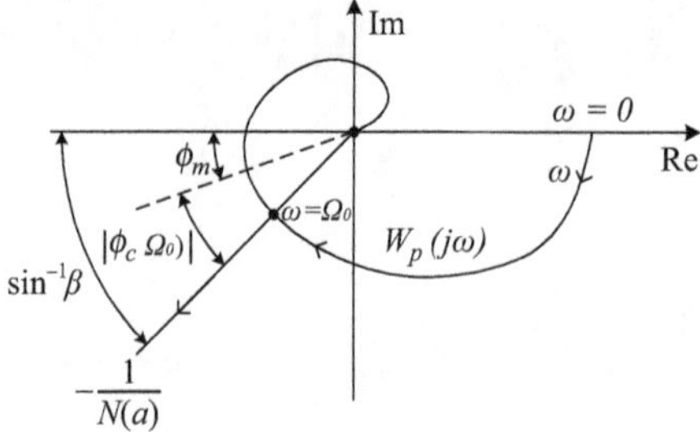

Figure 11.4 Phase margin specification in the MRFT-based auto-tuning

specified phase margin, ϕ_m. This is achieved by selecting (β, c_1, c_2, c_3) according to the following (different) set of constraints [15]:

$$\begin{aligned} c_1\sqrt{1+\xi^2} &= \quad 1 \\ \beta &= \sin\left(\phi_m - \tan^{-1}\xi\right) \end{aligned}, \text{ where } \xi = 2\pi c_3 - \frac{1}{2\pi c_2}. \tag{11.12}$$

Proof of this is given in [18], which can be understood by considering the diagram in Figure 11.4, which represents the state of sustained oscillations due to the MRFT (at the test frequency $\omega = \Omega_0$); $\phi_c(\Omega_0)$ represents the phase lag due to the controller at Ω_0. The constraints (11.12) are formulated so that introducing the PID controller results in a phase shift such that the phase margin exactly equals the specified ϕ_m.

Next, it is noted that criteria (11.11) for guaranteeing a specified gain margin, or criteria (11.12) for guaranteeing a specified phase margin, each have two equations, whereas in each case there are four parameters to select, namely β, c_1, c_2, and c_3. A degree of freedom therefore exists in the selection of these variables. This freedom is used to optimize the selection in order to meet a certain performance criterion, as explained in the following section.

11.3 Derivation of optimal tuning rules

In the MRFT tuning method, optimal tuning rules may be developed for a particular class of plant dynamics, whose so-called *situational parameters* are assigned ranges of variations. This is achieved through solving the problem of parametric optimization for (β, c_1, c_2, c_3) with a certain performance criterion while satisfying the constraints (11.11) or (11.12), as first suggested in [16,17]. γ_m or ϕ_m may be specified, but if their best values are not known (which is quite common), they may also be included as decision variables in the optimization, making it a set of five variables: β, c_1, c_2, c_3, and γ_m (or ϕ_m). But only three can be taken as decision variables, since the remaining are given by the two constraints of (11.11) or (11.12). The approach taken in this work is to set as decision variables the ones that require additional constraints to be placed on them. β is taken as one decision variable, as this makes it easier to constrain it to its allowed range of $(-1, 1)$. As for γ_m (or ϕ_m), it is desirable to have a value that is above a certain minimum, and so it is also beneficial to set γ_m (or ϕ_m) as a decision variable. Finally, c_2 is confined to [0.1, 100] in order to limit the integral action, and so c_2 is taken as the third decision variable.

11.3.1 Development of the model structure and parameter ranges

The first step is to develop a model of the considered class of plant dynamics, that being the class of digital voltage-mode dc–dc buck converters. The following control-to-output transfer function (TF) model is used, where d is the duty-cycle, v_o the output voltage, V_s the input voltage, L the filter inductor, C_o the total output capacitance (including C_1 and C_2 in Figure 11.2), and R_o the load:

$$G = \frac{v_o}{d} = e^{-\tau s}\frac{V_s}{LC_o s^2 + \frac{L}{R_o}s + 1}. \tag{11.13}$$

The model of (11.13) is considered accurate enough if the equivalent series resistance (ESR) of the output capacitance (represented by R_{C1} and R_{C2} in Figure 11.2) is low, especially in relation to R_o, such that all of the ESR can be ignored. This is true for most modern buck converters, which mainly use ceramic or film capacitors that have very low ESR. Also note that the inductor current sense resistor (R_s), and the inductor series resistance (R_L) are ignored as well. Let the sampling time period be T_s (which is also equal to one PWM time-period). τ represents the total delay in the system and is given by the sampling and control delay (one full T_s) plus a $0.5T_s$ delay due to the triangular PWM, for a total of $\tau = 1.5T_s$.

It is first desired to reduce the number of variables in the model structure of (11.13), which may be done through exploring standard design equations for L and C_o as they are likely to include common terms. First, the minimum inductance ($L_{\min}$) is expressed using the equation below [19], where V_o is the nominal output voltage, I_o the (maximum) load current, and f_s the switching frequency:

$$L_{\min} = \frac{V_o(1-D)}{K_L I_o f_s} = \frac{V_o(1-0.5)}{0.4 I_o f_s} = \frac{5}{4}\frac{R_o}{f_s}$$

The duty-cycle (D) above was set to the maximum, which is around 0.5 for practical buck converter applications, in order to get the minimum L. Similarly, the normalized inductor current ripple percentage (K_L) was set to 0.4, which is the maximum in typical designs. Note that the ratio of V_o to I_o is expressed as an equivalent (minimum) resistance (R_o). To accommodate designs using a larger L, the scaling factor (α_L) defined below is used and is assigned a range between 1 and 2:

$$L = \alpha_L L_{\min} = \alpha_L \frac{5}{4}\frac{R_o}{f_s}. \tag{11.14}$$

As for capacitor design the criterion of maximum overshoot ($\Delta V_{o\text{-}os}$) due to a load transient (ΔI_{step}) is used. The minimum capacitance ($C_{o\text{-}\min}$) using this criterion is given below [19], where a 20–80% load step is considered, and where K_L is now set to only 0.2 in order to get the minimum C_o. A scaling factor (α_C) is used for C_o:

$$C_{o-\min} = \frac{\left(\Delta I_{step} + 0.5K_L I_o\right)^2 L}{\left(V_o + \Delta V_{o-os}\right)^2 - V_o^2}$$

$$= \frac{((0.8 - 0.2)I_o + 0.5 \times 0.2 \times I_o)^2 \alpha_L L_{\min}}{(V_o + 0.05V_o)^2 - V_o^2}$$

$$C_{o-\min} = \frac{(0.6 + 0.1)^2 I_o^2}{\left(1.05^2 - 1\right) V_o^2} \alpha_L \frac{5}{4} \frac{R_o}{f_s} = 5.98 \frac{1}{R_o^2} \alpha_L \frac{R_o}{f_s} = \frac{5.98\alpha_L}{f_s R_o}$$

$$C_o = \alpha_C C_{o-\min} = \alpha_C \alpha_L \frac{5.98}{f_s R_o}. \tag{11.15}$$

α_C in (11.15) is restricted to the range [1, 1.5], since α_L (which has a range [1,2]) appears in the expression for $C_{o\text{-min}}$, which makes the effective scaling of C_o from 1 to 3. Using (11.14) and (11.15), the TF of (11.13) may be rewritten as follows:

$$G = \frac{v_o}{d} = e^{-\tau s} \frac{V_s}{7.475\alpha_L^2 \alpha_C \frac{s^2}{f_s^2} + 1.25\alpha_L \frac{s}{f_s} + 1}. \tag{11.16}$$

The dc scaling term (V_s) may be removed for simplicity as it only affects the value of K_c, but will not have an effect on the tuning rule coefficient c_1. Likewise, a time-scaled Laplace variable ($s' = s/f_s$) may be used to eliminate f_s, since such scaling only affects T_i & T_d and not c_2 or c_3. The only remaining parameters are thus α_L and α_C, as shown in the TF below:

$$G = \frac{v_o}{d} = e^{-1.5s'} \frac{1}{7.475\alpha_L^2 \alpha_C s'^2 + 1.25\alpha_L s' + 1}. \tag{11.17}$$

11.3.2 Procedure for obtaining the optimal tuning rules

The parameters α_L and α_C make up the *domain of situational parameters* over which the optimization is performed. Each situational parameter is assigned a discretized range as follows:

$$\alpha_L = [1, \quad 1.25, \quad 1.5, \quad 1.75, \quad 2], \text{ and}$$

$$\alpha_C = [1, \quad 1.125, \quad 1.25, \quad 1.375, \quad 1.5].$$

which results in 5×5 = 25 combinations that are denoted θ_n, $n = [1, 2, \ldots, 25]$. The θ_n thus define the space of the situational parameters. The plant models based on the TF of (11.17) that are formed from the different θ_n are denoted G_n, $n = [1, 2, \ldots, 25]$. Because the plant model is not fixed but is given by a set of models, a special formulation of optimization on the domain of situational parameters is used [16]. The situational parameters are considered as representing an uncertainty that must be compensated for in the computation of a cost function, so that various tuning rules can be weighed against each other. Such problem has been solved in [16] using a two-stage optimization process. Let $\mathbf{x}^T$ denote the tuning set (β, c_1, c_2, c_3, γ_m) for gain margin specification or (β, c_1, c_2, c_3, ϕ_m) for phase margin specification. In the first stage, a conventional optimization problem for $\mathbf{x}^T$ is solved for each G_n, based on a performance criterion f that is taken here as the "integral of time weighted absolute error" (ITAE). The ITAE is computed during the optimization through simulation of a setpoint-step test (i.e. a step in the reference output voltage) applied to the closed-loop system consisting of a certain G_n and a certain PID controller. Let $\mathbf{x}^*_n$ be the $\mathbf{x}$ that produces the minimum ITAE when used to tune a G_n formed from a single set of situational parameters θ_n. This first stage of optimization is expressed mathematically as follows:

$$f^*(\boldsymbol{\theta}_n) = \min\{f(\boldsymbol{\theta}_n, \mathbf{x})\}, \quad n = [1, 2, \ldots, 25], \tag{11.18}$$

$$\mathbf{x}^*_n := \{\mathbf{x} | f(\boldsymbol{\theta}_n, \mathbf{x}) = f^*(\boldsymbol{\theta}_n)\}, \quad n = [1, 2, \ldots, 25]. \tag{11.19}$$

An explanation of the steps of this stage is given below, where the following routine is repeated for each G_n, $n = [1, 2, \ldots, 25]$:

1. Initial values are assigned to β, c_2, and γ_m (or ϕ_m).
2. The corresponding c_1 and c_3 are calculated using (11.11) for gain margin specification or (11.12) for phase margin specification.
3. The test stage (the MRFT) is simulated for G_n using the initial/current β to obtain T_u and a_0; alternatively, since G_n is known, and given that β corresponds to a certain phase lag, (T_u, a_0) may be obtained analytically from G_n.
4. PID controller coefficients are obtained from (11.4) using the (T_u, a_0) pair from step 3 and the initial/current c_1, c_2, & c_3.
5. f is computed through a closed-loop simulation of a setpoint-step for the system consisting of the current G_n and the PID controller form step 4.
6. β, c_2, and γ_m (or ϕ_m) are updated and steps 2–5 are iterated until the minimum f, denoted $f^*(\boldsymbol{\theta}_n, \mathbf{x})$, is obtained; $\mathbf{x}^*_n$ is the $\mathbf{x}$ that results in this optimal f^* for the given G_n.

The first stage of the optimization is executed using the MATLAB® *fminsearch* function, which is based on the Nelder–Mead simplex direct search algorithm [20]. The output is an array of the optimal $\mathbf{x}$ for each G_n, i.e., $\mathbf{x}^*_n$, $n = [1, 2, \ldots, 25]$. In the second stage of the optimization, one of the $\mathbf{x}^*_n$ is selected as the best optimal set based on the principle of *least performance degradation*, where the chosen $\mathbf{x}^*_n$ is the one that results in the least-worst performance when used to tune all the other G_n. Following is an explanation of the steps of the second stage:

1. For each G_n, $n = [1, 2, \ldots, 25]$, a PID controller is tuned using all $\mathbf{x}^*_m$, $m = [1, 2, \ldots, 25]$. This is done as follows: For each G_n, (T_u, a_0) for every β from $\mathbf{x}^*_m$,

$m = [1, 2, \ldots, 25]$ are found either through an MRFT simulation or directly through the TF G_n. The (T_u, a_0) in each case are used along with the (c_1, c_2, c_3) from the corresponding $\mathbf{x}^*_m$ in the tuning rules (11.4) to design a PID controller. The output of this step is thus 25 PID controllers for each of the 25 G_n.

2. For each G_n, a closed-loop setpoint-step simulation is performed for each of its 25 corresponding PIDs from the previous step. The resulting ITAE is recorded in each case, thus resulting in a 25 × 25 matrix of ITAEs, where each ITAE is denoted $f(\boldsymbol{\theta}_n, \mathbf{x}^*_m)$, where $n = 25$ are the different G_n and $m = 25$ are the different tuning sets $\mathbf{x}^*_m$.
3. To allow for a direct comparison of the ITAE performances (due to the $m = 25$ $\mathbf{x}^*_m$) across the $n = 25$ different G_n, the effect of the situational parameters is removed by expressing all $m = 25$ ITAEs of a given G_n as a ratio to $f^*(\boldsymbol{\theta}_n)$, where $f^*(\boldsymbol{\theta}_n)$ is the ITAE obtained for that G_n with an optimal tuning $\mathbf{x}^*_n$. This quantity is an expression of degradation relative to the optimal tuning case, and it is defined as:

$$q_n^m = f(\boldsymbol{\theta}_n, \mathbf{x}^*_m)/f^*(\boldsymbol{\theta}_n), \tag{11.20}$$

$n = [1, 2, \ldots, 25]$ and $m = [1, 2, \ldots, 25]$.

4. Finally, the optimal tuning set is found as:

$$\mathbf{x_{opt}} := \left\{ \mathbf{x} \middle| \min_{n=1,\ldots,25} \left\{ \max_{m=1,\ldots,25} q_n^m \right\} \right\}. \tag{11.21}$$

In other words, $\mathbf{x_{opt}}$ is the tuning set that when used to tune all G_n, $n = [1, 2, \ldots, 25]$, results in a maximum relative degradation that is lower than any other maximum relative degradation obtained with the tuning of any G_n with any other $\mathbf{x}^*_m$. The two-stage optimization described above was executed for gain margin specification (i.e. using constraints (11.11)), and with the tuning rules being optimized for setpoint-step transients as mentioned above. γ_m was restricted to the range [1,3], and only positive values were allowed for c_1 and c_3. Also as mentioned earlier, β and c_2 were confined to the ranges $(-1, 1)$ and $[0.1, 100]$, respectively. The result of the optimization was the following:

$$\mathbf{x_{opt}}^T = (\beta, c_1, c_2, c_3, \gamma_m) = (-0.3, \quad 0.3, \quad 3.2, \quad 0.05, \quad 3) \tag{11.22}$$

Using these $\mathbf{x_{opt}}$, the maximum relative degradation (q_n^m) was around 1.22. This number can be used as a metric showing the performance of the tuning rules: with the use of the synthesized tuning rules with the coefficients $\mathbf{x_{opt}}$ of (11.22), it is guaranteed that the performance of any system from the considered class will not be worse than 22% of the best theoretical performance for this system, estimated in terms of the ITAE criterion. This fact allows us to term the results of the proposed tuning as *near-optimal*, and the tuning rules themselves as *optimal* for this class of plant dynamics.

11.4 Experimental results

A dc–dc buck converter prototype using a 32-bit microcontroller was used for experimental verification of the method above with the developed tuning rules. An LC filter of $L = 10$ μH and $C = 660$ μF was used. The switching frequency (and the sampling rate) were set to 200 kHz. The output voltage is 2 V, stepped down from an input of 9 V. The auto-tuning was prompted by the user's command, although it could be programmed to run according to a schedule, or even be triggered by a certain event. Using the value of $\beta = -0.3$ from the $\mathbf{x}_{\mathbf{opt}}$ of (11.22), the MRFT test stage was run and the T_u and a_0 of the oscillations were measured. The experimental MRFT oscillations are shown in Figure 11.5(a) and (b), where the A/D converter samples of the output voltage are marked by blue circles, with their values given on the left vertical axis. The red trace labeled "Relay" represents the output of the MRFT algorithm; +1 (on the right vertical axis) indicates an MRFT output of $D + h$, while −1 an MRFT output of $D–h$. An important aspect of the A/D output voltage samples that enables accurate measurement of T_u and a_0 is the absence of switching ripple, which besides the LC filtering is achieved by properly synchronizing the sampling with the PWM switching in order to avoid switching transitions. It is also noted from Figure 11.5 that the amplitude of the MRFT oscillations is relatively low, being < ±0.05 V around the nominal 2 V (i.e. less than 2%). This does not present any significant disturbance to the converter, and thus the test and tuning may be carried during normal operation of the converter. Another advantage noted from Figure 11.5 is the short duration of the test stage.

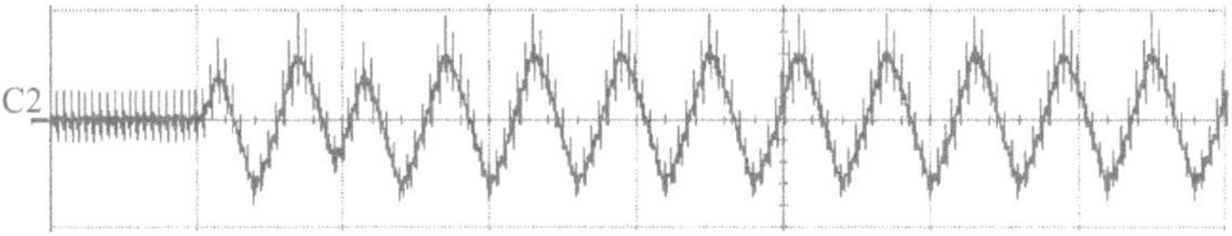

(a) MRFT oscillations as recorded in oscilloscope (100 mV/div, 100 μs/div)

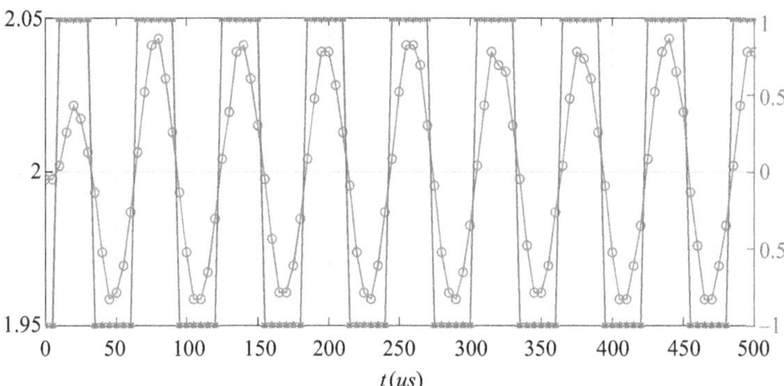

(b) A/D converter samples of MRFT oscillations from microcontroller memory

Figure 11.5 MRFT oscillations in output voltage

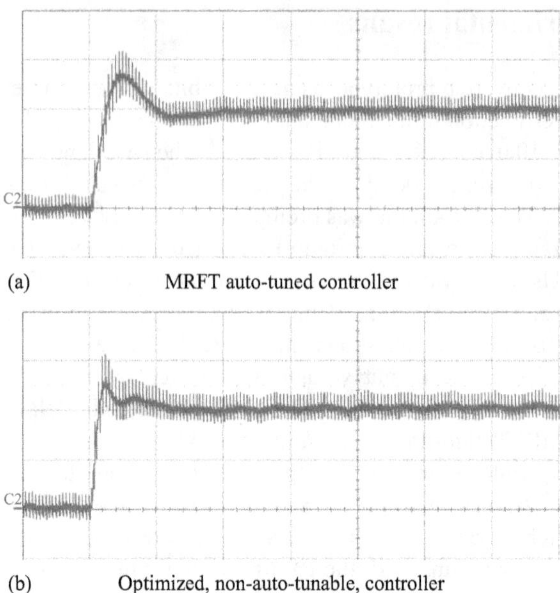

Figure 11.6 Response of output voltage to setpoint-step (50 mV/div, 100 μs/div)

Only a few stable oscillations are required to take reasonably accurate averages of T_u and a_0. For example, to get only five stable oscillations – plus the short initial and final transients – perhaps only 0.5 ms would suffice.

Using the measured T_u and a_0, the new PID controller parameters were computed using the tuning rules (11.4) with the optimal coefficients from (11.22). The performance of the MRFT-auto-tuned controller for a setpoint-step is shown in Figure 11.6, where it is compared to an optimized controller designed using full knowledge of the converter. While the performance of the optimized controller is expectedly better, the MRFT auto-tuned controller, which was tuned not using any knowledge of the converter's parameters, has only slightly higher overshoot and slightly longer settling time. And since the parameters of the test buck converter were randomly picked, the MRFT auto-tuning is expected to also perform well for other converter designs within the LC parameter ranges defined in the derivation of the tuning rules. Furthermore, it is noted that even for dc–dc buck converters that fall outside of the design range used in the tuning rules derivation, the MRFT auto-tuning still guarantees a gain margin of 3, and through that should provide an acceptable performance.

11.5 Conclusion

This chapter has presented the design and implementation of an auto-tuning based on the MRFT tuning method for the class of digitally controlled voltage-mode dc–dc buck converters. The presented auto-tuning is simple and consists of short test and tuning stages, and it guarantees stability through allowing the specification of the gain or phase margin – all without requiring any knowledge of the converter or load parameters. The chapter provides a theoretical background on the MRFT, as well as an explanation of how the test stage is adapted to dc–dc switching power converters. The derivation of optimal tuning rules for the class of digitally controlled voltage-mode dc–dc buck converters is also given in detail. Besides guaranteeing a specified gain margin, the tuning rules are also optimized for dynamic performance of the considered class of converters. It is shown through an experiment on a test converter that the MRFT auto-tuned controller performs closely to an optimized controller designed with full knowledge of the test converter's parameters. It is finally noted that even for dc–dc buck converter designs that have LC parameters outside the design range used in developing the tuning rules, the auto-tuning still guarantees a stable performance with the specified gain margin and through that provides an acceptable performance.

References

[1] J. G. Ziegler and N. B. Nichols, "Optimum settings for automatic controllers," *Trans. ASME*, vol. 64, pp. 759–768, 1942.

[2] K. J. Åström and T. Hägglund, "Automatic tuning of simple regulators with specifications on phase and amplitude margins," *Automatica*, vol. 20, pp. 645–651, 1984.

[3] R. W. Erickson and D. Maksimovic, *Fundamentals of Power Electronics,* New York, NY: Kluwer Academic/Plenum Publishers 2001.

[4] J. Li and F. C. Y. Lee, "Modeling of V^2 current-mode *control," IEEE Trans. Circ. Sys. I: Regular Papers*, vo. 57, no. 9, pp. 2552–2563, 2010.

[5] D. Maksimovic, R. Zane, and R. Erickson, "Impact of digital control in power electronics," in: *16th Int. Symp. Power Semiconductor Devices and ICs*, Kitakyushu, Japan, 2004, pp. 13–22.

[6] B. Miao, R. Zane, and D. Maksimovic, "Automated digital controller design for switching converters," in: *36th Power Electron. Specialists Conf. – PESC* 2005, Recife, Brazil, pp. 2729–2735.

[7] M. Shirazi, R. Zane, and D. Maksimovic, "An autotuning digital controller for DC–DC power converters based on online frequency-response measurement," *IEEE Trans. Power Electron.*, vol. 24, no. 11, pp. 2578–2588, 2009.

[8] A. Barkley and E. Santi, "Improved online identification of a DC–DC converter and its control loop gain using cross-correlation *methods," IEEE Trans. Power Electron.*, vol. 24, no. 8, pp. 2021–2031, 2009.

[9] M. Bhardwaj, S. Choudhury, R. Poley, and B. Akin, "Online frequency response analysis: a powerful plug-in tool for compensation design and health assessment of digitally controlled power converters," *IEEE Trans. Ind. App.*, vol. 52, no. 3, pp. 2426–2435, 2016.

[10] A. Shehada, Y. Yan, A. R. Beig, and I. Boiko, "Comparison of relay feedback tuning and other tuning methods for a digitally controlled buck

converter," in: *Proc. 45th Annu. Conf. IEEE Ind. Electron. Soc.* – IECON 2019, Lisbon, Portugal, pp. 1647–1652.

[11] W. Stefanutti, P. Mattavelli, S. Saggini, and M. Ghioni, "Autotuning of digitally controlled DC–DC converters based on relay feedback," *IEEE Trans. Power Electron.*, vol. 22, no. 1, pp. 199–207, 2007.

[12] L. Corradini, P. Mattavelli, and D. Maksimovic, "Robust relay-feedback based autotuning for DC–DC converters," *38th IEEE Power Electron. Specialists Conf. – PESC 2007*, Orlando, FL, pp. 2196–2202.

[13] Z. Zhao and A. Prodic, "Limit-cycle oscillations based auto-tuning system for digitally controlled DC–DC power supplies," *IEEE Trans. Power Electron.*, vol. 22, no. 6, pp. 2211–2222, 2007.

[14] J. Morroni, L. Corradini, R. Zane, and D. Maksimovic, "Adaptive tuning of switched-mode power supplies operating in discontinuous and continuous conduction modes," *IEEE Trans. Power Electron.*, vol. 24, no. 11, pp. 2603–2611, 2009.

[15] I. Boiko, "Loop tuning with specification on gain and phase margins via modified second-order sliding mode control algorithm", *Int. J. Syst. Sci.*, vol. 43, no. 1, pp. 97–104, 2012.

[16] I. Boiko, *Non-Parametric Tuning of PID Controllers: A Modified Relay-Feedback-Test Approach*, London, UK: Springer, 2013.

[17] I. Boiko, "Design of non-parametric process-specific optimal tuning rules for PID control of flow loops," *J. Franklin Inst.*, vol. 351, no. 2, pp. 964–985, 2014.

[18] A. Shehada, Y. Yan, A. R. Beig, and I. Boiko, "Auto-tuning of DC–DC buck converters through the modified relay feedback test," *IEEE Access*, vol. 9, pp. 62505–62518, 2021.

[19] Application Note AND9544/D, "Buck Converter External Components Selection," *ON Semiconductor,* Oct. 2017.

[20] J. C. Lagarias, J. A. Reeds, M. H. Wright, and P. E. Wright, "Convergence properties of the Nelder–Mead simplex method in low dimensions," *SIAM J. Optim.*, vol. 9, no. 1, pp. 112–147, 1998.

Chapter 12

Sliding mode control for DC–DC buck and boost converters

Igor Boiko[1] *and Ayman Ismail Al Zawaideh*[1]

Nomenclature

Symbol	Description	Unit
V_s	Source voltage	V
L	Inductance	mH
r_L	Inductance series resistance	mΩ
C	Capacitance	μF
R	Load resistance	Ω
V_{out}	Output voltage	V
x_1	Capacitor voltage (output voltage)	V
x_2	Inductor current	A
x_{1ref}	Output voltage reference	V
x_{2ref}	Inductor current reference	A
Σ	Sliding variable	V
B	Relay hysteresis	V
U	Relay control signal	
C	Relay amplitude	
$W_{2u\to x1}(s)$	Transfer function from the control signal to the output voltage	
$W_{2u\to x2}(s)$	Transfer function from the control signal to the inductor current	
$J(\omega)$	LPRS function	
k_n	Equivalent gain	
a_0	Amplitude of oscillations (ripple)	V
a_E	Amplitude of fluctuations	V
a_{AM}	Amplitude of oscillations (ripple) due to the presence of source voltage fluctuations	V
ω_E	Fluctuations frequency	rad/s
V_{s0}	DC term of source voltage	V
θ_1	Control signal positive pulse duration	s
θ_2	Control signal negative pulse duration	s
f	Switching frequency	kHz
Ω	Switching frequency of the self-excited oscillations	rad/s

(*Continues*)

[1]Electrical Engineering and Computer Science Department, Khalifa University, United Arab Emirates

(Continued)

Symbol	Description	Unit
$W_{r\to\tilde{x}_1}(s)$	Transfer function from the input signal to the deviation in the voltage	
$W_{r\to\tilde{x}_2}(s)$	Transfer function from the input signal to the deviation in the current	
$W_{r\to\tilde{y}}(s)$	Transfer function from the input signal to the deviation in the output signal	
δ_{vprop}	Source voltage fluctuations effect on the output voltage waveform	V
δ_{iprop}	Source voltage fluctuations effect on the inductor current waveform	A
ω_m	Modulated signal frequency	rad/s
$M(\omega)$	Linear dynamics magnitude characteristic	
$\varphi(\omega)$	Linear dynamics phase characteristic	
a_δ	Amplitude of voltage fluctuations due to propagation	V
a_H	Upper boundary due to the superposition principle (AM & prop.)	V
a_L	Lower boundary due to the superposition principle (AM & prop.)	V

DC–DC power converters are used in different industrial applications as DC power sources due to their fast dynamic response and high efficiency [1]. Power converters require an on–off switching action, which makes sliding mode (SM) controller a good candidate [2, 3]. SM controller provides high robustness with respect to external disturbances and parameters variations [4, 5]. The principle of SM was introduced in the 1960s [6] and was later developed in theory and used in applications [2, 5, 7–10]. In the present book chapter, the application of the Locus of a Perturbed Relay System (LPRS) method to analysis and design of DC–DC SM power converters is considered. The use of the LPRS method brings the power converter design methodology to a new level not attainable before.

12.1 Introduction

DC–DC power converters are widely used in electronics, power systems and other industrial applications as DC power sources due to their fast dynamic response and high efficiency [1]. One of the popular principles that is put into the foundation of the DC–DC converter control is the sliding mode (SM). The classical SM control theory describes the processes occurring in the converter dynamics as switching of infinite frequency, and the subsequent propagation of this discontinuous control to the converter output, which results in the zero amplitude of the voltage ripple at the controller output, ideal robustness with respect to the external disturbances and parameter variations. However, in real-life applications, the presence of additional (not accounted for in the model used, which is also often referred to as unmodeled or parasitic) dynamics precludes such operation and results in the appearance of a phenomena known as chattering [11–13]. Chattering causes the so-called sliding variable to oscillate at a finite amplitude and a high frequency which depends on the system's parameters. The attempt to implement the converter control based on the ideal (non-hysteretic) relay switching, which theoretically should ensure the ideal SM, in reality is manifested as oscillations of high but uncontrolled frequency, accompanied by high electromagnetic interferences and switching losses. Thus, in practical design, SM control is intentionally designed to have a lower but controlled frequency of switching. This results in a deviation of the actual converter performance from the performance under the ideal SM principle. Yet, despite this mismatch between the theory and practice, classical SM control principles and metrics are used in the SM converter design. In this book chapter, the Locus of a Perturbed Relay System (LPRS) method [8] is used to describe the dynamics of the processes in a SM power converter. The LPRS method allows one to analyze a real SM, which features finite-frequency switching and non-ideal robustness, and provide more accurate and realistic evaluation of system performance. The main concepts and principles of the LPRS-based analysis and design of SM power converters are presented in this chapter.

Analysis and design problems of SM buck and boost converters have been extensively studied in the literature. Different methods of analysis are presented in [14], converter design and modeling in [15–18], and reduction of the chattering in [19–27]. Digital SM current control DC–DC boost converter is designed and implemented in [28] to minimize the quasi-SM effect. Paper [29] proposes a design procedure for SM application to a maximum power point tracking in a photovoltaic system. A two-loop controller DC–DC buck converter having a PI controller for the inner loop and a SM controller for the outer loop is proposed in [30] to control a DC motor. Publication [31] proposes a two-loop controller for a boost converter, where the outer-loop uses a fuzzy logic controller implemented in an 8-bit microcontroller and the inner-loop uses an analog PI controller circuit. An adaptive control scheme is presented in [32] to tackle the change in the SM controller's switching frequency caused by a variation in the source voltage and the output load resistance. In [33], a systematic design procedure is proposed to regulate the voltage using a compensating network for SM boost converter. A digital SM control DC–DC boost converter mathematical analysis under a constant load is derived in [34]. In [35], a design of a continuous terminal SM control based on a finite time disturbance observer is proposed to transient response and tracking of the output voltage. A normalization voltage error sliding variable design for a cascade PI-SM control of DC–DC boost converter is presented in [36]. The designed sliding variable improves the transient response as it increases the design margin for the PI controller coefficients. A cascade control single ended primary inductor converter analysis and design is proposed in [37].

The available analysis tools for SM control system are based on solving the original differential equation, the describing function (DF) method [14], the classical SM control theory [15], or the Tsypkin locus [38]. The first method from the listed above can accurately describe all the processes occurring in the converter dynamics, but is hard to use to produce any explanation of these processes. The DF is an approximate method that does not give exact results for the problems of oscillations analysis and the servo aspect. This is especially relevant to SM power converters, where the sliding variable shape does not usually resemble the sinusoid, thus not meeting the condition necessary for a valid application of the DF method. The classical SM control theory deals with the reduced order system model, which does not allow one to explain the effect of slow signals propagation – the logical conclusion is that external disturbances are fully rejected, which is not supported by the practice. And the Tsypkin locus can provide an exact solution in the relay feedback system (RFS) for the periodic motion, but is totally incapable of analyzing propagation of control signals or external disturbances.

This chapter deals with the latest advancements in the field of design of SM DC–DC buck and boost converters, and offers a new approach to analysis of the converter dynamics and SM controller design, based on the LPRS method [8, 39, 40]. The LPRS is a frequency domain analysis method that provides an exact analysis of a periodic motion (self-sustained oscillations or chattering in a SM control system) and an external signal propagation (input or disturbance) in a RFS. The solution of the latter problem deals with the equivalent gain concept, which can be determined using the LPRS. The equivalent gain concept has been implicitly used in the SM converters theory. The hysteretic comparator in [16] was assumed to have infinite gain in SM operation, and to be a finite number in PWM operation. The LPRS can provide an exact value of the equivalent gain in SM converters, showing that this quantity is not infinite, which results in non-ideal disturbance rejection. It is, thus, an essential tool for the analysis of slow signals propagation through the system. The presented methodology (LPRS) can be used to solve multiple analysis and design problems, such as: chattering (ripple) analysis, solving for the amplitude and the frequency of chattering for given converter parameters, finding the effect of fluctuations in the source voltage on the output voltage waveform, or the design of the converter parameters to operate the system at a specified switching frequency, and possibly to design compensators to ensure specifications characterizing converter robustness (the latter is not addressed in this book chapter).

The present chapter deals with at least two main challenges arising from the use of the LPRS method. The first one is the transformation of the DC–DC buck and boost converters switching model to a RFS, which would be suitable for the LPRS analysis. The second problem is the analysis of the propagation of source voltage fluctuations to the current and voltage signals. The effect of source voltage fluctuations is attributed to the two effects, the amplitude modulation (AM) and the input signal propagation. Solution of these challenges is important to both theory and practice. Moreover, the LPRS is used in the design part, specifically for finding an exact value of the relay hysteresis to ensure the designed switching frequency.

This is demonstrated in and supported by the prototype design. The LPRS analysis for a DC–DC buck converter was reported in [41] and [42], and for an H-bridge inverter in [43]. The LPRS analysis for a DC–DC boost converter for the linearized model was presented in [44, 45], and for the nonlinear system model in [46]. The performance and stability analysis of a cascaded PI-SM control DC–DC boost converter through LPRS is presented in [47].

The chapter is organized as follows. In Section 12.2, the model of a SM DC–DC buck and boost converter is presented, and further transformed into a model suitable for the LPRS analysis in Section 12.3. Sections 12.4 and 12.5 deal with the LPRS analysis of the SM DC–DC buck and boost converters for the linear and non-linear models. In Sections 12.6–12.8, the analysis of source voltage fluctuations on the output voltage waveform is derived for both types of converters. Sections 12.9 and 12.10 present simulations, hardware implementation and experimental verification. Finally, a summary and a conclusion are presented in Section 12.11.

12.2 Modeling of DC–DC buck and boost converters

The schematics of the buck and the boost converters are shown in Figures 12.1 and 12.2, where Vs is the source voltage, L and r_L are the inductance and its series resistance, C is the capacitance, R is the load resistance, and u is the control signal which can be 0 and 1, corresponding to opening and closing of the controlled switch.

The state space representation of the buck converter is [41]:

$$\begin{aligned} \dot{x}_1 &= -\frac{1}{RC}x_1 + \frac{1}{C}x_2 \\ \dot{x}_2 &= -\frac{1}{L}x_1 - \frac{r_L}{L}x_2 + \frac{V_s}{L}u \end{aligned} \tag{12.1}$$

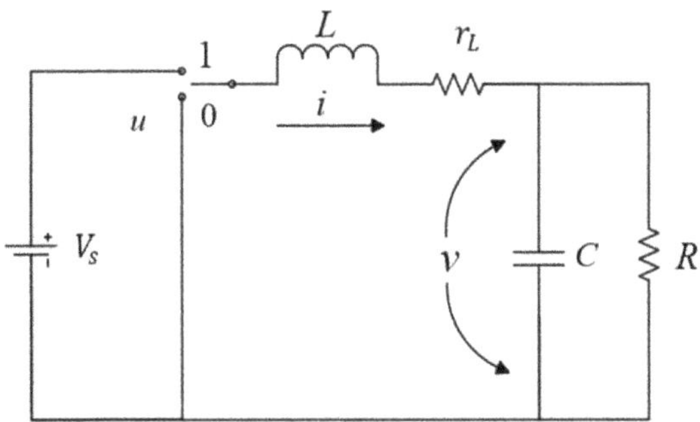

Figure 12.1 Buck converter schematic

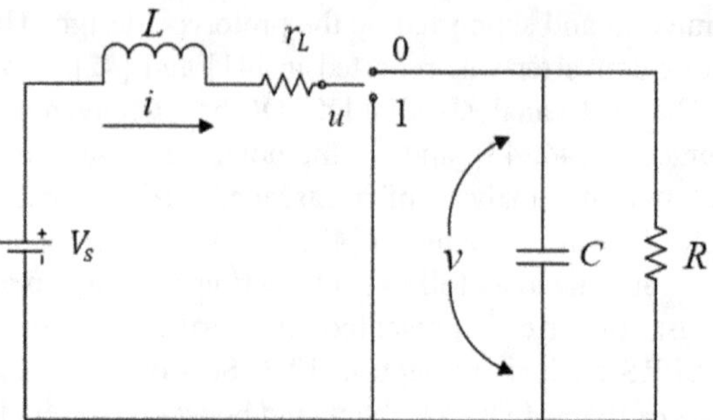

Figure 12.2 Boost converter schematic

where x_1 is the voltage across the capacitor and x_2 is the current through the inductor. The state-space representation for the boost converter is [15]:

$$\begin{aligned} \dot{x}_1 &= -\frac{1}{RC}x_1 + (1-u)\frac{1}{C}x_2 \\ \dot{x}_2 &= -(1-u)\frac{1}{L}x_1 - \frac{r_L}{L}x_2 + \frac{V_s}{L} \end{aligned} \tag{12.2}$$

The sliding variable is designed for both of the converters to be a weighted sum of the errors of the inductor current and the output voltage as:

$$\sigma = c_1\left(x_{1ref} - x_1\right) + c_2\left(x_{2ref} - x_2\right) \tag{12.3}$$

The designed operating switching frequency can be attained through the introduction of a hysteresis in the relay controller [23, 48, 49]. The implementation of the hysteresis band can be done using analog circuit following the bellow conditions:

$$u = \begin{cases} 1 & if \quad \sigma \geq b \quad or \quad (\sigma > -b, \quad u(t-0) = 1) \\ 0 & if \quad \sigma \leq -b \quad or \quad (\sigma < b, \quad u(t-0) = 0) \end{cases} \tag{12.4}$$

where $2b$ is the hysteresis value and $u(t-0)$ is the controller value at time immediately preceding the current time.

SM DC–DC buck and boost converters block diagrams are shown in Figures 12.3 and 12.4, where they represent a RFS with an asymmetric relay control action. The relay action needs to be transformed to symmetric for one to be able to analyze the plant dynamics. This can be done through the following control law:

$$u = \frac{1}{2}(1+\overline{u}) \tag{12.5}$$

where the controller $\overline{u}$ is defined as:

$$\overline{u} = \begin{cases} 1 & if \quad \sigma \geq b \quad or \quad (\sigma > -b, \quad \overline{u}(t-0) = 1) \\ -1 & if \quad \sigma \leq -b \quad or \quad (\sigma < b, \quad \overline{u}(t-0) = -1) \end{cases} \tag{12.6}$$

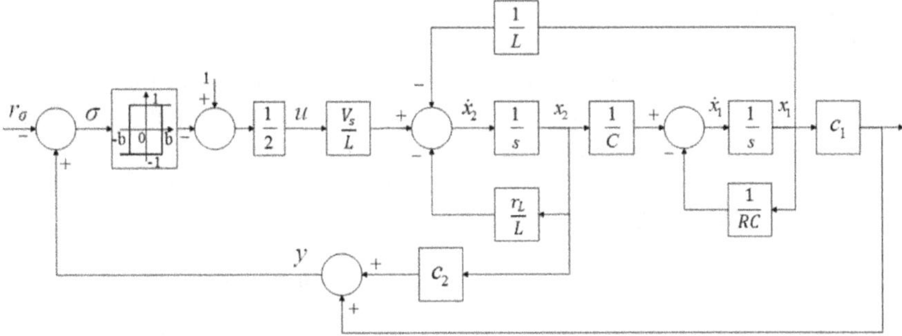

Figure 12.3 Block diagram of SM buck converter dynamic model

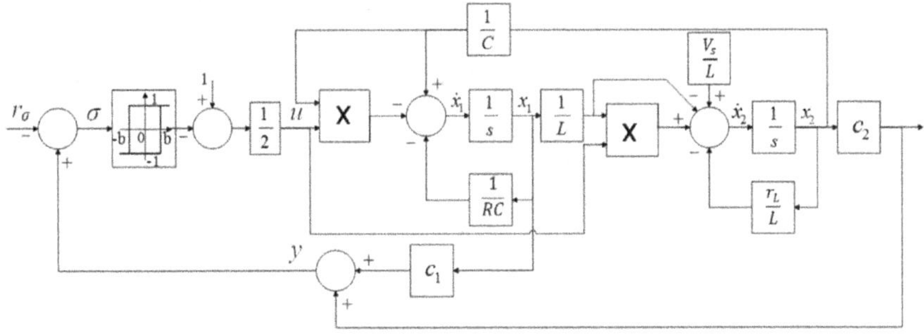

Figure 12.4 Block diagram of SM boost converter dynamic model

The dynamics of the buck converter in (12.1) can be rewritten as:

$$\begin{aligned}\dot{x}_1 &= -\frac{1}{RC}x_1 + \frac{1}{C}x_2 \\ \dot{x}_2 &= -\frac{1}{L}x_1 - \frac{r_L}{L}x_2 + \frac{V_s}{2L}\overline{u} + \frac{V_s}{2L}\end{aligned} \tag{12.7}$$

and for the boost converter in (12.2):

$$\begin{aligned}\dot{x}_1 &= -\frac{1}{RC}x_1 + \frac{1}{2C}x_2 - \frac{1}{2C}x_2\overline{u} \\ \dot{x}_2 &= -\frac{1}{2L}x_1 + \frac{1}{2L}x_1\overline{u} - \frac{r_L}{L}x_2 + \frac{V_s}{L}\end{aligned} \tag{12.8}$$

12.3 Transformation to a RFS suitable for the LPRS analysis

The SM DC–DC buck and boost converters need to be transformed from their original switching model to a RFS for analysis through the LPRS method. The steady-state operating mode of the switching model in (12.1) – (12.5) is not an equilibrium point but self-sustained oscillations. Therefore, this transformation, and if necessary linearization (in boost converter dynamics), still can be done but only at the point representing the averaged values of the output voltage and inductor's current. Let us refer to this point, which would encompass the constant values of the control, output voltage, and current equal to the respective average values under the control being equally spaced control u switching between 0 and 1, the *virtual equilibrium*.

12.3.1 Transformation of the buck converter model

The buck converter block diagram, drawn as per the original equations, is shown in Figure 12.3. It can be seen that the converter's model is a RFS with a small difference which arises from the presence of two input signals. Since the buck converter's model is linear, system transformation into a RFS can be done by propagating the bias "1" through it. Through rules of equivalent transformation of the closed loop, the system can be represented as a RFS shown in Figure 12.5.

The transfer function from the control signal to the output voltage and inductor current can be written as:

$$W_{2u\to x_1}(s) = \frac{V_s}{2L\left[C\left(s+\frac{r_L}{L}\right)\left(s+\frac{1}{RC}\right)+\frac{1}{L}\right]} \tag{12.9}$$

$$W_{2u\to x_2}(s) = \frac{V_sC\left(s+\frac{1}{RC}\right)}{2L\left[C\left(s+\frac{r_L}{L}\right)\left(s+\frac{1}{RC}\right)+\frac{1}{L}\right]} \tag{12.10}$$

and the propagation of "1" through the system dynamics results in an increment of the total input signal which can be evaluated as:

$$\Delta r_\sigma = \frac{V_s(c_1R+c_2)}{2(R+r_L)} \tag{12.11}$$

The overall input signal can be rewritten as:

$$\begin{aligned} r_\sigma - \Delta r_\sigma &= c_1x_{1ref} + c_2x_{2ref} - \frac{V_s(c_1R+c_2)}{2(R+r_L)} \\ &= c_1\left(x_{1ref}\frac{RV_s}{2(R+r_L)}\right) + c_2\left(x_{2ref} - \frac{V_s}{2(R+r_L)}\right) \end{aligned} \tag{12.12}$$

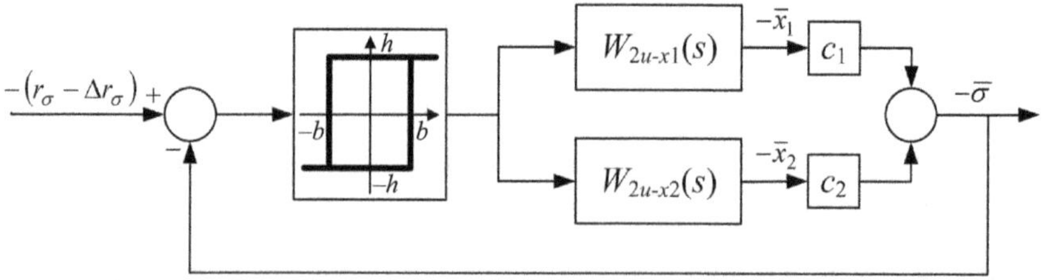

Figure 12.5 Equivalent block diagram of SM buck converter dynamic model

This effect will result in changing the state variables (without the DC components) to:

$$\tilde{x}_1 = x_1 - \frac{V_s R}{2(R + r_L)}$$
$$\tilde{x}_2 = x_2 - \frac{V_s}{2(R + r_L)} \tag{12.13}$$

It can be seen from (12.13) that by setting the reference state magnitudes to such values that the input signal is equal to zero, would result in having a symmetric periodic oscillations in the system.

12.3.2 Transformation of the boost converter model

The first step in the transformation involves representing the boost converter's dynamics in terms of deviations of the states from their averaged values. The steady-state values can be evaluated through setting $\overline{u}$ and $x_{1,2}$ to be equal to zero and solving for them, which yields:

$$x_1^* = \frac{2RV_s}{R + 4r_L}, x_2^* = \frac{4V_s}{R + 4r_L} \tag{12.14}$$

The new introduced state variables do not include the DC components:

$$\tilde{x}_1 = x_1 - x_1^* = x_1 - \frac{2RV_s}{R + 4r_L}$$
$$\tilde{x}_2 = x_2 - x_2^* = x_2 - \frac{4V_s}{R + 4r_L} \tag{12.15}$$

The plant switching model in (12.8) can be rewritten to include the new variable (12.15):

$$\dot{\tilde{x}}_1 = -\frac{1}{RC}\tilde{x}_1 + \frac{1}{2C}\tilde{x}_2 - \frac{1}{2C}\overline{u}\left(\tilde{x}_2 + x_2^*\right)$$
$$\dot{\tilde{x}}_2 = -\frac{1}{2L}\tilde{x}_1 - \frac{r_L}{L}\tilde{x}_2 + \frac{1}{2L}\overline{u}\left(\tilde{x}_1 + x_1^*\right) \tag{12.16}$$

The sliding variable can be expressed with the new variable as:

$$\tilde{\sigma} = c_1\left(x_{1ref} - x_1^*\right) + c_2\left(x_{2ref} - x_2^*\right) - \left(c_1\tilde{x}_1 + c_2\tilde{x}_2\right) \tag{12.17}$$

where the first two terms are the input signal $\left(c_1\left(x_{1ref} - x_1^*\right) + c_2\left(x_{2ref} - x_2^*\right)\right)$ and the output signal:

$$\tilde{y} = c_1\tilde{x}_1 + c_2\tilde{x}_2 \tag{12.18}$$

In (12.16), given that $x_1^* \gg \tilde{x}_1$ and $x_2^* \gg \tilde{x}_2$; the nonlinear terms, which are the product of the control signal and the state, can be disregarded, allowing to

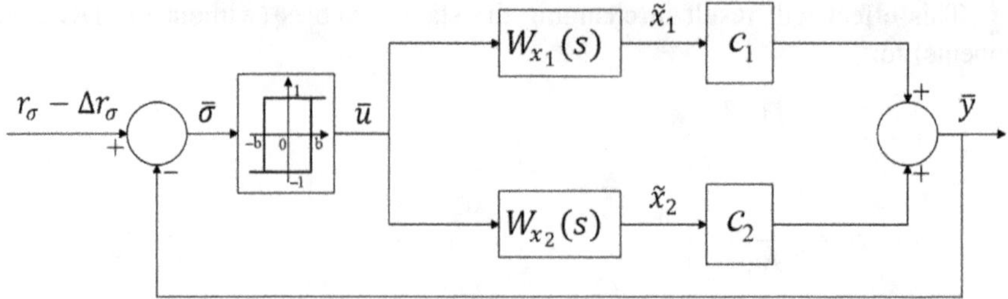

Figure 12.6 Equivalent block diagram of SM boost converter dynamic model

represent (12.16) as:

$$\begin{aligned}\dot{\tilde{x}}_1 &= -\frac{1}{RC}\tilde{x}_1 + \frac{1}{2C}\tilde{x}_2 - \frac{1}{2C}x_2^*\overline{u} \\ \dot{\tilde{x}}_2 &= -\frac{1}{2L}\tilde{x}_1 - \frac{r_L}{L}\tilde{x}_2 + \frac{1}{2L}x_1^*\overline{u}\end{aligned} \tag{12.19}$$

Therefore, the plant dynamics can be represented as:

$$\begin{aligned}\dot{\tilde{\mathbf{x}}} &= \mathbf{A}\tilde{\mathbf{x}} + \mathbf{B}\overline{\mathbf{u}} \\ \tilde{y} &= \mathbf{C}\tilde{\mathbf{x}}\end{aligned} \tag{12.20}$$

where $\tilde{\mathbf{x}} = [\tilde{x}_1 \quad \tilde{x}_2]^T$, and the output is $\tilde{y} = c_1\tilde{x}_1 + c_2\tilde{x}_2$. The block diagram of the linearized boost converter model is depicted in Figure 12.6, where the transfer functions from the relay output to the state variables signals are:

$$W_{x_1}(s) = \frac{x_1^* - 2x_2^*(Ls + r_L)}{4LC\left(s + \frac{1}{RC}\right)\left(s + \frac{r_L}{L}\right) + 1} \tag{12.21}$$

$$W_{x_2}(s) = \frac{x_2^* - 2x_1^*\left(Cs + \frac{1}{R}\right)}{4LC\left(s + \frac{1}{RC}\right)\left(s + \frac{r_L}{L}\right) + 1} \tag{12.22}$$

In this chapter, the boost converter will be analyzed through both the linearized (12.19) and the non-linear (12.16) models.

12.4 LPRS analysis of a linear plant model

This section presents the LPRS analysis for linear plant dynamics, which makes it applicable to both buck and linearized boost converters. The LPRS of a RFS with an arbitrary linear plant not containing pure integrators is derived in [8]. The LPRS of a linear plant can be represented as the following function of frequency:

$$J(\omega) = -0.5\mathbf{C}\left[\mathbf{A}^{-1} + \frac{2\pi}{\omega}\left(I - e^{\frac{2\pi}{\omega}\mathbf{A}}\right)^{-1} e^{\frac{\pi}{\omega}\mathbf{A}}\right]\mathbf{B}$$

$$+j\frac{\pi}{4}\mathbf{C}\left(\mathbf{I} + e^{\frac{\pi}{\omega}\mathbf{A}}\right)^{-1}\left(\mathbf{I} - e^{\frac{\pi}{\omega}\mathbf{A}}\right)\mathbf{A}^{-1}\mathbf{B} \quad (12.23)$$

where A, B, and C are the linear plant's matrices, A is invertible. The function $J(\omega)$ is defined as a characteristic of the response of the linear plant in the RFS to the unequally spaced control signal subject to the asymmetry of the control pulses approaching zero, as the frequency is varied from zero to infinity [8]. The real part of the LPRS is used to find the equivalent gain, and the imaginary part contains the necessary condition for the relay's switching; hence, information of the switching frequency.

In the analysis problem, the exact equivalent gain and the switching frequency can be evaluated from the point of intersection of the LPRS and the straight line at a distance equal to $\frac{\pi b}{4c}$ below the real axis for a positive hysteresis. For the design problem, the hysteresis value b can found for the system to produce switching of a desired frequency. The relay hysteresis can be evaluated by solving:

$$\mathrm{Im}(J(\Omega)) = -\frac{\pi b}{4c} \quad (12.24)$$

The relay's equivalent gain is obtained from:

$$k_n = -\frac{1}{2\mathrm{Re}(J(\Omega))} \quad (12.25)$$

The oscillations amplitude can be exactly evaluated as derived in [50]:

$$a_0 = \max_{t\in[0,T]}|y(t,\omega)| \quad (12.26)$$

where

$$y(t,\omega) = \mathbf{C}e^{\mathbf{A}t}\left(\mathbf{I} + e^{\frac{\pi}{\omega}\mathbf{A}}\right)^{-1}\left(\mathbf{I} - e^{\frac{\pi}{\omega}\mathbf{A}}\right)\mathbf{A}^{-1}\mathbf{B} + \mathbf{C}\mathbf{A}^{-1}\left(e^{\mathbf{A}t} - \mathbf{I}\right)\mathbf{B}c \quad (12.27)$$

The amplitude of chattering can also be approximately evaluated using the first harmonic of the respective signal:

$$a_0 = \frac{4c}{\pi}\left|W_{x_{1,2}}(j\Omega)\right| \quad (12.28)$$

MATLAB® function "lprs(A,B,C,ω)" found in [51] can be used for the LPRS analysis of the SM DC–DC buck and linearized boost converters.

12.5 LPRS analysis of a nonlinear plant model

The LPRS of the nonlinear boost converter model is presented in this section. Poincare map is used to analyze the periodic motion of the system (12.16) subject to an unequally spaced control signal due to a non-zero input signal. From (12.16), when the control signal $\overline{u} = 1$, the state-space equations can be written as:

$$\begin{aligned} \dot{\tilde{x}}_1 &= -\frac{1}{RC}\tilde{x}_1 - \frac{1}{2C}x_2^* \\ \dot{\tilde{x}}_2 &= -\frac{r_L}{L}\tilde{x}_2 + \frac{1}{2L}x_1^* \end{aligned} \tag{12.29}$$

and when $\overline{u} = -1$, the states can be written as:

$$\begin{aligned} \dot{\tilde{x}}_1 &= -\frac{1}{RC}\tilde{x}_1 + \frac{1}{C}\tilde{x}_2 + \frac{1}{2C}x_2^* \\ \dot{\tilde{x}}_2 &= -\frac{1}{L}\tilde{x}_1 - \frac{r_L}{L}\tilde{x}_2 - \frac{1}{2L}x_1^* \end{aligned} \tag{12.30}$$

The dynamics of the plant can be represented as $\dot{\tilde{\mathbf{x}}} = \mathbf{A}_i\tilde{\mathbf{x}} + \mathbf{B}\overline{u}$, where the matrices are defined as:

$$\begin{aligned} \mathbf{A}_1 &= \begin{bmatrix} -\frac{1}{RC} & 0 \\ 0 & -\frac{r_L}{L} \end{bmatrix}, \mathbf{A}_2 = \begin{bmatrix} -\frac{1}{RC} & \frac{1}{C} \\ -\frac{1}{L} & -\frac{r_L}{L} \end{bmatrix}, \mathbf{B} = \mathbf{B}_1 = -\mathbf{B}_2 \\ &= \begin{bmatrix} -\frac{1}{2C}x_2^* \\ \frac{1}{2L}x_1^* \end{bmatrix} \end{aligned} \tag{12.31}$$

where $(\mathbf{A}_1, \mathbf{B}_1)$ and $(\mathbf{A}_2, \mathbf{B}_2)$ are the dynamics of (12.29) and (12.30), respectively.

The linear plant response to a constant input is given by:

$$\mathbf{x}(t) = e^{\mathbf{A}t}\mathbf{x}(0) + \mathbf{A}^{-1}\left(e^{\mathbf{A}t} - \mathbf{I}\right)\mathbf{B}\mathbf{u} \tag{12.32}$$

Let us assume the existence of a periodic motion. When $\overline{u} = 1$, the plant response can be written as:

$$\mathbf{x}(\theta_1) = e^{\mathbf{A}\theta_1}\mathbf{x}(0) + \mathbf{A}_1^{-1}\left(e^{\mathbf{A}\theta_1} - \mathbf{I}\right)\mathbf{B}$$

and similarly when $\overline{u} = -1$:

$$\mathbf{x}(\theta_1 + \theta_2) = e^{\mathbf{A}\theta_2}\mathbf{x}(\theta_1) - \mathbf{A}_2^{-1}\left(e^{\mathbf{A}\theta_2} - \mathbf{I}\right)\mathbf{B}$$

For a periodic motion, $\mathbf{x}(0) = \mathbf{x}(\theta_1 + \theta_2)$ and θ_1 and θ_2 are the control signal's positive and negative pulse durations, respectively. SM boost converter periodic motion can be found through finding a fixed point of the Poincare return map:

$$\mathbf{x}(0) = \left[\mathbf{I} - e^{\mathbf{A}_2\theta_2}e^{\mathbf{A}_1\theta_1}\right]^{-1}\left[e^{\mathbf{A}_2\theta_2}\left(e^{\mathbf{A}_1\theta_1} - \mathbf{I}\right)\mathbf{A}_1^{-1} - \left(e^{\mathbf{A}_2\theta_2} - \mathbf{I}\right)\mathbf{A}_2^{-1}\right]\mathbf{B}$$

$$\mathbf{x}(\theta_1) = \left[\mathbf{I} - e^{\mathbf{A}_1\theta_1}e^{\mathbf{A}_2\theta_2}\right]^{-1}\left[e^{\mathbf{A}_1\theta_1}\left(\mathbf{I} - e^{\mathbf{A}_2\theta_2}\right)\mathbf{A}_2^{-1} - \left(e^{\mathbf{A}_1\theta_1} - \mathbf{I}\right)\mathbf{A}_1^{-1}\right]\mathbf{B}$$

Following the LPRS open-loop definition in [8] and the obtained Poincare map, the LPRS for the DC–DC boost converter will be derived next. The imaginary part of the LPRS can be found by taking the limits $\theta_1, \theta_2 \to \theta = \frac{T}{2}$:

$$\begin{aligned}
\mathrm{Im}J(\omega) &= \frac{\pi}{4}\mathbf{C} \lim_{\theta_1,\theta_2 \to \theta = \frac{T}{2}} \left(\frac{x(0) - x(\theta)}{2} \right) \\
&= \frac{\pi}{8}\mathbf{C}\big[\left[\mathbf{I} - e^{\mathbf{A}_2\theta}e^{\mathbf{A}_1\theta}\right]^{-1}\left[e^{\mathbf{A}_2\theta}\left(e^{\mathbf{A}_1\theta} - \mathbf{I}\right)\mathbf{A}_1^{-1} - \left(e^{\mathbf{A}_2\theta} - \mathbf{I}\right) - \mathbf{A}_2^{-1}\right]\mathbf{B} \\
&\quad -\left[\mathbf{I} - e^{\mathbf{A}_1\theta}e^{\mathbf{A}_2\theta}\right]^{-1}\left[e^{\mathbf{A}_1\theta}\left(\mathbf{I} - e^{\mathbf{A}_2\theta}\right)\mathbf{A}_2^{-1} + \left(e^{\mathbf{A}_1\theta} - \mathbf{I}\right) - \mathbf{A}_1^{-1}\right]\mathbf{B}\big]
\end{aligned}$$

The periodic motion switching conditions due to the feedback action can be written as:

$$\begin{aligned}
f_0 - y(0) &= b \\
f_0 - y(\theta_1) &= -b
\end{aligned}$$

Solving for the input signal (f_0) yields:

$$f_0 = \frac{y(0) + y(\theta_1)}{2}$$

where f_0 is the input signal required to ensure symmetric periodic motion ($u_0 = 0$). The constant term of the sliding variable (error signal) $\sigma(t)$ when $u_0 = 0$ can be found from:

$$\sigma_0 = \sigma(t)|_{\theta_1,\theta_2 \to \theta = \frac{T}{2}} = f_0 - y_0 = \frac{y_p(0) + y_p(\theta)}{2} - y_0$$

The real part of the LPRS can be written as:

$$\begin{aligned}
\mathrm{Re}J(\omega) &= -\frac{1}{2}\lim_{u_0 \to 0} \frac{\Delta\sigma}{u_0} \\
&= -\frac{1}{2}\lim_{u_0 \to 0} \frac{\sigma_{avg} - \sigma_0}{u_0} \\
&= -\frac{1}{2}\lim_{u_0 \to 0} \frac{\left(0.5\mathbf{C}[\mathbf{x}(0) + \mathbf{x}(\theta_1)] - y_{avg}\right) - \sigma_0}{u_0}
\end{aligned}$$

where $\Delta\sigma$ is the increment of σ caused by an increment of u_0. Let $\gamma = \frac{\theta_1}{\theta_1+\theta_2}$, $\theta_1 = \gamma T$, $\theta_2 = (1-\gamma)T$, and $u_0 = 2\gamma - 1$. The real part of the LPRS can be represented as:

$$\mathrm{Re}J(\omega) = -\frac{1}{2}\lim_{\gamma \to \frac{1}{2}} \frac{\left(0.5\mathbf{C}[\mathbf{x}(0) + \mathbf{x}(\theta_1)] - y_{avg}\right) - \sigma_0}{2\gamma - 1}$$

The real part of the LPRS can be evaluated by solving for the below limits:

$$\mathrm{Re}J(\omega) = -\frac{1}{8}\mathbf{C}\lim_{\gamma\to\frac{1}{2}}\left(\frac{d\mathbf{x}(0)}{d\gamma} + \frac{d\mathbf{x}(\theta_1)}{d\gamma}\right) + \frac{1}{4}\lim_{\gamma\to\frac{1}{2}}\frac{dy_{avg}}{d\gamma}$$

The LPRS of the SM DC–DC boost converter nonlinear model is:

$$J(\omega) = -\frac{1}{8}\mathbf{C}\lim_{\gamma\to\frac{1}{2}}\left(\frac{d\mathbf{x}(0)}{d\gamma} + \frac{d\mathbf{x}(\theta_1)}{d\gamma}\right) + \frac{1}{4}\lim_{\gamma\to\frac{1}{2}}\frac{dy_{avg}}{d\gamma} + j\frac{\pi}{8}\mathbf{C}(\mathbf{x}(0) - \mathbf{x}(\theta)) \tag{12.33}$$

Where the real and the imaginary parts limits are:

$$\lim_{\gamma\to\frac{1}{2}}\left(\frac{d\mathbf{x}(0)}{d\gamma}\right) = T\left[e^{-\mathbf{A}_2\theta} - e^{\mathbf{A}_1\theta}\right]^{-1}(\mathbf{A}_1 - \mathbf{A}_2)$$

$$\left[e^{-\mathbf{A}_1\theta} - e^{\mathbf{A}_2\theta}\right]^{-1}\left[e^{\mathbf{A}_2\theta}(e^{\mathbf{A}_1\theta} - \mathbf{I})\mathbf{A}_1^{-1} - (e^{\mathbf{A}_2\theta} - \mathbf{I})\mathbf{A}_2^{-1}\right]\mathbf{B}$$

$$+\left[\mathbf{I} - e^{\mathbf{A}_2\theta}e^{\mathbf{A}_1\theta}\right]^{-1}\left[Te^{\mathbf{A}_2\theta}\left((e^{\mathbf{A}_2\theta} + \mathbf{I}) - \mathbf{A}_2(e^{\mathbf{A}_1\theta} - \mathbf{I})\mathbf{A}_1^{-1}\right)\right]\mathbf{B}$$

$$\lim_{\gamma\to\frac{1}{2}}\left(\frac{d\mathbf{x}(\theta_1)}{d\gamma}\right) = T\left[e^{-\mathbf{A}_1\theta} - e^{\mathbf{A}_2\theta}\right]^{-1}(\mathbf{A}_1 - \mathbf{A}_2)$$

$$\left[e^{-\mathbf{A}_2\theta} - e^{\mathbf{A}_1\theta}\right]^{-1}\left[e^{\mathbf{A}_1\theta}(\mathbf{I} - e^{\mathbf{A}_2\theta})\mathbf{A}_2^{-1} + (e^{\mathbf{A}_1\theta} - \mathbf{I})\mathbf{A}_1^{-1}\right]\mathbf{B}$$

$$+\left[\mathbf{I} - e^{\mathbf{A}_1\theta}e^{\mathbf{A}_2\theta}\right]^{-1}\left[Te^{\mathbf{A}_1\theta}\left(\mathbf{A}_1(\mathbf{I} - e^{\mathbf{A}_2\theta})\mathbf{A}_2^{-1} + (e^{\mathbf{A}_2\theta} + \mathbf{I})\right)\right]\mathbf{B}$$

$$\lim_{\gamma\to\frac{1}{2}}\left(\frac{dy_{avg}}{d\gamma}\right) = \frac{\mathbf{C}}{T}\left[Te^{\mathbf{A}_1\theta}, \mathbf{x}(0)-, T, e^{\mathbf{A}_2\theta}, \mathbf{x}, (\theta_1), +\mathbf{A}_1^{-1}, (e^{\mathbf{A}_1\theta} - \mathbf{I})\frac{d\mathbf{x}(0)}{d\gamma}\right.$$

$$\left.+\mathbf{A}_2^{-1}(e^{\mathbf{A}_2\theta} - \mathbf{I})\frac{d\mathbf{x}(\theta)}{d\gamma}, +T\mathbf{A}_1^{-1}, (e^{\mathbf{A}_1\theta} - \mathbf{I}), \mathbf{B}, +, T\mathbf{A}_2^{-1}, (e^{\mathbf{A}_2\theta} - \mathbf{I})\right]\mathbf{B}$$

where $\mathbf{x}(0)$ and $\mathbf{x}(\theta_1)$ are:

$$\mathbf{x}(0) = \left[\mathbf{I} - e^{\mathbf{A}_2\theta}e^{\mathbf{A}_1\theta}\right]^{-1}\left[e^{\mathbf{A}_2\theta}(e^{\mathbf{A}_1\theta} - \mathbf{I})\mathbf{A}_1^{-1} - (e^{\mathbf{A}_2\theta} - \mathbf{I})\mathbf{A}_2^{-1}\right]\mathbf{B}$$

$$\mathbf{x}(\theta_1) = \left[\mathbf{I} - e^{\mathbf{A}_1\theta}e^{\mathbf{A}_2\theta}\right]^{-1}\left[e^{\mathbf{A}_1\theta}(\mathbf{I} - e^{\mathbf{A}_2\theta})\mathbf{A}_2^{-1} - (e^{\mathbf{A}_1\theta} - \mathbf{I})\mathbf{A}_1^{-1}\right]\mathbf{B}$$

12.6 Source voltage fluctuations effect on the buck converter

Analysis of source voltage fluctuations on the output voltage or inductor current in terms of the averaged values using the LPRS method is derived in this section. For instance, the rectification of a 50 Hz or 60 Hz AC voltage to be fed as a source voltage to a DC–DC converter, resulting in a source voltage with fluctuations having the frequency of 100 Hz or 120 Hz. The resulting frequency is much slower than the frequency of the self-oscillations, which permits the use of the equivalent gain concept [8]. The source voltage fluctuations can be modeled as the sum of a constant DC term and a periodic term:

$$V_s(t) = V_{s0} + a_E \sin(\omega_E t), \omega_E \ll \Omega \tag{12.34}$$

where V_{s0} is the DC term of source voltage, a_E is the amplitude of the fluctuations, and ω_E is the fluctuations frequency. The effect of fluctuations of the source voltage can be attributed as the combination of the effects of amplitude modulation (AM) and input signal propagation.

The effect of AM is due to the presence of the source voltage (V_s) as a factor in the numerators of $W_{2u\to x_1}(s)$ and $W_{2u\to x_2}(s)$ ((12.9) and (12.10)). The relay amplitude is directly multiplied with V_s. Thus, the total amplitude of chattering due to the presence of the fluctuations can be evaluated as:

$$a_{AM}(t) = a_0\left(1 + \frac{a_E \sin(\omega_E t)}{V_{s0}}\right), \tag{12.35}$$

where a_0 is the nominal amplitude of chattering of the output voltage or the inductor current for the DC (nominal) source voltage.

The propagation effect is due to the presence of V_s in the input signal in Figure 12.5. The total input signal due to the source voltage fluctuations is:

$$r_\sigma - \Delta r_\sigma = -a_E \sin(\omega_E t)\frac{c_1 + c_2 R}{2(R + r_L)} \tag{12.36}$$

The propagation effect of the averaged signals can be studied using the concept of the equivalent gain, which can be evaluated by the replacement of the relay action by k_n found through the LPRS. The described replacement transforms the RFS into a linear system which can be used for analysis of propagation of the slow signals through the relay system. The transfer function from the input signal to the deviation in the voltage is given by:

$$W_{r\to\tilde{x}_1}(s) = \frac{k_n W_{2u\to x_1}(s)}{1 + k_n(c_1 W_{2u\to x_1}(s) + c_2 W_{2u\to x_2}(s))} \tag{12.37}$$

and from the input signal to the deviation in the current:

$$W_{r\to\tilde{x}_2}(s) = \frac{k_n W_{2u\to x_2}(s)}{1 + k_n(c_1 W_{2u\to x_1}(s) + c_2 W_{2u\to x_2}(s))} \tag{12.38}$$

and from the input signal to the output signal:

$$W_{r\to\tilde{y}}(s) = \frac{k_n(c_1 W_{2u\to x_1}(s) + c_2 W_{2u\to x_2}(s))}{1 + k_n(c_1 W_{2u\to x_1}(s) + c_2 W_{2u\to x_2}(s))} \tag{12.39}$$

Therefore, the effect of fluctuations in the source voltage on the output voltage waveform can be evaluated as:

$$\delta_{vprop}(t) = a_E \sin(\omega_E t + arg W_{r\to x_1}(j\omega_E)) \frac{c_1 + c_2 R}{2(R + r_L)} |W_{r\to x_1}(j\omega_E)| \tag{12.40}$$

and of the current:

$$\delta_{iprop}(t) = a_E \sin(\omega_E t + arg W_{r\to x_2}(j\omega_E)) \frac{c_1 + c_2 R}{2(R + r_L)} |W_{r\to x_2}(j\omega_E)| \tag{12.41}$$

The overall effect of the source voltage fluctuations on the voltage or current states can be found by the principle of superposition of the AM effect and the propagation effect:

$$y(t) \approx a_0 \left(1 + \frac{a_E \sin(\omega_E t)}{V_{s0}}\right) \sin(\Omega t) + \delta_{(i,v)prop}(t) \tag{12.42}$$

The approximation is due the non-sinusoidal shape of the chattering, which can be observed in (12.27).

12.7 Source voltage fluctuations effect on the boost converter – analysis through linearized model

The derivation of the effect of source voltage fluctuations for the linearized boost converter model is similar to the buck converter, and the difference in the formulas will be evaluated in this section.

Since the source voltage parameter is present in the numerator of (12.21) and (12.22), the AM effect can be evaluated using (12.35). The propagation effect is due to having the source voltage in the input signal in Figure 12.6. The input signal can be rewritten to include this effect as:

$$\bar{r}_\sigma = r_\sigma - \Delta r_\sigma = c_1 \left(\frac{2V_{s0}R}{R + 4r_L} - \frac{2V_s R}{R + 4r_L}\right) + c_2 \left(\frac{4V_{s0}}{R + 4r_L} - \frac{4V_s}{R + 4r_L}\right)$$

$$r_\sigma - \Delta r_\sigma = -2a_E \sin(\omega_E t) \left[\frac{c_1 R + 2c_2}{R + 4r_L}\right] \tag{12.43}$$

Therefore, the effect of fluctuations in the source voltage on the output voltage waveform can be evaluated as:

$$\begin{aligned}\delta_{vprop}(t) &= \frac{2a_E}{R+4r_L} a_E \sin\left(\omega_E t + arg W_{\bar{r}_\sigma \to \tilde{x}_1}(j\omega_E)\right) \\ &(R - [c_1 R + 2c_2]|W_{\bar{r}_\sigma \to \tilde{x}_1}(j\omega_E)|) + \frac{2V_s R}{R+4r_L}\end{aligned} \tag{12.44}$$

and of the current:

$$\begin{aligned}\delta_{iprop}(t) &= \frac{2a_E}{R+4r_L} a_E \sin\left(\omega_E t + arg W_{\bar{r}_\sigma \to \tilde{x}_2}(j\omega_E)\right) \\ &(2 - [c_1 R + 2c_2]|W_{\bar{r}_\sigma \to \tilde{x}_2}(j\omega_E)|) + \frac{4V_s}{R+4r_L}\end{aligned} \tag{12.45}$$

The overall effect of both AM effect and propagation effect can be evaluated using (12.42).

12.8 Source voltage fluctuations effect on the boost converter – analysis through nonlinear model

The effect of the source voltage fluctuations of the source voltage on the nonlinear boost converter dynamics is derived in this section.

12.8.1 Effect of propagation

The state-space representation of the system in (12.16) can be rewritten for the analysis of slow signals propagation as follows:

$$\begin{aligned}\dot{\tilde{x}}_{10} &= -\frac{1}{RC}\tilde{x}_{10} + \frac{1}{2C}\tilde{x}_{20} - \frac{1}{2C}\overline{u}_0 x_2^* - \frac{1}{2C}\overline{u}_0\tilde{x}_{20} \\ \dot{\tilde{x}}_{20} &= -\frac{1}{2L}\tilde{x}_{10} - \frac{r_L}{L}\tilde{x}_{20} + \frac{1}{2L}\overline{u}_0 x_1^* + \frac{1}{2L}\overline{u}_0\tilde{x}_{10}\end{aligned} \tag{12.46}$$

where

$$\begin{aligned}\overline{u}_0 &= \alpha_u + a_u \sin(\omega_E t) \\ \tilde{x}_{10} &= \alpha_1 + a_1 \sin(\omega_E t + \phi_1) \\ \tilde{x}_{20} &= \alpha_2 + a_2 \sin(\omega_E t + \phi_2)\end{aligned} \tag{12.47}$$

The multiplication of the averaged state dynamics with the averaged control signal can be evaluated as:

$$\begin{aligned}\overline{u}_0\tilde{x}_{10} &= \frac{1}{2} a_u a_{x1}(\cos(\phi_1) - \cos(2\omega_E t + \phi_1)) \\ \overline{u}_0\tilde{x}_{20} &= \frac{1}{2} a_u a_{x2}(\cos(\phi_2) - \cos(2\omega_E t + \phi_2))\end{aligned} \tag{12.48}$$

The terms of $2\omega_E t$ are disregarded as they are filtered out and of a lower magnitude, the multiplied terms in (12.48) can be rewritten as:

$$M_1 = \overline{u}_0 \tilde{x}_{10} = \frac{1}{2} a_u a_{x1} \cos(\phi_1)$$
$$M_2 = \overline{u}_0 \tilde{x}_{20} = \frac{1}{2} a_u a_{x2} \cos(\phi_2) \qquad (12.49)$$

Rewriting (12.46) to include (12.49) yields:

$$\dot{\tilde{x}}_{10} = -\frac{1}{RC}\tilde{x}_{10} + \frac{1}{2C}\tilde{x}_{20} - \frac{1}{2C}\overline{u}_0 x_2^* - \frac{1}{2C}M_2$$
$$\dot{\tilde{x}}_{20} = -\frac{1}{2L}\tilde{x}_{10} - \frac{r_L}{L}\tilde{x}_{20} + \frac{1}{2L}\overline{u}_0 x_1^* + \frac{1}{2L}M_1 \qquad (12.50)$$

The constant terms in (12.50) can be transposed to the input signal through the introduction of new variables:

$$\overline{x}_{10} = \tilde{x}_{10} - \tilde{x}_{10}^*$$
$$\overline{x}_{20} = \tilde{x}_{20} - \tilde{x}_{20}^* \qquad (12.51)$$

where:

$$\tilde{x}_{10}^* = \frac{R(M_1 - 2r_L M_2)}{R + 4r_L}, \quad \tilde{x}_{20}^* = \frac{2M_1 + RM_2}{R + 4r_L}$$

Therefore, the system new state-space representation for the averaged motion including the introduced variables can be written as:

$$\dot{\overline{x}}_{10} = -\frac{1}{RC}\overline{x}_{10} + \frac{1}{2C}\overline{x}_{20} - \frac{1}{2C}x_2^*\overline{u}_0$$
$$\dot{\overline{x}}_{20} = -\frac{1}{2L}\overline{x}_{10} - \frac{r_L}{L}\overline{x}_{20} + \frac{1}{2L}x_1^*\overline{u}_0 \qquad (12.52)$$

and the new input signal is:

$$\begin{aligned}\overline{r}_\sigma = r_\sigma - \Delta r_\sigma - \left(c_1\tilde{x}_{10}^* + c_2\tilde{x}_{20}^*\right) &= c_1\left(\frac{2V_{s0}R}{R+4r_L} - \frac{2V_sR}{R+4r_L}\right) \\ &+ c_2\left(\frac{4V_{s0}}{R+4r_L} - \frac{4V_s}{R+4r_L}\right) - \left(c_1\tilde{x}_{10}^* + c_2\tilde{x}_{20}^*\right) \\ &= -2a_E\sin(\omega_E t)\left[\frac{c_1R + 2c_2}{R+4r_L}\right] - \left(c_1\tilde{x}_{10}^* + c_2\tilde{x}_{20}^*\right)\end{aligned} \qquad (12.53)$$

The dynamics of the plant can written as follows:

$$\dot{\overline{\mathbf{x}}}_0 = \mathbf{A}\overline{\mathbf{x}}_0 + \mathbf{B}\overline{\mathbf{u}}_0$$
$$\tilde{y}_0 = \mathbf{C}\overline{\mathbf{x}}_0$$

where the output signal is $\tilde{y}_0 = c_1\tilde{x}_{10} + c_2\tilde{x}_{20}$, and the matrices are:

$$\mathbf{A} = \begin{bmatrix} -\frac{1}{RC} & \frac{1}{2C} \\ -\frac{1}{2L} & -\frac{r_L}{L} \end{bmatrix}, \mathbf{B} = \begin{bmatrix} -\frac{1}{2C}x_2^* \\ \frac{1}{2L}x_1^* \end{bmatrix}, \mathbf{C} = \begin{bmatrix} c_1 & c_2 \end{bmatrix}$$

The propagation effect on the states $\tilde{x}_{10}$ and $\tilde{x}_{20}$ can be evaluated as follows:

$$\begin{aligned} \tilde{x}_{10} = & -2a_E\left[\frac{c_1R + 2c_2}{R + 4r_L}\right]|W_{\tilde{r}_\sigma\rightarrow\tilde{x}_{10}}(j\omega_E)|\sin\left(\omega_E t + argW_{\tilde{r}_\sigma\rightarrow\tilde{x}_1}(j\omega_E)\right) \\ & -\left(c_1\tilde{x}_{10}^* + c_2\tilde{x}_{20}^*\right)\frac{k_nW_{\tilde{x}_{10}}(0)}{1 + k_n(c_1W_{\tilde{x}_{10}}(0) + c_2W_{\tilde{x}_{20}}(0))} + \tilde{x}_{10}^* \end{aligned} \tag{12.54}$$

$$\begin{aligned} \tilde{x}_{20} = & -2a_E\left[\frac{c_1R + 2c_2}{R + 4r_L}\right]|W_{\tilde{r}_\sigma\rightarrow\tilde{x}_{20}}(j\omega_E)|\sin\left(\omega_E t + argW_{\tilde{r}_\sigma\rightarrow\tilde{x}_2}(j\omega_E)\right) \\ & -\left(c_1\tilde{x}_{10}^* + c_2\tilde{x}_{20}^*\right)\frac{k_nW_{\tilde{x}_{20}}(0)}{1 + k_n(c_1W_{\tilde{x}_{10}}(0) + c_2W_{\tilde{x}_{20}}(0))} + \tilde{x}_{20}^* \end{aligned} \tag{12.55}$$

and on the control signal:

$$\tilde{u}_0 = -k_n\left[2a_E\left(\frac{c_1R + 2c_2}{R + 4r_L}\right)\sin\left(\omega_E t\right) + \left(c_1\tilde{x}_{10}^* + c_2\tilde{x}_{20}^*\right)\right] \tag{12.56}$$

The variables in (12.47) can be found by solving (12.54)–(12.56). The propagation effect on the output voltage and the inductor current are:

$$\delta_{vprop}(t) = \tilde{x}_{10} + \frac{2V_sR}{R + 4r_L} \tag{12.57}$$

$$\delta_{iprop}(t) = \tilde{x}_{20} + \frac{4V_s}{R + 4r_L} \tag{12.58}$$

12.8.2 Propagation of amplitude-modulated signal through linear dynamics

The fluctuations of the source voltage give rise to another component, which will be referred to as AM effect. We will consider the effect of how the AM propagates through a linear system dynamics. Let a carrier signal $(u(t))$ be modulated by a signal $V(t)$, and the resulting signal propagates through a linear plant $(G(s))$, as depicted in Figure 12.7, where:

$$u(t) = a_c\sin\left(\Omega t\right) \tag{12.59}$$

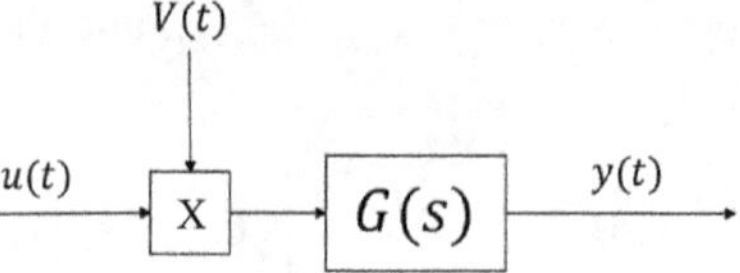

Figure 12.7 Propagation of amplitude modulated signal through linear dynamics

$$V(t) = 1 + a_m \sin(\omega_m t), \omega_m \ll \Omega \tag{12.60}$$

For simplicity of notation, which does not affect the overall results and conclusions, we assume that the carrier signal and the modulating signals are sinusoidal – as given by (12.59) and (12.60). The product of the modulation and the carrier signals yields:

$$\begin{aligned} u(t)V(t) &= a_c \sin(\Omega t)(1 + a_m \sin(\omega_m t)) \\ &= a_c \sin(\Omega t) + \frac{a_c a_m}{2}[\cos((\Omega - \omega_m)t) - \cos((\Omega + \omega_m)t)] \end{aligned} \tag{12.61}$$

Since $\Omega \gg \omega_m$, then $\Omega - \omega_m$ and $\Omega + \omega_m$ can be evaluated as small deviations from Ω, the linear approximation can be considered:

$$M(\Omega + \omega_m) = M(\Omega) + \left.\frac{\partial M}{\partial \omega}\right|_{\omega=\Omega} \omega_m \tag{12.62}$$

$$M(\Omega - \omega_m) = M(\Omega) - \left.\frac{\partial M}{\partial \omega}\right|_{\omega=\Omega} \omega_m \tag{12.63}$$

where

$$M(\omega) = |G(j\omega)| \tag{12.64}$$

is the linear dynamics magnitude characteristic, and

$$\varphi(\Omega + \omega_m) = \varphi(\Omega) + \left.\frac{\partial \varphi}{\partial \omega}\right|_{\omega=\Omega} \omega_m \tag{12.65}$$

$$\varphi(\Omega - \omega_m) = \varphi(\Omega) - \left.\frac{\partial \varphi}{\partial \omega}\right|_{\omega=\Omega} \omega_m \tag{12.66}$$

where

$$\varphi(\omega) = \arg(G(j\omega)) \tag{12.67}$$

is the linear dynamics phase characteristics. From (12.61), the propagation of the carrier frequency (Ω), lower side frequency ($\Omega - \omega_m$) and upper side frequency ($\Omega + \omega_m$) responses are separately calculated and summed together. The total output signal is:

$$y(t) = Ma_c \sin(\Omega t + \varphi) + \frac{a_c a_m}{2}[(M - \Delta M)\cos((\Omega - \omega_m)t + \varphi - \Delta\varphi) \\ -(M + \Delta M)\cos((\Omega - \omega_m)t + \varphi + \Delta\varphi)] \tag{12.68}$$

where

$$M = M(\Omega), \quad \Delta M = \left.\frac{\partial M}{\partial \omega}\right|_{\omega=\Omega} \omega_m$$

$$\varphi(\omega) = \arg(G(j\omega)), \quad \Delta\varphi = \left.\frac{\partial \varphi}{\partial \omega}\right|_{\omega=\Omega} \omega_m$$

In (12.68), using the formula for the cos of a sum to represent the angles as a sum of $\Delta\varphi$ and the rest of the expression yields:

$$\begin{aligned}
y(t) = {} & Ma_c \sin(\Omega t + \varphi) \\
& + \frac{a_c a_m}{2}\{M \cos((\Omega - \omega_m)t + \varphi)\cos(\Delta\varphi) \\
& + M \sin((\Omega - \omega_m)t + \varphi)\sin(\Delta\varphi) \\
& - \Delta M \cos((\Omega - \omega_m)t + \varphi)\cos(\Delta\varphi) \\
& - \Delta M \sin((\Omega - \omega_m)t + \varphi)\sin(\Delta\varphi) \\
& - M \cos((\Omega + \omega_m)t + \varphi)\cos(\Delta\varphi) \\
& + M \sin((\Omega + \omega_m)t + \varphi)\sin(\Delta\varphi) \\
& - \Delta M \cos((\Omega + \omega_m)t + \varphi)\cos(\Delta\varphi) \\
& + \Delta M \sin((\Omega + \omega_m)t + \varphi)\sin(\Delta\varphi)\} \\
= {} & Ma_c \sin(\Omega t + \varphi) \\
& + \frac{a_c a_m}{2}\{M \cos(\Delta\varphi)[\cos((\Omega - \omega_m)t + \varphi) \\
& - \cos((\Omega + \omega_m)t + \varphi] \\
& + M \sin(\Delta\varphi)[\sin((\Omega - \omega_m)t + \varphi) \\
& + \sin((\Omega + \omega_m)t + \varphi)] \\
& - \Delta M \cos(\Delta\varphi)[\cos((\Omega - \omega_m)t + \varphi) \\
& + \cos((\Omega + \omega_m)t + \varphi)] \\
& - \Delta M \sin(\Delta\varphi)[\sin((\Omega - \omega_m)t + \varphi) \\
& - \sin((\Omega + \omega_m)t + \varphi)]\sin(\Delta\varphi)
\end{aligned}$$

$$
\begin{aligned}
&= Ma_c \sin(\Omega t + \varphi) \\
&\quad + \frac{a_c \alpha_m}{2}\{M \cos(\Delta\varphi) 2 \sin(\Omega t + \varphi) \sin(\omega_m t) \\
&\quad + M \sin(\Delta\varphi) 2 \sin(\Omega t + \varphi) \cos(\omega_m t) \\
&\quad - \Delta M \cos(\Delta\varphi) 2 \cos(\Omega t + \varphi) \cos(\omega_m t) \\
&\qquad + \Delta M \sin(\Delta\varphi) 2 \cos(\Omega t + \varphi) \sin(\omega_m t)\} \\
&= Ma_c \sin(\Omega t + \varphi)\{1 + \alpha_m \sin(\omega_m t) \cos(\Delta\varphi)\} \\
&\quad + a_c \alpha_m \Delta M \cos(\Omega t + \varphi)\{\cos(\omega_m t) \cos(\Delta\varphi) + \sin(\omega_m t) \sin(\Delta\varphi)\} \\
&= y_p(t)\{1 + \alpha_m \sin(\omega_m t + \Delta\varphi)\} - a_c \alpha_m \Delta M \cos(\Omega t + \varphi) \cos(\omega_m t + \Delta\varphi)
\end{aligned} \tag{12.69}
$$

where $y_p(t) = Ma_c\cos(\Omega t + \varphi)$ is the output of propagation of $u(t)$ through the linear plant's dynamics ($G(s)$). Let $y_p^*(t)$ be a conjugate signal of $y_p(t)$, which is represented as:

$$y_p^*(t) = Ma_c\cos(\Omega t + \varphi) \tag{12.70}$$

Rewriting (12.69) to include $y_p^*(t)$ yields:

$$y(t) = y_p(t)\{1 + \alpha_m \sin(\omega_m t + \Delta\varphi)\} - \frac{\alpha_m \Delta M}{M} y_p^*(t) \cos(\omega_m t + \Delta\varphi) \tag{12.71}$$

Given that $y_p^*(t)$ is $y_p(t)$ shifted by $\frac{\pi}{2}$, the equivalent AM diagram can be presented as in Figure 12.8, where the AM is applied at the output of the linear plant. Thus, the resulting output modulated signal has two components:

- AM by $(1 + \alpha_m \sin(\omega_m t + \Delta\varphi))$ which is applied to the propagation of $u(t)$ through $G(s)$.

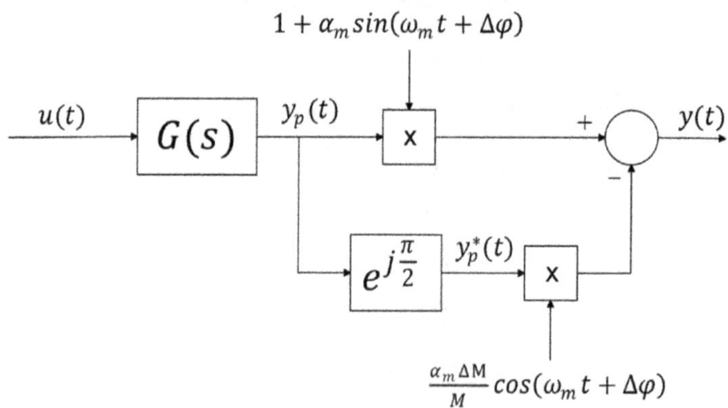

Figure 12.8 Equivalent amplitude modulation diagram

- AM by $\left(\frac{\alpha_m \Delta M}{M}\cos(\omega_m t + \Delta\varphi)\right)$ applied to $y_p^*(t)$.

Since $y_p(t)$ and $y_p^*(t)$ are shifted by $\frac{\pi}{2}$, the upper envelop of AM in (12.71) can be calculated as a quadratic summation of the two components as follows:

$$y_{env}(t) = \left\{ (1 + \alpha_m \sin(\omega_m t + \Delta\varphi))^2 + \left(\frac{\alpha_m \Delta M}{M} y_p^*(t) \cos(\omega_m t + \Delta\varphi) \right)^2 \right\}^{1/2}$$

The derived AM effect can be applied to the DC–DC boost converter by substituting $\alpha_m = \frac{a_E}{V_i}$, $\omega_m = 2\pi \cdot 100$ and Ma_c is equal to the amplitude of chattering.

12.9 Simulation

The performance of the designed controller and the verification of the analytical derivations are validated using MATLAB®/Simulink® simulations. Table 12.1 show the parameters and their values of the buck and the boost converters. Figure 12.9 shows the LPRS of the DC–DC buck and boost converters based on the converter parameters in Table 12.1. The inductor current, the output voltage, and the sliding variable waveforms of the buck are shown in Figure 12.10, and for the boost converters in Figure 12.11. The amplitude of ripple of the sliding variables (buck and boost) matches the designed hysteresis values; and the averaged simulation values of the output voltage and the inductor current are equal to the their references. Such values insured a symmetric of the period oscillations of the control signal at the designed switching frequency.

12.9.1 Simulation results of the source voltage fluctuations effect

The simulation results of the effect of source voltage fluctuations on the output voltage waveform are presented in this section. The source voltage fluctuations frequency is $\omega_E = 628.3$ rad/s (100 Hz), which is the result of rectifying of a 50 Hz

Table 12.1 Parameters of DC–DC buck and boost converters

Description	**Parameter**	**Nominal value (buck)**	**Nominal value (boost)**
Input voltage	V_s [V]	25.2	22
Capacitance	C [μF]	1.0	99
Inductance	L [mH]	2.0	0.334
Inductor series resistance	r_L [Ω]	0.02	0.58
Load resistance	R [Ω]	60	100
Desired output voltage	V_{out} [V]	12.6	43.0
Desired switching frequency	f [kHz]	40	50

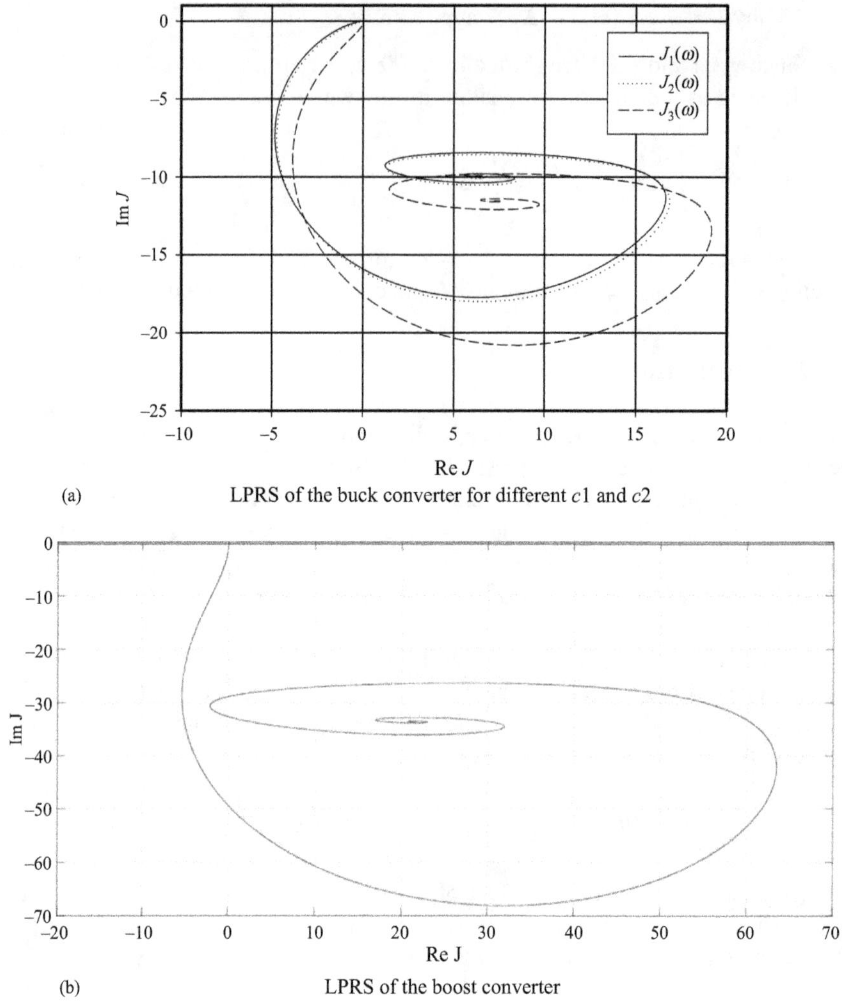

(a) LPRS of the buck converter for different $c1$ and $c2$

(b) LPRS of the boost converter

Figure 12.9 LPRS of the DC–DC buck and boost converters based on the circuit parameters

ac voltage source. The fluctuations amplitude for the buck converter $a_E = 1\text{V}$, and for the boost converter $a_E = 0.93\text{V}$. It can be observed from (12.35) that the upper boundary of the AM voltage is almost in the counter-phase with the propagation effect signal (12.40), and the signal of the lower boundary of the AM is nearly in phase with (12.40) for the buck converter (due to the phase lag being small $(arg W_{\bar{r}_o \to \bar{x}_1}(j\omega_E) \approx 0)$, and the reverse effect for the boost converter. Therefore, the resulting amplitude of the lower and the upper boundaries due to the superposition principle (12.42) can be found as: $a_L = 2a_\delta + \frac{a_0 a_E}{V_s}$, $a_H = 2a_\delta - \frac{a_0 a_E}{V_s}$ for the buck converter, and $a_L = 2a_\delta - \frac{a_0 a_E}{V_s}$, $a_H = 2a_\delta + \frac{a_0 a_E}{V_s}$ for the boost converter.

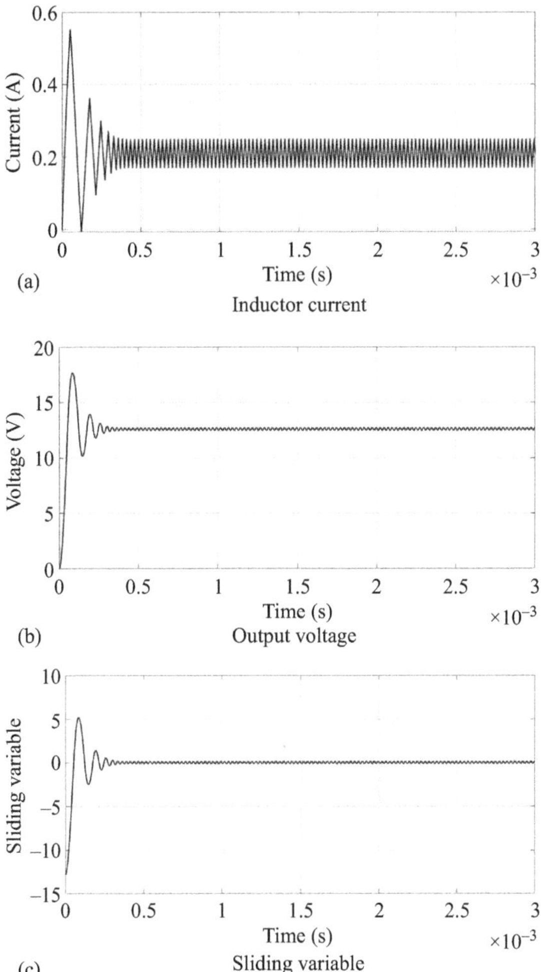

Figure 12.10 Buck converter inductor current, output voltage, and sliding variable waveforms, switching frequency 40 kHz

The resulting superposition effect on the output voltage waveform can be clearly seen for the buck converter in Figure 12.12 and for the boost converter in Figure 12.13. The simulation results show an exact match with the analytical derivations as presented in Table 12.2.

12.10 Hardware implementation

The presented analytical derivations and simulations are further validated using an experimental prototype. The SM boost converter circuit schematic is

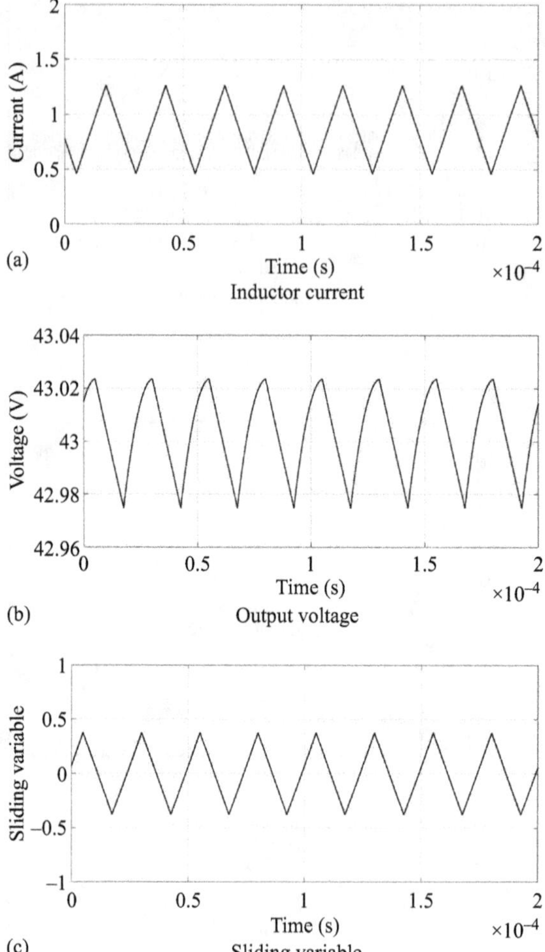

Figure 12.11 Boost converter inductor current, output voltage, and sliding variable waveforms at steady state operation, switching frequency 40 kHz

depicted in Figure 12.14, and its signals description is shown in Table 12.3. The proposed control circuit includes three operational amplifiers (LM318). The output of the first and the second op-amps are the voltage error and the current error signals, respectively. Both of these signals are summed in the third op-amp to generate the designed sliding variable in (12.3), which is then fed to a non-inverting Schmitt Trigger (LM311). The resulting signal drives the digital circuit consisting of an opto-coupler (6n137) and a MOSFET gate driver (MC34151).

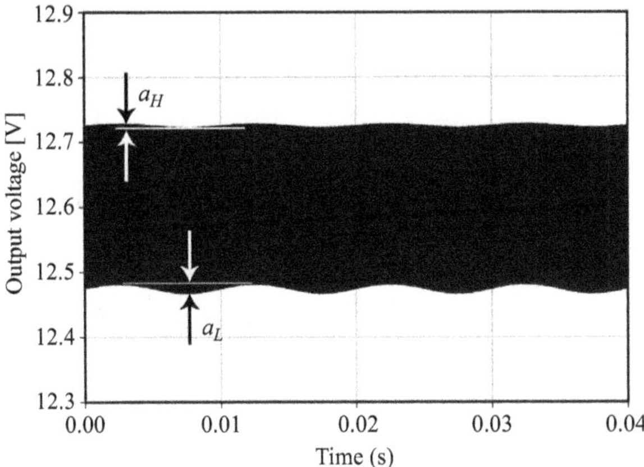

Figure 12.12 Output voltage response due to source voltage fluctuations (buck converter)

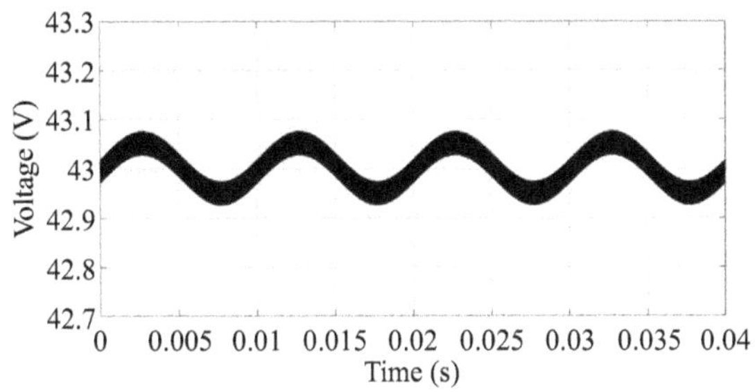

Figure 12.13 Output voltage response due to source voltage fluctuations (boost converter)

Table 12.2 Oscillations parameter of DC–DC buck and boost converters

Description	**Magnitude (buck)**	**Magnitude (boost)**
Amplitude of voltage ripple/chattering [V]	0.1277	0.0244
Amplitude of voltage AM component	0.0050	0.0010
Equivalent gain	6.0718	205.78
Voltage fluctuations amplitude due to propagation [V]	0.0032	0.0511

12.10.1 Experimental results

The performance of the designed circuit is evaluated in this section. The designed circuit prototype is shown in Figure 12.15, where the circuit parameters are based

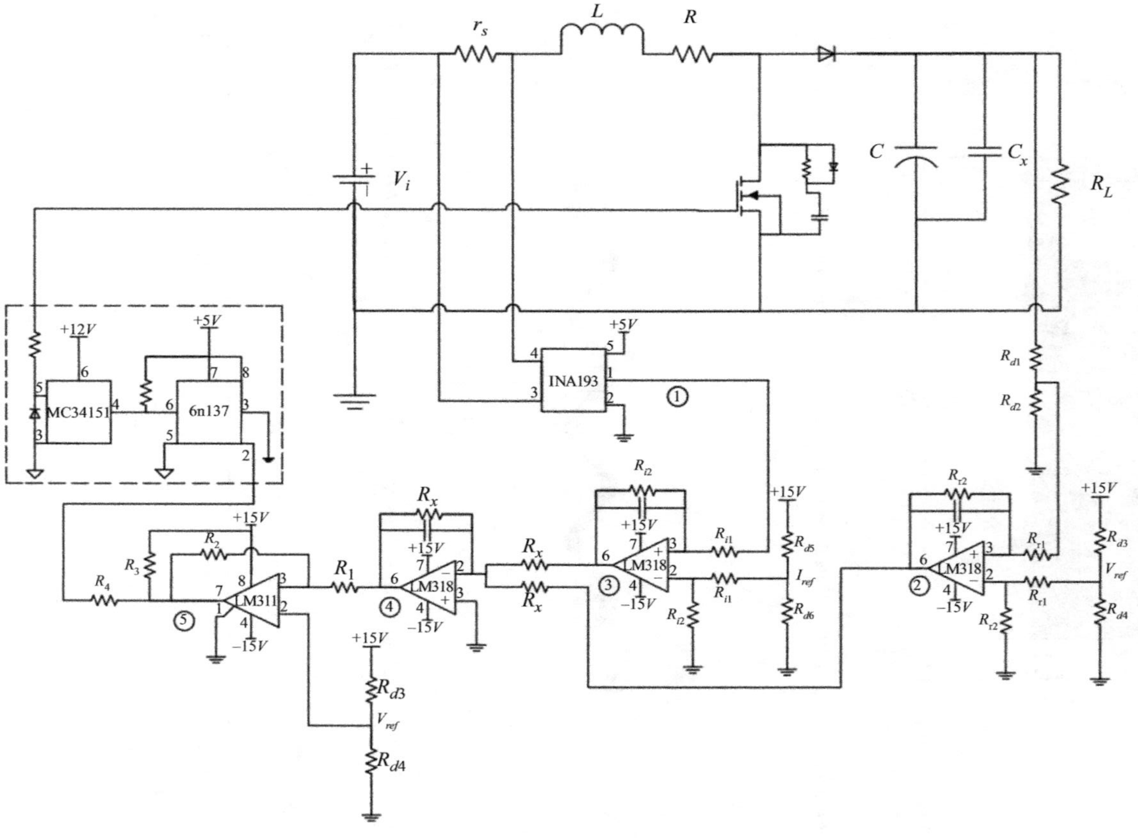

Figure 12.14 Circuit schematics of SM of boost converter

Table 12.3 Signals description

Test location	Description
1	i_L
2	$c_1(x_1 - x_{1ref})$
3	$c_2(x_2 - x_{2ref})$
4	$c_1(x_{1ref} - x_1) + c_2(x_{2ref} - x_2)$
5	U

Figure 12.15 PCB of SM boost converter

on Table 12.1. The circuit is designed to operate at 40 kHz. The steady-state oscillatory waveforms of the output voltage, the inductor current, and the gate signal are shown in Figure 12.16. The output voltage average value is 43.1 V, and the operating switching frequency is 40.09 kHz. The source voltage fluctuations (at 100 Hz) which is the result of a rectification of an AC voltage supply (at 50 Hz) is shown in Figure 12.17. Its effect on the output voltage waveform is depicted in Figure 12.18. Table 12.4 shows a comparison between the analytical and the experimental results. The small difference between analytical and the experimental results are due to having the source voltage fluctuations not purely sinusoidal as approximated in (12.34).

12.10.2 Load variation

The effect of varying the load resistance ($60\Omega < R < 140\Omega$) on the switching frequency and the average output voltage is depicted in Figure 12.19. It can be noticed that the effect of increasing the load resistance is an increase of both the

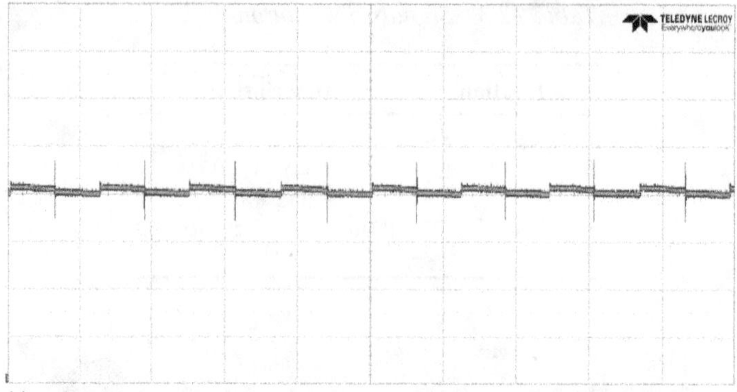

(a) *Output Voltage, average value* V_{out} *= 43.1V, each vertical division is 1 V*

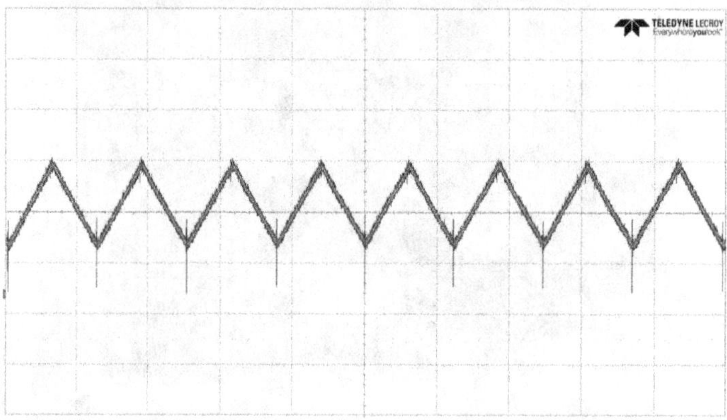

(b) *Inductor current, average value* I_L*= 0.86A, each vertical division is 0.5 V*

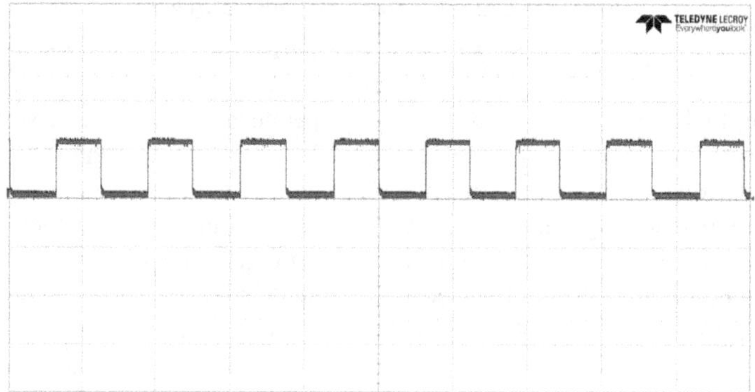

(c) *Mosfet gate signal, switching frequency* f *= 40.09 kHz, each vertical division is 10 V*

Figure 12.16 Experimental results of SM boost converter. Each horizontal division is 20 μs

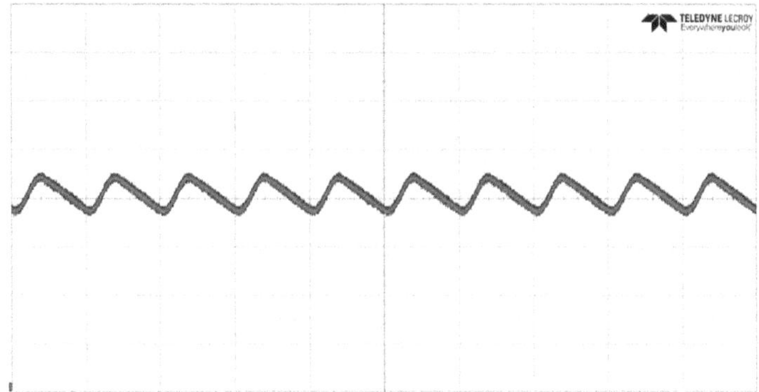

Figure 12.17 Source voltage fluctuations, frequency of fluctuations f = *100 Hz, amplitude* $a_E = 0.93$ *V. Each horizontal division is 10 ms, each vertical division is 2 V*

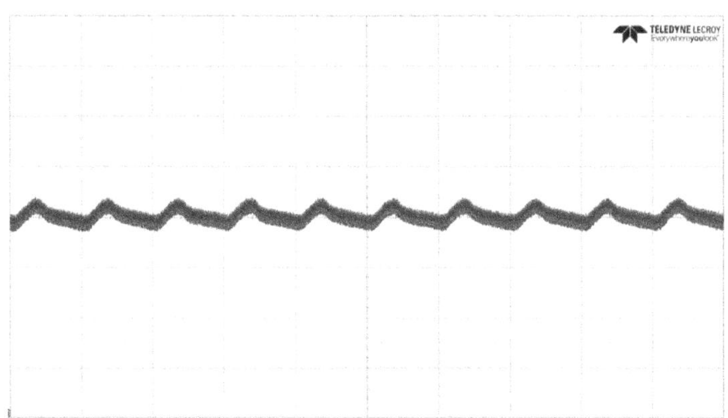

Figure 12.18 Output voltage due to fluctuating source voltage, fluctuations frequency of f = *100 Hz. Each horizontal division is 10 ms, each vertical division is 0.5 V*

Table 12.4 Comparison between experimental and analytical results

Description	**Analysis**	**Experimental**
Desired output voltage V_{out} [V]	43.0	43.074
Switching frequency f [kHz]	40	40.09
Upper boundary of the output voltage envelop (AM&prop) [V]	0.0521	0.054
Lower boundary of the output voltage envelop (AM&prop) [V]	0.0500	0.052

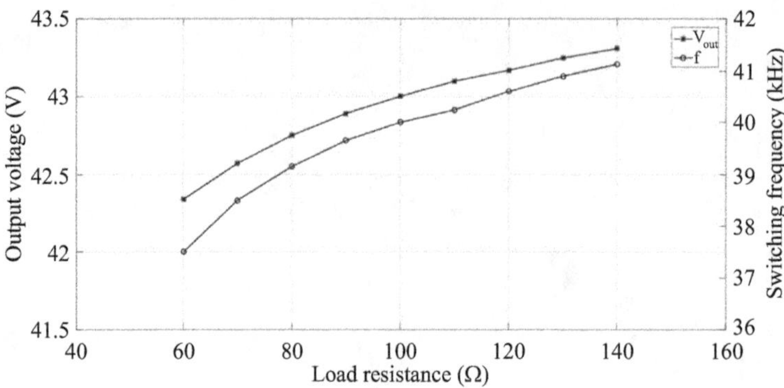

Figure 12.19 Experimental data of the effect of varying the load resistance on the switching frequency and the output voltage

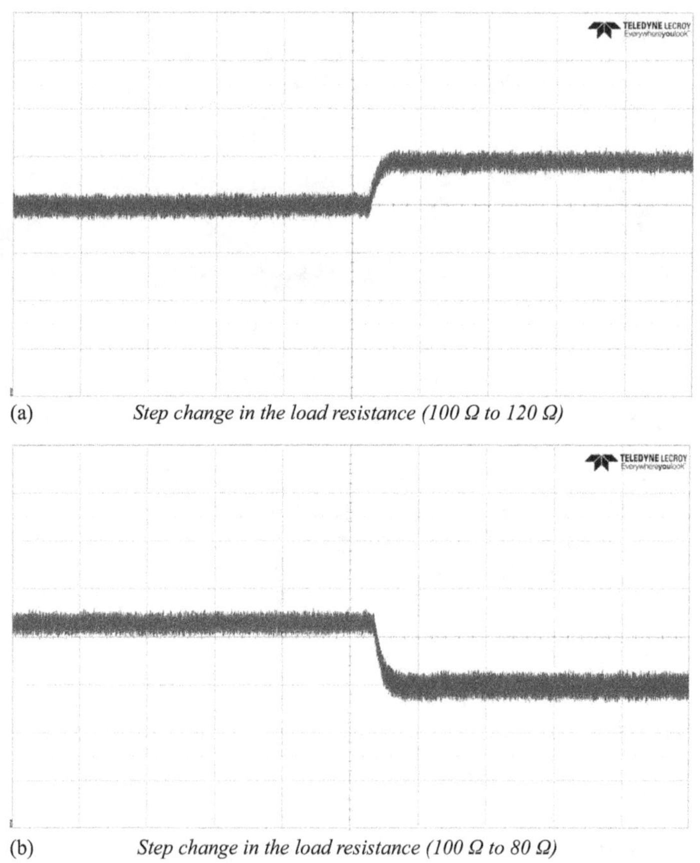

TELEDYNE LECROY
(a)
Step change in the load resistance (100 Ω to 120 Ω)
TELEDYNE LECROY
(b)
Step change in the load resistance (100 Ω to 80 Ω)

Figure 12.20 Experimental data of the output voltage transient response due to a step change in the load resistance. Each horizontal division is 2 ms, each vertical division is 0.2 V

switching frequency and the output voltage. The change in the load resistance (±40%) results in a deviation of the output voltage by only 0.97 V (2.26% of the nominal output voltage; further improvements can be and are usually achieved through adding another loop containing integral control), and a deviation of a 3.6 kHz for the operating switching frequency. The presented results demonstrate the robustness of the designed controller. A load step change from 100 Ω to 120 Ω and from 100 Ω to 80 Ω is shown in Figure 12.20. The results show a very fast speed response with a rise time of 0.9 ms and with no overshoot, which is a good feature of SM controlled converters. The authors believe that despite the SM converter was designed as means of the theory verification and demonstration, its characteristics would be suitable for many applications.

12.11 Conclusion

Analysis of a SM DC–DC buck and boost converters dynamics using the LPRS methodology is proposed. Some design aspects are considered too. The buck converter model can be directly transformed to a RFS, which allows one to directly apply the LPRS method for analysis of the SM converter dynamics and control design. The boost converter model was transformed from a switching model to a RFS too. However, the plant dynamics after this transformation are nonlinear. And there two different approaches to analysis: linearization of the nonlinear dynamics and the use of the formulation of the LPRS for a nonlinear plant, with subsequent derivation of the LPRS function. Both options are explored in this book chapter, and the analysis of the boost converter was carried for both the linear and the nonlinear dynamic models. The functionality of the LPRS method, which features exact determination of the frequency and amplitude of chattering and possibility of analysis of external signal propagation, was fully utilized in the presented development. Chattering (ripple) in inductor current and output voltage, as well as, the effect of source voltage fluctuations, which may come from rectification of AC voltage, were analyzed through the concepts of the LPRS method. It was possible to identify the reason and explain the effects of the source voltage fluctuations, which are a combination of AM and input signal propagation effects. The proposed analysis is supported by simulations and experimental results. The presented methodology makes it a superior tool of analysis of SM DC–DC buck and boost converters.

References

[1] M. H. Rashid, *Power Electronics: Devices, Circuits, and Applications*, 4th ed. London: Pearson, 2014.

[2] V. Utkin, J. Guldner, and J. Shi, *Sliding Mode Control in Electro-Mechanical Systems,* Boca Raton, FL: CRC Press, 2009.

[3] V. I. Utkin, *Sliding Modes in Control and Optimization*, Berlin, Germany: Springer Science & Business Media, 2013.

[4] J.-J. E. Slotine and W. Li, *Applied Nonlinear Control,* Englewood Cliffs, NJ: Prentice Hall, 1991.

[5] Y. Shtessel, C. Edwards, L. Fridman, and A. Levant, *Sliding Mode Control and Observation,* Cambridge, MA: Birkhauser, 2014.

[6] V. Utkin, "Equations of slipping regime in discontinuous systems," *Autom. Remote Control*, vol. 32, no. 12, pp. 1897–1907, 1971.

[7] C. Edwards and S. Spurgeon, *Sliding Mode Control: Theory and Applications*, London:Taylor & Francis, 1998.

[8] I. Boiko, *Discontinuous Control Systems: Frequency-Domain Analysis and Design*, Boston, MA: Birkhauser, 2009.

[9] A. Sabanovic, L. Fridman, and S. Spurgeon (eds.), *Variable Structure Systems: From Principles to Implementation*, New York, NY: IEE, 2004.

[10] F. Plestan, Y. Shtessel, V. Bregeault, and A. Poznyak, "New methodologies for adaptive sliding mode control," *Int. J. Control.,* vol. 83, no. 9, pp. 1907–1919, 2010.

[11] K. D. Young, V. I. Utkin, and U. Ozguner, "A control engineer's guide to sliding mode control," *IEEE Trans. Control Syst. Technol.*, vol. 7, no. 3, pp. 328–342, 1999.

[12] K. D. Young, "Controller design for a manipulator using theory of variable structure systems," *IEEE Trans. Syst., Man, Cybern.*, vol. 8, no. 2, pp. 101–109, 1978.

[13] V. I. Utkin, "Sliding mode control design principles and applications to electric drives," *IEEE Trans. Ind. Electron.*, vol. 40, no. 1, pp. 23–36, 1993.

[14] S. Kadwane, "Frequency-domain approach for sliding mode control of DC–DC buck converter," *Control Intell. Syst.,* vol. 40, no. 2, 2012.

[15] V. Utkin, "Sliding mode control of DC–DC converters," *J. Frankl. Inst.,* vol. 350, no. 8, pp. 2146–2165, 2013.

[16] A. Romero, L. Martinez-Salamero, J. Calvente, E. Alarcon, S. Porta, and A. Poveda, "Sliding mode control of switching converters: general theory in an integrated circuit solution," *HAIT J. Sci. Eng*., vol. B2, nos. 5–6, pp. 609–624, 2005.

[17] J. M. Olm, X. R. Oton, and Y. Shtessel, "Stable inversion-based robust tracking control in DC–DC non-minimum phase switched converters," *Automatica*, vol. 47, no. 1, pp. 221–226, 2011.

[18] V. M. Nguyen and C. Q. Lee, "Indirect implementation of sliding mode control law in buck-type converters," *APEC*, vol. 1, pp. 111–115, 1996.

[19] D. J. Perreault, K. Sato, R. L. Selders, and J. G. Kassakian, "Switching-ripple-based current sharing for paralleled power converters," *IEEE Trans. Circuits Syst.*, vol. 46, no. 10, pp. 1264–1274, 1999.

[20] A. Cid-Pastor, R. Giral, J. Calvente, V. Utkin, and L. Martinez-Salamero, "Interleaved converters based on sliding mode control in a ring configuration," *IEEE Trans. Circuit Syst.*, vol. 58, 10, pp. 2566–2577, 2011.

[21] H. Lee and V. Utkin, "Chattering reduction using multiphase sliding mode control," *Int. J. Control*., vol. 82, no. 9, pp. 1720–1737, 2009.

[22] D. Biel and E. Fossas, Some experiments on chattering suppression in power converters, MSC2009, St.-Petersburg, July, 2009.

[23] Y. M. Siew-Chong Tan, M. K. H. Lai, Cheung, and C. K. Tse, "On the practical design of a sliding mode voltage controlled buck converter," *IEEE Trans. Power Electron.*, vol. 20, no. 2, pp. 425–437, 2005.

[24] L. Martinez-Salamero, G. Garsia, M. Orellana, *et al.*, "Analysis and design of a sliding-mode strategy for start-up control and voltage regulation in a buck converter," *IET Trans. Power Electron.*, vol. 6, no. 1, pp. 52–59, 2013.

[25] H. El Fadil and F. Giri, "Reducing chattering phenomenon in sliding mode control of buck-boost power converters," in: *Proceedings of the IEEE International Symposium on Industrial Electronics*, 2008, pp. 287–292.

[26] X. Du, L. Zhou, and H.-M. Tai, "Double-frequency buck converter," *IEEE Trans. Ind. Electron.,* vol. 56, no. 5, pp. 1690–1698, 2009.

[27] D. Biel, E. Fossas, F. Guinjoan, E. Alarcon, and A. Poveda, "Application of sliding-mode control to the design of a buck-based sinusoidal generator," *IEEE Trans. Ind. Electron.,* vol. 48, no. 3, pp. 563–571, 2001.

[28] E. Vidal-Idiarte, C. E. Carrejo, J. Calvente, and L. Martinez-Salamero, "Two-loop digital sliding mode control of DC–DC power converters based on predictive interpolation," *IEEE Trans. Ind. Electron.*, vol. 58, no. 6, pp. 2491–2501, 2011.

[29] E. Mamarelis, G. Petrone, and G. Spagnuolo, "Design of a sliding-mode-controlled SEPIC for PV MPPT applications," *IEEE Trans. Ind. Electron.*, vol. 61, no. 7, pp. 3387–3398, 2014.

[30] R. Silva-Ortigoza, V. M. Hernandez-Guzman, M. Antonio-Cruz, and D. Munoz-Carrillo, "DC/DC buck power converter as a smooth starter for a DC motor based on a hierarchical control," *IEEE Trans. Power Electron.*, vol. 30, no. 2, pp. 1076–1084, 2015.

[31] E. Vidal-Idiarte, L. Martinez-Salamero, F. Guinjoan, J. Calvente, and S. Gomariz, "Sliding and fuzzy control of a boost converter using an 8-bit microcontroller," *IEE Proc. Electric Power Appl.*, vol. 151, no. 1, pp. 5–11, 2004.

[32] Siew-Chong Tan, Y. M. Lai, C. K. Tse, and M. K. H. Cheung, "Adaptive feedforward and feedback control schemes for sliding mode controlled power converters," *IEEE Trans. Power Electron.*, vol. 21, no. 1, pp. 182–192, 2006.

[33] L. Martinez-Salamero, A. Cid-Pastor, A. E. Aroudi, R. Giral, J. Calvente, and G. Ruiz-Magaz, "Sliding-mode control of DC–DC switching converters," *IFAC Proc. Volumes*, vol. 44, no. 1, pp. 1910–1916, 2011.

[34] A. El Aroudi, B. A. Martínez-Treviño, E. Vidal-Idiarte, and L. Martínez-Salamero, "Analysis of start-up response in a digitally controlled boost converter with constant power load and mitigation of inrush current problems," *IEEE Trans. Circuits Syst. I, Reg. Papers*, vol. 67, no. 4, pp. 1276–1285, 2020.

[35] Z.Wang, S. Li, and Q. Li, "Continuous nonsingular terminal sliding mode control of DC–DC boost converters subject to time-varying disturbances," *IEEE Trans. Circuits Syst. II, Exp. Briefs*, vol. 67, no. 11, pp. 2552–2556, 2020.

[36] S. H. Chincholkar, W. Jiang, and C.-Y. Chan, "A normalized output error-based sliding-mode controller for the DC–DC cascade boost converter," *IEEE Trans. Circuits Syst. II, Exp. Briefs*, vol. 67, no. 1, pp. 92–96, 2020.

[37] H. Komurcugil, S. Biricik, and N. Guler, "Indirect sliding mode control for DC–DC SEPIC converters," *IEEE Trans. Ind. Informat.*, vol. 16, no. 6, pp. 4099–4108, 2020.

[38] Z. Ya, *Tsypkin, Relay Control Systems*, Cambridge, England: Cambridge University Press, 1984.

[39] I. Boiko, "Oscillations and transfer properties of relay servo systems – the locus of a perturbed relay system approach," *Automatica*, vol. 41, no. 4, pp. 677–683, 2005.

[40] I. Boiko, "Analysis of closed-loop performance and frequency-domain design of compensating filters for sliding mode control systems," *IEEE Trans. Autom. Control*, vol. 52, no. 10, pp. 1882–1891, 2007.

[41] I. Boiko, "LPRS *analysis of sliding mode buck converter," J. Franklin Inst.*, vol. 353, no. 18, pp. 5137–5150, 2016.

[42] R. Rafiezadeh and R. Grino, "A relay controller with parallel feed-forward compensation for a buck converter feeding constant power loads," in *Proc. 24th IEEE Int. Conf. Emerg. Technol. Factory Autom.*, Sep. 2019, pp. 445–452.

[43] L. Benadero, F. Torres, E. Ponce, and A. E. Aroudi, "Dynamic analysis of self-oscillating H-bridge *inverters with state feedback," J. Franklin Inst.*, vol. 357, no. 1, pp. 494–521, 2020.

[44] A. AlZawaideh and I. Boiko, "LPRS analysis of sliding mode control of a boost converter," in *Proc. 15th Int. Workshop Variable Struct. Syst.*, Jul. 2018, pp. 461–463.

[45] A. AlZawaideh and I. Boiko, "Analysis of a sliding mode boost converter under fluctuating input source voltage, using LPRS method," *Control Eng. Pract.*, vol. 92, Art. no. 104132, 2019.

[46] A. AlZawaideh and I. Boiko, "Analysis of a sliding mode dc-dc boost converter through LPRS of a nonlinear plant," *IEEE Trans. Power Electron.*, vol. 35, no. 11, pp. 12 321–12 331, 2020.

[47] A. Al Zawaideh and I. M. Boiko, "Analysis of stability and performance of a cascaded PI sliding-mode control DC–DC boost converter via LPRS," *IEEE Trans. Power Electron.*, vol. 37, no. 9, pp. 10455–10465, 2022, doi:10.1109/TPEL.2022.3169000.

[48] B. J. Cardoso, A. F. Moreira, B. R. Menezes, and P. C. Cortizo, "Analysis of switching frequency reduction methods applied to sliding mode controlled DC–DC converters," in *Proc. '92 7th Annu. Appl. Power Electron. Conf. Expo.*, Feb. 1992, pp. 403–410.

[49] P. Mattavelli, L. Rossetto, G. Spiazzi, and P. Tenti, "General-purpose sliding-mode controller for DC/DC converter applications," in *Proc. IEEE Power Electron. Specialist Conf.*, Jun. 1993, pp. 609–615.

[50] I. Boiko, N. Kuznetsov, R. Mokaev, T. Mokaev, M. Yuldashev, and R. Yuldashev, "On counter-examples to aizerman and kalman conjectures," *Int. J.Control*, 95, pp. 1–8, 2020.
[51] I. Boiko, Discontinuous Control Systems, 2011. Available: http://www.mathworks.com/matlabcentral/fileexchange/30870-m-files-for–discontinuous-control-systems-?focused=5180508&tab=function

Chapter 13

Fractional-order controllers in power electronic converters

Allan G. S. Sánchez[1], *Francisco J. Pérez–Pinal*[1] and *Martín A. Rodríguez-Licea*[1]

Fractional calculus is considered nowadays as a true option for modelling and efficiency enhancement of power electronic converters. This chapter addresses the integration of fractional calculus into a control strategy to regulate the output of the basic configurations of buck, boost, or buck–boost converter from a practical perspective. The strategy considers a non-integer approach to investigate its effectiveness in controlling minimum and non-minimum phases systems. A standard PID controller structure is used to integrate the fractional-order derivative and integral through a biquadratic approximation of the Laplacian operator. The viability and effectiveness of the resulting controller is corroborated numerically and experimentally through its physical implementation. Fast response, stable regulation and good tracking characteristic were the most remarkable results.

Notation

$\frac{d}{dt}$	derivative operator
${}_aD_t^{\alpha}$	fractional-order operator
log	logarithm
Re(z), Im(z)	real/imaginary part of z
$\arg\{z\}$, $z = x + jy$	principal value of z, $\arctan\left(\frac{y}{x}\right)$
$\Gamma(\cdot)$	Gamma function
$\bar{a}$	average value of a
$\widehat{a}$	AC value of a
x_i	ith state
$\approx$	approximately equal to

[1]Instituto Tecnológico de Celaya, México

$G_p(s)$	plant transfer function
$G_{mp}(s)$	minimum-phase transfer function
$G_{nm}(s)$	non-minimum phase transfer function
$G_c(s)$	controller transfer function

Acronyms

AC/DC	alternating/direct current
ANN	artificial neural network
BOA	bat optimisation algorithm
BWOA	black widow optimisation algorithm
CCM	continuous conduction mode
DL	deep learning
EOA	evolutionary optimisation algorithm
FL	fuzzy logic
FOC	fractional-order control
FPGA	field programmable gate arrays
GOA	genetic optimisation algorithm
MOSFET	metal-oxide semiconductor field-effect transistor
MPPT	maximum power point tracking OA: optimisation algorithm
PEC	power electronic converter
PI	proportional-integral
PID	proportional-integral-derivative
PSOA	particle swarm optimisation algorithm
PWM	pulse width modulation
RHP	right-half plane
SMC	sliding mode control
WOA	whale optimisation algorithm

13.1 Introduction

Power electronic converters (PECs) have become fundamental electronic devices for numerous industrial and household applications due to their versatility to operate in low- and high-power ranges. Power electronics and control theory are combined to analyse, determine, and modify the transient and permanent dynamic characteristics of PECs. The interaction of both fields has derived a significant amount of knowledge describing a variety of new PECs topologies targeting different applications, improvements to existing control strategies and new proposals to manage energy storage and its delivery to the load in the most efficient ways.

Several linear and non-linear control techniques have been adapted and tested for voltage regulation of PECs. Three different approaches can be distinguished from the existing literature. On the one hand, the control designer chooses the control method based on its most prominent characteristic looking for it to prevail. The typical case is the sliding mode control (SMC) method, which fulfils the control objective whilst producing a robust response. The state feedback control method is effective and straightforward, although it requires additional considerations, such as system linearisation. Proportional-integral (PI) and proportional-integral-derivative (PID) have also been considered due to their simplicity, ease of implementation, acceptable performance and stability, in spite of their limited robustness.

On the other hand, a second approach seeks to enhance the performance of the control method through optimisation algorithms (OA) such as particle swarm (PSOA), Whale (WOA), Bat (BOA), Black Widow (BWOA), Evolutionary (EOA), Genetic (GOA) ones, or by combining a more sophisticated technique such as fuzzy logic (FL), deep learning (DL), or artificial neural networks (ANN) to mention the most relevant from artificial intelligence. Nature-based optimisation algorithms are preferred to improve controller performance due to their simplicity and physical interpretation. The common structure of the minimisation/maximisation criterion includes steady-state error, time constants and overshoot. The improvements include robustness, transient and permanent response enhancement; however, limitations on global minimum, optimal trapping and computational time can arise.

Lastly, it has been determined that fractional calculus represents an alternative to enhance modelling, increase efficiency and improve performance, mainly transient response and robustness of PECs through control strategies that adapted fractional derivative and integral in the strategy.

Fractional-order control (FOC) has been significantly benefited by the generalisation of the derivative and integral developed in fractional calculus, which allows considering integer, non-integer and even complex orders. This mathematical development allowed better flexibility in the controller structure, which produced many possible applications after A. Oustaloup suggested the first algorithm to synthesise a non-integer controller in 1991.

The main reasons that motivated the integration of fractional calculus to the control strategy are as follows [1,2]:

- Robustness.
- Reduced noise level due to reduction of derivative order.
- Additional degrees of freedom enhance the system model.
- Accurate description of frequency behaviour.
- Ladder/tree arrangements are used for implementation.
- Suitable describing long-term memory or fractal properties.
- Accurate modelling of lumped- or distributed-parameter systems.
- Flexibility for infinite-dimensional nature approximation.

Even though many interesting and relevant results have been derived from integrating fractional-order calculus into the PECs field, there still exist some drawback that have hindered its total acceptance from practical researchers and industry, mainly. Among the main limitation one can find [3]:

- Higher computational load.
- Fractional order has to be approximated through high-order transfer functions.
- Cost of implementation (more components or more sophistication).
- Lack of reliable tuning methods.

Notwithstanding, fractional calculus continues attracting the researchers' attention deriving in a significant amount of knowledge, mainly the one that relates PECs and FOC, deriving in remarkable improvements that include robustness [4], tuning flexibility [5] and smoother control signals [6], to mention the most relevant.

In this chapter, a fractional approximation of the PID controller is proposed as a true option to regulate output in basic DC–DC converters. Considering the described advantages of fractional calculus, a FOC approach is suggested to investigate its performance regulating the output of systems with a non-minimum phase classification. Since PECs are operated with high-speed control signals, the fractional-order approach is expected to endow the control strategy with the necessary dynamic to operate the plant efficiently and with high performance.

The controller structure will be generated with the fractional-order approximation of the Laplacian operator, which in turn will be synthesised with a biquadratic module that can generate a phase curve that exhibits flatness around its centre frequency. Frequency response of the controller resulting structure will be modified and shaped between derivative/integral effect as needed through two controller gains, which results in a simpler tuning method.

The physical realisation of fractional-order PID is possible by approximating its structure through rational functions of polynomials evaluated at a specific frequency band. This partial fraction expansion can be directly generated through a five-term electrical arrangement that resulted new and different, which can be implemented with *RC* networks and operational amplifiers in adder configuration. This represents a novelty due to the simplicity of proposed structure and simplification for analogue implementation, specially when fractional orders are commonly approximated through high-order transfer functions that require extensive ladder- or tree-like electrical circuits, or by using the fractional definition of the derivative/integral to be implemented in FPGA boards.

The content of this manuscript is distributed as follows: a brief revision of state of the art on FOC applied to PECs is provided in Section 13.2. Section 13.3 provides the preliminaries on two major topics of this manuscript: DC–DC converters and fractional calculus, which includes different methods to perform fractional-order approximation of derivatives/integral. In Section 13.4, the algorithm to synthesise the structure approximation, the tuning method and its effect on the frequency response are explained. In Section 13.5, numerical and experimental results are provided and described. A generalisation of the controller structure to regulate in either conversion mode of the described DC–DC converters, is also explained and tested in this section. Lastly, discussion on the presented results and the relevant conclusions are given in Sections 13.6 and 13.7, severally.

13.2 State of the art

The integration of fractional calculus in control of dynamic systems was well received in the last decade due to the theoretical development that made its interpretation, approximation and realisation possible. Power electronics is one of the fields that has benefited from the FOC development, which derived in alternative and more accurate representation of DC–DC converters [7–9], robust and flexible control strategies [10,11] or more precise controllers [12,13].

Some significant results on FOC applied to PECs can be summarised as follows: in [14,15], either the common fractional-order poles placement or the k-factor-based approaches were investigated to control the buck converter output. By using a commensurable transfer function or classical structure of compensator to represent the controller, stability region can be extended or simply guaranteed by establishing phase margin. By approximating the controller fractional order through the Oustaloup method and the Hankel model reduction, the fractional dominant poles were located based on three-term fractional transfer functions. The main improvement was robustness against load variations and parametric uncertainties. The PI/PID structures are one of the most used to test the fractional-order approach. In [16], a fractional PI was cascaded with a compensator of fuzzy-logic nature to regulate the output in a buck converter via capacitor current. This combination resulted in improvements on the capacity of disturbance suppression and good tracking characteristic. In a similar approach but with more sophistication, either fuzzy- or MPPT-based PID controllers were suggested in [17–19] to deal with disturbances and uncertainties when regulating output in a buck converter. In addition, optimisation algorithms were used to determine controller parameters. The combined strategies resulted in effective regulation and suppression of instability effects caused by a constant power load of microgrids. To addressed the same problem with similar objectives, a backstepping-based control, in combination with the ANN technique to estimate controller parameters, was investigated in [20]. Constant and stable voltage supply and good disturbance rejection were the improvements. A fractional PID approximation was investigated in [21] to achieve output voltage regulation of a buck converter. Control strategy considered both performance and robustness requirements. The resulting controller directly compensated the system frequency response through a lead or lag effect depending on the gains chosen. Robust performance of closed-loop system and a fast tracking characteristic were the improvements. In [22–24], fractional sliding and fractional-terminal sliding mode controllers with either back-stepping or reset control approach were investigated for regulating voltage in a buck converter. Designed controllers considered uncertainties rejection, non-linear loads and non-modelled dynamics. The system exhibited a reduction of steady-state error, enhanced perturbation rejection and L_2 stability in Lyapunov sense. A two-loop control diagram was used in [25] to generate a novel non-linear sliding surface to achieve voltage regulation in a buck converter. A compromise relating the start-up effect and transient of step response was considered in the problem formulation to ensure the terminal power. A fast response, robustness in the presence of load variations and convergence in finite time are the main results. A fractional PI-based anti-windup controller was investigated in [26]. The approach was proposed to control a motor for low speed applications. Benefits of the proposal were a softer switching pulse, superior tracking speed, steady-state reduction and wind up phenomena removal. Some other proposals have used classical structures to be combined with less known techniques, such as artificial ecosystems-, lion swarm- or hunger games-based optimisation algorithms. In [27–30], these techniques were used to find PI and PID controller generalisations parameters to achieve robustness, disturbance rejection, fast tracking and minimum steady-state error.

Using a similar approach but changing the control strategy, a sliding mode control of non-integer-order with adaptive properties was investigated in [31] to control output in a boost converter. The adaptive rules were designed for current tracking through state observer based on Lyapunov functions. Enhancement of step response transient and robust performance were the improvements. A standard fractional approximation of PID controller was investigated in [32] to control voltage in a boost converter. Control strategy consisted in approximating closed-loop transfer function by using characterisation of input inductance and output capacitance through a first-order system. A fast transient recovery and less overshoot were demonstrated through the system digital implementation. In [33], an optimised fractional PI controller was investigated to achieve output regulation in a DC–DC boost converter. The control strategy included augmentation of controller structure with a pre-stage non-linear error-modulator to enhance convergence rate during the transient. The closed-loop response exhibited short transits with less damping and good disturbance rejection property. Optimised typical PI controllers are suggested in [34] to control output of a boost converter that operates coupled to a photovoltaic system. Control strategy considered pollination and water cycle optimisation algorithms for controller parameter tuning. Smoother control efforts, increased robustness and improved transient response were the advantages of the approach. In [4], a typical fractional-order structure of PID controller was employed to achieve output regulation in a boost converter. Control strategy considered both robustness and performance requirements to address instability of the system caused by the feedback loop. The approach integrated the internal mode control concept to separate the undesired system dynamics form modelling it as a delay. Experimental validation showed fast response, good tracking characteristic and robustness against parameter uncertainties as improvements. With a similar idea, an optimised PID with fractional-order approach was suggested in [35] to control output of a boost converter. Control strategy consisted of criteria to be met to achieve the desired performance requirements related to the start-up response. The artificial bee colony algorithm was employed to compute controller parameters. The comparison with classical approach showed superiority of the fractional PID controller in transient stage. In [36], an Internal Mode Control and Direct Synthesis techniques were investigated to design a controller to regulate output in a boost converter. Main objectives were the enhancement of transient response and robustness. The combination resulted in a simple design process that produced satisfactory performance parameters and robustness, being those produced by the Internal Mode Control remarkable.

Lastly, to achieve regulation of buck–boost converter output, control strategies are similar to those that successfully achieved the control objective when controlling the DC–DC boost converter, since they have a similar closed-loop behaviour. Some examples are [10,12,37–40], where either standard fractional-order PID structure or fractional sliding mode control have been combined with either optimisation algorithms or artificial intelligence-based techniques to determine controller parameters. Others combine two-loop strategies with current mode control to achieve the objective [11] or the hard limiter approach [41], which introduces a device in the inductor branch to regulate its current. An entirely different approach was based on the ultracapacitor behaviour, whose characterisation of fractional dynamic was applied to buck–boost converter with state feedback techniques and used to stabilise its closed-loop behaviour in the sense of Lyapunov [42–45].

Definitely, it is hard to describe all the evolution of FOC applied to PECs; however, the most recent and relevant techniques/approaches have been described. For those readers interested in the topic, revise the provided references and references therein.

In the following section, necessary concepts on DC–DC converters and fractional calculus are provided.

13.3 Preliminaries

This section provides necessary preliminaries on converters basic configurations buck, boost and buck–boost, including the description of their mathematical models.

13.3.1 DC–DC converters

DC–DC converters are electronic devices that can produce, depending on the configuration, a variable but continuous source of voltage that can be lower or higher than the converter power supply. Buck configuration, which generates an output voltage lower than its power supply due to its conversion ratio less than one, is one of the two most used basic topologies to regulate DC motors velocity or as a regulated voltage source. On the other hand, boost configuration is the classic way to interconnect lower-voltage sources with loads that demand higher levels of voltage, mainly because its conversion ratio is characterised by a gain greater than one. Lastly, buck–boost converter topology gathers both characteristics, since it can produce a lower or higher voltage level than the one provided by its power supply with inverse polarity.

In the following, the mathematical models of these three DC–DC converters are provided.

13.3.1.1 Buck converter

The electronic representation of buck converter is depicted in Figure 13.1. Elements composing this electronic device include: a MOSFET Q and a diode D,

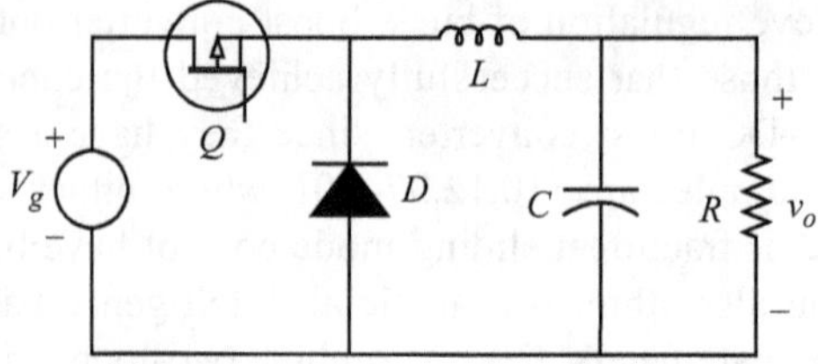

Figure 13.1 Electrical representation of buck converter

Table 13.1 Parameters and input/output operation values for buck converter from Figure 13.1 to ensure CCM

Parameter	Notation	Value	Parameter	Notation	Value
Input voltage	V_g	25 V	Frequency	f_{sw}	20 kHz
Output voltage	v_o	15 V	Resistance	R	10 Ω
Output power	P_o	10 W	Capacitor	C	7 μF
Voltage ripple	Δv_C	10%	Inductor	L	2.7 mH
Current ripple	Δi_L	10%			

which are operated as complementary switches to deliver the energy of the DC voltage source V_g to the inductance L, a capacitor C and a resistor R.

Assuming continuous conduction mode (CCM) and ideal operation, buck converter depicted in Figure 13.1 is described as follows [46],

$$\begin{aligned} L\frac{di_L}{dt} &= \bar{d}V_g - v_C \\ C\frac{dv_C}{dt} &= i_L - \frac{1}{R}v_C \end{aligned} \tag{13.1}$$

where $\bar{d} \in [0, 1]$ is computed over one period and represents the average value of duty cycle d, whose equilibrium point $[i_L, \ v_C] = [\bar{d}V_g/R, \ \bar{d}V_g]$ was obtained by setting $di_L/dt = dv_C/dt = 0$. The system output $y(t)$ is identified by considering that $v_o = v_C$, which is manipulated through $\bar{d}$, which corresponds to the control input $u(t)$, thus, transfer function from $\bar{d}$-to-v_c of system (13.1) will be given by,

$$G_p(s) = \frac{Y(s)}{U(s)} = \frac{\left(\frac{V_g}{CL}\right)}{s^2 + \left(\frac{1}{RC}\right)s + \left(\frac{1}{CL}\right)} \tag{13.2}$$

Parameters and input/output operation values shown in Table 13.1 were used to ensure CCM in the buck converter depicted in Figure 13.1. System frequency response is provided in Figure 13.2.

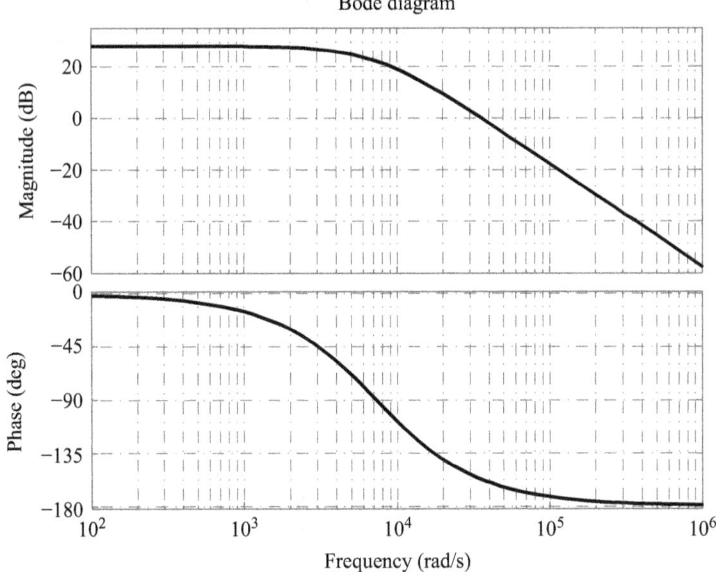

Figure 13.2 Frequency response for buck converter in Figure 13.1

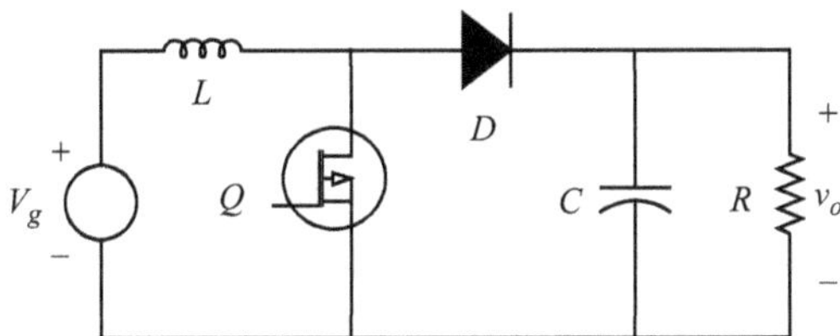

Figure 13.3 Electrical diagram of DC–DC boost converter

In the following section, the mathematical description of boost converter will be provided.

13.3.1.2 Boost converter model

Electrical representation of boost converter is presented in Figure 13.3. As previously done, ideal operation and CCM are considered; thus the continuous average model of the system will be given by [46]

$$\begin{aligned} L\frac{\mathrm{d}i_L}{\mathrm{dt}} &= V_g - (1-\bar{d})v_C \\ C\frac{\mathrm{d}v_C}{\mathrm{dt}} &= (1-\bar{d})i_L - \frac{1}{R}v_C \end{aligned} \tag{13.3}$$

which is a bilinear system. This implies that model non-linearities are given by the product of $x_i u$, where x_i is the ith state variable and u the control input.

Obtaining this converter transfer function requires linearisation of (13.3). Small-signal method enables us to generate the linear representation of the boost converter model. This analysis technique is commonly used to approximate behaviour around operating points by perturbing signals of the system to generate their DC and AC components [47]. Resulting model will be the small-signal representation, which is the AC equivalent where non-linearities $x_i u$ of the original model (13.3) are replaced by linear elements related to the operating or equilibrium point $[i_L, v_C] = \left[V_g/(R(1-\bar{d})^2),\ V_g/(1-\bar{d})\right]$.

Small-signal representation of boost converter can be transformed into the state-space model $\dot{x} = \mathbf{A}x + \mathbf{B}u$, $y = \mathbf{C}x$ given as follows [47]:

$$\begin{bmatrix} \dot{\hat{i}}_L \\ \dot{\hat{v}}_C \end{bmatrix} = \begin{bmatrix} 0 & \dfrac{-(1-\bar{D})}{L} \\ \dfrac{(1-\bar{D})}{C} & \dfrac{-1}{RC} \end{bmatrix} \begin{bmatrix} \hat{i}_L \\ \hat{v}_C \end{bmatrix} + \begin{bmatrix} \dfrac{V_g}{L(1-\bar{D})} & \dfrac{1}{L} \\ \dfrac{-V_g}{RC(1-\bar{D})^2} & 0 \end{bmatrix} \begin{bmatrix} \hat{\bar{d}} \\ \hat{v}_g \end{bmatrix} \tag{13.4}$$

and

$$y = \begin{bmatrix} 0 & 1 \end{bmatrix} \begin{bmatrix} \hat{i}_L \\ \hat{v}_C \end{bmatrix} \tag{13.5}$$

where $\bar{D}$ is the DC component of $\bar{d}$ and $\hat{v}_g$, $\hat{v}_C$, $\hat{i}_L$, $\hat{\bar{d}}$, are the AC or perturbation terms of V_g, v_C, i_L, $\bar{d}$, respectively.

Thus, system transfer function $G_p(s)$ will be given by,

$$G_p(s) = C(sI - A)^{-1}B_1 \tag{13.6}$$

which is the relation of control law to output $\hat{\bar{d}}$-to-$\hat{v}_c$, where vector $B_1 = \left[\dfrac{V_g}{L(1-\bar{D})} \dfrac{-Vg}{RC(1-\bar{D})^2}\right]^T$. Therefore, the DC–DC boost converter transfer function will be given by,

$$G_p(s) = \frac{\left(\frac{V_g}{CL}\right) - \left(\frac{V_g}{RC(1-\bar{D})^2}\right)s}{s^2 + \left(\frac{1}{RC}\right)s + \frac{(1-\bar{D})^2}{CL}} \tag{13.7}$$

Note that from system transfer function (13.7) a right-half plane (RHP) zero can be obtained from its numerator, thus it is a non-minimum phase system, which represents a challenging control problem because the RHP zero becomes an RHP pole in closed loop, thus producing the characteristic closed-loop instability of boost converter.

Recalling that non-minimum phase systems such as (13.7) typically respond as shown in Figure 13.4, where the delay phenomena can be identified, transfer function (13.7) can be separated as follows:

$$G_p(s) = G_{mp}(s)G_{nm}(s) \tag{13.8}$$

where

$$G_{mp}(s) = \frac{\left(\frac{V_g}{RC(1-\bar{D})^2}\right)\left(\frac{R(1-\bar{D})^2}{L} + s\right)}{s^2 + \left(\frac{1}{RC}\right)s + \frac{(1-\bar{D})^2}{CL}} \tag{13.9}$$

is the minimum phase part of $G_p(s)$ and

$$G_{nm}(s) = \frac{\frac{R(1-\bar{D})^2}{L} - s}{\frac{R(1-\bar{D})^2}{L} + s} \tag{13.10}$$

the normalized non-minimum phase part. It is important to highlight from (13.10) that $|G_{nm}(j\omega)| = 1$, thus it is considered an all-pass system which has the structure of Padé approximation [48]:

$$e^{-s\tau} \approx \frac{2/\tau - s}{2/\tau + s} \tag{13.11}$$

thus, it is responsible for the delay shown in the system response depicted in Figure 13.4. Therefore, one can consider that the boost converter dynamic will be given entirely by the minimum phase part $G_{mp}(s)$.

Parameters and input/output operation values shown in Table 13.2 were used to ensure CCM in the boost converter of Figure 13.3, whose frequency response is depicted in Figure 13.5.

In the remaining of this section, mathematical model of buck–boost converter, which also exhibits the non-minimum phase characteristic is provided.

13.3.1.3 Buck–boost converter model

DC–DC buck–boost converter results from combining in cascade buck and boost basic configurations. Without increasing the number of elements, this alternative arrangement can produce a reverse polarity output, which can be both lower or higher than the power supply.

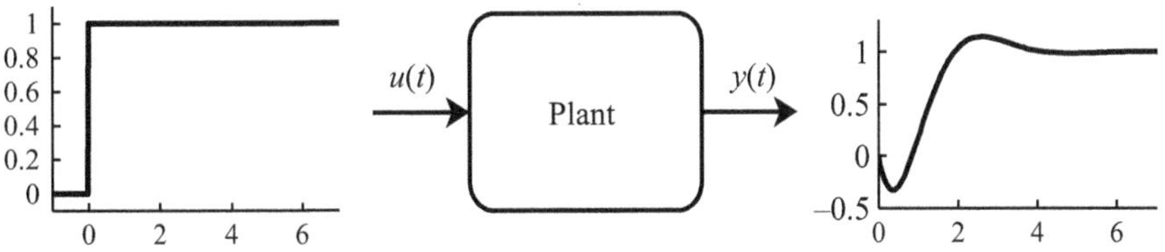

Figure 13.4 Typical response of non-minimum phase systems

Table 13.2 Parameters and input/output operation values for boost converter from Figure 13.3 to ensure CCM

Parameter	Notation	Value	Parameter	Notation	Value
Input voltage	V_g	15 V	Frequency	f_{sw}	20 kHz
Output voltage	v_o	30 V	Resistance	R	25 Ω
Output power	P_o	36 W	Capacitor	C	20 μF
Voltage ripple	Δv_C	5%	Inductor	L	3.7 mH
Current ripple	Δi_L	5%			

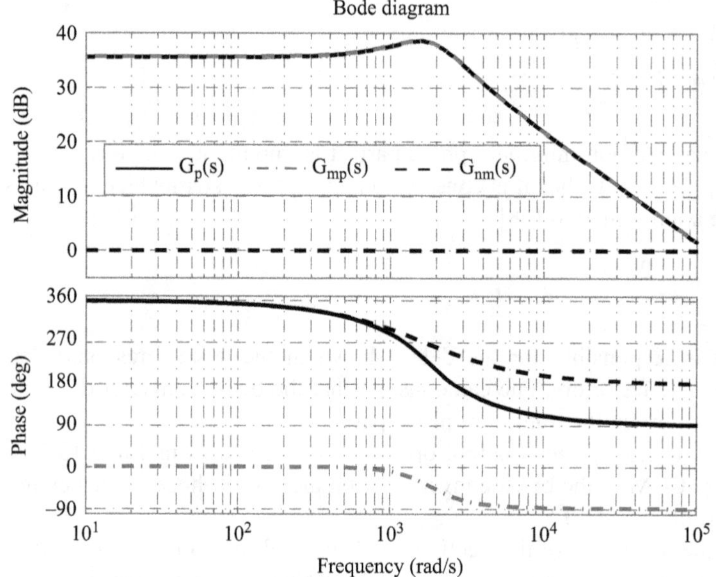

Figure 13.5 Frequency response for boost converter in Figure 13.3, where solid, dot-dashed and dashed lines stand for $G_p(s)$, $G_{mp}(s)$ and $G_{nm}(s)$, respectively

Averaged mathematical model of buck–boost converter from Figure 13.6, under ideal operating conditions and CCM, is given as follows [46]:

$$\begin{aligned} L\frac{di_L}{dt} &= V_g\bar{d} + (1-\bar{d})v_C \\ C\frac{dv_C}{dt} &= -(1-\bar{d})i_L - \frac{1}{R}v_C \end{aligned} \tag{13.12}$$

where reverse-polarity output voltage can be confirmed through the second equation of (13.12). As it happened in the previous case, the buck–boost converter is a bilinear system; thus, the same procedure done for the boost converter can be applied.

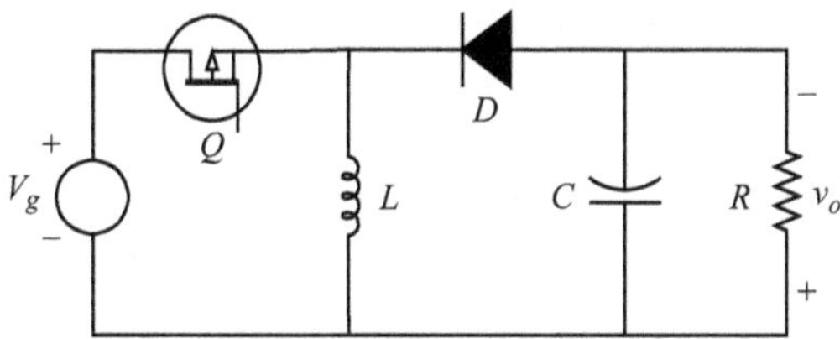

Figure 13.6 Electrical diagram of the DC–DC buck–boost converter

Linearised representation around the operating point or equilibrium $[i_L,\ v_C] = \left[V_g\bar{D}/(R(1-\bar{D})^2),\ V_g\bar{D}/(1-\bar{D})\right]$ of the resulting small-signal average model is given as follows [47],

$$\begin{bmatrix} \dot{\hat{i}}_L \\ \dot{\hat{v}}_C \end{bmatrix} = \begin{bmatrix} 0 & \frac{(1-\bar{D})}{L} \\ \frac{-(1-\bar{D})}{C} & \frac{-1}{RC} \end{bmatrix} \begin{bmatrix} \hat{i}_L \\ \hat{v}_C \end{bmatrix} + \begin{bmatrix} \frac{V_g}{L(1-\bar{D})} & \frac{\bar{D}}{L} \\ \frac{V_g\bar{D}}{RC(1-\bar{D})^2} & 0 \end{bmatrix} \begin{bmatrix} \hat{d} \\ \hat{v}_g \end{bmatrix} \tag{13.13}$$

and

$$y = \begin{bmatrix} 0 & 1 \end{bmatrix} \begin{bmatrix} \hat{i}_L \\ \hat{v}_C \end{bmatrix} \tag{13.14}$$

Buck–boost converter transfer function $G_p(s)$ is computed from (13.6), as in the previous case, and $B_1 = \left[\frac{V_g}{L(1-\bar{D})} \frac{V_g\bar{D}}{RC(1-\bar{D})^2}\right]^T$ as follows:

$$G_p(s) = \frac{\left(\frac{V_g\bar{D}}{RC(1-\bar{D})^2}\right)s - \left(\frac{V_g}{CL}\right)}{s^2 + \left(\frac{1}{RC}\right)s + \frac{(1-\bar{D})^2}{CL}} \tag{13.15}$$

which also represents a non-minimum phase transfer function. By extending (13.15) as in (13.8), the factors will be

$$G_{mp}(s) = \frac{\left(\frac{V_g\bar{D}}{RC(1-\bar{D})^2}\right)\left(s + \frac{R(1-\bar{D})^2}{L\bar{D}}\right)}{s^2 + \left(\frac{1}{RC}\right)s + \frac{(1-\bar{D})^2}{CL}} \tag{13.16}$$

which is the minimum phase part of $G_p(s)$ and

$$G_{nm}(s) = \frac{s - \dfrac{R(1-\bar{D})^2}{L\bar{D}}}{s + \dfrac{R(1-\bar{D})^2}{L\bar{D}}} \tag{13.17}$$

the normalized non-minimum phase part. Note that $G_{nm}(s)$ introduced a delay in the case of boost converter, but in this case it is also the cause of output voltage polarity inversion.

Parameters and input/output operation values shown in Table 13.3 were used to ensure CCM in the buck–boost converter of Figure 13.6, whose the frequency response is provided in Figure 13.7.

Next section provides a summarised review of fractional calculus and the methodology to approximate the fractional-order Laplacian operator will be described.

13.3.2 Fractional-Order Calculus

Fractional calculus is the generalisation of methods or algorithms devoted to predicting, analysing and solving variables, including ordinary calculus. This theory consists in the possibility of defining derivatives and integrals with integer, non-integer/rational or even complex orders [49,50]. A differential/integral equation containing fractional-order derivatives/integrals is considered a fractional-order equation. On the other side, if a system description is given by fractional-order differential/integral equations, thus it is said to be a fractional system [49].

Different ways have been suggested to define fractional-order derivatives/integrals, each one with the aim of describing systems/phenomena more accurately. Three are the most known, studied and used definitions of fractional-order derivative/integral, defined through ${}_aD_t^\alpha$, which is the fundamental operator used in fractional-order calculus, where α represents the operation order and a, t the operation limits for fractional-order differentiation/integration. Depending on the value of α, the integro-differential operator will behave as follows [49,50]:

$$ {}_aD_t^\alpha = \begin{cases} \dfrac{d^\alpha}{dt^\alpha} & \alpha > 0 \\ 1 & \alpha = 0 \\ \displaystyle\int_a^t (d\tau)^\alpha & \alpha < 0 \end{cases} \tag{13.18} $$

Table 13.3 Parameters and input/output operation values for buck–boost converter from Figure 13.6 to ensure CCM

Parameter	Notation	Value	Parameter	Notation	Value
Input voltage	V_g	25 V	Frequency	f_{sw}	20 kHz
Output voltage	v_o	15/35 V	Resistance	R	15 Ω
Output power	P_o	15/81 W	Capacitor	C	30 μF
Voltage ripple	Δv_C	10%	Inductor	L	10 mH
Current ripple	Δi_L	10%			

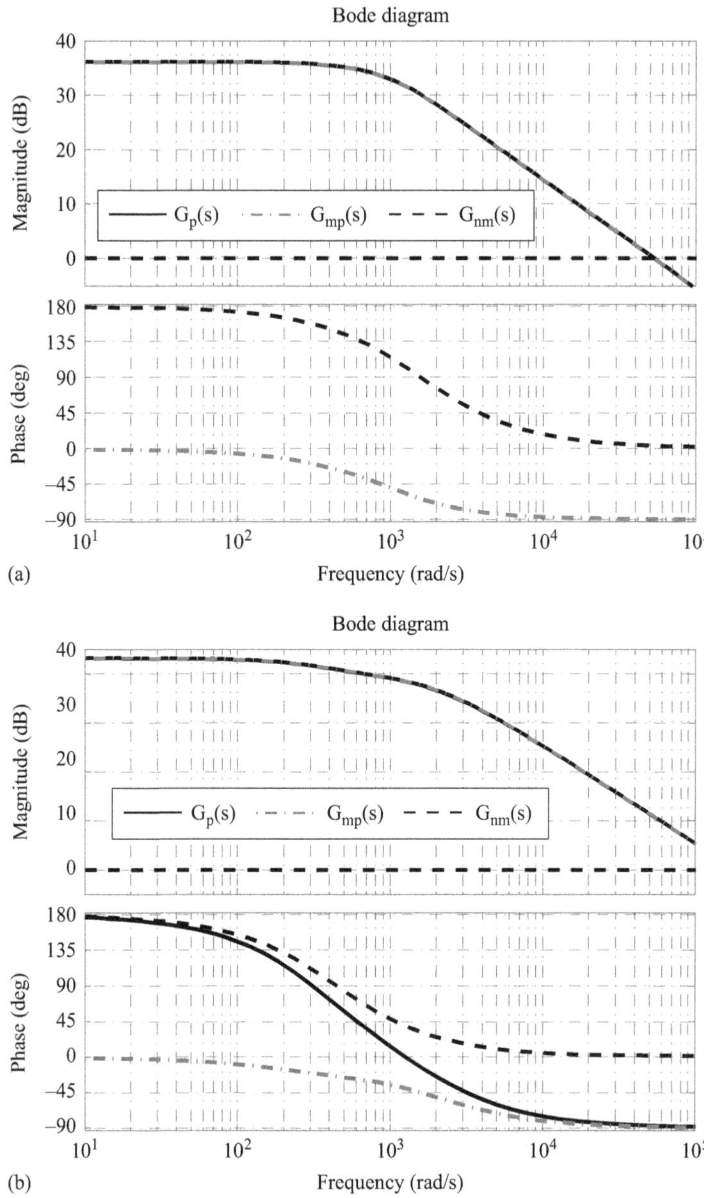

Figure 13.7 Frequency response for buck–boost converter in Figure 13.6, where solid, dot-dashed and dashed lines stand for $G_p(s)$, $G_{mp}(s)$ and $G_{nm}(s)$, respectively. (a) Buck mode. (b) Boost mode.

Riemann–Liouville fractional-order derivative, historically related to the beginnings of fractional calculus, is defined as follows [51]:

$$ {}_aD_t^{\alpha} f(t) = \frac{1}{\Gamma(m-\alpha)} \frac{\mathrm{d}^m}{\mathrm{d}t^m} \int_a^t (t-u)^{m-\alpha-1} f(u) du \tag{13.19} $$

where $m = \lceil \alpha \rceil$ is the celling rounding of α and $\Gamma(m-\alpha)$ the Euler gamma function. On the other hand, Caputo definition of non-integer order derivative is given as follows [51]:

$$ {}_aD_t^{\alpha} f(t) = \frac{1}{\Gamma(m-\alpha)} \int_a^t (t-u)^{m-\alpha-1} f^{(m)}(u) du \tag{13.20} $$

which is equivalent to Riemann–Liouville one under the assumption of Cauchy-type initial conditions [51].

Grünwald–Letnikov fractional-order definition for derivative is commonly used due to its practicality and mathematical interpretation, since it can be derived from the typical definition of the derivative

$$ \frac{d^n f}{dt^n} = \lim_{h \to 0} \frac{1}{h^n} \sum_{k=0}^{n} (-1)^k \binom{n}{k} f(t-kh) \tag{13.21} $$

where $\binom{n}{k}$ are the binomial coefficients. By considering $n = (t-a)/h$ for the integer part of the summation limit value [49,50],

$$ \lim_{\substack{h \to 0 \\ n=(t-a)/h}} f_h^{(n)}(t) = {}_aD_t^{\alpha} f(t) \tag{13.22} $$

where ${}_aD_t^{\alpha} f(t)$ operates on the function $f(t)$. Therefore, definition (13.21) will be given by,

$$ {}_aD_t^{\alpha} f(t) = \lim_{h \to 0} \frac{1}{h^{\alpha}} \sum_{k=0}^{(t-a)/h} (-1)^k \alpha k f(t-kh) \tag{13.23} $$

which is the so-called Grünwald–Letnikov fractional-order differintegral [51].

Relation commonly used to compute numerically fractional derivatives was derived from (13.23) as follows [49]:

$$ {}_{(k-L_m/h)}D_{t_k}^{q} f(t) \approx h^{-q} \sum_{j=0}^{k} c_j^{(q)} f(t_k - j) \tag{13.24} $$

where L_m refers to memory length, t_k time step and $c_j^{(q)}$ the binomial coefficients, which are given by [49]

$$ c_0^{(q)} = 1, \quad c_j^{(q)} = \left(1 - \frac{1+q}{j}\right) c_{j-1}^{(q)} \tag{13.25} $$

Although the above-described fractional-order derivative/integral definitions were proposed to enhance description of systems/phenomena, the main concern of applied researchers and industry is related to the way that fractional-order models/controllers/systems are implemented. This continues to be carried out by integer-order components that are approximated through higher integer-order transfer functions.

In the following, some of the most known fractional-order approximation approaches are briefly described. The selected methodology to synthesise the controllers for regulating voltage in DC–DC converters is also described.

13.3.2.1 Carlson fractional-order approximation

Laplace representation of fractional-order operator is s^α for $\alpha \in (0,\ 1)$, whose frequency response is given in magnitude and phase by $20\alpha \log \omega$ and $\alpha\pi/2$, respectively. There is no finite-order filter that fits straight lines generated for all the frequency spectrum; thus, it is advisable to locate frequency response within a finite range $(\omega_l,\ \omega_h)$.

The Carlson approximation is based on Newton iterative process and Continued Fraction Expansion [52] to approximate the Laplacian operator through the rational function $F_i(s) = s^{\pm\alpha}$, $\alpha \in (0,\ 1)$ as follows:

$$F_i(s) = F_{i-1}(s)\frac{\left(\frac{1-\alpha}{\alpha}\right)F_{i-1}^{\beta}(s) + \left(\frac{1+\alpha}{\alpha}\right)s}{\left(\frac{1+\alpha}{\alpha}\right)F_{i-1}^{\beta}(s) + \left(\frac{1-\alpha}{\alpha}\right)s} \tag{13.26}$$

where the initial value $F_0(s) = 1$ and $\beta = \lceil 1/\alpha \rceil$; thus, it can only assume integer values $\pm 2,\ \pm 3, \ldots$, which represents the main limitation of this approach. By using (13.26), Carlson's approximation of differentiators/integrators $s^{\pm\alpha}$ for $\alpha = 0.6$ produce the frequency responses shown in Figure 13.8.

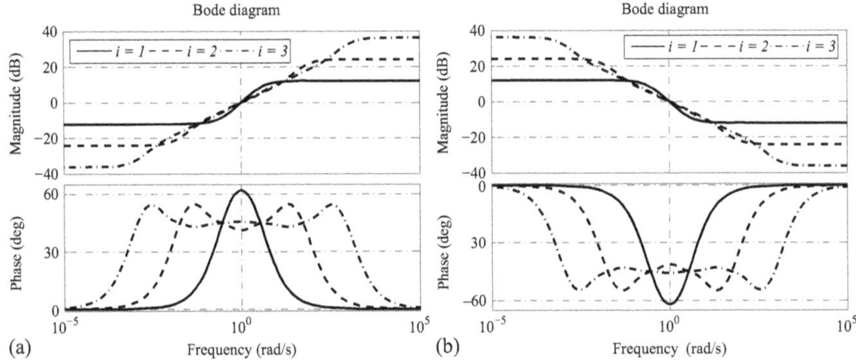

Figure 13.8 Frequency response of Carlson approximation (13.26) for iterations $i = 1,\ 2,\ 3$ *and fractional order* $\alpha = 0.6$, *where* $s^{0.6} = \frac{2.667s+0.6667}{0.6667s+2.667}$ *and* $s^{-0.6} = \frac{0.6667s+2.667}{2.667s+0.6667}$ *for* $i = 1$

13.3.2.2 Oustaloup and refined Oustaloup fractional-order approximation

Oustaloup fractional-order approximation is an integer-order polynomial quotient obtained by cascading N first-order transfer functions as follows [53]:

$$s^{\alpha} \approx \omega_h^{\alpha} \prod_{k=1}^{N} \frac{s + \omega_k'}{s + \omega_k} \tag{13.27}$$

where

$$\begin{aligned} \omega_k' &= \omega_l \left(\frac{\omega_h}{\omega_l} \right)^{\frac{2k-1-\alpha}{2N}} \\ \omega_k &= \omega_l \left(\frac{\omega_h}{\omega_l} \right)^{\frac{2k-1+\alpha}{2N}} \end{aligned} \tag{13.28}$$

for $0 < \alpha < 1$, $(\omega_l,\ \omega_h)$ is the operating frequency range of the approximation and N the approximation order. Note from (13.27) that poles and zeros composing the resulting quotient alternate, leading the zeros to the poles for s^{α} and vice versa for $s^{-\alpha}$, which indicates that the approximation can produce derivate or integral effect. In Figure 13.9, fourth- and tenth-order Oustaloup approximation frequency response is shown for derivative/integral effects when $\alpha = 0.6$ and $[\omega_l,\ \omega_h] = [0.01,\ 100]$ rad/s.

On the other hand, the refined Oustaloup approximation was proposed to overcome the frequency limitations of the original proposal. The refined procedure doubles the amount of zeros/poles than those introduced by the original definition

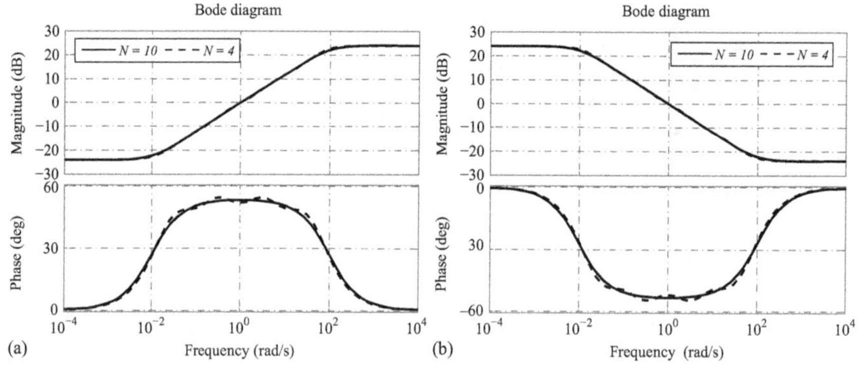

Figure 13.9 Frequency response of Oustaloup approximation (13.27) of fourth/tenth order and $\alpha = 0.6$, *where* $s^{0.6} = \frac{15.85s^4+279.1s^3+446.3s^2+70.1s+1}{s^4+70.1s^3+446.3s^2+279.1s+15.85}$ *and* $s^{-0.6} = \frac{s^4+70.1s^3+446.3s^2+279.1s+15.85}{15.85s^4+279.1s^3+446.3s^2+70.1s+1}$ *for* $N = 4$

as follows [6]:

$$s^{\alpha} \approx \left(\frac{d\omega_h}{b}\right)^{\alpha} \left(\frac{ds^2 + b\omega_h s}{d(1-\alpha)s^2 + b\omega_h s + d\alpha}\right) \prod_{k=-N}^{N} \frac{s + \omega_k'}{s + \omega_k} \tag{13.29}$$

where $b = 10$ and $d = 9$.

Main disadvantages of Oustaloup and refined Oustaloup approximations include the inability to fit the overall expected frequency range when implemented and high integer-order transfer functions [6].

13.3.2.3 Charef fractional-order approximation

Similarly, the Charef methodology approximates fractional-order systems of the form [54]

$$H(s) = \frac{1}{\left(1 + \frac{s}{\omega_l}\right)^{\alpha}} \approx \frac{1}{s^{\alpha}} \tag{13.30}$$

within the frequency range ω_l, ω_h, where $\alpha \in (0, 1)$, by cascading poles and zeros as Oustaloup method to generate a quotient of polynomials that are factorised as follows [54]:

$$\widehat{H}(s) = \frac{\prod_{i=0}^{N-1}\left(1 + \frac{s}{z_i}\right)}{\prod_{i=0}^{N}\left(1 + \frac{s}{p_i}\right)} \tag{13.31}$$

where ω_l is the corner frequency determined from the magnitude curve at -3 dB, $p_0 = \omega_l b$, $a = 10^{y/10(1-\alpha)}$, $b = 10^{y/10\alpha}$ and $ab = 10^{y/10\alpha(1-\alpha)}$; thus, the poles and zeros for $i \geq 1$ are computed as follows:

$$\begin{aligned} p_i &= p_0(ab)^i \\ z_i &= ap_0(ab)^i \end{aligned} \tag{13.32}$$

and

$$N = \left\lfloor \frac{\log\left(\frac{\omega_h}{p_0}\right)}{\log(ab)} \right\rceil + 1 \tag{13.33}$$

where $\lfloor \cdot \rceil$ is the rounding to nearest integer. In Figure 13.10, the frequency response of the Charef approximation (13.31) for $\alpha = 0.6$ and frequency range $[\omega_l, \omega_h] = [1 \times 10^{-2}, 1 \times 10^{6}]$ rad/s is shown. Note from Figure 13.10 that despite the shape, the approximation phase contribution is 54° as in the previous cases.

13.3.2.4 El-Khazali fractional-order approximation

As in the previous methodologies, the El-Khazali approach approximates the Laplacian operator within a specific frequency band through bi-quadratic modules.

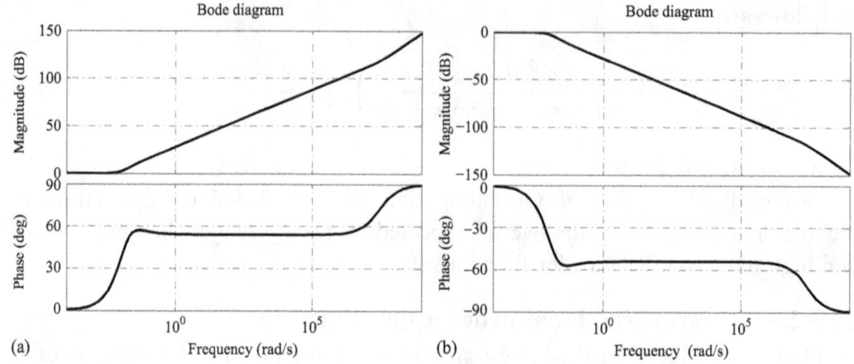

Figure 13.10 Frequency response of Charef approximation (13.31) for s^{α} and $s^{-\alpha}$, where α= 0.6. (a) Derivative effect. (b) Integral effect

Approximation of Laplacian operator s^{α}, $\alpha \in (0,1)$ according to El-Khazali methodology is given as follows [55,56]:

$$s^{\alpha} \approx T\left(\frac{s}{\omega_c}\right) = \frac{a_0\left(\frac{s}{\omega_c}\right)^2 + a_1\left(\frac{s}{\omega_c}\right) + a_2}{a_2\left(\frac{s}{\omega_c}\right)^2 + a_1\left(\frac{s}{\omega_c}\right) + a_0} \tag{13.34}$$

which represents a single bi-quadratic module capable of generating a flattened phase response, ω_c is the centre frequency and alpha-dependent real constants a_0, a_1, a_2 are given as follows:

$$\begin{aligned} a_0 &= \alpha^{\alpha} + 3\alpha + 2 \\ a_2 &= \alpha^{\alpha} - 3\alpha + 2 \\ a_1 &= 6\alpha \ \tan\frac{(2-\alpha)\pi}{4} \end{aligned} \tag{13.35}$$

If $\omega = \omega_c$ is assumed, the so-called integro-differential operator (13.34) will be given as

$$s^{\alpha} \approx T\left(\frac{j\omega_c}{\omega_c}, \alpha\right) = \frac{(a_2 - a_0) + ja_1}{-(a_2 - a_0) + ja_1} = \frac{-6\alpha + j6\alpha \tan\frac{(2-\alpha)\pi}{4}}{6\alpha + j6\alpha \tan\frac{(2-\alpha)\pi}{4}} \tag{13.36}$$

where $arg\{T(j1,\alpha)\} = \arctan\left(\frac{6\alpha \tan\left(\frac{(2-\alpha)\pi}{4}\right)}{-6\alpha}\right) - \arctan\left(\frac{6\alpha \tan\left(\frac{(2-\alpha)\pi}{4}\right)}{6\alpha}\right)$ is the phase contribution of s^{α}, thus the phase contribution of the approximation is given by [55,56],

$$\arg\left\{T(s/\omega_c)\big|_{j\omega_c}\right\} = \pm\alpha\pi/2, \tag{13.37}$$

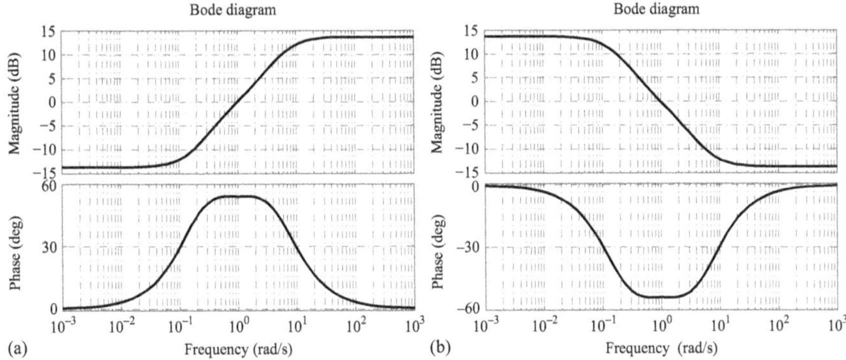

Figure 13.11 Frequency response of $s^{\pm\alpha}$, where $\alpha = 0.6$, $\omega_c = 1$ and $arg\{T(j1)\} = \pm 54^\circ$. (a) Derivative effect. (b) Integral effect

implying that the biquadratic approximation of Laplacian operator $s^{\pm\alpha}$ will have a phase contribution within the range $-90^\circ < \arg\{T(s/\omega_c)|_{j\omega_c}\} < 90^\circ$ depending on the value of α.

Derivative and integral effects will be produced by holding a condition over a_0 and a_2. For $a_0 > a_2 > 0$, (13.34) will exhibit derivative effect around ω_c, producing that $arg\{T(s/\omega_c)\} > 0$. In another way, for $0 < a_0 < a_2$, (13.34) exhibits integral effect, which implies $\arg\{T(s/\omega_c)|_{j\omega_c}\} < 0$. Analysing the location of zeros/poles that (13.34) produces, a zero appears more to the right in the plane, thus leading the poles for $a_0 > a_2$ while a pole appears more to the right leading zeros for $a_0 < a_2$, which is consistent with the derivative and integral effects.

Frequency response of El-Khazali approximation (13.34) for non-integer Laplacian operator of order $\alpha = 0.6$, where $a_0 = 4.536$, $a_1 = 7.0654$ and $a_2 = 0.9360$ is depicted in Figure 13.11.

By comparing the above-described approximations, it can be determined that the El-Khazali approach is the most convenient method to choose. This is due to it employs a second order polynomials in the transfer function for the approximation of Laplacian operator, whilst the other methods required greater orders to produce an acceptable one. In addition, the bi-quadratic module of El-Khazali produces a symmetrical wave-less flat-phase curve. Aiming to keep the controller design as simple as possible, the El-Khazali method was chosen to control voltage in the DC–DC converters described in Section 13.3.1.

The synthesis of controller structure and the steps to apply the control strategy, using the approximation of fractional-order for the Laplacian operator, are described in the following section.

13.4 Controller synthesis and control strategy

In this section, the design algorithm for the controller is provided. First, the procedure to synthesise the structure of a fractional PID controller is given. Then, the steps to apply the control strategy to regulate output in DC–DC converters are provided.

13.4.1 Synthesis of fractional-order approximation of PID controller

The fractional-order structure of a $PI^{\alpha}D^{\mu}$ controller is describe by [57]

$$G_c(s) = k_p\left(1 + \frac{1}{T_i s^{\alpha}} + T_d s^{\mu}\right) \tag{13.38}$$

where $\alpha, \mu \in (0, 1)$ stand for the integral and derivative mode orders, respectively. Moreover, proportional gain is represented by k_p, T_i stands for integral time constant and T_d for derivative time constant.

Note that the fractional definition (13.38) provides us with five parameters to be tuned, which is typically received as a good characteristic since the increase in the degrees of freedom may allow us to cover a wider dynamic. However, this also implies an increase in complexity, mainly when the resulting controller is implemented, without mentioning that the lack of reliable tuning methods for this approach is a big concern for industry [3]. Considering the importance of finding unique solutions for (13.38) that guarantee implementation simplicity, some conditions over T_i and T_d have to be made. By applying several optimisation strategies, it has been determined that constants T_i and T_d have to be linked with each other as $T_i = \zeta T_d$, where ζ is a constant. Preliminary results suggested that $\zeta = 4$ guaranteed a good relation between controller performance and implementation simplicity [58]. Complementary results determined that lower values of ζ produced further improvement [58,59].

Considering the relation above described and by slightly modifying the fractional-order controller structure (13.38), a perfect square trinomial can be created if it is set $\alpha = \mu$, $\zeta = 1$; thus, the controller (13.38) will be given by,

$$G_c(s) = \frac{k_c(T_i s^{\alpha} + 1)^2}{s^{\alpha}} \tag{13.39}$$

where controller gain is given by $k_c = k_p/T_i$.

If (13.34) is substituted into (13.39), the fractional approximation of the controller will be given by

$$G_c\left(\frac{s}{\omega_c}\right) = \frac{k_c\left(T_i\left[a_0\left(\frac{s}{\omega_c}\right)^2 + a_1\left(\frac{s}{\omega_c}\right) + a_2\right] + a_2\left(\frac{s}{\omega_c}\right)^2 + a_1\left(\frac{s}{\omega_c}\right) + a_0\right)^2}{\left(a_0\left(\frac{s}{\omega_c}\right)^2 + a_1\left(\frac{s}{\omega_c}\right) + a_2\right)\left(a_2\left(\frac{s}{\omega_c}\right)^2 + a_1\left(\frac{s}{\omega_c}\right) + a_0\right)} \tag{13.40}$$

By considering $s = j\omega_c$ and substituting (13.35) into (13.40), the controller will be given by,

$$G_c = \frac{k_c[(T_i - 1)(a_2 - a_0) + j(T_i + 1)a_1]^2}{-(a_0^2 + a_1^2 + a_2^2) + 2a_0 a_2} \tag{13.41}$$

where

$$\arg\left\{G_c(s/\omega_c)\big|_{j\omega_c}\right\} = \arctan\left(\frac{-2(T_i^2-1)\tan\frac{(2-\alpha)\pi}{4}}{(T_i-1)^2-(T_i+1)^2\tan^2\frac{(2-\alpha)\pi}{4}}\right) \tag{13.42}$$

Since that $\tan 2\theta = \frac{2\tan\theta}{1-\tan^2\theta}$, thus the controller phase contribution will be given by

$$\phi_c = \pm\alpha\pi/2 \tag{13.43}$$

as was stated previously through (13.37). Recalling that equation $\phi_c + \phi_p = -\pi + \phi_m$ relates phase margin ϕ_m, controller phase contribution ϕ_c and plant phase ϕ_p at gain crossover frequency ω_{gc}, therefore,

$$\alpha = \frac{(-\pi - \phi_p + \phi_m)}{(\pi/2)} \tag{13.44}$$

As the reader can see, by using the structure (13.39), the controller effect can be varied from integral ($T_i \to 0$) to derivative ($T_i \to \infty$) by appropriately selecting the value of T_i according to the required controller phase contribution, which will be given by α.

In the following section, the proposed steps to apply the control strategy are provided.

13.4.2 Control strategy: design algorithm

In this section, the control strategy is described. It had been taken into account that two of the DC–DC converter configurations are non-minimum phase systems; thus, the steps are designed for the minimum phase part of the three described topologies, but applied to the complete system.

The proposed algorithm to ensure voltage regulation in the DC–DC power converters is described as follows:

1. Consider the transfer function $G_p(s)$ of the DC–DC converter under consideration.
2. Determine if $G_p(s)$ of the converter is a minimum phase transfer function.
3. If the requirement is not met, consider it divided into $G_p(s) = G_{mp}(s)G_{nm}(s)$.
4. Consider $G_p(s)$ or $G_{mp}(s)$ as the transfer function of plant to be controlled.
5. Compute phase of plant to be controller ϕ_p.
6. Consider ϕ_{md} as the required phase margin for the closed-loop system.
7. Compute controller order α by using (13.44) to produce the required ϕ_{md}.
8. Synthesise through (13.34) the fractional-order structure for approximation of s^α.
9. Determine the controller $G_c(s)$ through (13.39).
10. Tune T_i and k_c gains to achieve the proper effect.
11. Analyse the closed-loop effectiveness of the controller.

13.4.2.1 Algorithm: step 1

First step of control design algorithm indicates to consider the transfer function $G_p(s)$ of the converter under consideration, which can be either

$$G_p(s) = \frac{\left(\frac{V_g}{CL}\right)}{s^2 + \left(\frac{1}{RC}\right)s + \left(\frac{1}{CL}\right)} \tag{13.45}$$

for the buck converter of Figure 13.1, the transfer function given by

$$G_p(s) = \frac{\left(\frac{V_g}{CL}\right) - \left(\frac{V_g}{RC(1-\bar{D})^2}\right)s}{s^2 + \left(\frac{1}{RC}\right)s + \frac{(1-\bar{D})^2}{CL}} \tag{13.46}$$

for the boost converter of Figure 13.3, or the transfer function described by

$$G_p(s) = \frac{\left(\frac{V_g\bar{D}}{RC(1-\bar{D})^2}\right)s - \left(\frac{V_g}{CL}\right)}{s^2 + \left(\frac{1}{RC}\right)s + \frac{(1-\bar{D})^2}{CL}} \tag{13.47}$$

for the buck–boost converter of Figure 13.6. Note that as previously stated, with the same amount of elements three configuration can be generated for two different modes of conversion.

13.4.2.2 Algorithm: step 2

The second step of control design algorithm requires to determine whether the transfer function of the converter under consideration $G_p(s)$ is a minimum phase one.

A transfer function $H(s)$ is considered of minimum phase if both $H(s)$ and $1/H(s)$ are causal and stable. This means that if we consider $H(s) = N_H(s)/D_H(s)$, all roots of $N_H(s)$ and $D_H(s)$, which are numerator and denominator polynomials, have negative real part, thus all zero and poles will be located at the stable region of s-plane.

From buck transfer function (13.45) one can see that $N_{G_p}(s) = V_g/(CL)$ and $D_{G_p}(s) = s^2 + 1/(RC)s + 1/(CL)$ from which is easy to determine that no zeros or poles are located in the right-half plane. By applying the same analysis on the boost transfer function (13.46), it can be found that the numerator $N_{G_p}(s) = V_g/(CL) - V_g/(RC(1-\bar{D})^2)s$ has a root in the right region of complex plane. Thus, transfer function (13.46) is a non-minimum phase one. A similar conclusion is deduced when analysing the buck–boost transfer function (13.47).

13.4.2.3 Algorithm: steps 3 and 4

From the analysis performed in step two, it was determined that transfer functions of boost (13.46) and buck–boost (13.47) require to be divided into the form $G_p(s) = G_{mp}(s)G_{nm}(s)$. Thus, transfer function considered as the plant to be controlled will be given by,

$$G_{mp}(s) = \frac{\left(\frac{V_g}{RC(1-\bar{D})^2}\right)\left(\frac{R(1-\bar{D})^2}{L} + s\right)}{s^2 + \left(\frac{1}{RC}\right)s + \frac{(1-\bar{D})^2}{CL}} \tag{13.48}$$

for boost converter case, which is the minimum phase part of transfer function (13.46). For buck–boost converter case, minimum phase part of its transfer function (13.47) is given by

$$G_{mp}(s) = \frac{\left(\frac{V_g\bar{D}}{RC(1-\bar{D})^2}\right)\left(s + \frac{R(1-\bar{D})^2}{L\bar{D}}\right)}{s^2 + \left(\frac{1}{RC}\right)s + \frac{(1-\bar{D})^2}{CL}} \tag{13.49}$$

Lastly, for the case of buck converter, the plant to be controlled will be given by (13.45) due to it is a minimum phase one.

13.4.2.4 Algorithm: steps 5 and 6

For step five of the algorithm, plant phase contribution can be obtained by either determining the value of ω that makes $|G_p(j\omega)| = 1$ when $s = j\omega$ is substituted in the transfer function of uncontrolled plant. For boost and buck–boost converters, the expression will be given by $|G_{mp}(j\omega)| = 1$. The plant phase contribution will then be described as follows:

$$\phi_p = \arctan\left(\frac{\mathrm{Im}(G_p(j\omega))}{\mathrm{Re}(G_p(j\omega))}\right) \tag{13.50}$$

where $\mathrm{Im}(G_p(j\omega))$ and $\mathrm{Re}(G_p(j\omega))$ are imaginary and real parts of $G_p(j\omega)$ or $G_{mp}(j\omega)$.

Alternatively, when using the frequency response through the equation $\phi_c + \phi_p = -\pi + \phi_m$ with no control effort. Thus, plant phase will be given by $\phi_p = -\pi + \phi_m$ at the crossover frequency ω_c.

On the other hand, for step 6, it is well known that acceptable stability margins are within the range $g_m \geq 10$ dB and $30° \leq \phi_{md} \leq 60°$, where g_m represents gain margin. Thus, for the three cases of DC–DC power converters described in Section 13.3.1, the desired phase margin will be $\phi_{md} = 60°$.

13.4.2.5 Algorithm: step 7

Note that the equation relating ϕ_p and ϕ_m is used in this stage of the design procedure. As previously stated, the required phase margin for the closed-loop system will be $\phi_{md} = 60°$ for the three cases of DC–DC converters.

Thus, to apply this step of the algorithm, it is only needed to define the converter to be controlled, whose phase contribution can be computed by either of the procedures described in step five. Once ϕ_p and ϕ_{md} are available, the fractional order α is derived from (13.44).

13.4.2.6 Algorithm: step 8

Since the fractional order α has been determined at this point, the Laplacian operator s^{α} can be synthesised through

$$s^{\alpha} \approx \frac{a_0\left(\frac{s}{\omega_c}\right)^2 + a_1\left(\frac{s}{\omega_c}\right) + a_2}{a_2\left(\frac{s}{\omega_c}\right)^2 + a_1\left(\frac{s}{\omega_c}\right) + a_0} \tag{13.51}$$

which depends on a_0, a_1 and a_2, that in turn depend on α as shown by

$$\begin{aligned} a_0 &= \alpha^{\alpha} + 3\alpha + 2 \\ a_2 &= \alpha^{\alpha} - 3\alpha + 2 \\ a_1 &= 6\alpha \tan\frac{(2-\alpha)\pi}{4} \end{aligned} \tag{13.52}$$

Thus, this step is directly applied once the fractional order α has been obtained.

13.4.2.7 Algorithm: step 9

In this step, the synthesis of controller structure is generated through

$$G_c(s) = \frac{k_c(T_i s^{\alpha} + 1)^2}{s^{\alpha}} \tag{13.53}$$

which requires the fractional-order approximation of Laplacian operator s^{α}. Note that at this point the approximation corresponding to a specific α must be available, thus, it is substituted in $G_c(s)$ to obtain the controller structure as function of T_i and k_c.

Computation of these two parameters is what step ten is about.

13.4.2.8 Algorithm: steps 10 and 11

Once the controller structure has been obtained with T_i and k_c as values to be determined, the approximation (13.51) can be manipulated as $s^{\alpha} \approx T(s) \equiv A(s)/B(s)$ and substituted into (13.53) to obtain that $G_c(s) = k_c A(s)/B(s)$ as $T_i \to \infty$. Conversely, $G_c(s) = k_c B(s)/A(s)$ as $T_i \to 0$, which confirms that the fractional-order operator s^{α} behaves as a differentiator/integrator subject to the value of integral time constant T_i.

In Figure 13.12, the transition between derivative and integral effects for the controller structure (13.53) is shown for fractional-order approximation (13.51) with $\alpha = 0.6$, where controller phase contribution is within the range $-54^\circ \leq \phi_c \leq 54^\circ$ as shown in Figure 13.11 originally.

Lastly, as a final step of this design algorithm, the synthesised controller is tested with the converter transfer function under consideration to determine its effect on the voltage regulation through performance parameters.

The following section provides and explains the results of applying the proposed algorithm to regulate voltage in the three DC–DC power converters described in Section 3.1.

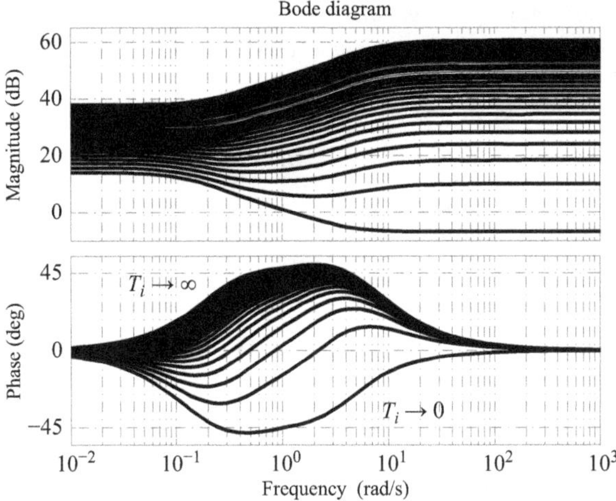

Figure 13.12 Transition between derivative and integral effects of controller $G_c(s)$ depending on the value of T_i

13.5 Results

Numerical and experimental results obtained from applying the steps described previously are provided in this section. Since the explained algorithm is applied to the described power converters, a generalization for the PID controller of fractional-order to operate in buck and boost modes will be generated.

13.5.1 Numerical results

The approximation synthesis of fractional PID (13.39) for the DC–DC power converters described in Section 13.3.1 will be computed under the following conditions:

- Consider the parameter values given in Tables 13.1, 13.2 and 13.3 for buck, boost and buck–boost converters depicted in Figures 13.1, 13.3 and 13.6, severally.
- Equilibrium points $[i_L,\ v_C]_{\text{buck}} = [1,\ 15]$, $[i_L,\ v_C]_{\text{boost}} = [2.4,\ 30]$ and $[i_L,\ v_C]_{\text{b-b}} = [1.6/5.6,\ 15/35]$, which correspond to $D = 0.6$, $\bar{D} = 0.5$ and $\bar{D} = 0.375/0.5833$ for buck, boost and buck–boost, respectively.
- Transfer functions (13.2), (13.9) and (13.16) with parameter values of Tables 13.1, 13.2 and 13.3 and condition described for equilibrium points of buck, boost and buck–boost, respectively.

With no control effort $\phi_c = 0$, the phase margin ϕ_m and the uncontrolled plant phase ϕ_p for each converter will be given by $\phi_p = -\pi + \phi_m$ as shown in Table 13.4 rows one and two. Thus, controller phase ϕ_c to fulfil the desired phase margin ϕ_{md} can be directly determined through $\phi_c + \phi_p = -\pi + \phi_{md}$. Controller phase ϕ_c and

Table 13.4 Plant phase ϕ_p, controller phase ϕ_c and fractional-order α for buck, boost and buck–boost converters, which correspond to steps 5–7 of the proposed algorithm

Parameter	Notation	Buck	Boost	Buck/boost
Phase margin	ϕ_m	22.7°	90.2°	90.7°/90.5°
Uncontrolled plant phase	ϕ_p	−157.3°	89.8°	−89.3°/−89.5°
Desired phase margin	ϕ_{md}	60°	60°	60°
Required controller contribution	ϕ_c	37.3°	−30.2°	−30.7°/−30.5°
Fractional-order	α	0.4148	−0.335	−0.3412/−0.3394

its corresponding fractional-order α are provided for the three DC–DC converters in Table 13.4 rows four and five.

Once the information in Table 13.4 has been computed, steps 8–11 of the proposed algorithm are applied to obtain the numerical simulations of voltage regulation in buck, boost and buck–boost converters as follows.

13.5.1.1 Voltage regulation in DC–DC buck converter

In the buck converter case, for steps one to four, the transfer function (13.2) is already a minimum phase one; thus, it is considered the uncontrolled plant. For steps five to seven and from Table 13.4, the system phase margin $\phi_m = 22.7°$ with no control effort is not the desired one; thus, the difference to achieve 60° has to be compensated by the controller. Therefore its contribution must be $\phi_c = 37.3°$, which implies a required fractional order for the approximation of $\alpha = 0.4148$, recalling from (13.43) that $\alpha = 2\phi_c/\pi$.

For steps 8–10 and using the computed fractional order $\alpha = 0.4148$ in (13.34), the frequency response of the approximation, depicted in Figure 13.13(a), confirms the designed phase contribution. Synthesis of PID controller is performed by substituting the obtained approximation $s^{0.4148}$ into (13.39). Once the controller structure has been obtained in terms of its gain and integral time constant, the required controller phase contribution of $\phi_c = 37.3°$ is achieved by setting $T_i = 5.1$ and $k_c = 0.18$, leading to the following controller transfer function

$$G_c(s) = k\frac{s^4 + \rho_1 s^3 + \rho_2 s^2 + \rho_3 s + \rho_4}{s^4 + \rho_5 s^3 + \rho_6 s^2 + \rho_7 s + \rho_8} \tag{13.54}$$

whose parameters are listed in Table 13.5.

The frequency response for fractional approximation of PID controller (13.54) is shown in Figure 13.13(b), which corroborates that desired phase margin is achieved with the controller.

Regulation characteristic produced by the fractional PID controller approximation is corroborated through the system's step response, shown in Figure 13.14 (a), where acceptable, stable and fast tracking characteristic can be corroborated. In

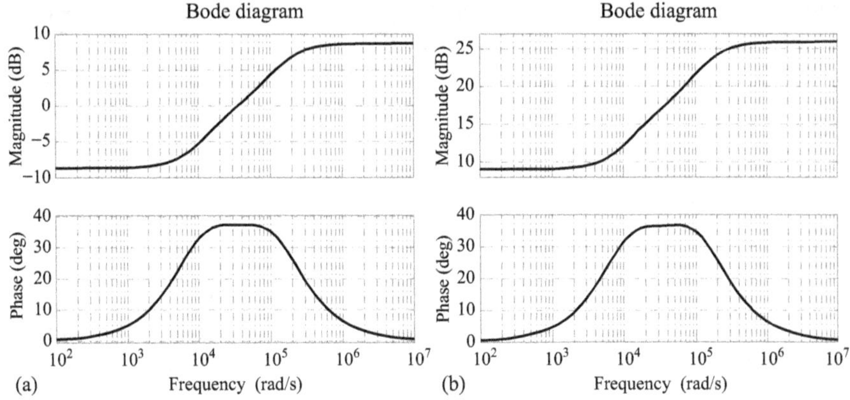

Figure 13.13 (a) Frequency response of fractional approximation $s^{0.4148}$. (b) Frequency response of fractional PID controller for output regulation in buck converter

Table 13.5 Coefficients for approximation of the fractional PID controller (13.54)

Coefficient	Value	Coefficient	Value
ρ_1	1.489×10^5	ρ_5	2.481×10^5
ρ_2	6.885×10^9	ρ_6	1.603×10^{10}
ρ_3	9.98×10^{13}	ρ_7	3.159×10^{14}
ρ_4	4.491×10^{17}	ρ_8	1.622×10^{18}
k	14.621		

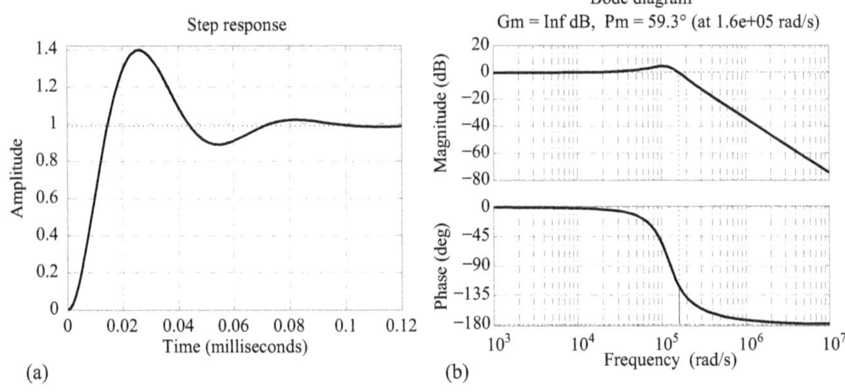

Figure 13.14 (a) Closed-loop step response of buck converter. (b) Frequency response of closed-loop system corroborating $\phi_{md} \approx 60°$

addition, closed-loop frequency response is provided in Figure 13.14(b) to validate that the desired phase margin has been reached.

It is a common practice to determine controller effectiveness regulating output voltage in the presence of load variations and disturbances, thus, in addition to step response, those tests were also performed on the closed-loop system.

In Figure 13.15, controller effectiveness regulating output voltage in the presence of load variations is shown. Figure 13.15 (top) depicts buck converter output voltage v_o, where regulation disruptions by the load variations can be identified. Figure 13.15 (bottom) shows the dynamic of control voltage v_c, where it can be seen that the controller successfully restored output regulation.

On the other hand, disturbance rejection is related to controller ability of achieving regulation in the presences of additive bounded unknown dynamics at the input of the plant. To determine controller effectiveness regulating the system output, a disturbance of about 10% of the maximum value that can be reached by the control law was applied.

In Figure 13.16, output and control voltage are shown. One can see from Figure 13.16 (top) that once the disturbance was applied, the controller was not able to completely eliminate its effect, since the desired value for the output was not restored. However, by looking at the control voltage v_c shown in Figure 13.16 (bottom), it can be determined that the controller partially rejects the effect of the applied signal since the dashed line approaches to its value before the disturbance.

Note that this represents the main disadvantage of the proposed approach. Even though it is a non-desirable characteristic in a controller, it can be preliminary attributed to the simplicity of tuning process, which did not include any consideration related to disturbance rejection. This can be addressed through multi-objective optimisation where a criterion to maximise robustness and disturbance rejection can be included.

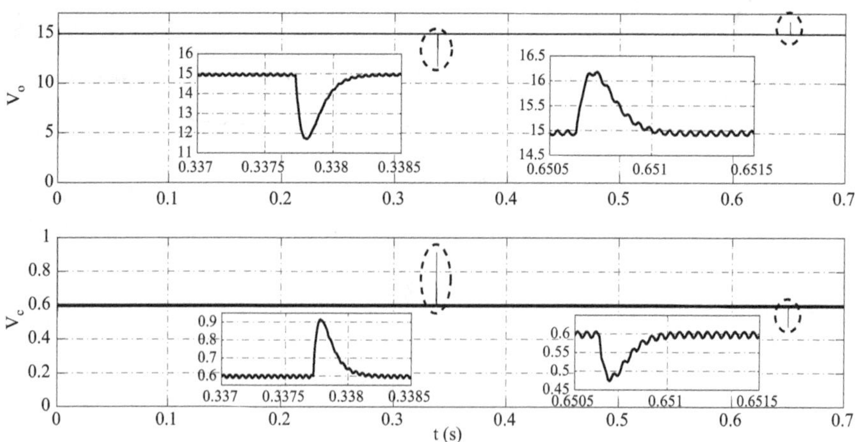

Figure 13.15 Dynamic of output (top) and control voltage (bottom) in the presence of load variations

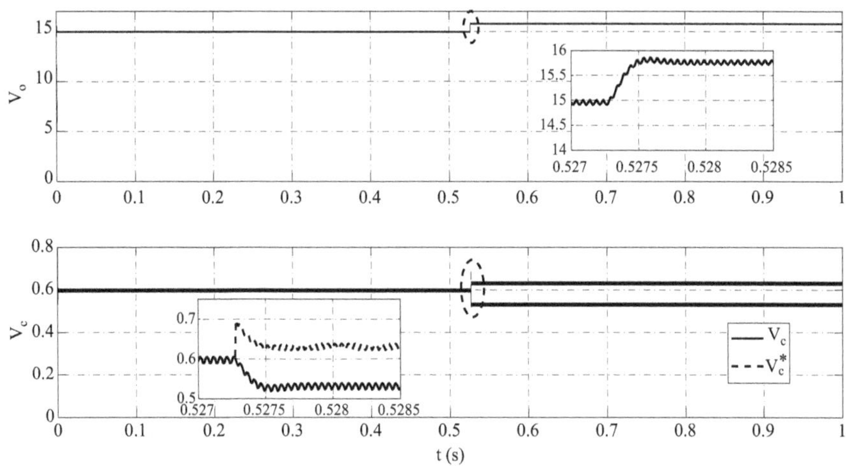

Figure 13.16 Regulation characteristic in the presence of additive disturbance at the plant input, where V_c is the control voltage after the disturbance was applied and V_c^* includes its effect

For the purpose of illustrating the effectiveness of proposed approach, a comparison with a classical PID structure is shown. For a fair comparison, the integer-order PID controller was tuned to meet the same requirements. This means that a closed-loop phase margin of $\phi_{md}= 60°$ is expected for the fractional-order case.

The tuning process was performed by using the MATLAB® algorithm through the function *pidtool*, where gain crossover frequency and closed-loop phase margin had to be adjusted only to achieve the requirements. Resulting parameters for the standard representation of typical PID controller are $[k_p,\ T_i,\ T_d] = [10.3033,\ 4.7953 \times 10^{-4},\ 3.0459 \times 10^{-6}]$.

In Figure 13.17(a), system regulation characteristic is shown. It can be seen by comparing Figures 13.14(a) and 13.17(a) that the overshot M(%) and settling time t_s of the latter are bigger than the curve corresponding to fractional PID controller. In fact, performance parameters of step response with the typical PID controller are $t_r= 9.64\ \mu s$, $t_p= 24\ \mu s$, M=50.4%, $t_s= 141\ \mu s$, from which superiority of the proposed approach can be determined if those are compared to M = 40% and $t_s= 100\ \mu s$ produced by the fractional PID controller. Figure 13.17(b) depicts frequency response of the system when using typical PID controller, where desired closed-loop phase margin $\phi_{md}= 61°$ can be corroborated.

It is important for the reader to know that by sacrificing closed-loop bandwidth, i.e., moving the gain crossover frequency to the left in the bode diagram, smoother step responses can be achieved with the typical PID controller, however, this impacts directly in the response velocity making it way slower and therefore not comparable to the one obtained with the fractional PID controller.

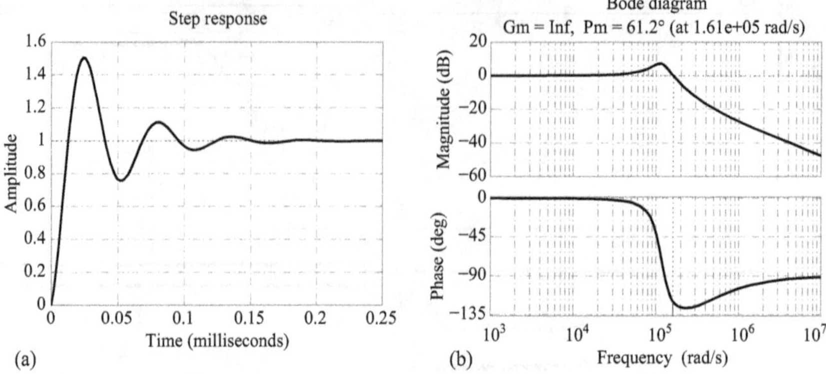

Figure 13.17 (a) Step response of buck converter regulated with a typical PID controller. (b) System frequency response corroborating $\phi_{md} = 60°$

13.5.1.2 Voltage regulation in boost converter

In boost converter case, for steps 1–4 of the proposed algorithm, its transfer function (13.7) is not of minimum phase; thus, considering the converter transfer function divided into $G_p(s) = G_{mp}(s)G_{nm}(s)$, $G_{mp}(s)$ will represent the uncontrolled plant given by (13.9).

For steps 5–7, the system phase margin with no control effort has been provided in Table 13.4, where $\phi_m = 90.2°$. Even when the value is superior to the desired one, the phase margin will be limited to the upper value of the described acceptable range to make it achievable and to keep the system realistic and implementable; thus, to reach the 60° of phase margin the controller has to exhibit a predominant integral effect; therefore, its contribution must be $\phi_c = -30.2°$, which implies a required fractional order $\alpha = -0.335$.

As in the previous case, after using computed fractional order $\alpha = -0.335$ in (13.34), the frequency response of the approximation, depicted in Figure 13.18(a), confirms that the approximation exhibit a phase curve typical of an integrator. The fractional PID controller is obtained by substituting the approximation $s^{-0.335}$ into (13.39). By setting integral time gain $T_i = 0.001$ and gain $k_c = 1.5$, the controller phase contribution is $\phi_c = -30.2°$, leading to the same structure given by (13.54), whose parameters to regulate output in a boost converter are given in Table 13.6.

The frequency response of fractional PID approximation given by (13.54) with parameter in Table 13.6 is depicted in Figure 13.18(b). Note that phase curve exhibits the integral effect with the required phase contribution. By comparing Figure 13.18(a) and (b), the approximation and controller phase contribution can be corroborated. It is important to highlight that, as predicted in Figure 13.12, the integral time constant has to be $T_i \approx 0$ to generate in the controller an integral effect.

Figure 13.19(a) shows the controller effectiveness through the step response. Converter exhibits a fast and stable regulation response with a good tracking

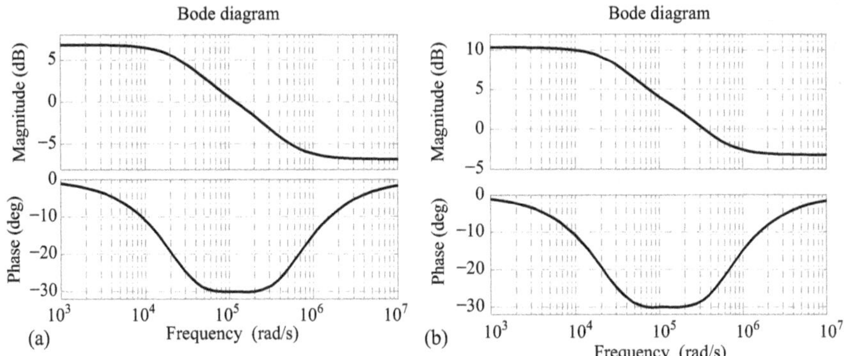

Figure 13.18 *(a) Frequency response of fractional approximation* $s^{-0.335}$. *(b) Frequency response of fractional PID for voltage regulation in boost converter*

Table 13.6 *Coefficients for approximation of the fractional PID (13.54) for output regulation in boost converter*

Coefficient	Value	Coefficient	Value
ρ_1	2.179×10^6	ρ_5	1.269×10^6
ρ_2	1.367×10^{12}	ρ_6	2.837×10^{11}
ρ_3	1.96×10^{17}	ρ_7	1.829×10^{16}
ρ_4	8.094×10^{21}	ρ_8	2.075×10^{20}
k	0.2417		

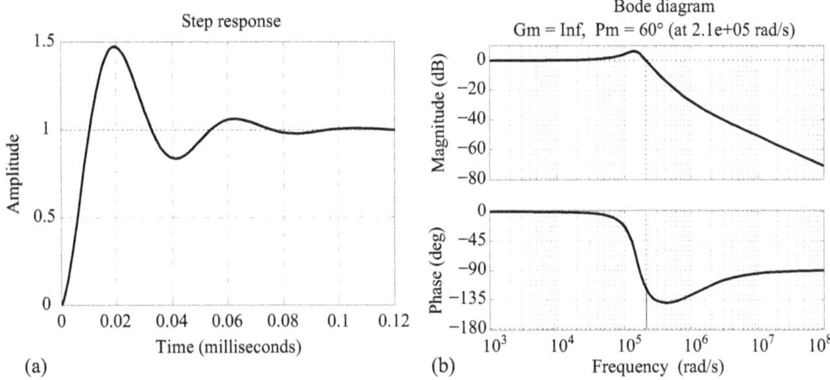

Figure 13.19 *(a) Step response of boost converter. (b) Closed-loop frequency response corroborating* $\phi_{md} = 60°$

characteristic. This can be concluded from steady-state of output, which converges to the reference in a considerable short time. On the other hand, system frequency response is provided in Figure 13.19(b). The achievement of desired phase margin can be corroborated in this figure.

As in the buck converter case, a typical PID controller was tuned to regulate the output in a boost converter to be compared with a fractional PID controller. To make both results comparable the same requirements were met. Parameters of typical PID controller resulting from the tuning process are $[k_p,\ T_i,\ T_d] = [0.51173,\ 2.5638 \times 10^{-6},\ 6.4095 \times 10^{-7}]$.

In Figure 13.20(a), system closed-loop step response is provided. The curve exhibits a fast regulation and stable behaviour. In order to determine superiority of the proposed approach Figures 13.19(a) and 13.20(a) are compared. The typical PID controller produced a step response that presents an overshoot $M = 55.5\%$, which compared to the one produced by the fractional-order PID approximation $M = 49\%$ results bigger.

On the other side, the step response obtained with the typical PID controller settles at a time $t_s = 128\ \mu s$, which compared to the $t_s = 80\ \mu s$ of the proposed approach results bigger as well. Rise and peak time for the integer-order PID case were computed at $t_r = 7.04\ \mu s$ and $t_p = 18.6\ \mu s$, respectively. Figure 13.20(b) shows frequency response to corroborate that the closed-loop phase margin ϕ_{md} was achieved and that it occurs at the same gain crossover frequency that in the case of fractional PID, thus, the comparison done is valid.

As in the previous case, the possibility of smoothing the step response obtained with the typical PID controller was revised. However, conclusion was similar as in the buck converter, improving one aspect of the response worsen the other.

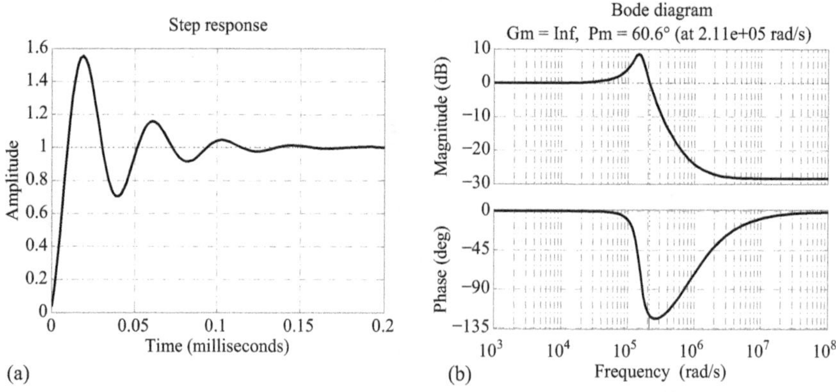

Figure 13.20 (a) Step response of boost converter regulated with a typical PID controller. (b) Closed-loop frequency response corroborating $\phi_{md} = 60°$

13.5.1.3 Voltage regulation in buck–boost converter

When dealing with output regulation of buck–boost converter, for steps 1–4 of the explained procedure, the converter transfer function (13.15) is not of minimum phase one. It is necessary to consider it divided into $G_p(s) = G_{mp}(s)G_{nm}(s)$; thus, $G_{mp}(s)$ can be regarded as the plant to be controlled given by (13.16). Since both buck or boost conversion modes can be achieved, the procedure is applied to both cases.

For steps 5–5, the system phase margin with no control effort has been provided in Table 13.4 as in the previous cases, where ϕ_m= 90.7° for buck mode and ϕ_m= 90.5° for boost mode. Phase margin is limited to the upper value to keep the control objective achievable, the system realistic and the controller implementable. Therefore to reach the 60° of phase margin, the controller has again to exhibit a predominant integral effect in both operation modes; thus, the contribution must be of $\phi_c = -30.7°$ when the converter is operating in buck mode and $\phi_c = -30.5°$ when the operation is in boost mode. This implies that the fractional order for the approximation will be $\alpha = -0.3412$ for buck mode and $\alpha = -0.3394$ for boost mode.

By using the fractional orders obtained for both cases, $\alpha = -0.3412$ and $\alpha = -0.3394$ in the approximation (13.34), the frequency response for both buck and boost modes are shown in Figure 13.21(a), where the integral effect is confirmed in the approximation. Note that both phase curves look very similar, since the approximation is computed for almost the same phase; however, the frequency band is different.

The structure of fractional PID will be obtained by substituting the approximation of both orders $s^{-0.3412}$ and $s^{-0.3394}$ into (13.39). By setting integral time constant T_i= 0.002 and k_c= 2 for the buck mode and T_i= 0.001 and k_c= 2 for the boost mode, the controller exhibits the required effect in each case with a phase contribution of $\phi_c = -30.7°$ and $\phi_c = -30.5°$ with the same structure given by (13.54), with parameters provided in Table 13.7 for both conversion modes.

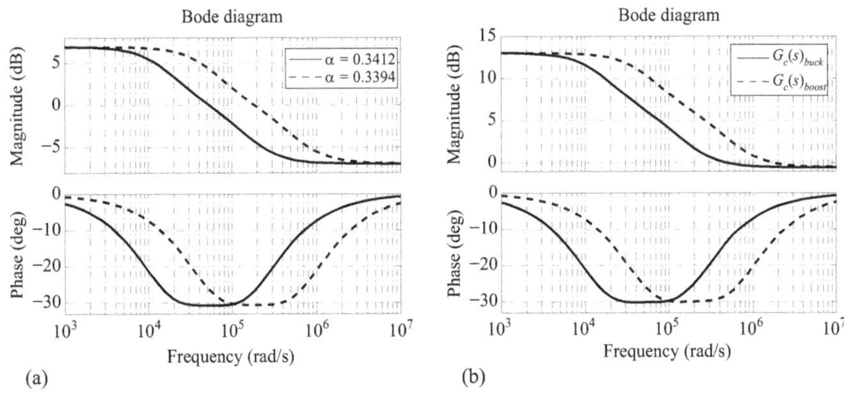

Figure 13.21 (a) Frequency response of fractional-order approximations $s^{-0.3412}$ and $s^{-0.3394}$. (b) Frequency response of fractional-order PID controllers for voltage regulation in buck–boost converter

Table 13.7 Coefficients for approximation of fractional PID (13.54) for both operation modes

Coefficient	**Buck**	**Boost**
ρ_1/ρ_5	$9.813 \times 10^5/5.73 \times 10^5$	$3.434 \times 10^6/1.996 \times 10^6$
ρ_2/ρ_6	$2.77 \times 10^{11}/5.694 \times 10^{10}$	$3.391 \times 10^{12}/6.941 \times 10^{11}$
ρ_3/ρ_7	$1.777 \times 10^{16}/1.629 \times 10^{15}$	$7.607 \times 10^{17}/6.954 \times 10^{16}$
ρ_4/ρ_8	$3.28 \times 10^{20}/8.092 \times 10^{18}$	$4.907 \times 10^{22}/1.214 \times 10^{21}$
k	0.3183	0.3166

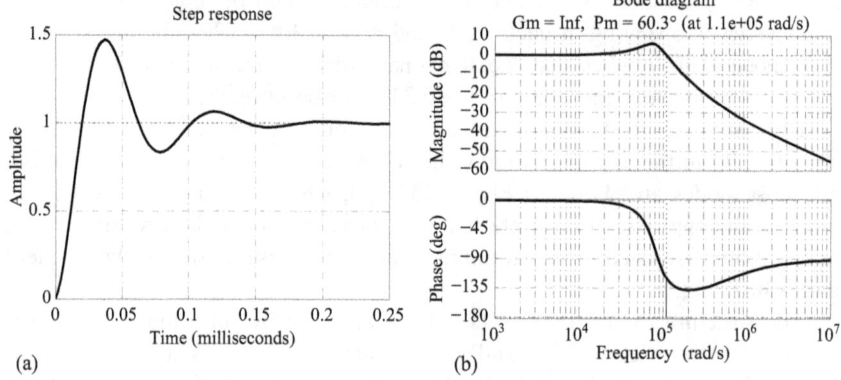

Figure 13.22 (a) Step response of DC–DC buck–boost converter operating in buck mode. (b) Frequency response of closed-loop system corroborating ϕ_{md}= 60° in buck mode

The structure for fractional approximation PID is similar to that of the previous cases given by (13.54) with parameters provided in Table 13.7, whose the frequency response is given in Figure 13.21(b) for both conversion modes. Note from the phase curve that the controller exhibits the integral effect as was expected with the required phase contribution. Similarly with the approximation, the main difference is the operating frequency band.

Effectiveness of the controller is corroborated through the step and frequency response. Figure 13.22(a) depicts step response of buck–boost converter whilst operating in buck mode. Stable, fast and good tracking characteristic can be confirmed, since the output reaches the reference value in about 0.15 ms with no steady-state error. Closed-loop frequency response shown in Figure 13.22(b) allows us to corroborate that the proposed controller achieved the desired phase margin.

On the other side, Figure 13.23(a) provides step response of buck–boost converter operating in boost mode. The system output evolves fast, reaching the reference value of about 0.05 ms. The output exhibits a good tracking characteristic with zero steady-state error. Closed-loop frequency response, provided in

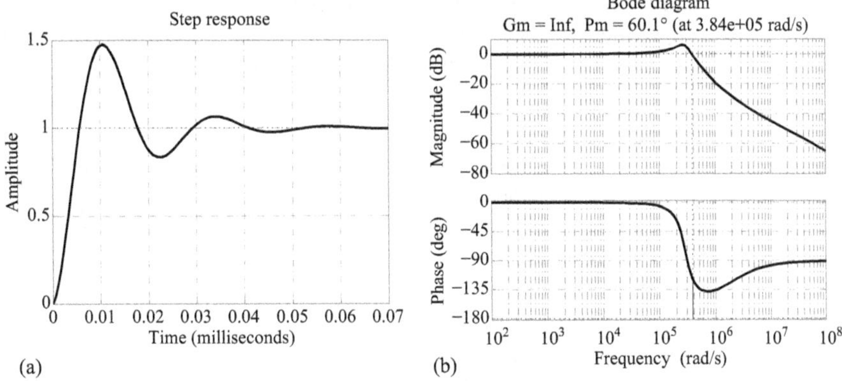

Figure 13.23 (a) Step response of buck–boost converter operating in boost mode. (b) System frequency response corroborating ϕ_{md}= 60° in boost mode

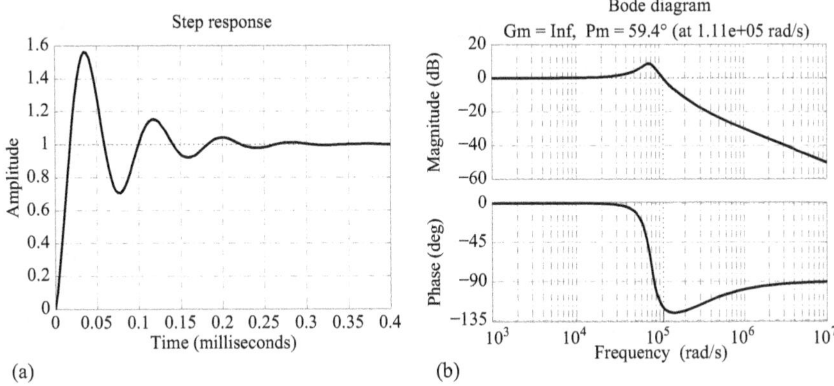

Figure 13.24 (a) Step response of buck–boost converter operating in buck mode with a typical PID controller. (b) System frequency response corroborating $\phi_{md} \approx 60°$ in buck mode

Figure 13.23(b), allows us to confirm that the fractional PID controller achieved the desired phase margin.

Lastly, two additional integer-order PID controllers were tuned to determine their effectiveness regulating the output of buck–boots converter. A comparison with results of fractional PID controller will be performed. As described for both previous cases, the typical PID controller was tuned to meet the same requirements that were considered in the fractional-order PID design.

For buck–boost converter case operating in buck mode, PID controller parameters are $[k_p,\ T_i,\ T_d] = [0.58572,\ 5.0909 \times 10^{-6},\ 1.2266 \times 10^{-6}]$. Corresponding closed-loop step response is provided in Figure 13.24(a). The effectiveness of typical

PID controller can be corroborated since it generates a fast regulation characteristic. However, by comparing Figures 13.22(a) and 13.24(a) can be determined that the response obtained with the fractional PID controller exhibits smaller both overshoot M = 49% and settling time t_s= 150 μs than the one obtained with the integer PID controller. Performance parameters of closed-loop step response of a buck–boost converter operating in buck conversion mode when using a typical PID controller are t_r= 13.7 μs, t_p= 34.7 μs, M = 56.3% and t_s= 246 μs. Figure 13.24(b) depicts frequency response of the system to corroborate that closed-loop phase margin ϕ_{md} is achieved around the same gain crossover frequency, thus, this comparison is valid.

A similar procedure was done for the typical PID controller to regulate the buck–boost converter operating in boost conversion mode. The controller parameters are $[k_p,\ T_i,\ T_d] = [0.61048,\ 1.5405 \times 10^{-6},\ 3.5184 \times 10^{-10}]$. The step response is depicted in Figure 13.25(a). As expected, the typical PID controller effectively regulated the system's output. However, by comparing again Figures 13.23(a) and 13.25(a) superiority of the proposed approach can be corroborated since, overshoot M = 49% and settling time t_s= 50 μs are both smaller than the ones obtained with the integer PID, whose performance parameters are given by t_r= 3.92 μs, t_p= 10.3 μs, M = 55.2% and t_s= 61.4 μs.

It is relevant to preliminary discuss at this point results presented. The most remarkable aspect of the proposed approach is the improvement achieved in response velocity. For the three cases of DC–DC converters described, the fractional-order PID controller effectively regulated the system output with a fast and stable characteristic. Critical aspects of a response to determine controller effectiveness are rise time t_r, settling time t_s and overshoot M (%).

For all conversion modes described, even though rise time t_r was similar for both approaches, settling time t_s and overshoot M (%) were smaller when considering the fractional PID controller. Note that the amount of overshoot obtained by the proposed approach might be seen as a disadvantage that could be avoided by

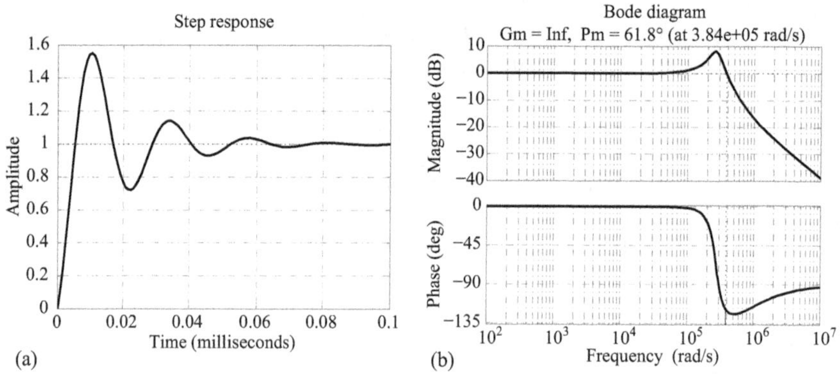

Figure 13.25 (a) Step response of buck–boost converter operating in boost mode with a typical PID controller. (b) System frequency response corroborating ϕ_{md}= 61° in boost mode

using a typical PID controller, however, it has been shown that this is not the case, since results obtained with the classical approach resulted in a bigger value of it. Even when smoother curves can be achieved by sacrificing bandwidth of the closed-loop system, this modification directly impacts the response velocity, making it way slower than the one produced by the fractional PID. We attribute the incapacity of matching the results obtained with the proposed approach from the classical PID to controller structures, since the synthesised fractional-order PID is of higher order that the one produced by the integer-order PID controller.

The amount of overshoot produced by the proposed approach remains as an area of opportunity that can be addressed through an optimisation approach. This has been thought of as the future direction of the this work.

A controller generalisation to operate in buck and boost mode is described in the following section. Experimental results only on output regulation in buck–boost converter are provided since it operates in both conversion modes and also exhibits the closed-loop stability of the boost converter.

13.5.2 Experimental results

To ease the controller implementation, a structure generalization of the fractional PID has to be derived. Since buck–boost converter operation depends on the duty cycle *D*, the controller structure must be investigated in both conversion modes and the transition from one to the other. Aiming to achieve the controller proper regulation in either mode, a generalization is determined and tested.

Electrical realization of the proposed fractional PID (13.40) requires a mathematical simplification. This can be done by representing (13.39) or (13.40) through its partial fraction expansion to derive a simpler mathematical expression that can be represented electrically by well-known circuitry.

The roots type is required before performing the partial fraction expansion. By considering the controller structure (13.39) and $s^{\alpha} \approx N(s)/D(s)$, the fractional PID approximation will be described by

$$G_c(s) = \frac{k_c\left(T_i^2 N(s)^2 + 2T_i N(s)D(s) + D(s)^2\right)}{N(s)D(s)} \tag{13.55}$$

which can be simplified even more if the effect of T_i is considered, as previously described through Figure 13.12. Therefore, since controller $G_c(s)$ is determined by quadratic polynomials $N(s)$ and $D(s)$, the roots of partial fraction expansion of $G_c(s)$ will be real as long as $a_1^2 > 4a_2a_0$.

Recalling that $N(s)$ and $D(s)$ depend on a_0, a_1 and a_2, which in turn depend on α, condition $a_1^2 > 4a_2a_0$ will hold $\forall\ \alpha \in (0, 1)$ as shown in Figure 13.26, thus the PID controller approximation partial fraction expansion will be described by real poles as follows:

$$G_c(s) = \left(\frac{A_1}{\gamma_1 s + 1}\right) + \left(\frac{A_2}{\gamma_2 s + 1}\right) + \left(\frac{A_3}{\gamma_3 s + 1}\right) + \left(\frac{A_4}{\gamma_4 s + 1}\right) + A_5 \tag{13.56}$$

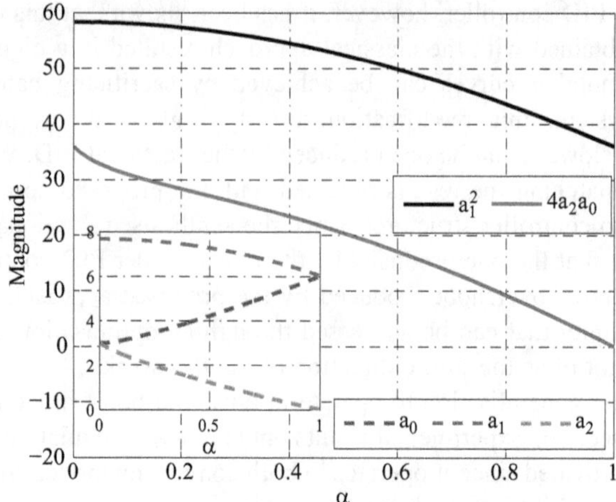

Figure 13.26 Approximation parameters $a_0(\alpha)$, $a_1(\alpha)$, $a_2(\alpha)$ *and values* a_1^2, $4a_2a_0$ *that ensure* $a_1^2 > 4a_2a_0$

which can be generated electrically by using *RC* networks and operational amplifiers as shown in Figure 13.27.

After applying partial fraction expansion to the fractional PID controller, the resulting gains and parameter values are provided in Table 13.8.

By revising columns four to six, it is determined that the proposed fractional PID approximation is implementable, since all the parameter values are commercial. Implementation results allowed us to determine the proposed method viability. The experiment was performed by using elements with the following technical characteristics: a 30 μF polypropylene metallized capacitor with V_{maxDC} = 500 V, 5% tolerance and R = 3.5 mΩ. A 10 mH inductor 1140-103K-RC with R_{DC} = 2.76 Ω and I_{max} = 10 A. A SR504 R0 diode with V_D = 0.55 V. A MOSFET NTP5864NG with V_{DS} = 60 V, current I_{DS} = 63 A and 12.4 mΩ of R_{ON}. A 4 MHz operational amplifier LF347N and a TL494 for the PWM signal. Tolerance of ±5% for capacitors and ±1% for resistances of computed values.

In Figure 13.28, the electrical scheme of the closed-loop system is depicted. As can be seen, the experiment comprises the plant (buck–boost converter), the comparator, the fractional PID approximation and the pulse width modulator (PWM), which delivers the control law to the final control element (MOSFET *Q*). The comparator stage produces the signal error through an operational amplifier in a differential configuration, which is given by

$$e(t) = \left(\frac{r}{r+r}\right)\left(1+\frac{R_f}{R_i}\right)V_r - \left(\frac{R_f}{R_i}\right)\left(\frac{r_2}{r_1+r_2}\right)v_o \tag{13.57}$$

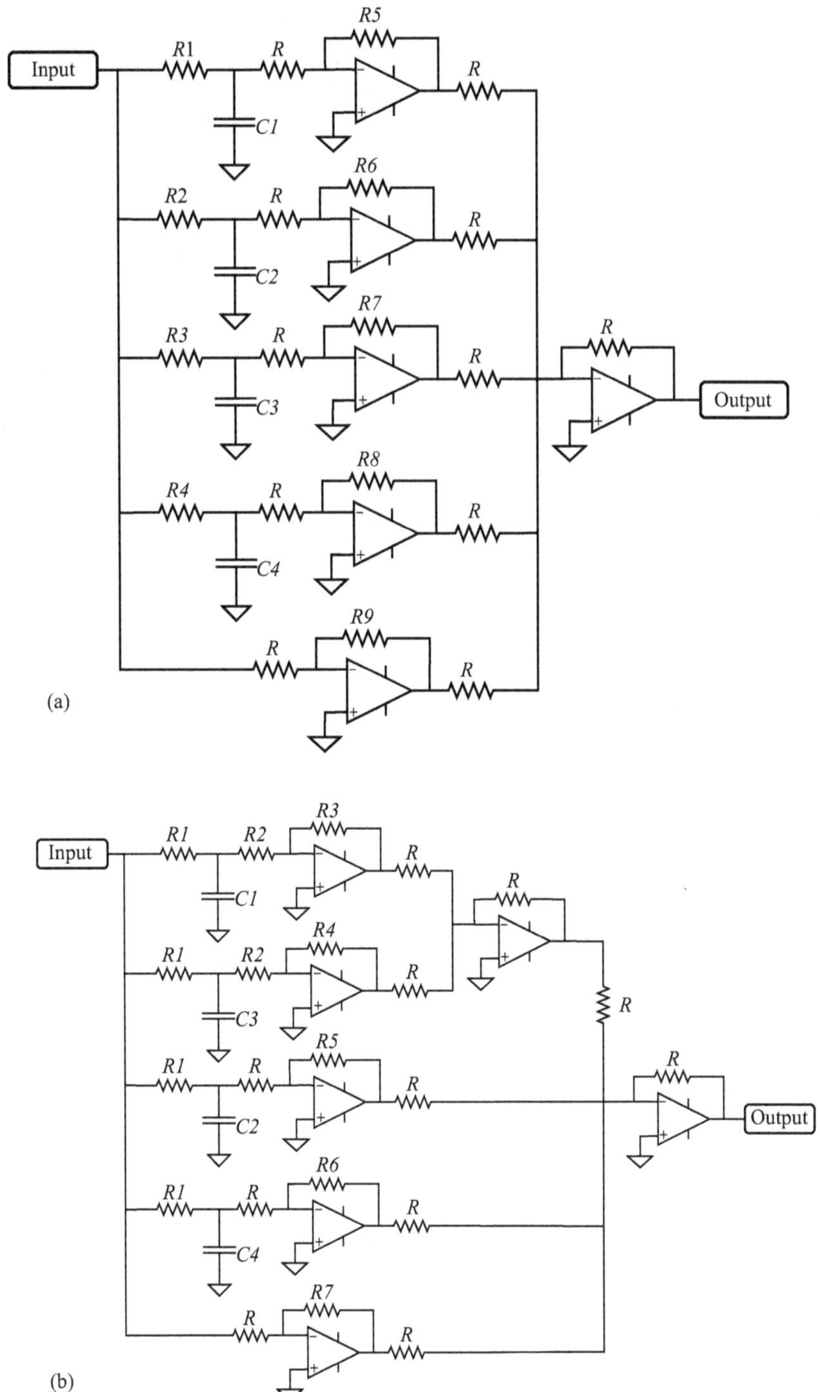

Figure 13.27 (a) Electrical circuit to generate the partial fraction expansion of controller (13.56). (b) Electrical arrangement of controller (13.56) for parameter values of Table 13.7

Table 13.8 Constant values for A's and γ's of fractional-order PID controller (13.56) and parameters for electrical implementation of the arrangement in Figure 13.27(b) for both conversion modes

Constant	Buck	Boost	Element	Buck	Boost
A_1	-4.7134×10^{-5}	-1.1682×10^{-5}	R_1	100 Ω	10 Ω
A_2	0.7987	0.8012	R_2	100 kΩ	100 kΩ
A_3	-3.1949×10^{-6}	-8.0119×10^{-7}	R_3	4.7 Ω	1.16 Ω
A_4	11.7835	11.6822	R_4	3.2 Ω	0.08 Ω
A_5	0.3183	0.3166	R_5	798 Ω	801 Ω
γ_1	2.19×10^{-6}	0.63×10^{-6}	R_6	11.8 kΩ	11.7 kΩ
γ_2	14.14×10^{-6}	4.04×10^{-6}	R_7	318 Ω	317 Ω
γ_3	24.86×10^{-6}	7.11×10^{-6}	R	1 kΩ	1 kΩ
γ_4	160.22×10^{-6}	45.51×10^{-6}	C_1	0.022 μF	0.063 μF
			C_2	0.142 μF	0.404 μF
			C_3	0.25 μF	0.712 μF
			C_4	1.602 μF	4.55 μF

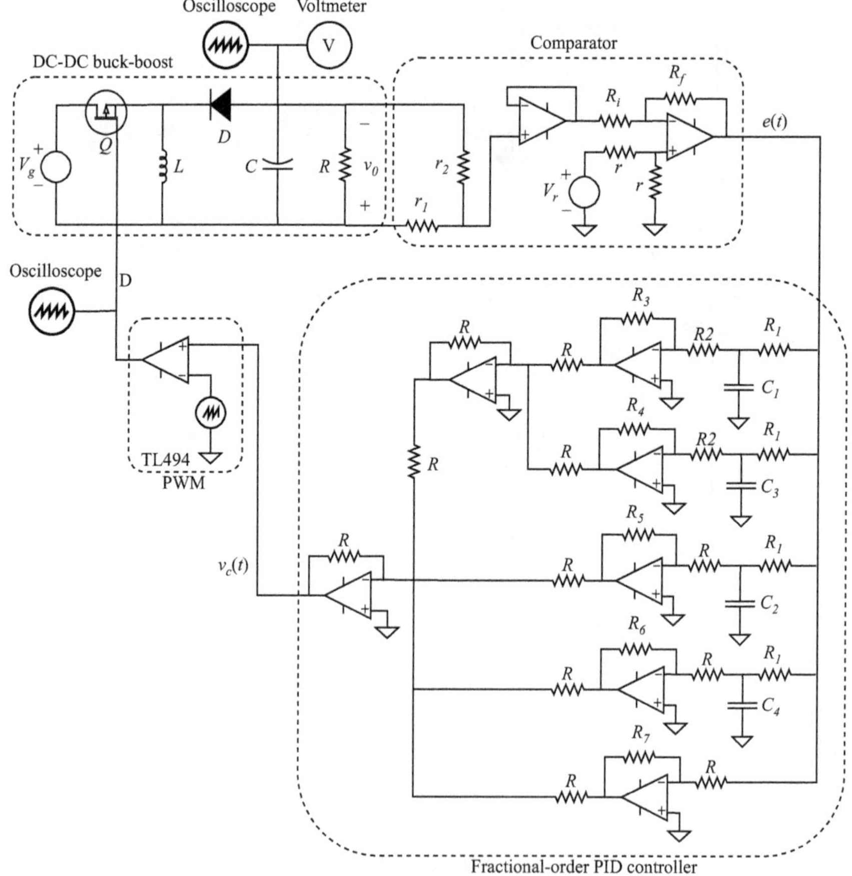

Figure 13.28 Electrical diagram of the implemented closed-loop system

Assuming that $R_f = R_i$, the error signal derived by the comparator will be

$$e(t) = V_r - \left(\frac{r_2}{r_1 + r_2}\right)v_o \tag{13.58}$$

which let us know the importance of the scaling applied to the converter output voltage.

The experimental results are shown in Figures 13.29–13.32. In Figure 13.29, the buck–boost converter voltage regulation in buck conversion mode was verified. In Figure 13.29(a), the screenshot of results obtained directly from the oscilloscope is shown. Technical aspects of measurement as the scale of horizontal/vertical axes

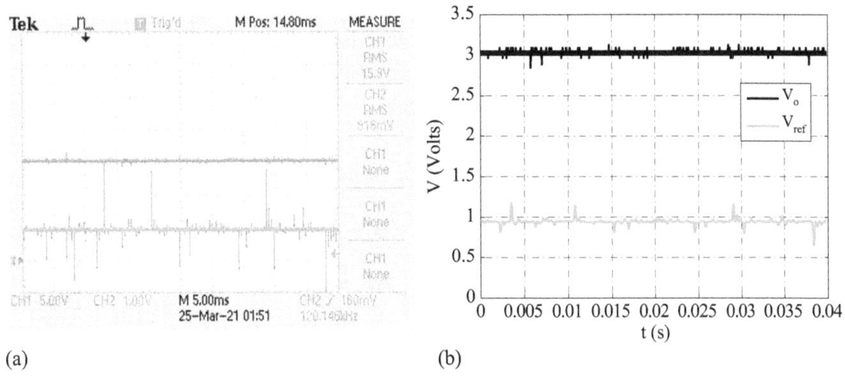

Figure 13.29 Voltage regulation to v_o= 15 V of buck–boost converter operating in buck conversion mode. (a) Oscilloscope perspective. (b) Alternative view of exported experimental data preserving scale of 5 V per unit in the output voltage

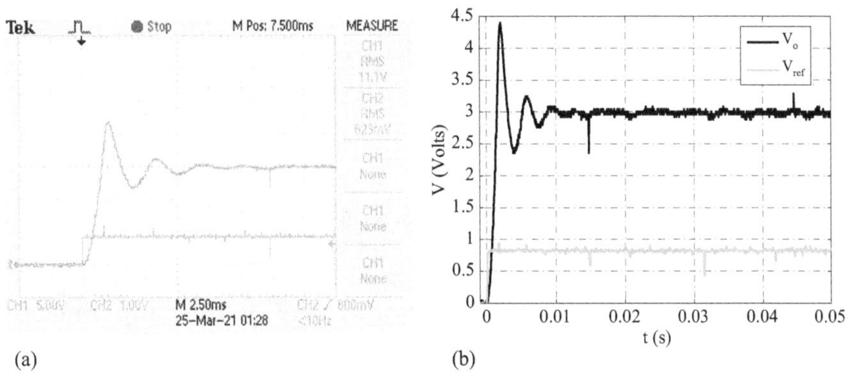

Figure 13.30 Tracking characteristic of buck–boost converter. (a) Oscilloscope perspective. (b) Alternative view of experimental data preserving 5 V per unit in the output voltage

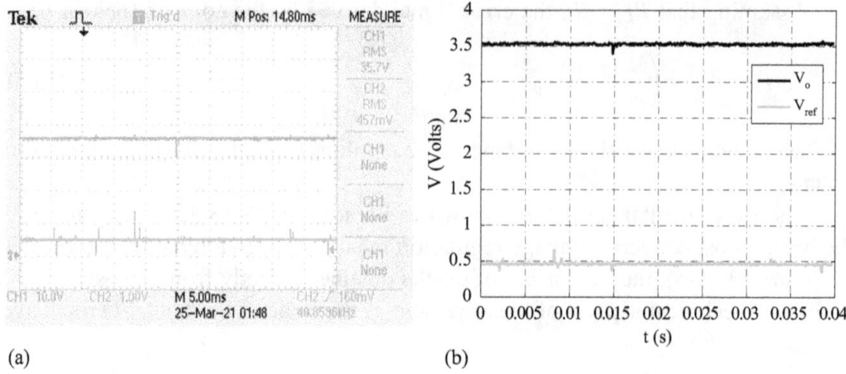

Figure 13.31 Voltage regulation to v_o= 35 V of buck–boost converter operating in boost conversion mode: (a) Oscilloscope perspective. (b) Alternative view of exported experimental data preserving 10 V per unit in the output voltage

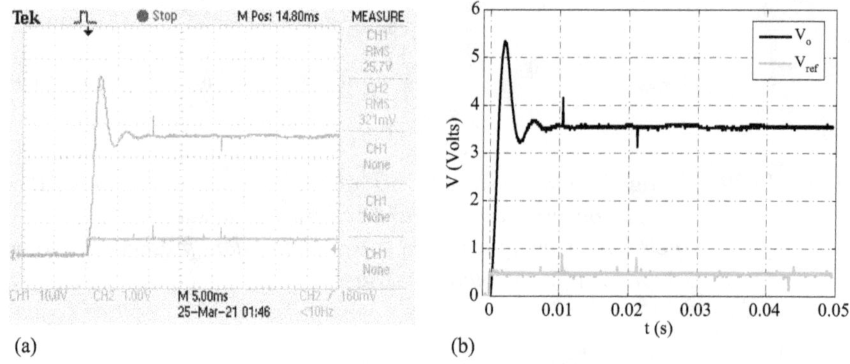

Figure 13.32 Tracking characteristic of buck–boost converter. (a) Boost mode. (b) Alternative view of exported experimental data preserving 10 V per unit in the output voltage

and final value of the output can be consulted in this figure. Experimental data of both output and reference voltages were exported and plotted externally to overcome oscilloscope quality shortage. Figure 13.29(b) depicts the output voltage for buck conversion mode (v_o= 15 V), where scales were preserved for comparative reasons.

Figure 13.30 verifies the tracking characteristic of buck–boost converter operating in buck conversion mode, which exhibits a fast and stable behaviour. Thus, it is determined that the fractional PID effectively regulates the converter output voltage even in the presence of reference changes.

In Figure 13.31, the buck–boost converter voltage regulation in boost conversion mode can be verified. Figure 13.31(a) depicts the screenshot of results obtained during the measurement. As in the previous case, experimental data of both output and reference voltages in this conversion mode were exported and plotted externally. Figure 13.31(b) depicts the output voltage for boost conversion mode (v_o= 35 V), where scales used during the measurement with the oscilloscope were preserved for comparative reasons.

Lastly, Figure 13.30 verifies that the buck–boost converter operating in boost conversion mode exhibits a fast and stable tracking characteristic. Thus, the conclusion that the fractional PID effectively regulates the converter output even in the presence of reference changes is confirmed.

Viability of the proposed approach was verified experimentally; thus, the fractional PID approximation can be considered as an alternative to regulate output of DC–DC converters. Taking into account that the system exhibited a considerably fast response, high-speed applications are the number one candidates for testing this proposal.

In the remaining of this chapter, a discussion on the described results and some relevant conclusions are given.

13.6 Discussion

This chapter investigated viability of a fractional approximation of PID controller to regulate output in DC–DC buck, boost and buck–boost converters. These converters relevance rests in their importance for systems that require stable/fast-response power sources or appropriate energy management strategy.

It can be highlighted from the proposed technique the combination of both requirements, closed-loop response performance and robustness and the integration of fractional calculus, which has proven higher accuracy describing systems and enhanced robustness against parameter variations or uncertainties.

Viability of proposed approach was investigated to determine its possibilities to become an alternative to control highly efficient converters among which are Silicon-Carbide ones or high-speed applications. Output voltage regulation was confirmed through experimental results in a buck–boost converter, due to it gathers the most significant characteristics of buck and boost converters, using a single control loop. Aiming to enhance the regulation velocity and response speed, alternative approaches such as the current control mode can be considered as future work of this investigation.

Although the approach of fractional-order control in DC–DC converters is not new, approximating the controller structure with a lower order module (quadratic polynomials) and synthesising a five-term electrical arrangement that can be implemented through *RC* networks resulted new and different, especially when fractional orders are commonly approximated through high-order transfer functions, such as some of the ones described in Section 13.3.2, which are realizable by extensive ladder- or tree-like electrical circuits, or by using the fractional definition of the derivative/integral, which required digital implementation due to the complexity of their mathematical descriptions.

Superiority of the described approach was determined by comparing performance of both typical PID and fraction-order PID approximation. Critical parameter such as rise time t_r, settling time t_s and overshoot M (%) favoured the results obtained by fractional-order PID. The possibility of improving step response of integer PID to match performance of fractional approach was investigated. We determined that smoother step responses were possible by sacrificing bandwidth which directly impacts response velocity, making worse the performance of typical PID.

On the other side, the proposed approach exhibited a limited capacity to reject the effect of disturbances. As previously mentioned, this is attributed to the tuning process, which did not consider any criterion of disturbance rejection. This disadvantage might be addressed by employing optimisation methods that include not only performance requirements but also robustness and disturbance rejection, resulting in a multi-objective approach.

13.7 Conclusion

A fractional PID approximation was designed, synthesised and tuned to achieve output regulation in a buck–boost converter. The system model resulted bilinear of varying parameter; thus, a linearisation around the equilibrium point was performed. The small-signal technique was used for this purpose. Resulting plant was considered divided into its minimum and non-minimum phase parts, allowing a controller design simplification.

The proposed algorithm took into consideration a combination of performance to modify transient response, and robustness. The resulting structure simplified the tuning process, since required parameters to achieve stability requirements were directly determined by shaping the controller effect through T_i and k_c. Synthesised controllers regulated output in both conversion modes with a fast and stable tracking characteristic.

Viability was verified experimentally through a controller structure generalisation to operate in both conversion modes. The electrical arrangement was implemented through *RC* circuits and operational amplifiers, confirming the fractional-order controller effectiveness and viability.

The main opportunity area of presented results lies in the limited ability of fractional-order PID controller to reject disturbances at the plant input, attributed preliminary to the lack of any criterion related disturbance rejection in the design and tuning method. The proposal of a strategy that considers performance, robustness and disturbance rejection through a multi-objective optimisation scheme can be the future direction of this proposal.

The authors would like to thank CONACYT México for catedras 6782 and 4155 and student Josué Soto–Vega for his contribution on the experiment implementation.

References

[1] Hidalgo-Reyes, J., Gómez-Aguilar, J., Escobar-Jiménez, R., Alvarado-Martínez, V., and López-López, M., 'Classical and fractional-order modeling of equivalent electrical circuits for supercapacitors and batteries, energy management strategies for hybrid systems and methods for the state of charge estimation: a state of the art review', *Microelectron. J.* 2019, **85**(1), pp. 109–128.

[2] Tarasov, V.E., 'Review of some promising fractional physical models', *Int. J. Mod. Phys. B* 2013, **27**(9), p. 1330005.

[3] Warrier, P. and Shah, P., 'Fractional order control of power electronic converters in industrial drives and renewable energy systems: a review', *IEEE Access* 2021, **9**(1), pp. 58982–59009.

[4] Sánchez, A.G.S., Pérez-Pinal, F.J., Rodrguez-Licea, M.A., and Posadas-Castillo, C., 'Non-integer order approximation of a pid-type controller for boost converters', *Energies* 2021, **14**(11), p. 3153.

[5] Sánchez, A.G.S., Soto-Vega, J., Tlelo-Cuautle, E., and Rodríguez-Licea, M. A., 'Fractional-order approximation of pid controller for buck–boost converters', *Micromachines* 2021, **12**(6), p. 591.

[6] Monje, C.A., Chen, Y., Vinagre, B.M., Xue, D., and Feliu-Batlle, V., *Fractional-Order Systems and Controls: Fundamentals and Applications,* New York, NY: Springer Science & Business Media, 2010.

[7] Radwan, A.G., Emira, A.A., AbdelAty, A.M., and Azar, A.T., 'Modeling and analysis of fractional order dc–dc converter', *ISA Trans.* 2018, **82**(1), pp. 184–199.

[8] Wei, Z., Zhang, B., and Jiang, Y., 'Analysis and modeling of fractional-order buck converter based on riemann-liouville derivative', *IEEE Access* 2019, **7** (1), pp. 162768–162777.

[9] Fang, S. and Wang, X., 'Modeling and analysis method of fractional-order buck–boost converter', *Int. J. Circuit Theory Appl.* 2020, **48**(9), pp. 1493–1510.

[10] Devaraj, S.V., Gunasekaran, M., Sundaram, E., *et al.*, 'Robust queen bee assisted genetic algorithm (qbga) optimized fractional order pid (fopid) controller for not necessarily minimum phase power converters', *IEEE Access* 2021, **9**(1), pp. 93331–93337.

[11] Mollaee, H., Ghamari, S.M. Saadat, S.A., and Wheeler, P., 'A novel adaptive cascade controller design on a buck–boost dc–dc converter with a fractional-order pid voltage controller and a self-tuning regulator adaptive current controller', *IET Power Electron.* 2021, **14**(11), pp. 1920–1935.

[12] Aseem, K. and Kumar, S.S., 'Hybrid k-means grasshopper optimization algorithm based fopid controller with feed forward dc–dc converter for solar-wind generating system', *J. Ambient Intell. Human. Comput.* 2021, **1** (1), pp. 1–24.

[13] Tepljakov, A., Alagoz, B.B., Yeroglu, C., *et al.*, 'Towards industrialization of fopid controllers: a survey on milestones of fractional-order control and pathways for future developments', *IEEE Access* 2021, **9**(1), pp. 21016–21042.

[14] Junior, F.A.d.C.A., Bessa, I., Pereira, V.M.B., *et al.*, 'Fractional order pole placement for a buck converter based on commensurable transfer function', *ISA Trans.* 2020, **107**(1), pp. 370–384.

[15] Cengelci, E., Garip, M., and Elwakil, A.S., 'Fractional-order controllers for switching dc/dc converters using the k-factor method: analysis and circuit realization', *Int. J. Circuit Theory Appl.* 2022, **50**(2), pp. 588–613.

[16] Saleem, O., Shami, U.T., and Mahmood-ul Hasan, K., 'Time-optimal control of dc–dc buck converter using single-input fuzzy augmented fractional-order pi controller', *Int. Trans. Electric. Energy Syst.* 2019, **29** (10), p. e12064.

[17] Farsizadeh, H., Gheisarnejad, M., Mosayebi, M., Rafiei, M., and Khooban, M. H., 'An intelligent and fast controller for dc/dc converter feeding cpl in a dc microgrid', *IEEE Trans. Circuits Syst. II: Express Briefs* 2019, **67**(6), pp. 1104–1108.

[18] Ghamari, S.M., Narm, H.G., and Mollaee, H., 'Fractional-order fuzzy pid controller design on buck converter with antlion optimization algorithm', *IET Control Theory Appl.* 2022, **16**(3), pp. 340–352.

[19] Karad, S. G. and Thakur, R., 'Fractional order controller based maximum power point tracking controller for wind turbine system', *Int. J. Electron.* 2022, **109**(5), pp. 875–899.

[20] Sorouri, H., Sedighizadeh, M., Oshnoei, A., and Khezri, R., 'An intelligent adaptive control of dc–dc power buck converters', *Int. J. Electr. Power Energy Syst.* 2022, **141**, p. 108099.

[21] Soriano-Sánchez, A.G., Rodrguez-Licea, M.A., Pérez-Pinal, F.J., and Vázquez-López, J.A., 'Fractional-order approximation and synthesis of a pid controller for a buck converter', *Energies* 2020, **13**(3), p. 629.

[22] Delavari, H. and Naderian, S., 'Backstepping fractional sliding mode voltage control of an islanded microgrid', *IET Gener. Transm. Distrib.* 2019, **13**(12), pp. 2464–2473.

[23] Mohadeszadeh, M., Pariz, N., and Ramezani-al, M.R., 'A fractional reset control scheme for a dc–dc buck converter', *Int. J. Dyn. Control.* 2022, **10**, pp. 2139–2150.

[24] Jia, Z., Liu, L., and Liu, C., 'Dynamic analysis and fractional-order terminal sliding mode control of a fractional-order buck converter operating in discontinuous conduction mode', *Int. J. Bifurc. Chaos* 2022, **32**(04), p. 2250045.

[25] Wang, J., Xu, D., Zhou, H., Bai, A., and Lu, W., 'High-performance fractional order terminal sliding mode control strategy for dc–dc buck converter', *Plos One* 2017, **12**(10), p. e0187152.

[26] Paul, R., 'Fractional order modified awpi based dc–dc converter controlled sedc motor', In: *International Conference on Computational Techniques and Applications*, New York, NY: Springer, 2022, pp. 547–555.

[27] Izci, D., Hekimoğlu, B., and Ekinci, S., 'A new artificial ecosystem-based optimization integrated with Nelder-Mead method for pid controller design of buck converter', *Alexandria Eng. J.* 2022, **61**(3), pp. 2030–2044.

[28] Wang, T., Wang, H., Hu, H., Qing, J., and Wang, C., 'Lion swarm optimisation-based tuning method for generalised predictive fractional-order pi to control the speed of brushless direct current motor', *IET Elect. Power Appl.* 2022, **16**(8), pp. 879–895.

[29] Izci, D. and Ekinci, S., 'A novel improved version of hunger games search algorithm for function optimization and efficient controller design of buck converter system', *e-Prime Adv. Electr. Eng. Electron. Energy* 2022, **2**, p. 100039.

[30] Fathy, A., Yousri, D., Rezk, H., Thanikanti, S.B., and Hasanien, H. M., 'A robust fractional-order pid controller based load frequency control using modified hunger games search optimizer', *Energies* 2022, **15**(1), p. 361.

[31] Wang, J., Xu, D., Zhou, H., and Zhou, T., 'Adaptive fractional order sliding mode control for boost converter in the battery/supercapacitor hess', *PLoS One* 2018, **13**(4), p. e0196501.

[32] Seo, S.-W. and Choi, H.H., 'Digital implementation of fractional order pid–type controller for boost dc–dc converter', *IEEE Access* 2019, **7**(1), pp. 142652–142662.

[33] Saleem, O., Rizwan, M., Khizar, A., and Ahmad, M., 'Augmentation of fractional-order pi controller with nonlinear error-modulator for enhancing robustness of dc–dc boost converters', *J. Power Electron.* 2019, **19**(4), pp. 835–845.

[34] Mohamed, A.T., Mahmoud, M.F., Swief, R., Said, L.A., and Radwan, A.G., 'Optimal fractional-order pi with dc–dc converter and pv system', *Ain Shams Eng. J.* 2021, **12**(2), pp. 1895–1906.

[35] Merrikh-Bayat, F. and Jamshidi, A., 'Comparing the performance of optimal pid and optimal fractional-order pid controllers applied to the nonlinear boost converter', arXiv preprint arXiv:1312.7517 2013, 1(1).

[36] Jha, K.K., Anwar, M.N., Shiva, B.S., and Verma, V., 'A simple closed-loop test based control of boost converter using internal model control and direct synthesis approach', *IEEE J. Emerg. Sel. Topic Power Electron*. 2022, **10**(5), pp. 5531–5540.

[37] Rajeswari, C. and Santhi, M., 'Modified flower pollination algorithm for optimizing fopid controller and its application with the programmable n-level inverter using fuzzy logic', *Soft Comput.* 2021, **25**(4), pp. 2615–2633.

[38] Sahin, E. and Altas, İ.H., 'A pso tuned fractional-order pid controlled non-inverting buck–boost converter for a wave/uc energy system', *Int. J. Intell. Syst. Appl. Eng*. 2016, **1**(1), pp. 32–37.

[39] Xie, L., Liu, Z., Ning, K., and Qin, R., 'Fractional-order adaptive sliding mode control for fractional-order buck-boost converters', *J. Electr. Eng. Technol.* 2022, **17**(3), pp. 1693–1704.

[40] Doostinia, M., Beheshti, M.T., Alavi, S.A., and Guerrero, J.M., 'Distributed event-triggered average consensus control strategy with fractional-order local controllers for dc microgrids', *Electr. Power Syst. Res.* 2022, **207**, p. 107791.

[41] Yan, B., Wang, S., and He, S., 'Complex dynamics and hard limiter control of a fractional-order buck-boost system', *Math. Prob. Eng.* 2021, **1**(1).

[42] Bolea Monte, Y., Grau Saldes, A., and Martínez García, H., 'Buck-boost converter with fractional control for electric vehicles', *Renew. Energy Power Quality J.* 2010, **1**(8), pp. 1–5.

[43] Parreño, A., Roncero-Sánchez, P., del Toro Garca, X., Feliu, V., and Castillo, F., 'Analysis of the fractional dynamics of an ultracapacitor and its application to a buck-boost converter', In: *New Trends in Nanotechnology and Fractional Calculus Applications*, New York, NY: Springer, 2010, pp. 97–105.

[44] Martínez, R., Bolea, Y., Grau, A., and Martínez, H., 'Buck-boost converter with fractional control for electric vehicles', In: *International Conference on Renewable Energies and Power Quality*, 2010.

[45] Lopez, M. M., Moreno-Valenzuela, J., and He, W., 'A robust nonlinear pi–type controller for the dc–dc buck–boost power converter', *ISA Trans.* 2022, **129**(Part A), pp. 687–700.

[46] Erickson, R.W. and Maksimovic, D., *Fundamentals of Power Electronics,* New York, NY: Springer Science & Business Media, 2007.

[47] Ang, S., Oliva, A., Griffiths, G., and Harrison, R., 'Continuous-time modeling of switching converters', *In: Power-Switching Converters,* 3rd ed., London: CRC Press, 2010.

[48] Åström, K.J. and Murray, R.M., 'Frequency domain design', *In: Feedback Systems: An Introduction for Scientists and Engineers,* 1st ed., Princeton, NJ: Princeton University Press, 2010.

[49] Podlubny, I., *Fractional Differential Equations*, vol. 198, London: Academic Press, 1998.

[50] Petráš, I., *Fractional-Order Nonlinear Systems: Modeling, Analysis and Simulation,* New York, NY: Springer Science & Business Media, 2011.

[51] Garrappa, R., 'Grünwald–Letnikov operators for fractional relaxation in Havriliak–Negami models', *Commun. Nonlinear Sci. Numer. Simul.* 2016, **38**(1), pp. 178–191.

[52] Carlson, G. and Halijak, C., 'Approximation of fractional capacitors (1/s)^ (1/n) by a regular newton process', *IEEE Trans. Circuit Theory*. 1964, **11**(2), pp. 210–213.

[53] Oustaloup, A., Levron, F., Mathieu, B., and Nanot, F.M., 'Frequency-band complex noninteger differentiator: characterization and synthesis', *IEEE Trans. Circuits Syst. I: Fundam. Theory Appl.* 2000, **47**(1), pp. 25–39.

[54] Charef, A., Sun, H., Tsao, Y., and Onaral, B., 'Fractal system as represented by singularity function', *IEEE Trans. Automat. Control.* 1992, **37**(9), pp. 1465–1470.

[55] El-Khazali, R., 'Fractional-order $PI^{\lambda}D^{\mu}$ controller design', *Comput. Math. Appl.* 2013, **66**(5), pp. 639–646.

[56] El-Khazali, R., 'On the biquadratic approximation of fractional-order Laplacian operators', *Analog Integr. Circuits Signal Process.* 2015, **82**(3), 503–517.

[57] Podlubny, I., Petráš, I., Vinagre, B.M., O'leary, P., and Dorčák, L., 'Analogue realizations of fractional-order controllers', *Nonlinear Dyn.* 2002, **29**(1–4), pp. 281–296.
[58] Wallén, A., ÅRström, K., and Hägglun, T., 'Loop-shaping design of PID controllers with constant Ti/Td ratio', *Asian J. Control.* 2002, **4**(4), pp. 403–409.
[59] Monje, C. A., Vinagre, B.M., Feliu, V., and Chen, Y., 'Tuning and auto-tuning of fractional order controllers for industry applications', *Control Eng. Pract.* 2008, **16**(7), pp. 798–812.

Chapter 14

Adjustable speed drive systems for industrial applications

Apparao Dekka[1], *Deepak Ronanki*[2], *Ricardo Lizana Fuentes*[3] *and Venkata Yaramasu*[4]

The adjustable speed drives reduce the energy loss over fixed speed drives leading to a significant amount of energy cost savings in various industry applications. Particularly, the demand for high-power adjustable speed drives in medium-voltage capacity are continuously growing in industries due to the increase in production demand, low operating cost, and economy of scale. Nevertheless, the development of medium-voltage drives involves various requirements and challenges on the line-side or grid-side (e.g., total harmonic distortion, power factor, *LC* resonance), motor-side (e.g., motor winding derating, *dv/dt*, common-mode voltage, *LC* resonance), and semiconductor devices (e.g., switching frequency, reliability, series connection). The power converters and control methods play a key role in addressing the medium-voltage adjustable speed drive challenges. In this chapter, the latest developments in medium-voltage drive technologies including applications, semiconductor devices, power converter topologies, and control methods are presented. Finally, the future trends of medium-voltage drives, and conclusions are summarized in this chapter.

14.1 Introduction

The medium-voltage (MV) drives play a key role in modernization of almost all industrial sectors. Over the past few decades, the MV drives have a significant growth in their capacity reaching to a voltage rating of 13.8 kV and a power rating of 40 MW [1]. It is expected to reach their operating voltages to 15 kV or greater in the near future. Figure 14.1 illustrates the commercial applications of MV drives including fans, compressors, conveyors, and pumps. These applications dominantly use single or multi-motor standard drive structures of 1–4 MW at 3.3–6.6 kV

[1]Department of Electrical Engineering, Lakehead University, Canada
[2]Department of Energy Science and Engineering, Indian Institute of Technology Delhi, India
[3]Department of Environment and Energy, Universidad Catolica de la Santisima Concepcion, Chile
[4]School of Informatics, Computing, and Cyber Systems, Northern Arizona University, USA

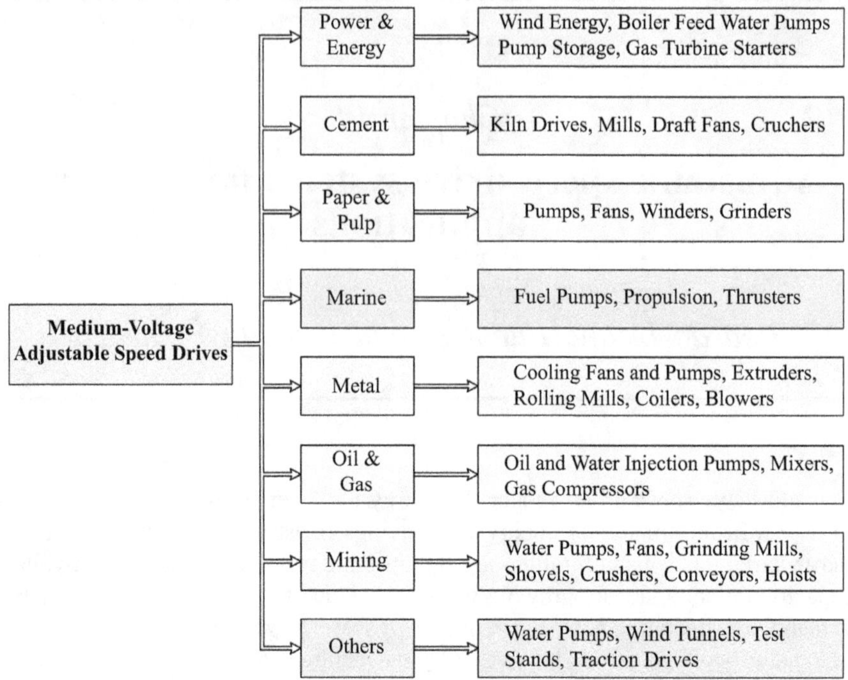

Figure 14.1 Applications of MV adjustable speed drives

voltage level [2]. Furthermore, the MV drives are designed with either fixed speed or adjustable speed operation to meet the application requirements. Among them, the adjustable speed operation is highly preferred in industries due to their ability to increase the productivity while saving the energy cost significantly [3]. In addition, there is a steady demand for adjustable speed drives (ASDs) in MV capacity with a global market value of $17 billion in 2020 and it is expected to reach $19.9 billion by 2026. The importance of MV drives continuously growing globally with the innovation of new high-voltage switching devices and power converters.

The traditional high-voltage switching devices are such as IGCT, GTO, and IGBT are designed with silicon (Si) material [4]. Some of the main features of these devices are ease of gate control, snubber less operation, and lesser device power losses [5]. Currently, the silicon carbide (SiC) material is used in the development of MOSFET and yet to commercialize in MV drive applications. These switching devices are available in low-voltage (LV) capacity and still under research stage to reach MV rating [6, 7]. The LV SiC devices are used to design the cascaded power converters for MV drives. The power converters used in MV drive applications are classified according to the type of DC-bus filter employed in them. They are named as voltage source converters (VSCs), current source converters (CSCs), and power converters without DC-link (e.g., matrix converters, cycloconverters) [8–10]. Among them, the VSCs are quite popular due to their flexible in design and control, and easy to meet the drive requirements compared to other topologies. They have a larger share of MV drives market and commercialized by various industrial manufacturers (e.g., ABB, Siemens, GE, Alstom, Hitachi, Rockwell Automation, etc.).

In addition, the MV drives require high-performance closed-loop control methods to regulate the motor variables (e.g., speed, torque/flux, currents). In MV drives, the classical control methods such as field-oriented control (FOC) and direct torque control (DTC) are commonly used and they are easy to implement in digital controllers [11, 12]. These methods use *dq*-frame proportional-integral (PI) controllers to regulate the motor variables and pulse width modulation (PWM) schemes to produce the converter gating pulses. Furthermore, the popular PWM schemes including multi-carrier and space vector PWM (SVPWM) methods are designed such that the converter switching frequency and drive common-mode voltage can be minimized while improving its power quality [13]. However, the classical control methods have poor transient response due to the limited bandwidth of PI-regulators and semiconductor device switching frequency [14].

Currently, the model predictive control (MPC) also known as predictive control has gained attention for power electronics and MV drive applications [15]. The MPC schemes for MV drives are designed from the philosophy of FOC and DTC approaches. In FOC approach, the inner current control PI-regulators are replaced with predictive algorithm resulting in predictive current control (PCC) [16, 17]. On the other hand, the torque and flux PI-regulators in DTC approach are replaced with predictive algorithm resulting in predictive torque control (PTC) for MV drives [18]. In these methods, the converter switching states are directly manipulated so that the drive control objectives are achieved. The implementation of MPC is quite easy and follows the intuitive approach. Furthermore, it is simple to incorporate the non-linearities of the system into their implementation. The MPC uses a simple cost function to achieve all of the converter control objectives and it exhibits a fast transient response compared with classical control methods [19]. However, the sampling time, variable switching frequency operation, selection of weighting factors, computational complexity, and accuracy of system models are major issues in MPC [20, 21]. Several new MPC methods such as long-horizon MPC [22] to handle sampling issues, two-stage MPC [23], and sequential MPC [24] to reduce the computational complexity, modulated MPC [25] to achieve the fixed switching frequency operation, and MPC without weighting factors [26] are developed. Currently, the model-free predictive control schemes and MPC with artificial intelligence (AI) are the main focus of the research.

Considering the importance of ASDs and its growth, this book chapter focused on the state-of-the-art developments in ASDs for industrial applications including semiconductor devices, high-power converter topologies, classical control methods, and model predictive control methods. This chapter is structured as follows:

- The overview of MV motor drive is presented in Section 14.2. Particularly, the evolution of semiconductor devices and power converter topologies are emphasized in this section.
- The multilevel converters for MV motor drives are highlighted in Section 14.3.

- The implementation and case studies of classical control methods including FOC, and DTC are presented in Section 14.4.
- The MPC methods for MV drives are discussed in Section 14.5.
- The future trends in MV drive technologies are given in Section 14.6.
- Section 14.7 describes the conclusions.

14.2 Overview of MV motor drives

The typical MV drive is a two-stage power conversion system, which involves the power conversion from fixed AC (constant magnitude and frequency) to variable AC (varying both magnitude and frequency) through an intermediate DC power conversion stage as shown in Figure 14.2. In the MV drive, the fixed AC-grid power is first converted to a constant DC power and is referred as a rectification stage. This stage comprises power grid, grid-side *LC* filter, transformer, and AC/DC converter. The induction motor receives variable AC power that is transformed from constant DC power. This stage is known as the inversion stage, and it consists of a DC/AC converter, a motor-side *LC* filter, and an induction motor. As illustrated in Figure 14.2, the DC-bus filter is a common link for both rectification and inversion stages and it is realized with either a capacitor or an inductor [1]. The transformer and grid-side filters minimize the total harmonic distortion (THD) in grid-side current to meet the IEEE-519 and IEC 1000-3-2 standards and eliminate/block the common-mode current entering the system [27, 28]. The motor-side filters are designed to improve the motor-side power quality including total harmonic distortion (THD) and *dv/dt*.

The power converters in rectification and inversion stages play a key role in achieving several objectives on grid-side and motor-side. The converters on grid-side are intended to improve the grid power factor, minimize the grid current THD, and regulate the DC-bus voltage/current magnitude. These converters are either passive or active type converters. In the case of passive type, the diode bridge converters are most popular choice due to their simplicity and ease of use. These converters limit the MV drive operation to two-quadrants only. Furthermore, the external power factor correction capacitors and multi-winding transformer are needed to achieve grid-side objectives [1]. Alternatively, the grid-side objectives can be easily fulfilled with the help of active type converters and closed loop control methods. The use of active type converters leads to a four-quadrant

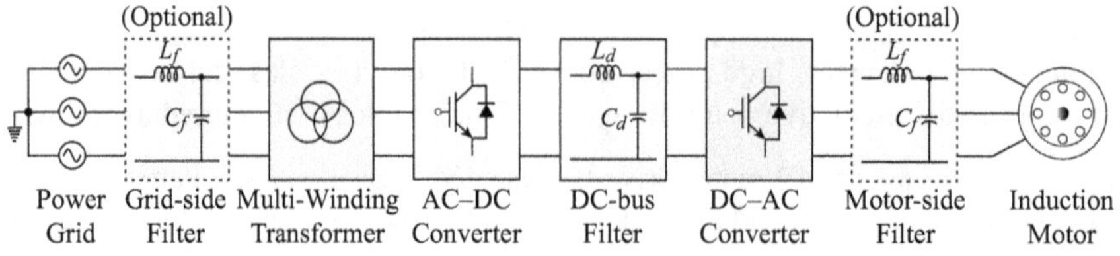

Figure 14.2 MV drive configuration

operation of MV drive. The motor-side converters are intended to reduce the current ripple to improve the drive performance, reduce the *dv/dt* stress on winding insulation, and reduce the common-mode voltage (CMV) stress thereby the bearing currents and shaft failures can be eliminated. These converters can also change their output voltage magnitude and frequency. Aforementioned objectives can be achieved with active type converters alone without any external components/devices [29].

The active type converters used in the rectification and inversion stages are broadly categorized based on the type of DC-bus filter employed in them. If the converters are designed with capacitive DC-bus filter, then such converters are referred as voltage source converters (VSCs), and they are renamed as voltage source inverter (VSI) and voltage source rectifier (VSR) based on their functionality [29]. If the converters are designed with inductive DC-bus filter then such converters are known as current source converters (CSCs), and they are renamed as current source inverter (CSI) and current source rectifier (CSR) based on their power conversion process [30, 31]. The VSCs follow the principle of voltage-to-voltage conversion process (AC voltage to DC voltage in VSR or DC voltage to AC voltage in VSI). Similarly, the CSCs follow the principle of current-to-current conversion process (AC current to DC current in CSR or DC current to AC current in CSI) [32].

The MV drives are available in a wide range of voltages of 2.3, 3.3, 4.16, 6.6, and 13.8 kV. To handle aforementioned voltages, the power converters employed in MV drives should be designed with similar voltage rating. Furthermore, the power converter topologies should have higher power conversion efficiency and less control complexity. Over the past decade, the efforts have been made to develop semiconductor switching devices with less power losses, higher switching frequency operation, and simple gating circuits for MV power converters [1]. The advancement of switching devices outpaces that of high-power converters. These converters are used in a variety of industries, propelling semiconductor technology toward greater power capacity, increased reliability, and lower costs. The most recent advancements in semiconductor devices and power converter topologies are described below.

14.2.1 Evolution of semiconductor devices

Currently, the semiconductor switching devices based on thyristor and transistor technologies are available for application in high-power converters [33, 34]. SCR, GTO, and GCT fall into the first group, whereas IGBT and IEGT fall into the second group. Other devices, such as power MOSFETs, ETO thyristors, MCTs, and SITs, have not proven to be useful in high-power applications [1, 33, 34].

The high-power switching devices are available in press-pack and module package and uses silicon (Si) material in their manufacturing process. Table 14.1 lists the commercial availability of important switching devices used in MV converters. The power diode is a basic device that is commonly used in MV converters as free-wheeling diodes, anti-parallel diodes, and clamping diodes etc. These

Table 14.1 Semiconductor device rating [1, 2, 8]

Device	Voltage (Max.)	Current (Max.)
Power diode	8.5 kV/1.2 kA	9.6 kA/1.8 kV
Thyristor	12 kV/1.5 kA	6.1 kA/1.8 kV
GTO	6 kV/6 kA	6 kA/6 kV
GCT	10 kV/1.7 kA	5 kA/4.5 kV
IGBT	6.5 kV/0.75 kA	3.6 kA/1.7 kV

devices have an 8.5 kV voltage rating and a 9.6 kA current rating. The maximum voltage and current ratings of thyristor are 12 kV and 6.1 kA, respectively. AC–DC converters, line-commutated inverters, and cycloconverters all make extensive use of these devices. Similarly, the GCTs are available in 6 kV and 6 kA capacity [1, 2, 8]. Other variants of the GCT technology are symmetrical, asymmetrical, and reverse conducting types. The symmetrical GCT has identical forward and reverse voltage blocking capability, and widely used in current source converters (CSCs). The asymmetrical (with external antiparallel diode) and reverse conducting (in-built antiparallel diode) GCTs are suitable for voltage source converters (VSCs). IGBTs have low-voltage and current rating capacity, however they can go as high as voltage rating of 6.5 kV and a current rating of 3.6 kA [1, 2, 8]. Currently, the IGBT is most popular switching device used in high-power converters.

However, the high-power silicon (Si) devices have poor thermal characteristics, resulting in poorer efficiency and the necessity for an effective cooling system. Also, the Si-based devices have higher on-state resistance leading to higher conduction losses and capable of operating at lower switching speed only [7, 35]. To address these challenges, the silicon carbide (SiC) devices have been explored for high-power applications. When compared to Si devices like IGBTs, the SiC devices can handle higher switching speeds and temperatures and have lesser conduction losses [1, 35]. In 2013, a hybrid SiC module with voltage rating of 1.7 kV and a current rating of 1.2 kA is developed. Also, a traction inverter was produced using 3.3 kV/1.5 kA SiC modules [7, 35]. Thanks to the technological advancements, the high-voltage/current SiC devices will be ready for commercialization in the near future. SiC devices offer a lot of potential in MV drives because they can reduce power losses and enhance performance, which is a future trend in MV drive research.

14.2.2 High-power converter topologies

As indicated in Figure 14.3, the MV drive converter topologies are divided into two families, named as power converters with DC-bus filters and power converters without DC-bus filters. In the former group, the power conversion takes place in two stages (AC–DC and DC–AC) between the AC source and the induction motor. This family of converters consists of a DC-bus filter, and it is realized with either a capacitor or an inductor [1, 8]. The converters with capacitive DC-bus filter are

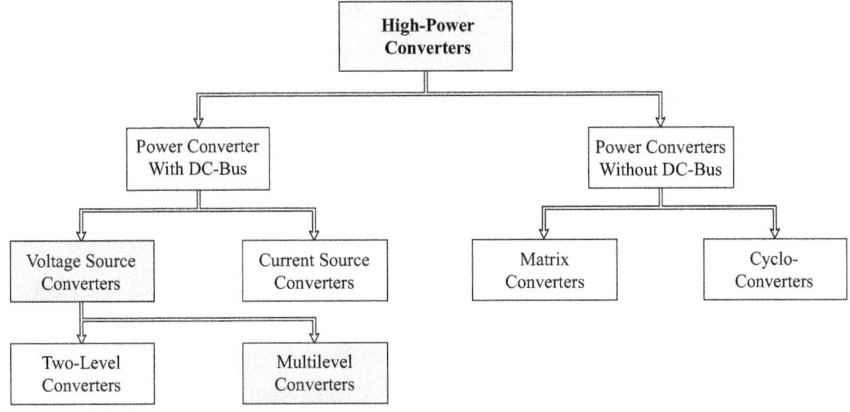

Figure 14.3 Classification of high-power converters

named as VSCs, whereas the converters with inductive DC-bus filter are named as CSCs. In the latter group, the power conversion takes in a single stage (AC–AC) between AC source and induction motor. This family of converters is designed without any DC-bus filters, which improves the reliability of the converter. This group includes cycloconverters (CCV) and matrix converters, in which the CCV uses an array of thyristors to connect the AC-grid directly to an induction motor. The CCV enables power flow in both ways, but it has poor transient response, limited range of voltage and frequency control, and poor power factor [9]. The matrix converter is a revolutionary topology that can generate sinusoidal input/output voltages, suitable for a wide range of voltage and frequency control and operates with good input factor [1].

VSC technology is the most developed of the high-power converters, with a higher market penetration than other high-power converters. As indicated in Figure 14.3, VSC technology is divided into two groups: two-level (2L) and multilevel converters. As illustrated in Figure 14.4(a), the two-level converter is matured VSC that is commonly utilized for low power applications at low-voltage (LV). To enhance the power capacity, the series connection of devices technique is employed in 2L-VSC to manage the MV operation. Alternatively, the parallel connection of devices is used to enhance the current carrying capacity at LV operation. Considering the present MV semiconductor device technology, the series connection of devices is highly preferred to achieve MV operation with 2L-VSC as illustrated in Figure 14.4(b). This strategy does not improve motor performance including THD, *dv/dt*, and CMV. In addition, they require output *LC* filters to reduce the THD, however this generates a *LC* resonance issue. In addition, voltage equalization circuits are required to ensure that the devices share the same voltage during blocking mode. The power losses in power converters are increased by these additional circuits [1, 2, 5]. As a result, 2L-VSCs are rarely used in high-power applications.

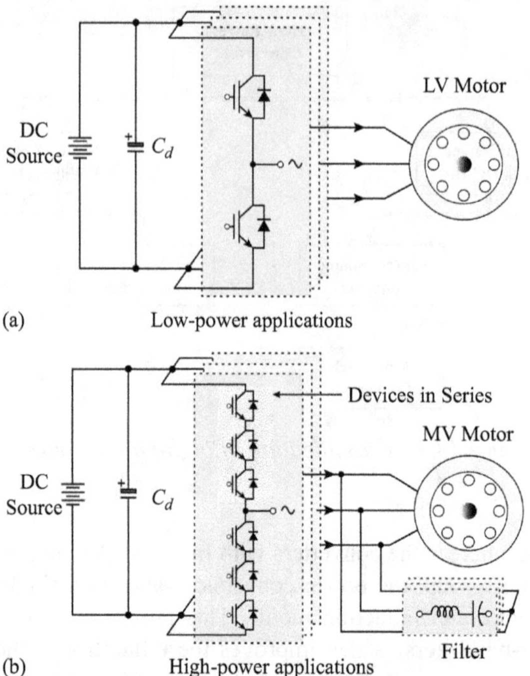

Figure 14.4 Two-level converter-fed motor drive

MV operation can also be achieved with power converter that are utilizing low-voltage and low-cost semiconductor device technology, and these converters are known as multilevel converters. Compared with 2L-VSC, the multilevel converters can generate stepped voltage waveform which significantly reduces the motor current THD without using external filters, *dv/dt*, and CMV. Moreover, the multilevel converters are designed without using series connection of devices thereby the device voltage sharing circuits can be eliminated [2, 5]. These converter topologies can operate at low switching frequencies without affecting the motor THD. By doing so, the switching power losses can be minimized leading to a higher converter efficiency compared with 2L-VSC. Some of the most recent developments in the multilevel converter technologies for MV motor drives are described in the following section.

14.3 Multilevel converters for MV drives

Multilevel converters can generate a stepped voltage waveform with minimal *dv/dt* and THD using an array of LV/MV devices and DC capacitors. The research on multilevel converters for MV drive applications are rigorously conducted over the years and a variety of converter topologies are produced as indicated in Figure 14.5.

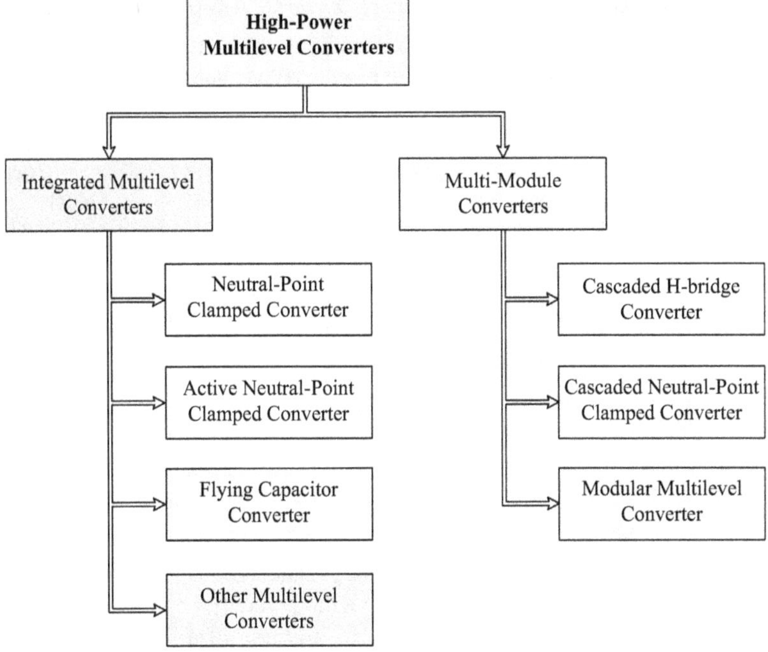

Figure 14.5 Classification of multilevel converters for MV drives

Multilevel converters are classified as integrated and multi-module converters based on their construction [1, 2, 5, 8, 29]. The integrated multilevel converters have an operating voltage range of 2.3–4.16 kV and comes in three-level (3L) to five-level (5L) configurations. To increase their operational voltage, these converters require considerable modifications or a step-up transformer, which is not cost-effective. In addition, they experience lengthier downtime due to failures, resulting in a loss of productivity and revenue in a variety of industrial processes [2, 5, 29]. Table 14.2 illustrates the commercial applications of neutral-point clamped (NPC), active neutral-point clamped (ANPC) converter, flying capacitor (FC) converters. Some of the new developments such as neutral-point piloted (NPP), nested neutral-point piloted (NNPP), and nested neutral-point clamped (NNPC) converters are summarized in other multilevel converters category and yet to find their way into commercial MV drive applications [1, 8].

Aforementioned problems can overcome with the multi-module converters as shown in Figure 14.5. The multi-module converters are modular in design and can operate at a voltage of up to 13.8 kV. Furthermore, with the identical modules in cascades, the output voltage levels of multi-module converters can be easily increased. Also, even when there are defects, they can continue to run at a reduced capacity [2]. The cascaded H-bridge (CHB) converter, cascaded neutral-point clamped (CNPC) converter, and modular multilevel converter (MMC) belongs to the multi-module converters family. The CHB and CNPC converters are among

Table 14.2 Market overview of multilevel converter fed MV drives [1, 2, 8]

Topology	Manufacturer	Product model	Power (MVA)	Voltage (kV)	Device
3L-NPC	ABB	ACS 1000	0.3–5	2.3–4.16	IGCT
	Converteam	VDM 7000	7–9.5	3.3	GTO/MV-IGBT
	Siemens	Sinamics GM150	0.6–10.1	2.3–6.6	MV-IGBT
	TMEIC-GE	Dura-Bilt5i MV	0.3–2.4	4.16	IGBT
	INGERDRIVE	Ingedrive MV100	0.8–15	3.15–4.16	HV-IGBT
3L-ANPC	ABB	PCS 8000	6–100	6–220	IGCT
4L-FC	Alstom	VDM 6000	0.3–8	2.3–4.2	MV-IGBT
CHB	Siemens	Perfect Harmony	0.3–30	2.3–13.8	LV-IGBT
	Hitachi	Hivectol-HVI-E	0.31–10	2.3–11	IGBT
	Rockwell	PF 6000	0.2–5.6	3–11	IGBT
5L-CNPC	ABB	ACS 5000	1.7–27	6–6.9	IGCT
	TMEIC-GE	TMdrive-XL85	30–120	7.2	IGCT
MMC	Siemens	Sinamics SM120	6–13.7	3.3–7.2	IGBT
	Benshaw	M2L 3000	0.23–7.5	2.3–6.6	IGBT

those that can operate at odd voltage levels. Isolated DC sources are required for these topologies, which are generated using a diode rectifier converter and multiple secondary winding transformer together. The addition of a transformer increases the system size and cost dramatically [36]. However, it will also assist in the enhancement of grid-side power quality. MMC is one of the most recent additions to the multi-module converter family, and it is capable of attaining a 400 kV operating voltage without the use of isolated DC sources. MMC also has fault tolerant capability and minimizes THD and *dv/dt* without filters [8]. However, given the same device voltage stress, it necessitates a large number of switching devices and flying capacitors. Only the commercially valuable power converter topologies are highlighted in this section.

14.3.1 Neutral-point clamped converter

The neutral-point clamped (NPC) converter is a well-known multilevel converter topology in both academia and industry. It is first commercialized in three-level operation for 2.3–3.3 kV operational voltage. There are several research studies on developing four-level (4L) and five-level (5L) NPC converters and their control methods [37, 38]. However, due to their control complexity and inherent topologies issues, the 4L and 5L-NPCs are not commercialized for MV drives. Figure 14.6 shows the 3L-NPC fed non-regenerative MV drive system in which the DC-bus voltage is generated using multipulse diode bridge rectifier system.

On the other hand, the identical 3L-NPC converter is employed on grid-side to achieve rectification as illustrated in Figure 14.7. Thereby, the rectifier and inverter systems will have a common DC-bus. This type of connection is referred as a back-to-back connection and it is widely employed in regenerative MV drives. The clamping diodes in 3L-NPC creates a DC-bus neutral point that splits the entire DC-bus voltage into two halves. The power devices only need to block half of the

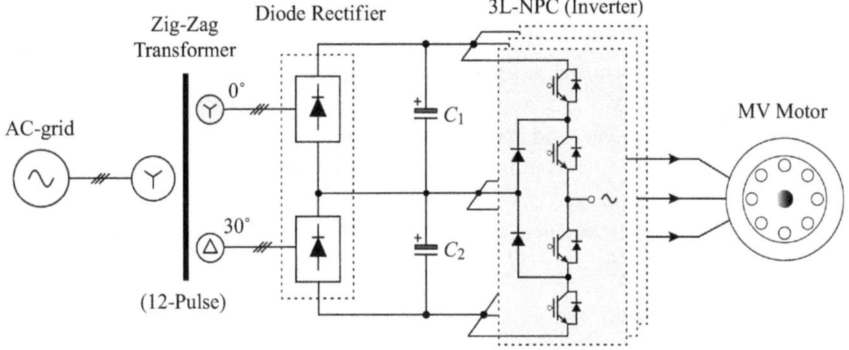

Figure 14.6 3L-NPC converter fed non-regenerative MV drive

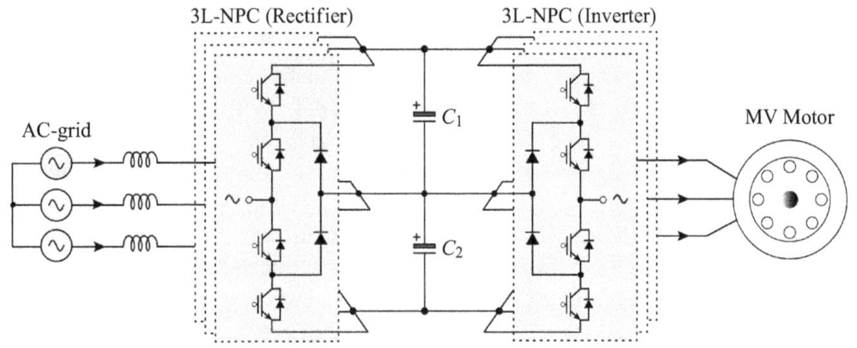

Figure 14.7 3L-NPC converter-fed regenerative MV drive

DC-bus voltage [36]. The voltage levels of 3L-NPC are 0, $+V_{dc}/2$, and $-V_{dc}/2$, which are measured with respect to the DC-bus neutral point. The multi-stepped waveforms lead to low voltage THD and *dv/dt* at the output [5]. The system voltage rating determines which semiconductor devices are used in 3L-NPC. To handle the system operation without series connection of devices, 3.3 kV devices are needed for 2.3 kV 3L-NPC, whereas a 3.3 kV 3L-NPC requires 4.5 kV devices [39]. Without connecting the devices in series, the 3L-NPC can provide an output voltage of up to 4.16 kV using today's semiconductor device technology. As indicated in Table 14.2, multiple manufacturers currently produced 3L-NPC with various system voltages and device technology. Siemens produces 2.3–6.6 kV 3L-NPC with 3.3 kV and 6.5 kV IGBTs, and ABB produces 3.3 kV 3L-NPC with 4.5 kV IGCTs, etc.

DC-bus neutral-point voltage control and DC-bus capacitor voltage ripples are major issues in 3L-NPC [2]. These ripples increase the voltage stress and affect the device reliability. Increasing the DC-bus capacitance has traditionally been used to reduce the DC-bus capacitor voltage ripples. Alternatively, modified multi-carrier PWM and SVPWM schemes are used to reduce the DC-bus capacitor voltage ripple [1]. The NPC requires a greater number of output voltage levels for higher operating voltages to suit the motor-side requirements. As a result, DC-bus capacitors, clamping diodes, and switching devices have increased significantly, resulting in higher system cost and control complexity. Furthermore, the loss distribution across the inner and outer switching devices is not equal. Hence, the NPC with more than three-levels is less attractive for high-power applications [8].

14.3.2 Active neutral-point clamped converter

In 3L-NPC, the inner devices conduct for a longer duration compared with outer devices leading to an asymmetrical distribution of power losses and their respective junction temperatures. This process has an impact on the designing of cooling mechanism and limits the device switching frequency and output power handling capability of the converter [29, 40]. To alleviate aforementioned difficulty, the 3L-NPC clamping diodes are replaced with active switching devices, resulting in a unique architecture known as a three-level active neutral-point clamped (3L-ANPC) converter as illustrated in Figure 14.8. With active switches, the neutral-point current can be easily changed to give an equal loss distribution amongst the devices [2, 5]. To improve neutral-point voltage controllability, the active switches additionally enable redundancy switching states. As indicated in Table 14.2, A 6-100 MVA 3L-ANPC is commercialized in different voltages of 6–220 kV. Through a transformer, the 3L-ANPC is connected to a 220 kV AC-grid. The ANPC concept is expanded to five levels by adding a 3L-FC cell to each phase of a 3L-ANPC converter, resulting in a 5L-ANPC converter. It is commercialized in 0.4–1 MVA at 6–6.9 kV voltage rating [36].

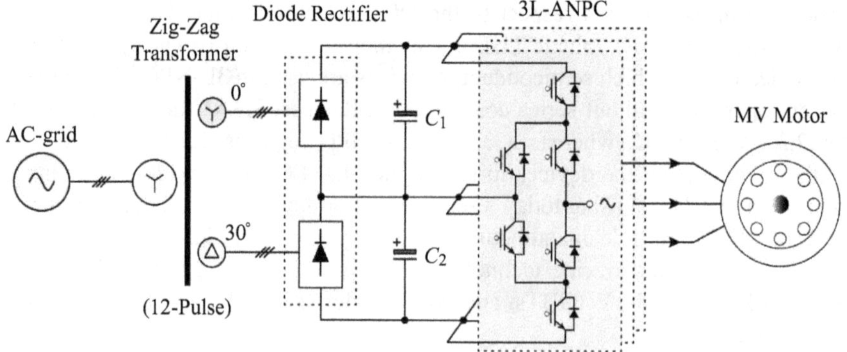

Figure 14.8 3L-ANPC converter-fed non-regenerative MV drive

14.3.3 Flying capacitor converter

In the place of clamping diodes in NPC and active switches in ANPC, the flying capacitors can be used to produce a stepped voltage waveform. Figure 14.9 illustrates the configuration of a flying capacitor (FC) converter, which can produce four-level (4L) voltage at their output terminals. In this topology, each flying capacitor together with a pair of switching devices produce an FC cell [41]. Each phase of this topology requires two FCs with different voltage ratings and six switching devices with identical voltage rating to produce a voltage waveform with four-levels [2, 5]. By simply adding more FC cells to an existing converter, the output voltage levels of FC converters can be easily increased. Furthermore, the usage of FCs eliminates the DC-bus neutral point, which forms a common DC-bus terminals for back-to-back connection.

The FCs have a zero initial voltage and require pre-charging circuits. In addition, the 4L-FC requires an additional voltage balancing technique to maintain the FCs nominal voltage during steady-state and dynamic operation. It is also possible to achieve natural balancing of FCs with carrier phase-shifted PWM (CPS-PWM). With CPS-PWM, the 4L-FC configuration is extremely appealing for high-speed MV drive applications due to the symmetrical loss distribution across the devices [1, 42]. However, the low-speed operation leads to a larger FC voltage ripple which in turn affects the device blocking voltage and voltage/current THD. The use of larger capacitance is the simplest method to reduce the voltage. This philosophy enhances the system cost and its size. Alternatively, the converter switching frequency can be increased to reduce the ripple. This philosophy leads to higher losses, resulting in poor conversion efficiency. Thereby, the 4L-FC is limited to medium switching frequency applications and used in traction and water pump systems. Table 14.2 indicates the power (0.3–8 MVA) and voltage (2.3–4.2 kV) specifications of the commercially available 4L-FC converter [8].

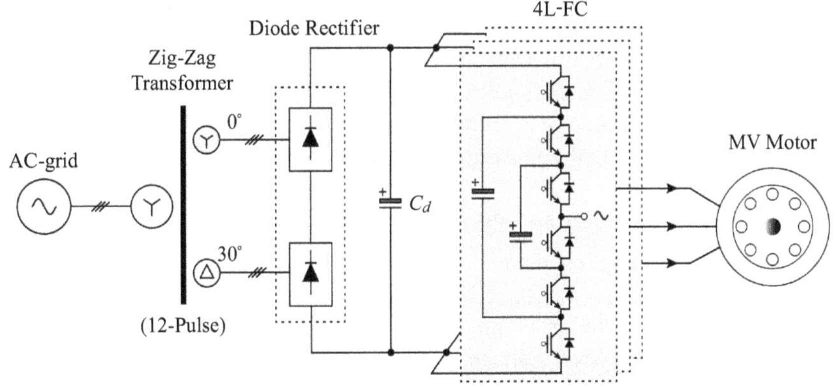

Figure 14.9 4L-FC converter-fed MV drive

14.3.4 Cascaded H-bridge converter

Without connecting devices in series, the integrated multilevel converters may attain an operational voltage of 4.16 kV. For higher operating voltages above 6.6 kV, these topologies are less appealing and cost-effective [2, 36]. Furthermore, the lack of fault-tolerant capability has limited their applications to higher voltage and power levels. Alternatively, employing low-cost and LV devices, multi-module converters can handle voltages up to 13.8 kV. These converters have a modular design as well as power/voltage scaling capability. Also, the multi-module converters can operate with reduced power capacity under fault conditions. The cascaded H-bridge (CHB) converter is commonly used multi-module converters in MV motor drives. Several manufacturers commercialized the CHB topology for various industry applications as shown in Table 14.2 [8]. Each phase of CHB converter is designed with a cascade connection of identical H-bridge modules and each leg of these modules are designed with 2L-VSC legs. This type of structure is named as modular structure. The voltage rating of a CHB converter can also be scaled up or down depending on how many H-bridge modules are used in each phase [43]. In the case of faults, the CHB converter takes lesser time to repair and often can continuously operate with reduced capacity to minimize the system downtime. Hence, its manufacturing and maintenance cost is very low.

As illustrated in Figure 14.10, the CHB converter uses multi-winding transformer to produce the isolate DC source for each H-bridge. The transformer together with rectifier forms a multi-pulse rectifier operation on grid-side, which cancels the dominant harmonics from the grid-side current. In regenerative MV drives, the active front-end (AFE) rectifier is used in place of diode bridge rectifier [1]. The voltage levels of CHB can be increased with addition of identical H-bridge modules in each phase. This process leads to a significant demand on the isolated DC sources and complex transformer structure. Hence, the maximum voltage rating of commercial CHB is limited to 13.8 kV and available in 17-levels as shown in Table 14.2 [44]. In CHB topology, the H-bridge modules in each phase are designed with an identical DC-bus voltage capacity and such converter is named as a symmetrical CHB. On the other hand, the H-bridges belongs to same phase of CHB are designed with unequal DC-bus voltages to rise the converter voltage levels without increasing the number of H-bridge modules and transformer complexity. This type of structure is known as an asymmetrical CHB [1]. However, it is not popular in commercial applications due to the loss of modular feature.

14.3.5 Cascaded neutral-point clamped converter

The cascaded neutral-point clamped (CNPC) converter is an adapted version of CHB converter in which the 2L-VSC legs of H-bridge modules are replaced with 3L-NPC legs as illustrated in Figure 14.11. Thereby, each 3L-NPC based H-bridge module generates five-level voltage waveform compared with 2L-VSC based H-bridge modules. Furthermore, by cascading numerous 3L-NPC-based H-bridge modules, the output voltage levels can be raised. By doing so, there is an increment in voltage and power capacity of CNPC converter while preserving the modular

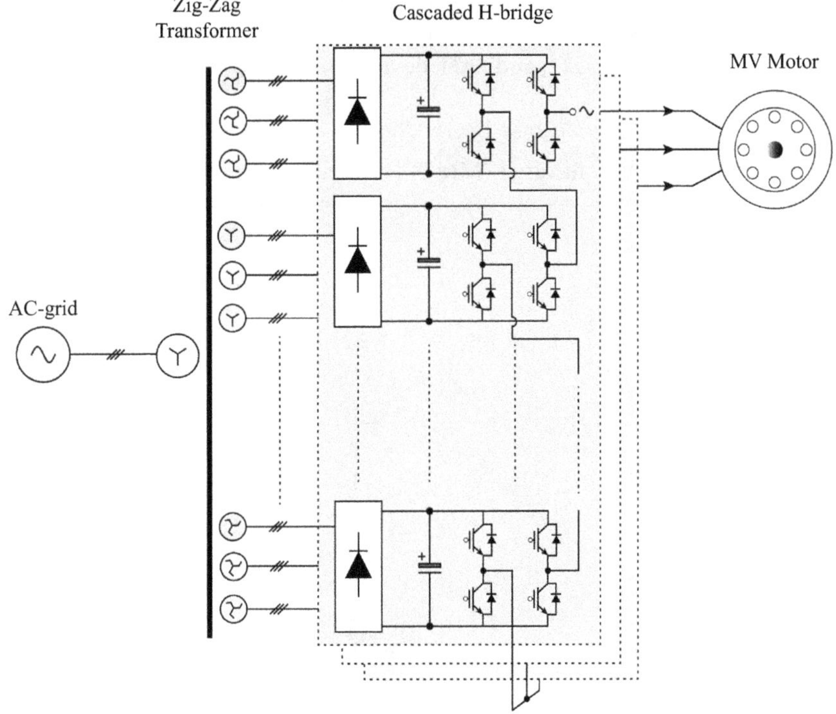

Figure 14.10 CHB converter fed MV drive

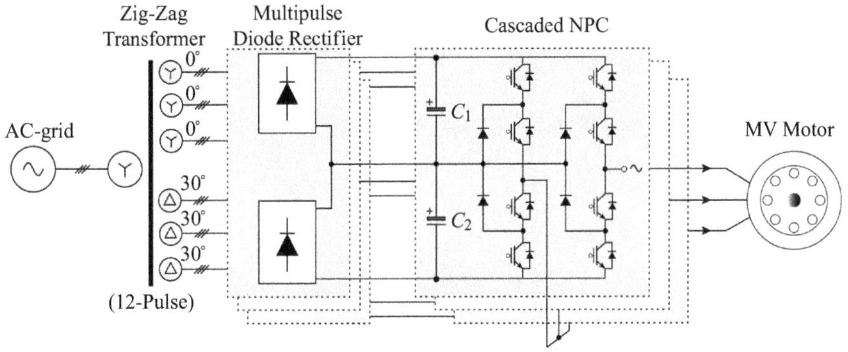

Figure 14.11 Cascade NPC (CNPC) converter-fed MV drive

feature of CHB. However, the CNPC converter requires a complex zig-zag transformer compared to CHB converter to meet the need of isolated DC sources [2, 36]. As a result, without series connection of devices, the CNPC operating voltage is limited to 6.6–7.2 kV. For compressors and conveyor applications, the system is designed with a 36-pulse rectifier by using an IGCT device [2, 36]. As illustrated in Table 14.2, several novel topologies with a voltage capacity of 7.2 kV were created by using MV-IGBTs, IEGTs, and GCT devices.

14.3.6 Modular multilevel converter

The modular multilevel converter (MMC) is a relatively new addition to the multi-module converter family, and its structure is depicted in Figure 14.12(a) [45]. The

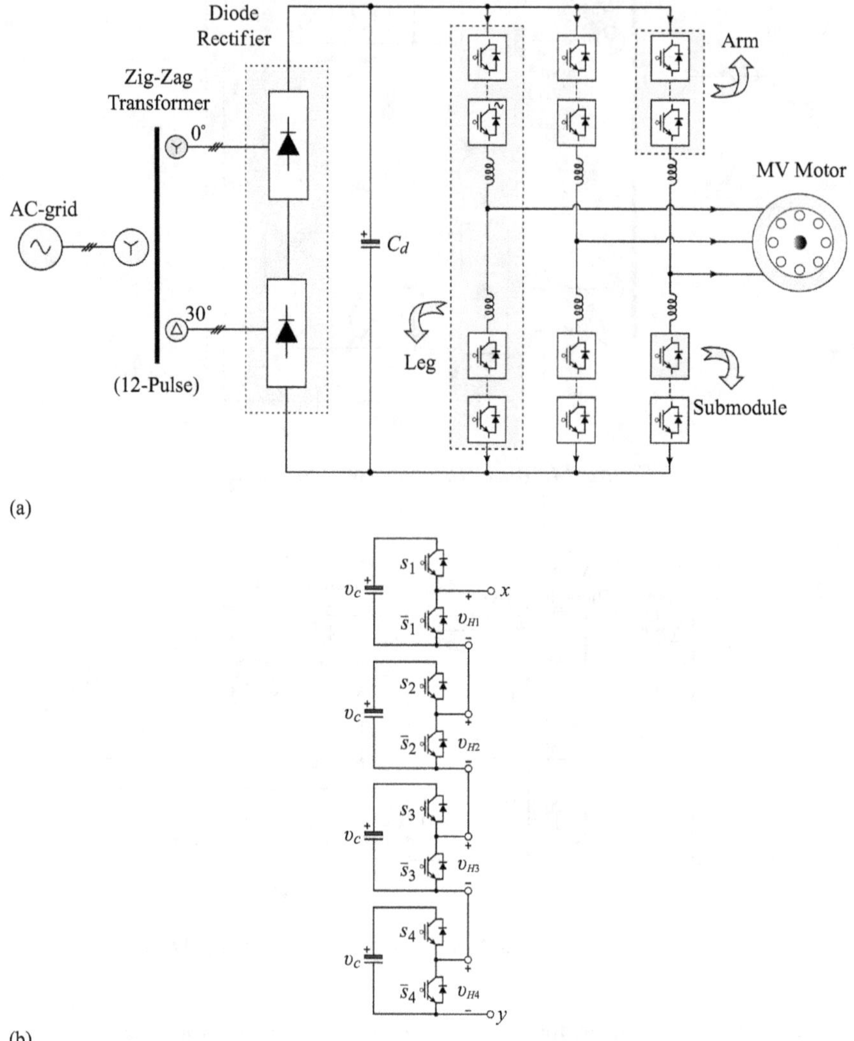

Figure 14.12 MMC and arm configuration: (a) MMC fed MV drive and (b) connection diagram of 4 HB-SMs in an arm

MMC retains the characteristics of CHB and CNPC converters while obviating the need for isolated DC sources. As a result, MMC may operate at any voltage and power rating, ranging from low (2.3–13.8 kV) to high (33–400 kV). These characteristics drew the attention of researchers from both academia and industry, who went on to commercially develop a number of products for a variety of applications, including MV drives, offshore wind farms, multi-terminal high-voltage direct current (HVDC) transmission systems, electrified transportation, HVDC transmission systems, and static synchronous compensators (STATCOM) [46]. Each phase of MMC is identical in construction and they are made up of cascading similar submodules (SMs) structure. Unlike CHB and CNPC converters, these SMs do not need isolated DC source. Furthermore, the SMs structure used in MMC varies with the application [45]. Figure 14.12(b) shows the cascade connection of half-bridge (HB) SMs in each arm of MMC. These modules are easy to manufacture and control unlike other modules.

The operation and control of MMC are limited by design issues, SM capacitor voltage control, minimization of circulating currents (CC), SM capacitor voltage ripple, and SM capacitor pre-charging procedure [8]. As a result, MMC necessitates a complicated control approach to meet those goals. Furthermore, MMC necessitates a greater number of switching devices, flying capacitors, and inductors, raising costs and compromising system reliability and efficiency. Furthermore, the MMC suffers from low-frequency fluctuation in the SM floating capacitors when used in MV drives [45]. To reduce the low-frequency fluctuations in SM capacitors, a high-frequency of CMV and CC are introduced into the system [21]. This results in increased device current stress and power losses, as well as higher voltage stress on the motor winding.

14.4 Classical control methods for MV drives

The induction motors (IMs) are available in medium-voltage and high-power capacity. Particularly, the squirrel-cage induction motors (SCIMs) became popular due to their rugged construction and maintenance free operation. Several classical control methods are available for controlling speed and torque/flux of IMs. The most popular and widely used control methods are shown in Figure 14.13. These methods are broadly categorized into scalar and vector control approaches. The scalar control methods can directly control the magnitude and rotating speed of the voltage/current/flux space vectors without any position information [12]. Hence, these methods are open-loop and have poor dynamic performance. Among them, the volt per hertz (*v/f*) control is a simple and commonly used method, where the ratio of stator voltage vector to frequency is maintained constant. As a result, under steady-state conditions, the stator flux vector magnitude is kept roughly constant at its nominal value. However, it cannot be maintained constant under transient conditions or dynamic speed adjustments. Also, the torque developed by the IM cannot be controlled with this method. Hence, the drives controlled by *v/f* method are not an ideal choice for applications which demands fast transient response [1].

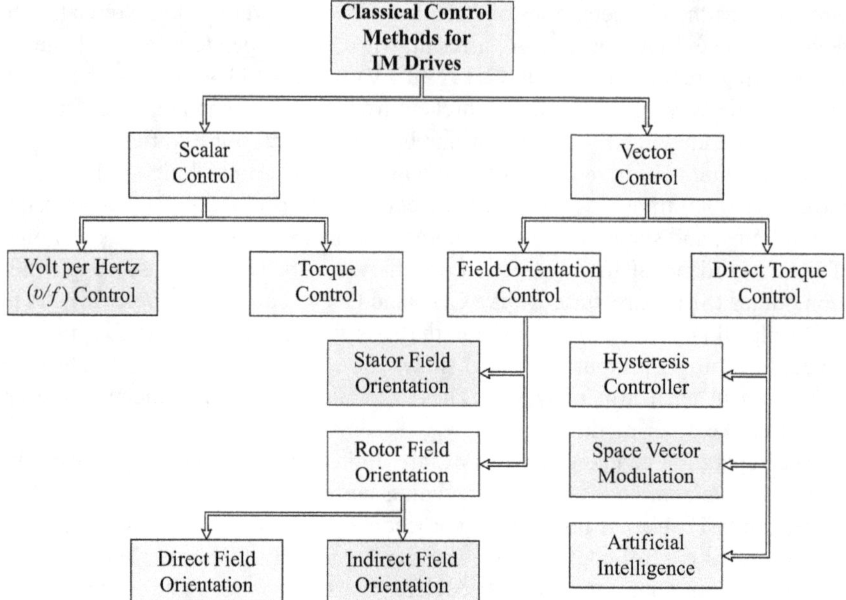

Figure 14.13 Classical control methods for MV drives

On the other hand, the vector control methods can control the position of voltage/current/flux space vectors along with their magnitude and frequency. Hence, these methods have very good steady-state and transient response. Among them, the FOC and DTC methods are highly sought for high-power MV drive applications [11]. These methods work on the principle of decoupling the flux and torque producing current components from motor stator current vector, thereby the motor flux and torque can be controlled independently. These methods are easily implemented in digital controllers with low-cost [1]. This section emphasizes on FOC and DTC implementation for IM drives only.

14.4.1 Field-oriented control

The philosophy of DC motor control is the foundation for FOC control of induction motor drive, in which the flux and torque are regulated independently. With the help of field orientation, the flux and torque producing current components are extracted from three-phase motor stator currents. These current components are used to control the flux and torque independently [11]. The field orientation can be achieved by using any one of the flux vectors from motor stator, rotor, and air-gap. However, the flux vector on rotor-side is commonly used in the field orientation for induction motor control [1]. The generalized expression for induction motor torque is given as

$$T_e = \frac{3PL_m}{2L_r}\left(\lambda_{dr}i_{qs} - \lambda_{qr}i_{ds}\right) \tag{14.1}$$

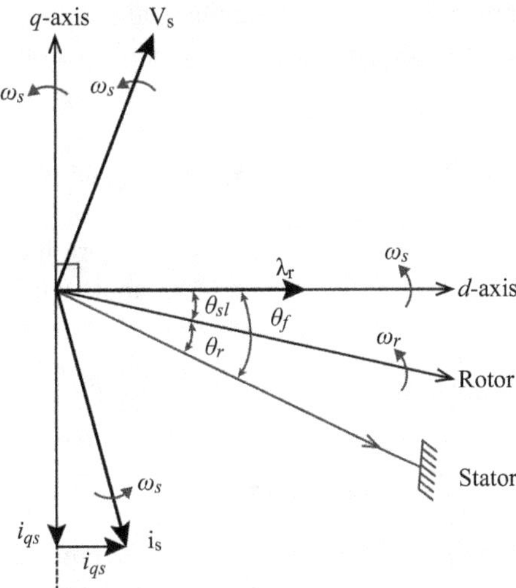

Figure 14.14 Rotor flux field orientation phasor diagram

where P is the stator pole pairs, L_m is the magnetizing inductance, L_r is the rotor self-inductance, i_{ds} and i_{qs} are the stator current components in synchronous dq-rotating reference frame, λ_{dr} and λ_{qr} are the rotor flux components in synchronous dq-rotating reference frame.

According to the principle of rotor flux field orientation, the rotor flux vector λ_r and the synchronous dq-reference frame d-axis are aligned together as shown in Figure 14.4. The rotating speed of dq-axis is given as,

$$\omega_s = 2\pi f_s \tag{14.2}$$

where ω_s is the synchronous speed and f_s is the stator frequency.

With the field orientation, the magnitude of d-axis component of rotor flux vector λ_{dr} is equal to λ_r, whereas the magnitude of rotor flux vector q-axis component λ_{qr} becomes zero. The motor torque expression is simplified as,

$$\left.\begin{aligned} T_e &= K_t \lambda_r i_{qs} \\ K_t &= 3PL_m/2L_r \end{aligned}\right\} \tag{14.3}$$

where K_t is the motor torque constant.

From Eq. (14.3), it is observed that the motor torque T_e varies linearly with the magnitude of q-axis component of stator current vector i_{qs} provided that the magnitude of rotor flux vector λ_r is maintained constant. From the phasor diagram shown in Figure 14.14, the stator current vector $\boldsymbol{i}_s$ can be decomposed into d- and q-axis components. The d-axis component of stator current vector i_{ds} is responsible for production of flux in the motor, whereas the q-axis component i_{qs} is responsible for production of motor torque. These current components are employed in the

FOC of SCIM. The rotor flux vector magnitude λ_r and its position θ_f are required to implement FOC. The rotor flux vector position can be estimated as follows:

$$\theta_f = \theta_r + \theta_{sl} \tag{14.4}$$

where θ_r is the position of rotor and θ_{sl} is the slip speed.

Depending on the type of method employed in finding the magnitude of rotor flux vector λ_r and its position θ_f, the FOC methods are broadly categorized into direct and indirect methods. In the direct FOC approach, the rotor flux vector position θ_f is calculated from the feedback stator voltages and currents. On the other hand, the indirect FOC approach follows the philosophy given in Eq. (14.4) in which the rotor position angle θ_r is directly obtained from speed/position sensor and the slip angle θ_{sl} is calculated from the feedback stator currents [1]. The general structure of FOC control block diagram is illustrated in Figure 14.15. The speed controller minimizes the difference between reference and measured speed (ω^*_r and ω_r) and it generates the reference torque command T^*_e. When the drive operates below the rated speed then the reference rotor flux command (λ^*_r) is set to the nominal value of motor rotor flux. When the drive speed above the rated speed then the rotor flux reference is weakened so that the stator voltages and currents will not be exceed their rated values [12]. The reference flux and torque variables are given to their respective flux and torque controllers.

These controllers minimize the difference between reference and estimated flux and torque variables, and generates modulation signals (v^*_{as}, v^*_{bs}, v^*_{cs}) as shown in Figure 14.15. The PWM modulator produce the switching pulses from the comparison of modulation and carrier signals. These pulses are applied to the multilevel inverter (MLI) switching devices, thereby the magnitude of MLI output voltage and its frequency will be adjusted while keeping their ratio constant. The main functions of flux/torque are (i) estimate the position of rotor flux vector θ_f to achieve the field orientation, (ii) estimate the magnitude of rotor flux vector λ_r or d-axis component of stator current vector i_{ds} responsible for flux production, (iii) estimate the magnitude of motor torque T_e or stator current vector q-axis component i_{qs} responsible for torque production, and (iv) estimate the angular rotational

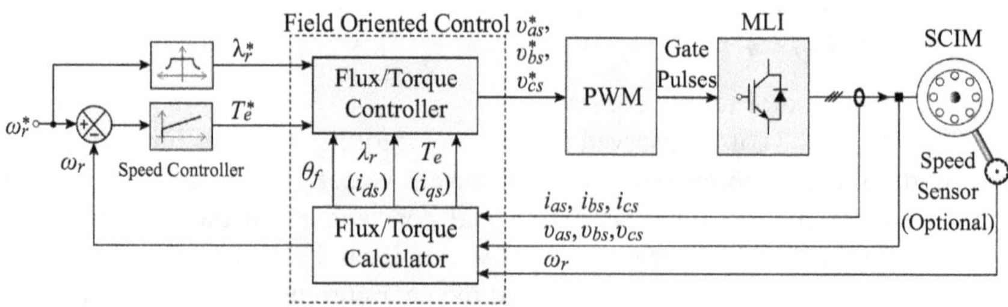

Figure 14.15 Generalized control block diagram of FOC with rotor flux orientation

speed of rotor ω_r [1]. The rotor angular speed ω_r can be directly measured using speed sensor or estimated from measured stator voltages and currents.

14.4.1.1 Direct FOC

The internal structure of FOC flux/torque controller and flux/torque calculator under DFOC philosophy is shown in Figure 14.16. The generation of reference flux and torque signals follows the philosophy shown in Figure 14.15. The DFOC features (i) rotor flux λ_r control loop, (ii) d-axis component of stator current vector i_{ds} control loop, and (iii) q-axis component of stator current vector i_{qs} control loop. The three-phase stator voltages and current signals are used to calculate the magnitude of rotor flux vector λ_r and its position θ_f as shown in Figure 14.16. The flux controller minimizes the difference between the reference and estimated rotor flux vector magnitudes (λ^*_r and λ_r) and it generates the reference stator current vector d-axis component i^*_{ds}. On the other hand, the reference torque command T^*_e produces the reference stator current vector q-axis component i^*_{qs} [1, 11]. The reference currents are compared with their respective feedback stator current components (i_{ds} and i_{qs}), and their difference is minimized by using the current controllers as shown in Figure 14.16. These controllers produce the reference stator voltage vector dq-axis components v^*_{ds} and v^*_{qs}, respectively. These dq-axis voltages are converted to abc-axis voltages v^*_{as}, v^*_{bs}, v^*_{cs} and given to PWM block to generate the gating pulses to the MLI. The rotor flux vector position θ_f is employed in the conversion process from abc/dq and vice versa [8]. Furthermore, the estimation of rotor flux vector magnitude λ_r is also very important to achieve the closed loop control of SCIM drive.

The induction motor space vector model in the arbitrary rotating reference frame is shown in Figure 14.17. From the model, the voltage vectors of stator and

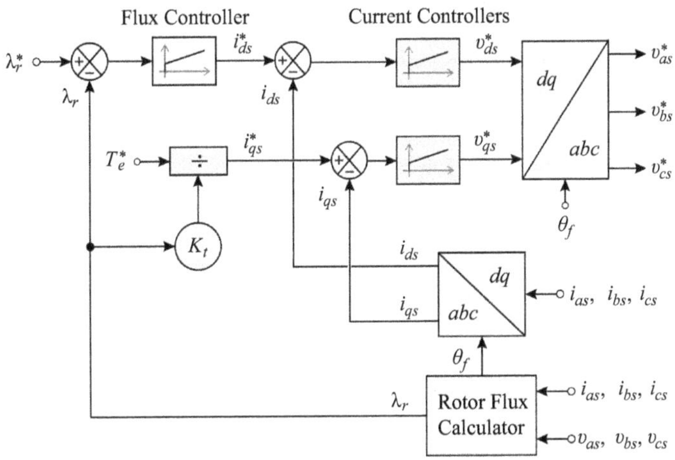

Figure 14.16 DFOC with rotor flux orientation

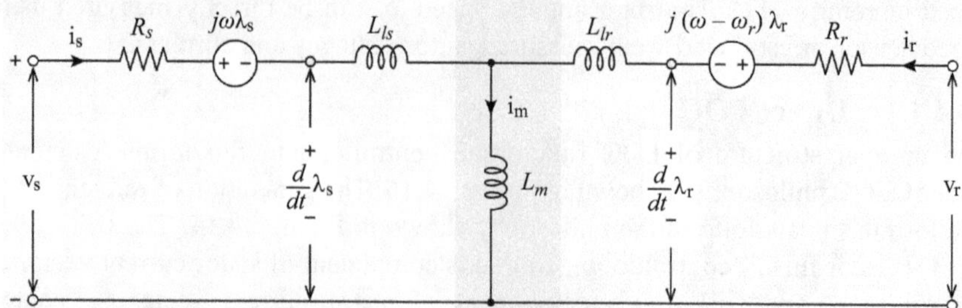

Figure 14.17 Induction motor space vector model in the arbitrary rotating reference frame

rotor side circuit are expressed as [1, 11]

$$\left.\begin{aligned} \boldsymbol{v}_s &= R_s\boldsymbol{i}_s + j\omega\boldsymbol{\lambda}_s + \frac{d\boldsymbol{\lambda}_s}{dt} \\ \boldsymbol{v}_r &= R_r\boldsymbol{i}_r + j(\omega-\omega_r)\boldsymbol{\lambda}_r + \frac{d\boldsymbol{\lambda}_r}{dt} \end{aligned}\right\} \tag{14.5}$$

where, $\boldsymbol{v}_s$, $\boldsymbol{i}_s$, and λ_s are the voltage, current, and flux vectors of stator circuit, $\boldsymbol{v}_r$, $\boldsymbol{i}_r$, and λ_r are the voltage, current, and flux vectors of rotor circuit, R_s and R_r are the winding resistance of stator and rotor circuits, respectively, ω is the arbitrary reference frame rotating speed, and ω_r is the rotating speed of rotor.

The flux vectors of stator and rotor sides are expressed in terms of winding inductances as,

$$\left.\begin{aligned} \boldsymbol{\lambda}_s &= \boldsymbol{i}_s L_s + \boldsymbol{i}_r L_m \\ L_s &= L_{ls} + L_m \\ \boldsymbol{\lambda}_r &= \boldsymbol{i}_s L_m + \boldsymbol{i}_r L_r \\ L_r &= L_{lr} + L_m \end{aligned}\right\} \tag{14.6}$$

where L_s and L_{ls} are the self and leakage inductances of stator winding, and L_r and L_{lr} are the self and leakage inductances of rotor winding.

The value of ω is set to zero in Eq. (14.5), which results in the induction motor model in the stationary $\alpha\beta$-reference frame as

$$\left.\begin{aligned} \boldsymbol{v}_s &= R_s\boldsymbol{i}_s + \frac{d\boldsymbol{\lambda}_s}{dt} \\ \boldsymbol{v}_r &= R_r\boldsymbol{i}_r - j\omega_r\boldsymbol{\lambda}_r + \frac{d\boldsymbol{\lambda}_r}{dt} \end{aligned}\right\} \tag{14.7}$$

In this analysis, the induction motor with cage-type rotor construction is considered. This results in zero rotor voltage ($\boldsymbol{v}_r = 0$) and Eq. (14.7) becomes

$$\left.\begin{aligned} \boldsymbol{v}_s &= R_s\boldsymbol{i}_s + \frac{d\boldsymbol{\lambda}_s}{dt} \\ 0 &= R_r\boldsymbol{i}_r - j\omega_r\boldsymbol{\lambda}_r + \frac{d\boldsymbol{\lambda}_r}{dt} \end{aligned}\right\} \tag{14.8}$$

From Eq. (14.8), the stator flux vector is written as,

$$\boldsymbol{\lambda}_s = \int (\boldsymbol{v}_s - R_s \boldsymbol{i}_s) dt \tag{14.9}$$

From Eqs (14.6) and (14.9), the rotor flux vector is given as,

$$\left.\begin{aligned} \boldsymbol{\lambda}_r = L_r \frac{\boldsymbol{\lambda}_s - L_s \boldsymbol{i}_s}{L_m} + L_m \boldsymbol{i}_s = \frac{L_r}{L_m}(\boldsymbol{\lambda}_s - \sigma L_s \boldsymbol{i}_s) \\ \sigma = 1 - \frac{L_m^2}{L_s L_r} \end{aligned}\right\} \tag{14.10}$$

where σ is the leakage factor.

The rotor flux vector λ_r given in Eq. (14.10) consists of two orthogonal components. These components are expressed in stationary $\alpha\beta$-reference frame as

$$\left.\begin{aligned} \lambda_{\alpha r} &= \frac{L_r}{L_m}(\lambda_{\alpha s} - \sigma L_s i_{\alpha s}) \\ \lambda_{\beta r} &= \frac{L_r}{L_m}(\lambda_{\beta s} - \sigma L_s i_{\beta s}) \end{aligned}\right\} \tag{14.11}$$

From Eq. (14.11), the magnitude of rotor flux vector and its position are expressed as,

$$\left.\begin{aligned} \lambda_r &= \sqrt{\lambda_{\alpha r}^2 + \lambda_{\beta r}^2} \\ \theta_f &= \tan^{-1}\left(\frac{\lambda_{\beta r}}{\lambda_{\alpha r}}\right) \end{aligned}\right\} \tag{14.12}$$

From Eqs (14.5) to (14.12), it is observed that

- The feedback signals of stator voltages $\boldsymbol{v}_s$, stator currents $\boldsymbol{i}_s$, and induction motor parameters (R_s, L_s, L_m, and L_r) are required to estimate the rotor flux vector magnitude λ_r and its position θ_f.
- The rotor flux calculator uses stationary $\alpha\beta$-reference frame, hence all the variables such as $\lambda_{\alpha r}$, $\lambda_{\beta r}$, $\lambda_{\alpha s}$, $\lambda_{\beta s}$, $i_{\alpha s}$, and $i_{\beta s}$ are vary in one cycle over a time.

Figure 14.18 shows the implementation block diagram of rotor flux calculator in stationary $\alpha\beta$-frame, which is designed based on Eqs (14.5)–(14.12). The sensors are used to measure the stator voltages and currents of the induction motor, respectively. Considering the principle of balanced system operation $v_{as} + v_{bs} + v_{cs} = 0$, two phase voltages of the induction motor are measured and the third phase voltage is estimated. Similarly, this philosophy is used in the stator current measurements. Thereby, the required number of voltage/current sensors for the implementation of DFOC will be reduced. The stator voltages and currents feedback in stationary *abc*-frame are transformed to stationary $\alpha\beta$-frame. The transformed $\alpha\beta$ voltage and current components are used along with Eqs (14.5)–(14.12) to estimate the magnitude of rotor flux vector λ_r and its position θ_f as shown in Figure 14.18.

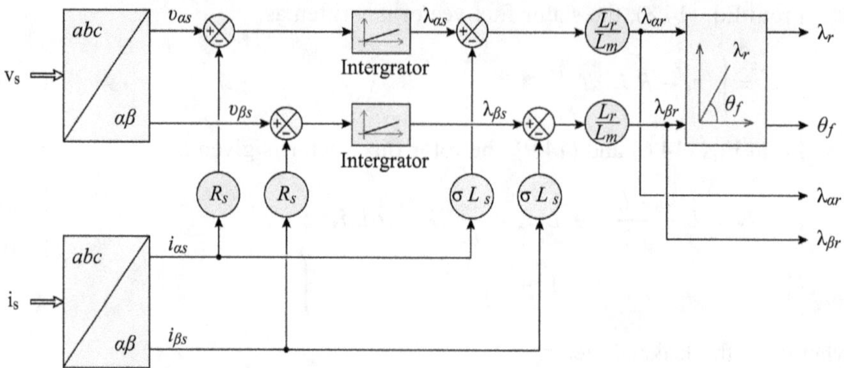

Figure 14.18 Block diagram of rotor flux calculator in DFOC

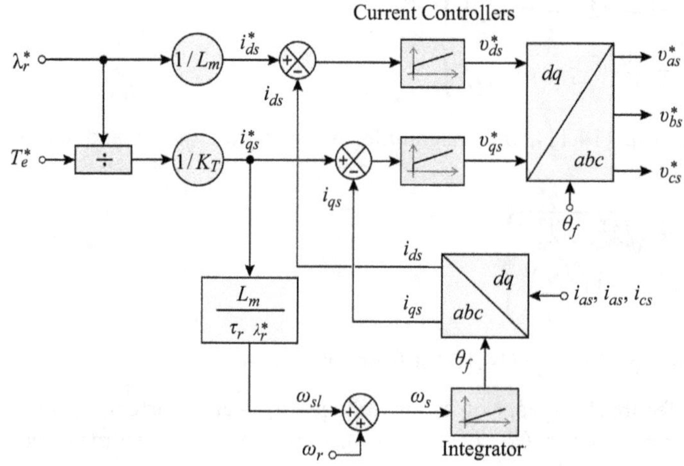

Figure 14.19 Block diagram of IFOC approach

14.4.1.2 Indirect FOC

The IFOC approach requires a speed sensor to obtain the position of rotor flux vector θ_f and it is expressed as follows:

$$\theta_f = \int (\omega_r + \omega_{sl})dt \tag{14.13}$$

where ω_{sl} is the slip angular frequency or slip angular speed.

From Eq. (14.13), it is observed that the position of rotor flux vector θ_f is obtained by integrating the summation of measured rotor speed ω_r and estimated slip speed ω_{sl}. The internal structure of flux/torque controller and flux/torque calculator in IFOC is shown in Figure 14.19 [1, 11]. In IFOC, the generation of reference rotor flux and torque follows the same principle illustrated in Figure 14.15. In addition, the IFOC consists of two feedback loops to control the flux and torque producing current components independently. The reference rotor flux vector magnitude λ^*_r produces the d-axis component of reference stator current vector i^*_{ds}, whereas the reference torque command T^*_e produces the reference q-axis component of stator current vector i^*_{qs} as illustrated in Figure 14.19. The feedback stator currents in dq-frame i_{ds} and i_{qs} are obtained from the measured stationary abc-reference frame stator currents i_{as}, i_{bs} and i_{cs}. The reference and feedback stator current vector dq-axis components are compared with each other, and resultant error current components are given to the current controllers to generate dq-axis reference voltages v^*_{ds} and v^*_{qs}, respectively. These voltages are converted to stationary abc-frame voltages with the help of rotor flux vector position angle θ_f as shown in Figure 14.19. These voltages are fed to the PWM block to produce the switching pulses to the MLI.

In IFOC approach, the position of rotor flux vector is employed to transform the variables from abc to dq-reference frame and vice versa as shown in Figure 14.19. To obtain θ_f, the slip speed or slip frequency is necessary and it is estimated from the measured stator current components and motor parameters. To estimate the slip speed ω_{sl}, the induction motor equivalent circuit shown in Figure 14.17, is employed. In which, the dq-axis rotating speed ω is set to ω_s, then the equivalent model given in Eq. (14.5) becomes [1, 11]:

$$\left.\begin{aligned} \boldsymbol{v}_s &= R_s\boldsymbol{i}_s + j\omega_s\boldsymbol{\lambda}_s + \frac{d\boldsymbol{\lambda}_s}{dt} \\ \boldsymbol{v}_r &= R_r\boldsymbol{i}_r + j\omega_{sl}\boldsymbol{\lambda}_r + \frac{d\boldsymbol{\lambda}_r}{dt} \\ \omega_{sl} &= \omega_s - \omega_r \end{aligned}\right\} \tag{14.14}$$

where ω_s is the synchronous speed of induction motor or rotating speed of the dq-reference frame axis.

Considering the induction motor with cage-type rotor construction then the rotor voltage in Eq. (14.14) becomes zero ($\boldsymbol{v}_r = 0$). The resultant induction motor model is given as

$$\left.\begin{aligned} \boldsymbol{v}_s &= R_s\boldsymbol{i}_s + j\omega_s\boldsymbol{\lambda}_s + \frac{d\boldsymbol{\lambda}_s}{dt} \\ 0 &= R_r\boldsymbol{i}_r + j\omega_{sl}\boldsymbol{\lambda}_r + \frac{d\boldsymbol{\lambda}_r}{dt} \end{aligned}\right\} \tag{14.15}$$

The angular slip speed (ω_{sl}) is derived from the rotor voltage equation in (14.15),

$$\frac{d\boldsymbol{\lambda}_r}{dt} = -R_r\boldsymbol{i}_r - j\omega_{sl}\boldsymbol{\lambda}_r \tag{14.16}$$

The expression for the rotor current vector is derived from Eq. (14.6)

$$\boldsymbol{i}_r = \frac{\boldsymbol{\lambda}_r - L_m \boldsymbol{i}_s}{L_r} \tag{14.17}$$

From Eqs (14.16) and (14.17), the derivative of rotor flux vector is expressed as follows:

$$\frac{d\boldsymbol{\lambda}_r}{dt} = -\frac{R_r}{L_r}(\boldsymbol{\lambda}_r - L_m \boldsymbol{i}_s) - j\omega_{sl}\boldsymbol{\lambda}_r \tag{14.18}$$

from which

$$\left.\begin{aligned} \boldsymbol{\lambda}_r(1 + \tau_r(p + j\omega_{sl})) &= L_m \boldsymbol{i}_s \\ \tau_r &= \frac{L_r}{R_r} \end{aligned}\right\} \tag{14.19}$$

where τ_r is the time constant of rotor and p is the mathematical derivative operator ($p = d/dt$).

The d- and q-axis components of stator current vector and rotor flux vector are separated from Eq. (14.19). In which, the rotor flux vector q-axis component λ_{qr} is set to zero, whereas the magnitude of d-axis component λ_{dr} is equal to λ_r due to the field orientation principle. The resultant d- and q-axis equations are given as,

$$\left.\begin{aligned} \lambda_r(1 + p\tau_r) &= L_m i_{ds} \\ \omega_{sl}\tau_r\lambda_r &= L_m i_{qs} \end{aligned}\right\} \tag{14.20}$$

The angular slip speed is obtained from the q-axis equation in (14.20) as,

$$\omega_{sl} = \frac{L_m}{\tau_r \lambda_r} i_{qs} \tag{14.21}$$

From Eq. (14.20), the reference stator current vector d-axis component i^*_{ds} is given as,

$$i^*_{ds} = \frac{(1 + p\tau_r)}{L_m}\lambda^*_r \tag{14.22}$$

The rotor flux vector magnitude λ_r is kept constant at their reference value of λ^*_r at steady state. As a result, the $p\lambda^*_r$ in Eq. (14.22) becomes zero and it results in

$$i^*_{ds} = \frac{1}{L_m}\lambda^*_r \tag{14.23}$$

From Eq. (14.3), the reference stator current vector q-axis component in terms of reference torque command T^*_e is given as

$$i^*_{qs} = \frac{1}{K_t \lambda^*_r} T^*_e \tag{14.24}$$

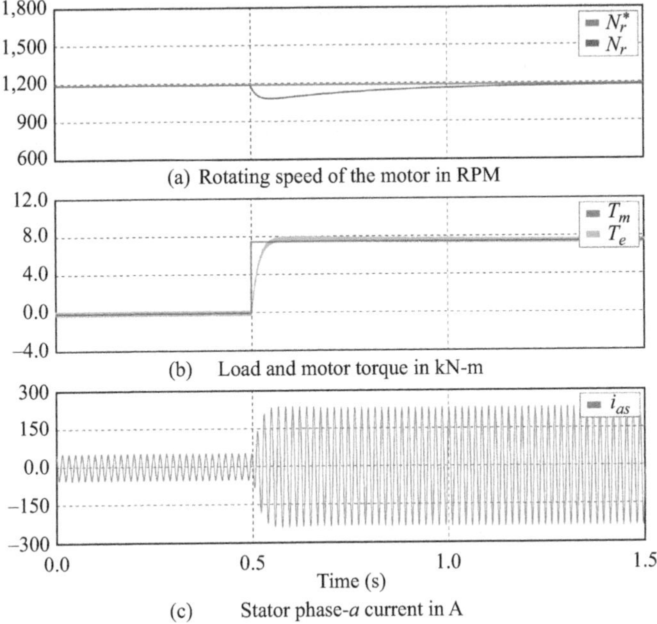

Figure 14.20 Simulated waveforms of IFOC drive operating at rated speed

Table 14.3 Three-phase SCIM parameters

Motor rating	SI unit	Motor parameter	SI unit
Output power (P_s)	1,475 hp	Stator resistance (R_s)	0.0726 Ω
Line-to-line voltage (V_s)	4,000 V	Rotor resistance (R_r)	0.1182 Ω
Stator current (I_s)	159 A (rms)	Stator leakage inductance (L_{ls})	3.9 mH
Speed (N_r)	1,189 RPM	Rotor leakage inductance (L_{lr})	3.9 mH
Torque (T_e)	7,935 N m	Magnetizing inductance (L_m)	173.4 mH
Stator flux linkage (λ_s)	9.0 Wb (peak)	Moment of inertia (J)	10 kg-m^2
Rotor flux linkage (λ_r)	8.35 Wb (peak)	Number of pole pairs (P)	3

The performance of induction motor drive with IFOC approach is illustrated in Figure 14.20. The motor nameplate details are given in Table 14.3. The sine-triangle PWM scheme with a triangular signal frequency of 1 kHz is used to generate the gating pulses for the inverter. In this study, the reference rotor flux vector magnitude λ^*_r is set to 8.35 Wb, whereas the motor reference speed N^*_r is set to 1,189 RPM. The actual rotating speed of the motor is controlled at their rated value as shown in Figure 14.20(a). During this period, the load torque demand T_m on the motor is set to zero. Figure 14.20(b) shows that the motor developed torque T_e is equal to the load torque demand T_m. Under this condition, the current drawn by the motor is only consists of magnetizing current component to setup the air-gap flux in the motor as illustrated in Figure 14.20(c). At $t = 0.5$ s, the load torque demand is suddenly changed from 0 N m to 7,935 N m. This step change affects the speed controller performance as shown in Figure 14.20(a). On the other hand, the motor developed torque follows the load torque demand and it is maintained at a torque of 7,935 N m as illustrated in Figure 14.20(b). The current drawn by the motor consists of both flux and torque producing current components as illustrated in Figure 14.20(c).

14.4.2 Direct torque control

As the name implies, the direct torque control (DTC) directly controls the torque and stator flux without any intermediate current controllers. Thereby, the DTC has superior dynamic performance compared with FOC. The DTC features are easy of implementation in digital control platforms and robust system operation [11, 12]. The motor developed torque is given as follows:

$$T_e = \frac{3P}{2}\frac{L_m}{\sigma L_s L_r}\lambda_s \lambda_r \sin\theta_T \tag{14.25}$$

where θ_T is referred as torque angle and it is measured between the stator and rotor flux vectors (λ_s and λ_r).

From the above equation, the torque angle θ_T can be used as a control variable to adjust the motor developed torque T_e. During this process, the stator flux vector magnitude λ_s is kept constant at their rated value. In DTC, the stator flux vector λ_s is estimated by using the system parameters and the stator voltages and current feedback signals. The derivative of stator flux vector is obtained from Eq. (14.8) as

$$\frac{d\boldsymbol{\lambda}_s}{dt} = \boldsymbol{v}_s - R_s\boldsymbol{i}_s \tag{14.26}$$

From Eq. (14.26), it is observed that the changes in stator voltage vector $\boldsymbol{v}_s$ are directly affect the stator flux vector λ_s. This voltage vector is also referred as an inverter output voltage vector and it is formulated by using various inverter stationary voltage vectors. The available stationary voltage vectors are varied with the inverter output voltage levels (n) and they are equal to n^3. The magnitude and position of stator flux vector λ_s can be directly adjusted with the help of n^3 stationary vectors [1, 11, 12]. Depending on the selection process of stationary voltage vectors, the DTC methods can be categorized into off-line and real-time DTC methods. The off-line methods use lookup or switching tables, whereas the online methods use PWM blocks in the selection process of stationary voltage vectors.

14.4.2.1 DTC with hysteresis control

The control block diagram of DTC with hysteresis controller is shown in Figure 14.21. This approach has an outer speed controller and inner torque/flux comparators to control the magnitude of stator flux vector and motor torque independently. The speed controller ensures the actual rotating speed of the motor follows their reference speed command and it generates the reference torque

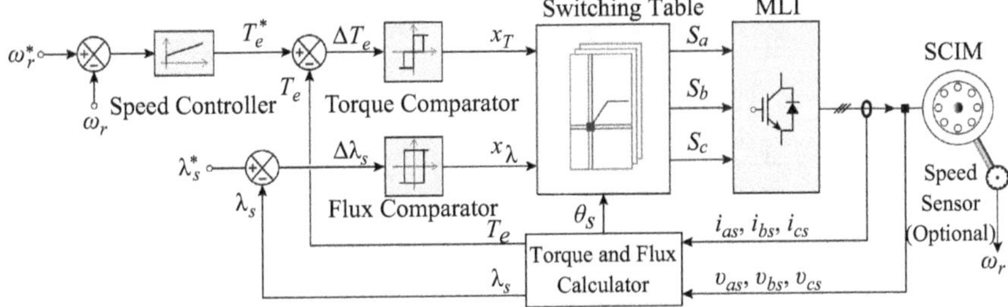

Figure 14.21 Block diagram of DTC scheme with hysteresis controller

command T^*_e. This command is compared with actual motor torque T_e and resultant error ΔT_e is fed to the torque comparator as shown in Figure 14.21. On the other hand, the magnitude of reference stator flux vector λ^*_s is compared with the actual stator flux vector magnitude λ_s. The resultant error $\Delta\lambda_s$ is fed to the flux comparator. In this approach, the magnitude of actual stator flux vector and motor torque signals are calculated by using stator voltage and current feedback signals as shown in Figure 14.21.

The flux and torque comparators are designed with hysteresis band, and it generates the signals x_λ and x_T based on the flux and torque errors, respectively. In which, the flux comparator output signal has two levels of $x_\lambda = +1$ and -1, whereas the torque comparator output has three levels of $x_T = +1$, 0, and -1. These levels represent the required control action need to be taken so that the error between reference and actual control signals can be minimized. For example, the "+1" represents the rise in λ_s or θ_T, "-1" represents the fall in λ_s or θ_T, and "0" signifies no change. The signals x_λ and x_T are fed to the switching table or look-up table as shown in Figure 14.21. The inverter stationary switching vectors are analyzed for different values of flux and torque comparators output and those vectors stored in the switching table [1, 11, 12]. Finally, the switching vector based on the value of x_λ and x_T is selected and applied to the inverter. However, the presence of hysteresis band causes variable switching frequency operation of the system.

The magnitude of stator flux vector and torque feedback signals are calculated from stator voltages and currents feedback signals, and motor parameters. From stationary $\alpha\beta$-reference frame models given in Eq. (14.9), the stator flux vector is defined as

$$\boldsymbol{\lambda}_s = \lambda_{\alpha s} + j\lambda_{\beta s} \tag{14.27}$$

The stator flux vector $\alpha\beta$-axis components are expressed as,

$$\left.\begin{aligned} \lambda_{\alpha s} &= \int (v_{\alpha s} - R_s i_{\alpha s})dt \\ \lambda_{\beta s} &= \int (v_{\beta s} - R_s i_{\beta s})dt \end{aligned}\right\} \tag{14.28}$$

from which its magnitude and angle are calculated as

$$\left.\begin{aligned} \lambda_s &= \sqrt{\lambda_{\alpha s}^2 + \lambda_{\beta s}^2} \\ \theta_s &= \tan^{-1}\left(\frac{\lambda_{\beta s}}{\lambda_{\alpha s}}\right) \end{aligned}\right\} \tag{14.29}$$

The developed motor torque in stationary $\alpha\beta$-reference frame is given as

$$T_e = \frac{3P}{2}\left(i_{\beta s}\lambda_{\alpha s} - i_{\alpha s}\lambda_{\beta s}\right) \tag{14.30}$$

From Eqs (14.28), (14.29), and (14.30), it is observed that the stator voltage and current feedback signals in stationary $\alpha\beta$-axis are needed to estimate the stator flux vector (magnitude and position) and motor torque. The motor stator winding resistance R_s is only parameter needed in the calculations. This is one of the major differences compared with DFOC where motor winding inductances are needed along with their resistances. The torque and flux calculator in DTC is designed by using Eqs (14.28), (14.29), and (14.30), and its structure is shown in Figure 14.22. The stator voltage and current feedback signals in stationary *abc*-reference frame are converted to stationary $\alpha\beta$-reference frame. These voltage and current components are used to calculate the magnitude of stator flux vector (λ_s) and its position (θ_s), and motor torque (T_e) as shown in Figure 14.22.

The performance of induction motor drive with DTC-hysteresis approach is shown in Figure 14.23. The motor nameplate details are listed in Table 14.3. In this study, the motor is operating at 1,189 RPM. During this operation, the load torque demand T_m on the motor is set to zero. Figure 14.23(a) shows that the motor developed torque T_e follows the load torque demand T_m. At $t = 0.1$ s, the load torque is suddenly increased to 7,935 N m and it is further reduced to 1,000 N m at $t = 0.3$ s as shown in Figure 14.23(a). The motor responds to these changes quickly and develops the torque T_e as per the load demand. Furthermore, the torque ripple

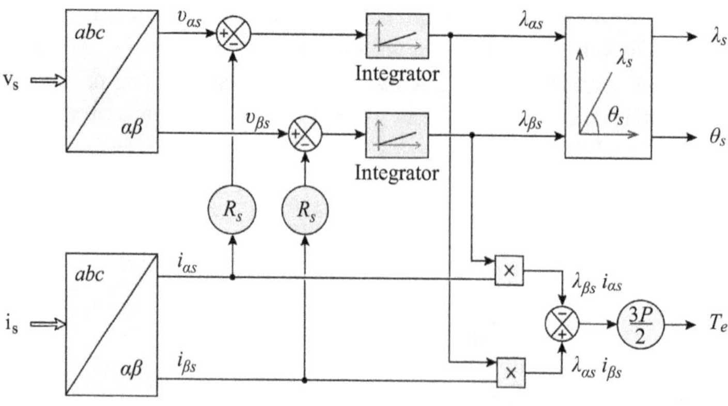

Figure 14.22 Block diagram of torque and flux calculator in DTC

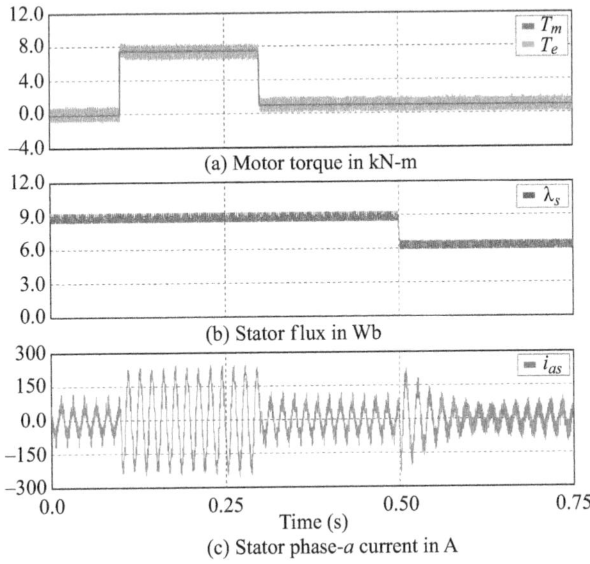

Figure 14.23 Simulated waveforms of induction motor drive with DTC at rated speed

depends on the torque comparator hysteresis band. With DTC, the motor torque and stator flux vector can be controlled independently as shown in Figure 14.23(b). From $t = 0$ to 0.5 s, the stator flux vector magnitude is maintained constant at their rated value of 9.0 Wb. The stator flux vector magnitude is suddenly reduced to 6.3 Wb to prove the effectiveness of the decoupled control of DTC at $t = 0.5$ s. During no-load condition, the current drawn by the motor is only consists of magnetizing current component to setup the air-gap flux in the motor as shown in Figure 14.23(c). From $t = 0.1$ to 0.3 s, the current drawn by the motor consists of both flux and torque producing current components. As it is further observed that the motor current magnitude varies with motor developed torque and stator flux magnitude as shown in Figure 14.23(c).

14.4.2.2 DTC with space vector modulation

The motor torque and stator flux vector hysteresis comparators in the DTC-hysteresis control approach are replaced with PI-controllers along with a PWM stage is added [11, 12]. By adding the PWM stage, the inverter switching frequency can be kept constant while keeping the current and torque ripple minimum. Also, the stationary vectors can be selected in real-time to achieve superior steady-state performance. Figure 14.24 shows the structure of DTC-SVM approach, in which the calculation of stator flux vector and torque is identical to the method used in the DTC-hysteresis approach. The flux and torque controllers are used to minimize their errors and gives the stator *dq*-axis reference voltages. These *dq*-axis voltages are transformed to stationary $\alpha\beta$-axis voltages by using the stator flux vector

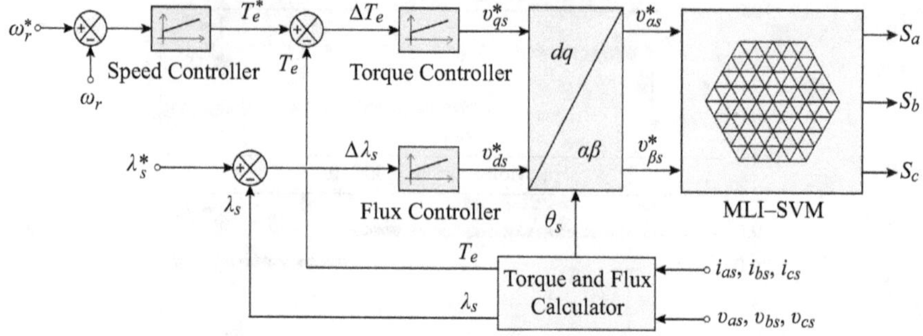

Figure 14.24 Block diagram of DTC with MLI-SVM

position θ_s as shown in Figure 14.24. The stationary $\alpha\beta$-axis reference voltages are applied to the SVM modulator, and it generates the gate pulses based on the selected stationary vectors and dwell times to the MLI. The SVM stage is designed based on the type of MLI used in the inverter stage [1].

The PI-regulators are very common in the classical control methods due to their simple structure, easy to design, and satisfactory performance. However, the changes in the system parameters, system non-linearities, and constant PI gains lead to poor control performance and unable to meet the control objectives with classical control methods [47]. Aforementioned issues are addressed through adaptive control techniques such as sliding mode control, self-tuned PI-regulators, and artificial intelligence (AI) techniques including genetic algorithms, fuzzy logic methods, and neural networks [47]. However, these methods are yet to become popular to be used in the commercial MV drive applications.

14.5 Model predictive control for MV drives

The use of a system model to forecast the behavior of the control variables to be managed distinguishes model predictive control (MPC). As demonstrated in Figure 14.25, MPC implementation comprises several components, including reference variables generation and extrapolation, control variable prediction using a system model, and system optimization utilizing cost function. External control loops generate the reference control variables $y^*(k)$ [48]. To reduce the tracking error between the reference and actual control variables, the reference variables at (k) sampling instant are projected to the $(k+1)$ sampling instant. When the sampling time (T_s) is smaller ($T_s < 20$ μs), then the extrapolation is not required [49]. The system model is used to forecast the behavior of the control variables at the $(k+1)$ sampling instant using the measured control variables ($y^m(k)$) and system parameters. The system model in discrete-time domain is used in the prediction process, and its accuracy is highly dependent on the sampling time. The difference between reference and predicted control variables is minimized by using a cost function. It is evaluated for all of the converter switching states. Finally, the state that gives the

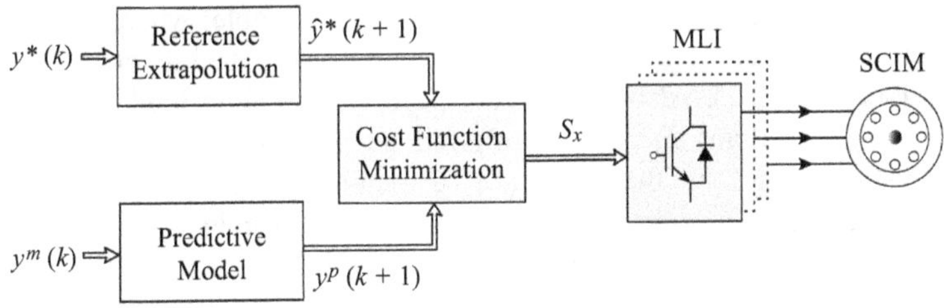

Figure 14.25 General approach of model predictive control

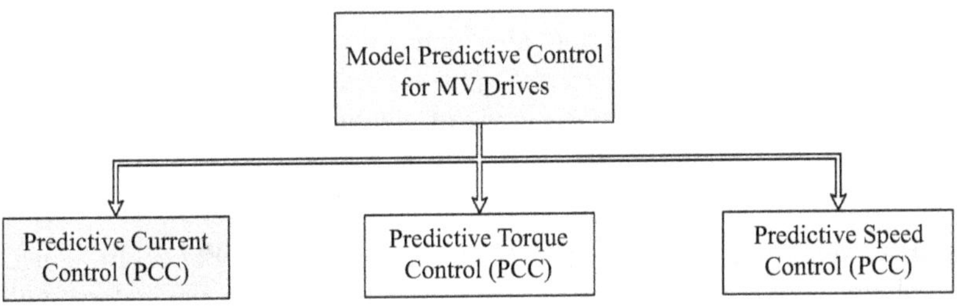

Figure 14.26 Classification of MPC for MV drives

least error between the reference and anticipated control variables is chosen and applied to the MLI [21].

The MPC is rigorously investigated for MV drives, renewable energy, power quality, and power supplies. In general, the MPC methods are broadly categorized into direct and indirect MPCs depending on the methodology employed in achieving the control objectives. In direct MPC, a single cost function is enough to achieve all of the system objectives [22, 50]. In some cases, more than one cost function is employed, and such methods are referred as multi-stage MPCs. On the other hand, the indirect MPC uses cost function along with classical control philosophy to achieve all the control objectives [21]. This approach minimizes the control complexity and weighting factor dependency. However, the system dynamic performance become sluggish compared with direct MPC. The direct and indirect MPC methods either directly manipulate the switching states or use a modulator to generate the switching states to the converter. The former approach causes variable switching frequency operation, whereas the latter approach leads to a constant switching frequency operation [51].

The MPC methods are further categorized based on the application and selected primary control variables in the optimization process. Particularly, the MPC methods for MV drives are derived from classical control methods such as FOC and DTC techniques and are listed in Figure 14.26. The stator current controllers in FOC are replaced with a predictive algorithm and the resultant method is referred as predictive current control (PCC) [50]. To manage the stator flux and torque, the predictive algorithm is used instead of torque and flux controllers in

DTC. In this method, the torque is primary control variable, whereas the flux has secondary importance. Hence, the resultant method is named as predictive torque control (PTC) [50, 52]. Also, the motor speed can be controlled with predictive algorithm and resultant method is referred as predictive speed control (PSC) [50, 53, 54]. Among them, the PCC and PTC methods are widely used in MV drive applications and they are emphasized in this section for induction motors. Despite its popularity, the digital implementation of MPC involves a number of challenges such as:

System models accuracy: The discrete-time system models are required to implement MPC. These models are created by discretizing continuous-time domain models, and their correctness varies depending on the approach used in the conversion process from continuous to discrete-time domain. In addition, the derived models consist of measured variables, predicted variables, and system parameters. Hence, the MPC performance vary with the accuracy of measurements and parameter's variations. Finally, the sampling time is the key factor in MPC implementation as it directly affects the system performance and its controllability [20, 48].

Tuning of weighting factors: The MPC can achieve a variety of control objectives by employing the cost function. Hence, the MPC is highly suited for multilevel converters control and its applications. Normally, the cost function consists of primary and secondary objectives. The relative value of secondary objectives compared to primary objectives is determined using weighting variables. Hence, the tuning of weighting factors is a key factor in achieving high-performance MPC [20, 48].

Computational burden: The MPC cost function is optimized for all of the converter's switching states. With more converter output levels, the number of switching states increases dramatically, putting a greater computational burden on the prediction algorithm. In addition, the MPC requires a longer sampling time to run the predictive algorithm, which has a significant impact on the system performance. As a result, MPC approaches without weighting factors are in high demand [20, 48].

Switching frequency: The direct MPC manipulates the switching states directly without using a modulator leading to a variable switching frequency operation. Higher device and converter switching losses, resonance issues, and a changeable harmonic profile result from this process. Whereas the employment of modulator causes fixed switching frequency operation, but it leads to a poor dynamic performance. In overall, it is required to make a tradeoff between dynamic performance and switching frequency during the selection of MPC methodology [20, 48].

14.5.1 Predictive current control

The predictive current control (PCC) is one of variations in MPC and widely used in power electronics and its applications. In MV drive applications, the PCC is designed to control the stator currents and it is derived from the FOC method [16, 20]. The control philosophy of PCC for MV drives is illustrated in Figure 14.27.

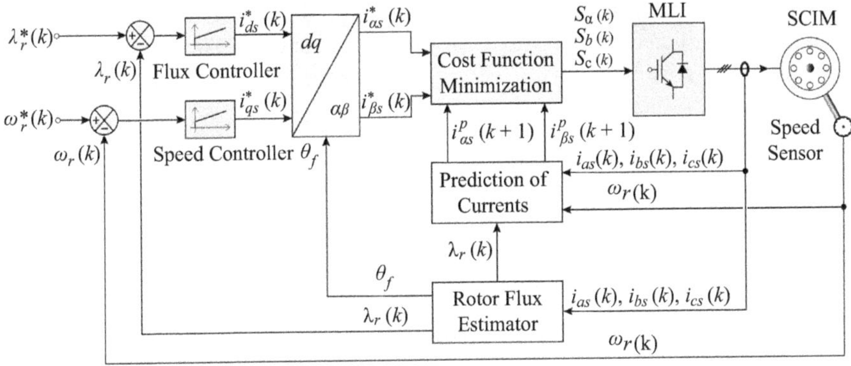

Figure 14.27 Block diagram of PCC for MV drive

The PCC follows the decoupled control of flux and torque producing current component's philosophy of FOC. The inner current controllers in FOC are replaced with predictive algorithm results in PCC. The rotor flux is controlled using the flux controller, which creates a reference stator current vector d-axis component i^*_{ds} as illustrated in Figure 14.27. The feedback flux information and its angle are estimated from the three-phase current feedback signals similar to the FOC philosophy as illustrated in Figure 14.27.

The reference stator current vector q-axis component i^*_{qs} is generated by the speed controller and its value is proportional to the motor reference torque command T^*_e. The feedback speed information ω_r can be obtained from digital speed sensor. The stator dq-axis reference current components are transformed to stationary $\alpha\beta$-frame components ($i^*_{\alpha s}$ and $i^*_{\beta s}$) as shown in Figure 14.27 [16, 20]. The stator currents are predicted by using the stator current model obtained from Eqs (14.6) and (14.8), and it is given as

$$\left.\begin{aligned} \frac{d\boldsymbol{i_s}}{dt} &= -\frac{R_\sigma}{\sigma L_s}\boldsymbol{i}_s + \frac{k_r}{\sigma L_s}\left(\frac{1}{\tau_r} - j\omega_r\right)\boldsymbol{\lambda}_r + \frac{1}{\sigma L_s}\boldsymbol{v}_s \\ \sigma = 1 - k_r k_s, \quad k_r &= \frac{L_m}{L_r}, \quad k_s = \frac{L_m}{L_s}, \quad R_\sigma = R_s + k_r^2 R_r \end{aligned}\right\} \tag{14.31}$$

where σ is the leakage factor, and k_s and k_r are the stator and rotor magnetic coupling factors, respectively.

The popular Euler's approach is used to transform the continuous-time domain model given in Eq. (14.31) to discrete-time domain as,

$$\boldsymbol{i}_s^p(k+1) = \left(1 - \frac{T_s R_\sigma}{\sigma L_s}\right)\boldsymbol{i}_s(k) + \frac{T_s k_r}{\sigma L_s}\left(\frac{1}{\tau_r} - j\omega_r(k)\right)\boldsymbol{\lambda}_r(k) + \frac{T_s}{\sigma L_s}\boldsymbol{v}_s(k) \tag{14.32}$$

where T_s is the sampling time.

In the stationary $\alpha\beta$-reference frame, the predicted stator current vector is divided into predicted α- and β-axis current components ($i^P_{\alpha s}(k+1)$ and $i^P_{\beta s}(k+1)$). The following cost function includes the reference and predicted stator current components as

$$g = \left(i^*_{\alpha s} - i^p_{\alpha s}(k+1)\right)^2 + \left(i^*_{\beta s} - i^p_{\beta s}(k+1)\right)^2 \tag{14.33}$$

For all conceivable switching states, the cost function is assessed. Finally, the converter's switching state with the lowest error is chosen and applied.

14.5.2 Predictive torque control

In PCC, the rotor flux and the torque are controlled through stator currents. On the other hand, a predictive algorithm can be used to control the motor torque and magnitude of stator flux vector directly, and the resultant method is referred as predictive torque control (PTC). In this method, the torque is primary control variable, whereas the stator flux is secondary variable. The PTC is derived from the conventional DTC in which the flux and toque controllers/comparators are replaced with predictive algorithm [16, 20]. The control block diagram of PTC for MV drives is shown in Figure 14.28. The reference stator flux vector magnitude λ^*_s is kept constant, whereas the reference torque command T^*_e comes from the speed controller [16, 20]. From Eqs (14.6) and (14.8), the predicted stator flux vector λ^P_s (k+1) is given as

$$\boldsymbol{\lambda}^p_s(k+1) = \boldsymbol{\lambda}_s(k) + T_s\mathbf{v}_s(k) - T_sR_s\boldsymbol{i}_s(k) \tag{14.34}$$

which is written in terms of rotor flux vector as follows:

$$\boldsymbol{\lambda}^p_s(k+1) = \sigma L_s\boldsymbol{i}_s(k+1) + k_r\boldsymbol{\lambda}_r(k) \tag{14.35}$$

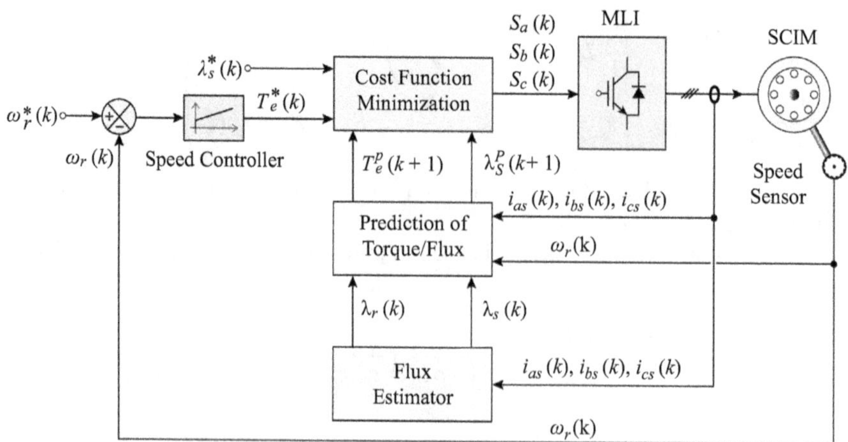

Figure 14.28 Block diagram of PTC for MV drive

The predicted torque $T^p{}_e\,(k{+}1)$ is given as

$$T_e^p(k+1) = \frac{3P}{2}\,\mathrm{Im}\left(\overline{\boldsymbol{\lambda}_s^p}(k+1)\,\boldsymbol{i}_s^p(k+1)\right) \tag{14.36}$$

where $\lambda^p{}_s$ with over the line represents the complex conjugate of stator flux vector.

The predicted stator current and rotor flux vectors are expressed in terms of measured stator currents, estimated rotor/stator flux, and predicted stator voltages as follows:

$$\left.\begin{aligned} \boldsymbol{i}_s^p(k+1) &= \left(1 - \frac{T_s R_\sigma}{\sigma L_s}\right)\boldsymbol{i}_s(k) + \frac{T_s k_r}{\sigma L_s}\left(\frac{1}{\tau_r} - j\omega_r(k)\right)\boldsymbol{\lambda}_r(k) + \frac{T_s}{\sigma L_s}\boldsymbol{v}_s(k) \\ \boldsymbol{\lambda}_r^p(k+1) &= \frac{T_s}{L_m \tau_r}\boldsymbol{i}_s(k) + \left(1 - T_s\left(\frac{1}{\tau_r} - j\omega_r(k)\right)\right)\boldsymbol{\lambda}_r(k) \end{aligned}\right\} \tag{14.37}$$

The predicted stator voltages and measured stator currents are used along with Eqs (14.6)–(14.12) in the calculation of stator and rotor flux vectors magnitude. The reference and predicted torque and stator flux variables are used to formulate the cost function as follows:

$$g = \left(T_e^* - T_e^p(k+1)\right)^2 + k_\lambda\left(\lambda_s^* - \lambda_s^p(k+1)\right)^2 \tag{14.38}$$

where k_λ is the weighting factor for flux variable and λ_s represents the stator flux vector magnitude. The cost function will be examined for all of the converter switching states, and the optimum state corresponding to the lowest value is selected.

14.5.3 Predictive speed control

As the name implies, the rotor speed is the primary control variable in predictive speed control (PSC) method. In PSC method, the rotor speed variable is included in a cost function along with current variables in PCC or torque/flux variables in PTC. The resultant cost function for PSC method becomes

$$\left.\begin{aligned} g &= \left(\omega_r^* - \omega_r^p(k+1)\right)^2 + k_i\left(i_{\alpha s}^* - i_{\alpha s}^p(k+1)\right)^2 + \\ &\quad k_i\left(i_{\beta s}^* - i_{\beta s}^p(k+1)\right)^2 \text{ (with PCC)} \\ g &= \left(\omega_r^* - \omega_r^p(k+1)\right)^2 + k_T\left(T_e^* - T_e^p(k+1)\right)^2 + \\ &\quad k_\lambda\left(\lambda_s^* - \lambda_s^p(k+1)\right)^2 \text{ (with PTC)} \end{aligned}\right\} \tag{14.39}$$

where k_i is the weighting factor for current variable and k_T is the weighting factor for torque variable.

According to the cost function, the PSC method prioritizes the motor speed and ensures accurate speed control. On the other hand, the current variables or torque/flux variables are handled with weighting factors leading to a poor response and steady-state error. Therefore, it is difficult to achieve high-performance drive using PSC method. Hence, the PSC method is not very popular in MV drive applications.

14.6 Future trends

The high-power adjustable speed drives at the MV level are preferred choice in modern industries to increase the productivity and its quality, and to save the energy cost. There is a steady demand for these drives with a global market value of $17 billion in 2020 and it is expected to reach $19.9 billion by 2026. To develop high-performance and low-cost industrial drives which can compete in the global market, innovation in research, development and manufacturing is essential. Over the years, there are several new developments in the MV drive technologies, including semiconductor devices, various multilevel converter topologies, modulation schemes, and current/torque control schemes.

With the available silicon (Si) semiconductor devices, it is difficult to improve the drive performance including efficiency and switching frequency operation. Hence, the silicon carbide (SiC) devices became an alternative due to their low on-state resistance and can reach higher switching frequency operation. Thereby, comparatively lower thermal generation during operation. Consequently, it can achieve higher power density and cost-effectiveness of the MV drive. Hence, the SiC devices will play a key role in developing new power converters for MV drive applications in the future. Even though, these devices have gained attention for high voltage applications, but some issues hinder their mass production including the manufacturing process, electromagnetic interference, thermal management, reliability, busbar layout, and gate driver circuit design.

Furthermore, the faults in MV drive power converters lead to a significant loss of production and revenue in various industry applications. This demands the incorporation of the fault-tolerant control schemes including fault isolation, detection, and reconfiguration approaches with the existing control schemes. Also, the condition monitoring of converter components can improve the reliability of the MV drive-by predictive maintenance. Another issue is the conventional control approaches demand a significant number of sensors (voltage, current, position, torque sensors) and communication ports for signal monitoring and system control. Therefore, it is imperative to develop sensorless control methods and parameter estimation methods, which bring down the MV drive system cost and enhance its reliability. Furthermore, special considerations such as reliability analysis-based design and incorporation of active thermal management systems can be an added advantage.

In addition, the demand for MV drive power converters with a higher power capacity is continuously growing due to the rapid industrialization and increased demand for high-quality products. Hence, the modular and scalable power converters along with fault-tolerant control strategies will become a trend in future MV drive developments. Also, the industry applications demand high-performance MV drives with faster transient response time and smaller torque/flux ripple. Currently, the MPC methods became popular and widely used to meet these requirements.

However, the existing MPC methods have various issues such as accuracy of system models, the effect of system parameters variation, requires a larger sampling time, variable switching frequency operation, selection of weighting factors, and computational complexity. Considering these issues, the current research focused on the development of model-free predictive control methods combined with artificial intelligence and machine learning techniques for MV drives.

14.7 Conclusions

An overview of state-of-the-art developments in MV drives including semiconductor devices, power converters, and control methods are presented in this chapter. Over the years, there are significant developments in semiconductor devices from thyristor to SiC MOSFET for MV drive power converters. In early days, the two-level VSCs are designed with series connection of devices to handle the MV operation. However, these converters didn't become popular due to the device voltage sharing problems and unable to meet the drive power quality standards. Alternatively, the multilevel converters become popular and capable of handling MV operation without devices in series while meeting the power quality standards. The basic functionality, features, and challenges associated with various multilevel power converter topologies for MV drives are presented. With the available semiconductor devices, the integrated multilevel converters are able to reach a voltage capacity of 4.16 kV, whereas the multi-module converters are able to reach a voltage capacity up to 13.8 kV.

Furthermore, the multi-module converters have fault-tolerant capability, which will improve the drive reliability. The MV drives requires closed-loop control methods to regulate the motor flux and torque. The classical control methods such as FOC and DTC philosophy and their implementation are presented. The performance of these methods is discussed with the help of simulation studies. However, the dynamic performance of FOC and DTC methods are affected due to the limitations on linear controller's bandwidth and device switching frequency. Alternatively, the MPC methods can achieve superior dynamic performance while achieving multiple objectives with the help of a cost function. The MPC methods for MV drives are categorized into PCC, PTC, and PSC methods based on the primary control variables used in the cost function. The operating principle and implementation of PCC, PTC, and PSC methods are presented. Finally, the future trends in MV drives are discussed.

References

[1] Wu B and Narimani M. *High-Power Converters and AC Drives*. New York, NY: John Wiley & Sons, 2017.

[2] Kouro S, Rodriguez J, Wu B, Bernet S, and Perez M. 'Powering the future of industry: high-power adjustable speed drive topologies'. *IEEE Industry Applications Magazine*. 2012;18(4):26–39.

[3] Abu-Rub H, Bayhan S, Moinoddin S, Malinowski M, and Guzinski J. 'Medium-voltage drives: challenges and existing technology'. *IEEE Power Electronics Magazine*. 2016;3(2):29–41.

[4] Stemmler HE. 'High-power industrial drives'. *Proceedings of the IEEE*. 1994;82(8):1266–1286.

[5] Rodriguez J, Franquelo LG, Kouro S, *et al.* 'Multilevel converters: an enabling technology for high-power applications'. *Proceedings of the IEEE*. 2009;97(11):1786–1817.

[6] Morya AK, Gardner MC, Anvari B, *et al.* 'Wide bandgap devices in AC electric drives: opportunities and challenges'. *IEEE Transactions on Transportation Electrification*. 2019;5(1):3–20.

[7] Ronanki D and Williamson SS. 'Evolution of power converter topologies and technical considerations of power electronic transformer-based rolling stock architectures'. *IEEE Transactions on Transportation Electrification*. 2017;4(1):211–219.

[8] Du S, Dekka A, Wu B, and Zargari N. *Modular Multilevel Converters: analysis, Control, and Applications*. New York, NY: John Wiley & Sons, 2018.

[9] Wu B, Pontt J, Rodríguez J, Bernet S, and Kouro S. 'Current-source converter and cycloconverter topologies for industrial medium-voltage drives'. *IEEE Transactions on Industrial Electronics*. 2008;55(7):2786–2797.

[10] Wheeler PW, Rodriguez J, Clare JC, Empringham L, and Weinstein A. 'Matrix converters: a technology review'. *IEEE Transactions on Industrial Electronics*. 2002;49(2):276–288.

[11] Casadei D, Profumo F, Serra G, and Tani A. 'FOC and DTC: two viable schemes for induction motors torque control'. *IEEE Transactions on Power Electronics*. 2002;17(5):779–787.

[12] Buja GS and Kazmierkowski MP. 'Direct torque control of PWM inverter-fed AC motors – a survey'. *IEEE Transactions on Industrial Electronics*. 2004;51(4):744–757.

[13] Leon JI, Kouro S, Franquelo LG, Rodriguez J, and Wu B. 'The essential role and the continuous evolution of modulation techniques for voltage-source inverters in the past, present, and future power electronics'. *IEEE Transactions on Industrial Electronics*. 2016;63(5):2688–2701.

[14] Kouro S, Perez MA, Rodriguez J, Llor AM, and Young HA. *'Model predictive control: MPC's role in the evolution of power electronics'. EEE Industrial Electronics Magazine*. 2015;9(4):8–21.

[15] Vazquez S, Leon JI, Franquelo LG, *et al.* 'Model predictive control: a review of its applications in power electronics'. *IEEE Industrial Electronics Magazine*. 2014;8(1):16–31.

[16] Rodriguez J, Garcia C, Mora A, *et al.* 'Latest advances of model predictive control in electrical drives—Part I: basic concepts and advanced strategies'. *IEEE Transactions on Power Electronics*. 2021;37(4):3927–3942.

[17] Rodriguez J, Garcia C, Mora A, *et al.* 'Latest advances of model predictive control in electrical drives—Part II: applications and benchmarking with

classical control methods'. *IEEE Transactions on Power Electronics*. 2021;37(5):5047–5061.

[18] Geyer T and Mastellone S. 'Model predictive direct torque control of a five-level ANPC converter drive system'. *IEEE Transactions on Industry Applications*. 2012;48(5):1565–1575.

[19] Rodriguez J, Kazmierkowski MP, Espinoza JR, *et al.* 'State of the art of finite control set model predictive control in power electronics'. *IEEE Transactions on Industrial Informatics*. 2012;9(2):1003–1016.

[20] Karamanakos P, Liegmann E, Geyer T, and Kennel R. *'Model predictive control of power electronic systems: methods, results, and challenges'. IEEE Open Journal of Industry Applications*. 2020;1:95–114.

[21] Dekka A, Wu B, Yaramasu V, Fuentes RL, and Zargari NR. 'Model predictive control of high-power modular multilevel converters—an overview'. *IEEE Journal of Emerging and Selected Topics in Power Electronics*. 2018;7(1):168–183.

[22] Karamanakos P, Geyer T, Oikonomou N, Kieferndorf FD, and Manias S. 'Direct model predictive control: a review of strategies that achieve long prediction intervals for power electronics'. *IEEE Industrial Electronics Magazine*. 2014;8(1):32–43.

[23] Zhou D, Ding L, and Li Y. 'Two-stage optimization-based model predictive control of 5L-ANPC converter-fed PMSM drives'. *IEEE Transactions on Industrial Electronics*. 2020;68(5):3739–3749.

[24] Zhang Y, Zhang B, Yang H, Norambuena M, and Rodriguez J. 'Generalized sequential model predictive control of IM drives with field-weakening ability'. *IEEE Transactions on Power Electronics*. 2018;34(9):8944–8955.

[25] Yang Y, Wen H, Fan M, Xie M, Chen R, and Wang Y. 'A constant switching frequency model predictive control without weighting factors for T-type single-phase three-level inverters'. *IEEE Transactions on Industrial Electronics*. 2018;66(7):5153–5164.

[26] Rojas CA, Rodriguez J, Villarroel F, Espinoza JR, Silva CA, and Trincado M. 'Predictive torque and flux control without weighting factors'. *IEEE Transactions on Industrial Electronics*. 2012;60(2):681–690.

[27] 'Institute of Electrical and Electronics Engineers. IEEE recommended practice and requirements for harmonic control in electric power systems'. New York, NY: IEEE, 2014.

[28] Garcia O, Cobos JA, Prieto R, Alou P, and Uceda J. 'Simple AC/DC converters to meet IEC 1000-3-2'. In *APEC 2000. Fifteenth Annual IEEE Applied Power Electronics Conference and Exposition (Cat. no. 00CH37058),* 2000 Feb 6 (vol. 1, pp. 487–493). New York, NY: IEEE.

[29] Abu-Rub H, Holtz J, Rodriguez J, and Baoming G. 'Medium-voltage multilevel converters—state of the art, challenges, and requirements in industrial applications'. *IEEE Transactions on Industrial Electronics*. 2010;57 (8):2581–2596.

[30] Dai H, Torres RA, Gossmann J, Lee W, Jahns TM, and Sarlioglu B. 'A seven-switch current-source inverter using wide bandgap dual-gate

bidirectional switches'. *IEEE Transactions on Industry Applications*. 2022;58(3):3721–3737.

[31] Phillips KP. 'Current-source converter for AC motor drives'. *IEEE Transactions on Industry Applications*. 1972;6:679–683.

[32] Rodriguez J, Wu B, Bernet S, *et al*. 'Design and evaluation criteria for high power drives'. In *2008 IEEE Industry Applications Society Annual Meeting* 2008 Oct 5 (pp. 1–9). New York, NY: IEEE.

[33] Moguilnaia NA, Vershinin KV, Sweet MR, Spulber OI, De Souza MM, and Narayanan EM. 'Innovation in power semiconductor industry: past and future'. *IEEE Transactions on Engineering Management*. 2005;52(4):429–439.

[34] Bose BK. 'Evaluation of modern power semiconductor devices and future trends of converters'. *IEEE Transactions on Industry Applications*. 1992; 28(2):403–413.

[35] Ronanki D, Singh SA, and Williamson SS. 'Comprehensive topological overview of rolling stock architectures and recent trends in electric railway traction systems'. *IEEE Transactions on Transportation Electrification*. 2017;3(3):724–738.

[36] Zargari NR, Cheng Z, and Paes R. 'A guide to matching medium-voltage drive topology to petrochemical applications'. *IEEE Transactions on Industry Applications*. 2017;54(2):1912–1920.

[37] Attique QM, Wang K, Zheng Z, Zhu H, Li Y, and Rodriguez J. 'A generalized simplified virtual vector PWM to balance the capacitor voltages of multilevel diode-clamped converters'. *IEEE Transactions on Power Electronics*. 2022;37(8):9377–9391.

[38] Monteiro AP, da Costa Bahia FA, Jacobina CB, and de Sousa RP. 'Cascaded transformers-based multilevel inverters with NPC'. *IEEE Transactions on Industrial Electronics*. 2021;69(8):7879–7889.

[39] Sayago JA, Bruckner T, and Bernet S. 'How to select the system voltage of MV drives—a comparison of semiconductor expenses'. *IEEE Transactions on Industrial Electronics*. 2008;55(9):3381–3390.

[40] Rech C, and Castiblanco WAP. 'Five-level switched-capacitor ANPC inverter with output voltage boosting capability'. *IEEE Transactions on Industrial Electronics*. 2023;70(1):29–38.

[41] Faraji F, Ghias AM, Guo X, Chen Z, Lu Z, and Hua C. 'A split-inductor flying capacitor converter for medium-voltage application'. *IEEE Journal of Emerging and Selected Topics in Power Electronics*. 2022, doi:10.1109/JESTPE.2022.3164130.

[42] Wu M, Tian H, Wang K, Konstantinou G, and Li YW. 'Generalized low switching frequency modulation for neutral-point-clamped and flying-capacitor four-level converters'. *IEEE Transactions on Power Electronics*. 2022;37(7):8087–8103.

[43] Kant P and Singh B. 'Multiwinding transformer fed CHB inverter with on-line switching angle calculation based SHE technique for vector controlled induction motor drive'. *IEEE Transactions on Industry Applications*. 2020;56(3):2807–2815.

[44] Singh B and Kant P. 'Multi-winding transformer for 18-pulse AC–DC converter fed 7-level CHB-inverter with fundamental switching based VCIMD'. *IEEE Open Journal of the Industrial Electronics Society*. 2020;1:1–9.

[45] Dekka A, Yaramasu V, Fuentes RL, and Ronanki D. 'Modular multilevel converters'. In Kabalci K. (ed.). *Multilevel Inverters. London: Academic Press*, 2021. p. 147–179.

[46] Ronanki D and Williamson SS. 'Modular multilevel converters for transportation electrification: challenges and opportunities'. *IEEE Transactions on Transportation Electrification*. 2018;4(2):399–407.

[47] Gadoue SM, Giaouris D, and Finch JW. 'Artificial intelligence-based speed control of DTC induction motor drives—a comparative study'. *Electric Power Systems Research*. 2009;79(1):210–219.

[48] Dekka A, Yaramasu V, Fuentes RL, and Ronanki D. 'Model predictive control of modular multilevel converters'. In Kabalci K. (ed.). *Multilevel Inverters*. London: Academic Press, 2021. p. 129–153.

[49] Rodriguez J and Cortes P. *Predictive Control of Power Converters and Electrical Drives*. New York, NY: John Wiley & Sons, 2012.

[50] Dekka A, Bahrami A, and Narimani M. 'Direct predictive current control of a new five-level voltage source inverter'. *IEEE Transactions on Industry Applications*. 2021;57(3):2941–2953.

[51] Yaramasu V, Dekka A, Dragicevi T, Zheng C, and Rodriguez J. 'Modulated model predictive control of an LC-filtered neutral-point clamped converter'. *In* 2020 IEEE 21st Workshop on Control and Modeling for Power Electronics *(*COMPEL*) 2020 Nov 9* (pp. 1–6). New York, NY: IEEE.

[52] Zhang Y, Yang H, and Xia B. 'Model-predictive control of induction motor drives: torque control versus flux control'. *IEEE Transactions on Industry Applications*. 2016;52(5):4050–4060.

[53] Wang F, Mei X, Rodriguez J, and Kennel R. 'Model predictive control for electrical drive systems – an overview'. *CES Transactions on Electrical Machines and Systems*. 2017;1(3):219–230.

[54] Vazquez S, Rodriguez J, Rivera M, Franquelo LG, and Norambuena M. 'Model predictive control for power converters and drives: advances and trends. *IEEE Transactions on Industrial Electronics*. 2016;64(2):935–947.

Index

Printed in the USA
CPSIA information can be obtained
at www.ICGtesting.com
JSHW061418280823
47398JS00003B/43